CLIMATE CHANGE: BIOLOGICAL AND HUMAN ASPECTS

In recent years climate change has become recognised as the foremost environmental problem of the twenty-first century, and a subject of considerable debate. Not only will climate change affect the multi-billion dollar energy strategies of countries worldwide, but it could also seriously affect many species, including our own. Written in an accessible style, this textbook provides a broad review of past, present and likely future climate change from the viewpoints of biology, ecology and human ecology. It is thoroughly referenced, allowing readers, if they wish, to embark on their own more specialist studies.

A fascinating introduction to the subject, this textbook will be of interest to a wide range of people, from students in the life sciences who need a brief overview of the basics of climate science, to atmospheric science, geography and environmental science students who need to understand the biological and human ecological implications of climate change. It will also be a valuable reference for those involved in environmental monitoring, conservation, policy-making and policy lobbying.

JONATHAN COWIE has spent many years conveying the views of biological science learned societies to policy-makers. His earlier postgraduate studies related to energy and the environment, and he is a former Head of Science Policy and Books at the Institute of Biology (UK). He is author of *Climate and Human Change: Disaster or Opportunity* (Parthenon Publishing, 1998).

CLIMATE CHANGE

Biological and Human Aspects

JONATHAN COWIE

CAMBRIDGE
UNIVERSITY PRESS

CAMBRIDGE UNIVERSITY PRESS
Cambridge, New York, Melbourne, Madrid, Cape Town, Singapore, São Paulo

Cambridge University Press
The Edinburgh Building, Cambridge CB2 8RU, UK

Published in the United States of America by Cambridge University Press, New York

www.cambridge.org
Information on this title: www.cambridge.org/9780521696197

First published 2007

Printed in the United Kingdom at the University Press, Cambridge

A catalogue record for this publication is available from the British Library

ISBN 978-0-521-87399-4 hardback
ISBN 978-0-521-69619-7 paperback

Contents

Introduction

This book is about biology and human ecology as they relate to climate change. Let's take it as read that climate change is one of the most urgent and fascinating science-related issues of our time and that you are interested in the subject: for if you were not you would not be reading this now. Indeed, there are many books on climate change but nearly all, other than the voluminous Intergovernmental Panel on Climate Change (IPCC) reports, tend to focus on a specialist aspect of climate, be it weather, palaeoclimatology, modelling and so forth. Even books relating to biological dimensions of climate change tend to be specialist, with a focus that may relate to agriculture, health or palaeoecology. These are, by and large, excellent value provided that they cover the specialist ground which readers seek. However, the biology of climate change is so broad that the average life-sciences student, or specialist seeking a broader context in which to view their own field, has difficulty in finding a wide-ranging review of the biology and human ecology of climate change. Non-bioscience specialists with an interest in climate change (geologists, geographers, atmospheric chemists, etc.) face a similar problem. This also applies to policy-makers and policy analysts, or those in the energy industries, getting to grips with the relevance of climate change to our own species and its social and economic activities.

In addition, specialist texts mainly refer to specialist journals. Very few university or research-institute libraries carry the full range. Fortunately the high-impact factor and multi-disciplinary journals such as *Science* and *Nature* do have specialist climate papers (especially those relating to major breakthroughs) and virtually all academic libraries, at least in the anglophone world, carry these publications. It is therefore possible to obtain a grounding in the biology (in the broadest sense) of climate-change science from these journals provided, that is, one is prepared to wade through several years' worth of copies.

This book hopefully scores with its broad biological approach, its tendency to cite the high-impact journals (although some specialist citations are also included) and its level of writing (hopefully appropriate for junior under-graduates and specialists reading outside their field). It should also be accessible to bioscientists as well as those outside of the life sciences. However, here is a quick word of advice. Familiarise yourself with the appendices at the back before you start reading!

Even so, this book can only be an introduction to the biology and human ecology – past, present and future – of climate change. Readers seeking more specialist knowledge on any particular aspect should seek out the references, at least as a starting point.

This book's style is also different to many textbooks. Reading it straight through from start to finish one may get the feeling that it is a little repetitious. This is only *partly* true. It is true in the sense that there are frequent references to other chapters and subsections. This is for those looking at a specific dimension, be they specialists putting their own work into a broader climate context, students with essays to write, or policy analysts and policy-makers looking at a special part of the human–climate interface. In short, this book is written as much, if not more, for those dipping into the topic as it is as a start-to-finish read.

There is another sense in which this book appears repetitious, although in reality it is not. It stems from one particular problem scientists have had in persuading others that human activity really is affecting our global climate. This is that there is no single piece of evidence that by itself proves such a hypothesis conclusively. Consequently those arguing a contrary case have been able to cite seemingly anomalous evidence, such as that a small region of a country has been getting cooler in recent years or that the Earth has been warmer in the past, or that there have been alternating warm and cool periods. All of this may be true individually but none of it represents the current big picture. So, instead of a single, all-powerful fact to place at the heart of the climate-change argument, there is a plethora of evidence from wide-ranging sources. For instance, there is a wealth of quite separate geological evidence covering literally millions of years of the Earth's history in many locations across the globe. This itself ranges from ice cores and fossils to isotopic evidence of a number of elements from many types of sediment. There is also a body of biological evidence from how species react to changes in seasons to genetic evidence from when species migrated due to past climate change. Indeed, within this there is the human ecological evidence of how we have been competing with other species for resources and how this relates observed changes in both human and ecological communities with past climate change.

This vast mass of evidence all points to the same big picture of how changes in greenhouse gases and/or climate have affected life in the past. Then again there is the present and the evidence used to build up a likely picture of what could well happen in the future. Here again, the evidence seems to be very largely corroborative. Therefore, to readers of this book it can seem as if the same ground is being covered when in fact it is a different perspective being presented each time that leads to the same concluding picture.

Indeed, because there is so much evidence contributing to the big picture, some may well find that evidence from their own specialist area of work is not included, or, if it is included, is covered briefly. This is simply because the topic is so huge and not due to a lack of recognition on my part of the importance of any particular aspect of climate-change science.

That there are similar themes running through specialist areas of climate-change science and the relating biology is in once sense comforting (we seem to be continually improving our understanding and coming to a coherent view) but in another it is frustrating. Over the years I have spoken to a large number of scientists from very disparate disciplines. Part of this has been due to my work (policy analysis and science lobbying for UK learned societies and before that in science journal and book management) and in part because I enjoy going to biosphere science as well as energy-related symposia. (There is nothing quite like looking over the shoulders of a diverse range of scientists and seeing what is happening in the laboratory and being discovered in the field.) The key thing is that these individual specialist, climate-related scientists all tend to say similar things, be they involved with ocean circulation, the cryosphere (ice and ice caps), tropical forests and so forth. They say the same as their colleagues in other specialist areas but equally do not appear to really appreciate that there is such a commonality of conclusion. For example, a common emerging theme is that matters are on the cusp. Change is either happening or clearly moving to a point where (frequently dependent on other factors) marked change could well happen. It is perhaps a little disappointing that more often than not such specialists seem to have a limited awareness of how their counterparts in other disciplines view things. (I should point out that, in my view, this has more to do with pressures from how science is undertaken these days, and not due to the high level of competence these specialists have within their own field. Scientists simply are not afforded the time to take several steps back from their work and view the larger scientific panorama.) That science is so compartmentalised tends to limit wide-ranging discussions, yet these, when properly informed by sound science, can be exceptionally fruitful.

By now you may be beginning to suspect what has been motivating my researching and writing of this book. The question that remains for me is whether this book will have any effect on your own motivations and under-standing? As it is quite likely that I will encounter at least some of you over the coming years, I dare say I will find out. Meanwhile, I hope you find this topic as fascinating as I do.

Jonathan Cowie
www.science-com.concatenation.org

Acknowledgements

I would very much like to thank all those in UK bioscience with whom I have interacted in some way or other on climate-change matters. In particular, I should thank a good number who have been on the various Institute of Biology science committees since the 1990s. This also goes to a score or two of my fellow members of the British Ecological Society and the Geological Society of London. A special thank you goes to those who have alerted (and, as often as not, invited) me to workshops and symposia on climate and energy issues as well as on biosphere science. I have found every one useful in at least one way: many provided a number of new insights and all gave me a reality check. Thank you.

This book also owes a lot to some research bodies. In the UK we are quite bad at making data from tax-payer-funded research publicly available (even for education and policy purposes). This is not so in the USA and so I greatly valued the open access that the National Oceanic and Atmospheric Administration give to their palaeoclimate-related data (which I have used to generate a number of the figures). Interested readers can visit their website at www.ncdc.noaa.gov/oa/ncdc.html. I am also extremely appreciative of the UK Environment Agency's current (2006) Chief Executive, without whom Figure 6.5 simply would not have been presented! Then there are the many who sent paper off-prints (e-mailed pdf files). There are too many to mention but be assured all are referenced.

Talking of references, as mentioned in the Introduction as far as possible I have taken either major reports, many of which are available on the Internet, or used high-impact-factor journals that can be found in most university libraries (these in turn cite papers in more specialist publications). However, I have also used a number of World Health Organization (WHO) press releases. This comes from my background in science policy, and the WHO have been sending me these for the best part of two decades. You will not find these in university libraries but fortunately you too can seek these out, on www.who.int/mediacentre/news/en.

A mention also has to go to the friendly and helpful librarians of Imperial College London, whose work really is appreciated. Then there are all those who have facilitated my site and field visits in the UK and abroad, be they to power stations (fossil, hydroelectric and nuclear), sites of special scientific interest (in the literal and not just the UK technical sense of the term) and educational institutions.

A thank you also goes to Peter Tyers for the cover picture. This is the second time he has done this for me, but then he is a good photographer.

Finally I must specifically thank Cambridge University Press and freelance copy-editor Nik Prowse for work on the manuscript. I like to think that I have long since found my feet with words, but any capability for editorial spit and polish has always eluded me. Nik has also greatly helped standardise the referencing and presentation. I therefore really do value good editors (and so should you) and especially those who appreciate those who try to do things a little differently. With luck you will notice.

1

An introduction to climate change

In most places on this planet's terrestrial surface there are the signs of life. Even in those places where there is not much life today, there are frequently signs of past life, be it fossils, coal or chalk. Further, it is almost a rule of thumb that if you do discover signs of past life, either tens of thousands or millions of years ago, then such signs will most likely point to different species to those found there today. Why? Here there are a number of answers, not least of which is evolution. Yet a key feature of why broad types of species (be they broad-leaved tree species as opposed to narrow needle-leaved ones) live in one place and not another is to do with climate. Climate is a fundamental factor influencing biology. Consequently a key factor (among others) as to why different species existed in a particular place 5000, 50 000, 500 000 or even 5 000 000 years ago (to take some arbitrary snapshots in time) is due to different climatic regimens existing at that place in those times.

It is also possible to turn this truism on its head and use biology to ascertain aspects of the climate, and biological remains are aspects of past climates. Furthermore, biology can influence climate: for example, an expanse of rainforest transpires such a quantity of water, and influences the flow of water through a catchment area, that it can modify the climate from what it otherwise would have been in the absence of living species. Climate and biology are interrelated.

Look at it another way. All living things flourish within a temperature range as well as have certain temperature tolerances for aspects of their life cycle. Furthermore, all living things require a certain amount of water and the availability of water, terrestrially, is again driven by climate. Given this essential connection of temperature and water to life, it is not difficult to see how important climate is in determining where different species, and assemblages thereof (ecosystems), can be found.

From this we can easily deduce that if climate is so important, then climate change is absolutely critical if we are to predict the likely fate of species in a certain region. It is also possible to use the reverse in an applied sense to note the presence (or past presence) of different species and then use this as an indicator of climate, both in the past and in the present. This interrelationship between life and climate is fundamental. It affects all species, which includes, we sometimes forget, our own – *Homo sapiens*. Here we also tend to forget that on every continent except Antarctica there are examples of deserted settlements and evidence of long-extinct civilisations. These are societies that once flourished but have now gone, due primarily to a change in climate.

If it is not sufficiently significant that living things, including human societies, are subject to the vagaries of climate change, there is now convincing evidence that our modern global society is currently altering the global climate in a profound way that also has regional, and indeed global, biological implications that will impact heavily on human societies. For these reasons there is currently considerable interest in the way living things interact with the climate, and especially our own species. As we shall see in the course of this book, biology, and the environmental sciences relating to ecology and climate, can provide us with information as to past climates and climate change (palaeoclimatology) which in turn can illuminate policy determining our actions affecting future climate. This will be invaluable if we are to begin to manage our future prospects.

1.1 Weather or climate

Any exploration of the biology of climate change needs to clarify what is meant by climate as distinct from weather. In essence the latter is the day-to-day manifestation of the former. The climate of a region is determined by long-term weather conditions including seasonal changes. The problem is that weather is in its own right a variable phenomenon: if it were not we would have less difficulty in arriving at more accurate long-term forecasts. Consequently, if the climate of a region changes we can only discern this over a long period of time once we have disentangled possible climate change from weather's natural background variability. Analogously, physicists and engineers refer to what they call the signal-to-noise ratio, and this they apply to electrical currents or an electromagnetic signal, be it a commercial radio broadcast or that from a stellar body. Similarly with climate change, the problem is to disentangle a small climatic-change signal from considerable background weather noise. For example, one very hot summer (or drought, or heavy monsoon, or whatever...) by itself does not signify climate change. On the other hand, a decade or more of these in succession may well be of climatic significance.

Before we explore climate change and especially current problems, we first need to be aware of some terms and the phenomena driving current global warming.

1.2 The greenhouse effect

The greenhouse effect is not some peripheral phenomenon only of importance to global warming. The greenhouse effect is at the heart of the Earth's natural climatic systems. It is a consequence of having an atmosphere, and of course the atmosphere is where climates are manifest.

The French mathematician Jean-Baptiste Joseph Fourier (not to be confused with the contemporary chemist of the same name) is generally credited with the discovery of the greenhouse effect. He described the phenomenon, in 1824 and then again in a very similar paper in 1827 (Fourier, 1824, 1827), whereby an atmosphere serves to warm a planet. These papers almost did not get written as Fourier was very nearly guillotined during the French Revolution and only escaped when those who condemned him were ultimately guillotined themselves.

Perhaps the best way to illustrate the greenhouse effect is to consider what it would be like if the Earth had no atmosphere. This is not as difficult as it might first seem. We only have to travel 384 400 km (238 856 miles) to the Moon and see the conditions there. On that airless world (its atmosphere is barely above vacuum at one trillionth (10^{-12}) of the Earth's) the daytime temperature is 390 K (117 °C), while at night it drops to 100 K (-173 °C), giving a median of some 245 K (-28 °C). During the lunar day, sunlight is either reflected off the Moon's rocky surfaces or is absorbed, warming the rocks that then re-radiate the energy. The total amount of incoming radiation equals that outgoing. However, at the Earth's surface the average global temperature is higher, at about 288 K (15 °C). The Earth's atmosphere keeps the planet warmer than it would otherwise be by some 43 K (43 °C). This 43-K warming is due to the Earth's atmospheric greenhouse. It is perfectly natural. This warming effect has (albeit to a varying extent) always existed. It occurs because not all the thermal radiation from the Sun falling on our planet's surface gets reflected back out into space. The atmosphere traps some of it just as on the Moon rock is warmed. However, more is trapped on Earth because the atmosphere is transparent to some frequencies (the higher frequencies) of thermal radiation, while opaque to some other, lower, frequencies. Conversely, rock on the Moon is not at all transparent so only the surface of the rock warms and not the strata deep beneath.

The reason why some of the light reflected from the Earth's surface, or radiated as infrared radiation from the lower atmosphere, becomes trapped is because it has changed from being of the sort to which the atmosphere as a whole is transparent to that to which the atmosphere is opaque. There are different types of light because photons of light can be of different energy. This energy (E) of electromagnetic radiation (light, thermal radiation and other rays) is proportional to its frequency (ν) or colour, with the constant of proportionality being Planck's constant (h, and which is estimated to be 6.626×10^{-34} J s). And so the atmosphere is transparent to some frequencies of light but not others. This transparency mix allows some higher-energy light into the blanket of atmosphere surrounding our planet, but hinders other, especially lower-energy infrared (heat-level), wavelengths from getting out. The exact mathematical relationship between the energy of a photon of light (or any other electromagnetic radiation) was elucidated, long after Fourier, in 1902 by the German physicist Max Planck. It can be expressed in the following simple equation.

$$E = h\nu.$$

E (energy) is measured in joules and ν (frequency) in hertz.

When sunlight or solar radiation is either reflected off dust particles and water droplets in the atmosphere, or alternatively off the ground, it loses energy. As a result of the above relationship between energy and frequency, this reflected light is now at a lower energy, hence lower frequency. As stated, the atmosphere, while transparent to many higher frequencies, is opaque to many of the lower thermal frequencies. The atmosphere traps these and so warms. Consequently the atmosphere acts like a blanket trapping lower-frequency radiation (see Figure 1.1). It functions just as the glass of a greenhouse does by allowing in higher-frequency light, but trapping some of the lower-frequency heat; hence the term greenhouse effect. This is why those constituents of the atmosphere that strongly exhibit these properties are called greenhouse gases. The Irish polymath John Tyndall described the greenhouse role of some gases in 1861 (Tyndall, 1861) and succeeded in quantifying their heat-absorbing properties.

There are a number of greenhouse gases. Many of these occur naturally at concentrations determined by natural, as opposed to human, factors. Water vapour (H_2O) is one, methane (CH_4) another, as is nitrous oxide (N_2O), but the one most frequently talked about is carbon dioxide (CO_2). Others do not occur naturally. For example, halocarbons such as CFCs (chlorofluorocarbons) are completely artificial (human-made), being products from the chemical industry that are used as coolants and in foam blowing. Then again, today there are the naturally occurring greenhouse gases, like carbon dioxide, whose atmospheric concentrations are further enhanced by human action.

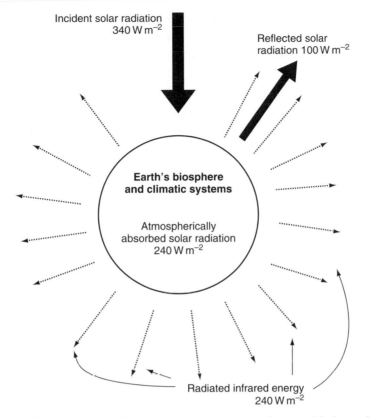

Incident solar radiation
340 W m^{-2}

Reflected solar
radiation 100 W m^{-2}

**Earth's biosphere
and climatic systems**

Atmospherically
absorbed solar radiation
240 W m^{-2}

Radiated infrared energy
240 W m^{-2}

Figure 1.1 A summary of the principal solar-energy flow and balance in the Earth's atmosphere. Not all the high-energy infrared radiation falling on the Earth is reflected back out into space. Some is converted into lower infrared energy in the atmosphere. The result is atmospheric warming. Note: the Sun radiates 1370 W m^{-2} to the Earth's distance. However, the Earth is a rotating sphere not a flat surface, so the average energy falling on the Earth's surface is just 340 W m^{-2}.

Tyndall not only recognised that there were greenhouse gases, he also speculated what would happen if their concentration in the atmosphere changed. He considered what it would be like if their warming effect did not take place (as on the Moon). Indeed, he contemplated that a reduction in greenhouse gases might throw the Earth into another ice age. Strangely though, he never considered what might happen if the concentration of greenhouse gases increased. Consequently he *never* asked what would happen if human action contributed additional greenhouse gases. In other words, what would happen if there was the addition of an anthropogenic contribution to the natural greenhouse effect?

It is this difference, between the natural greenhouse effect and the additional human-generated (anthropogenic) effect, which is at the heart of the current

issue of global warming. The Swedish chemist and Nobel laureate Svante
August Arrhenius first proposed that the human addition of carbon dioxide
to the atmosphere would result in warming in 1896, although he himself did not
use the term greenhouse but hothouse.

Today the atmosphere is indeed changing, as Arrhenius thought it might,
with the concentration of carbon dioxide increasing in recent terms largely due
to the burning of fossil fuels. In 1765, prior to the Industrial Revolution, the
Earth's atmosphere contained 280 ppm (parts per million) of carbon dioxide.
By 1990 (which is, as we shall see, a key policy date) it contained 354 ppm and
was still rising. By 2005 it had topped 380 ppm and was still climbing.

Over this time the Earth has also warmed. The warming has not been as
regular as the growth in greenhouse gas but, from both biological and abiotic
proxies (of which more later) as well as some direct measurements, we can
deduce it has taken place. Furthermore, we now know that Tyndall was right.
With less greenhouse gas in the atmosphere the Earth cools: there are ice ages.
As we shall see (in Chapter 3) we have found that during the last glacial period,
when the Earth was cooler, there was less atmospheric carbon dioxide.

Nonetheless there has been much debate as to whether the current rise in
atmospheric carbon dioxide has caused the Earth to warm. An alternative
view is that the warming has been too erratic and is due to random climate
variation. To resolve this issue the United Nations (UN), through the UN
Environment Programme (UNEP) and World Meteorological Organization
(WMO), established the Intergovernmental Panel on Climate Change (IPCC).
Its three main reports or assessments (Intergovernmental Panel on Climate
Change, 1990, 1995, 2001a, 2001b) have concluded that the 'emissions of
greenhouse gases and aerosols due to human activities continue to alter the
atmosphere in ways that are expected to affect the climate'.

The current rise in atmospheric greenhouse gases (over the past three
centuries to date) is well documented and is summarised in Table 1.1.

Table 1.1. *Summary of principal greenhouse gases (with the exception of
tropospheric ozone (O_3) due to lack of accurate data). Atmospheric lifetime
is calculated as content/removal rate.*

Greenhouse gas...	CO_2	CH_4	CFC-11	CFC-12	N_2O
Atmospheric concentration					
Late 18th century	280 ppm	0.7 ppm	0	0	288 ppb
2001	371 ppm	1.75 ppm	252 ppt	480 ppt	315 ppb
Atmospheric lifetime (years)	50–200	12	45	130	114

ppb, parts per billion; ppm, parts per million; ppt, parts per trillion.

As we shall see, each of the above greenhouse gases contributes a different proportion to the human-induced (anthropogenic) warming, but of these the single most important gas, in a current anthropogenic sense, is carbon dioxide.

There are two reasons for the different warming contributions each gas makes. First, the concentrations and human additions to the atmosphere of each gas are different. Second, because of the physicochemical properties of each gas, each has a different warming potential.

With regards to changes to the various present-day concentrations of the different gases, they are due to the post-Industrial Revolution increases in each gas: human influences on the global atmosphere were very different before the Industrial Revolution. The changes in the concentration of these key greenhouse gases each largely arise from different sets of human actions. For instance, part of the increase in carbon dioxide comes from the burning of fossil fuels and part from deforestation and changes in land use. Again, some of the increase in methane comes from paddy fields, while part of the rest comes from the fossil-fuel industry and biomass burning. We shall examine this in more detail in the next section when looking at the carbon cycle, but other methane increases (or, in the pre historic past, decreases) are due to more complex factors such as the climate itself, which can serve to globally increase, or decrease, the area of methane-generating wetlands.

Both carbon dioxide and methane are part of the global carbon cycle (see the following section). Nitrous oxide (N_2O) forms part of the nitrogen cycle and, like carbon dioxide and methane, has both natural and human origins. Naturally, nitrous oxide is given off by the decomposition of organic matter in soils, in particular by tropical forest soils that have high nutrient-cycling activity, as well as by oceans. Human sources include biomass burning and from the use of fertilisers. The principal agent removing nitrous oxide from the atmosphere is photolysis – removal by the action of sunlight – ultimately resulting in nitrogen (N_2) and oxygen (O_2).

As to the second factor determining the different warming contribution each gas makes, each has different physicochemical properties. These are quantified for each gas in what is called their global warming potential (GWP). GWPs are a comparative index for a unit mass of each gas measured against the warming potential of a unit mass of carbon dioxide *over a specific period of time*. Carbon dioxide has, therefore, a defined warming potential of 1. A complicating factor is that because different greenhouse gases have different atmospheric residence times (see Table 1.1) GWPs *have* to relate to a specific time frame. A GWP expressed without a time frame is nonsense. This can be understood by considering methane, which only has an average atmospheric

Table 1.2. *Global warming potentials (GWPs) for some of the principal greenhouse gases over three time frames (IPCC, 2001a).*

Gas	Atmospheric lifetime (years)	GWP		
		Time horizon ... 20 years	100 years	500 years
Carbon dioxide	50–200	1	1	1
Methane	12	62	23	7
Nitrous oxide	114	275	296	156

residence time of a dozen years. Nearly all of a kilogram of methane will still be in the atmosphere after a year. Roughly half of it will be in the atmosphere after 12 years and, assuming exponential decay, a quarter or less after 24 years. Conversely nitrous oxide has an average residence time of over a century. So, clearly, comparing the GWPs of nitrous oxide and methane over a decade will give different warming figures compared with the same comparison over a century. Finally, because of uncertainties, not least with carbon dioxide's own atmospheric residence times, different researchers have different GWP estimates. This can be especially frustrating, as estimates 'improve' with time or as different theories as to the dominating effect of, for example an aspect of the carbon cycle, come into vogue, it means that GWPs often vary both with research team and with time. Even the IPCC's GWP estimates vary a little from report to report. Furthermore, because the IPCC is science by committee – where uncertainty is resolved through consensus of opinion – one cannot simply dismiss one research team's estimates as being completely out of hand. Instead, when looking at a research team's climatic model, you need to see what GWP estimates are used as well as the model itself and then make your own judgement on its results compared to those of another team. Table 1.2 summarises the IPCC's 2001a estimates for GWPs for carbon dioxide, methane and nitrous oxide. CFCs (chlorofluorocarbons) and HFCs (hydrofluorocarbons) are not included as there are so many different ones. However, typically most have GWPs of a few thousand (compared to carbon dioxide's GWP of 1) for time horizons up to 500 years. Fortunately because of their low atmospheric concentration, human-made chemicals such as CFCs and HFCs contribute less than a quarter of current warming (see Figure 1.2).

There is one important greenhouse gas that has only briefly been mentioned so far, and that is water vapour. Water vapour is a powerful greenhouse gas contributing a significant proportion of the natural (as opposed to the human-induced) greenhouse effect. There is sufficient water vapour above the troposphere for it to absorb much of the infrared radiation at its absorptive

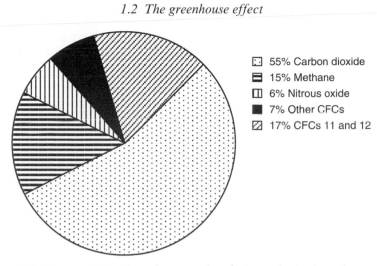

☐ 55% Carbon dioxide
☐ 15% Methane
☐ 6% Nitrous oxide
■ 7% Other CFCs
☒ 17% CFCs 11 and 12

Figure 1.2 The contribution from each of the principal anthropogenic greenhouse gases due to the change in warming (radiative forcing) from 1980 to 1990 (excluding ozone, which may or may not be significant and is difficult to quantify). Data from IPPC (1990).

frequencies. Indeed, if we were to look at the Earth from space, solely in water-vapour frequencies, our planet would appear as mist-veiled as Venus. This is true even over the dry Sahara Desert. But the concentration of water vapour is not consistent throughout the entirety of the atmospheric column. Tropospheric water vapour, in the atmospheric layer closest to the ground, varies considerably over the surface. In the first 1 2 km of the atmosphere (in the lower part of the troposphere), the amount of water vapour in a unit volume increases with temperature. In the troposphere above this point, the water-vapour greenhouse effect is most important and harder to quantify. Furthermore, current computer models of the global climate account for water-vapour feedback, whereby a warmer world sees more evaporation, hence more water vapour, and this tends to double the warming that one would expect from just a fixed-water-vapour model. The ability of current (early twenty-first century) global climate computer models to reproduce the likely effect of water vapour over a period of warming was given credence in 2005 by a US team of atmospheric scientists led by Brian Soden. They compared satellite observations between 1982 and 2004 at the 6.3 μm wavelength, which is part of water's absorption spectrum and especially useful for measuring its presence in the upper troposphere, and climate models. The satellite measurements and the models showed a good correlation.

Clouds (the suspension of fine water droplets in regions of saturated air) complicate the picture further still. Being reflective they tend to cool the surface during the day and at night act as an effective greenhouse blanket.

However, there are clouds and there are clouds. The picture is complex and our understanding incomplete, hence climate models are only an approximation of what is going on, but revealing approximations nonetheless. (We will return to climate change and the water cycle later in this chapter.)

Given that overall the Earth's atmosphere is broadly conferring a 43 °C greenhouse warming effect (since, as we have seen, the airless Moon is cooler), the question remains as to how much warming has been conferred anthropogenically since the Industrial Revolution, due to the human addition of greenhouse gases. We shall come to this in Chapter 5. Nonetheless it is worth noting for now that mathematicians Cynthia Kuo and colleagues from the Bell Laboratory, New Jersey, USA, statistically compared instrumentally determined changes in atmospheric CO_2 concentrations between 1958 and 1989 and global temperature (Kuo *et al.*, 1990). This confirmed that carbon dioxide and global temperature over that period were significantly correlated to over 99.99%. This is to say that were 10 000 alternative copies of the Earth similarly measured that only one would give similar results due to sheer chance and 9999 would give results because there is a link between carbon dioxide concentrations and global temperature. But before looking at how the human addition of carbon dioxide to the atmosphere affects climate we need a better understanding of atmospheric carbon dioxide's natural sources and sinks. Fundamental to this is the carbon cycle.

1.3 The carbon cycle

Carbon is one of the fundamental elements necessary for life. It is found in virtually all molecules (but not quite every molecule) associated with life. These include all carbohydrates, all proteins and all nucleic acids. As such, carbon is fundamental to biological structures, of both micro- and macro-organisms, including plants and animals; for example, lignin in plants and cartilage and bone in animals. Indeed biomolecules, as we shall see (Chapter 2), can be of great use to palaeoclimatologists as some of them (and hence the remains of species in which they are found) can be used as climatic indicators.

The carbon cycle itself refers to the circulation of carbon in the biosphere. The circulation is driven primarily (but not solely) by biological processes. A planet that does not have any biological processes sees carbon flows through its geosphere driven solely by geophysical processes. On Earth carbon, in the form of carbon dioxide, is fixed by photosynthesis into organic compounds in plants and photosynthetic algae and returned to the atmosphere mainly by the respiration of plants, animals and micro-organisms in the form of carbon dioxide, but also by the decay of organic material in the form of both methane and carbon dioxide. Abiotic drivers include the burning of organic material, be it natural (e.g. forest fires) or through human action (e.g. the burning of firewood or fossil fuels). Another abiotic driver is that of plate-tectonic movement. This contributes to the so-called deep carbon cycle operating on a scale of millions of years. Here plates in the process of subduction carry with them organic sediments down into the Earth's mantle. This plate movement results in volcanic activity that in turn converts the organic sediments to carbon-containing gases (again mainly carbon dioxide but also other volatile compounds) that come to the surface via volcanic and related activity. There are a variety of other abiotic processes including the chemical oxidation of methane in the atmosphere to carbon dioxide as well as processes (which frequently also accompany biotic ones) in organic sediments. An estimate of the principal carbon movements within the carbon cycle is given in Figure 1.3.

The carbon cycle is at the centre of biology's relationship with the global climate (and hence global climate change). More than this, it demonstrates the

Figure 1.3 Broad estimates of the principal carbon sources and sinks in gigatonnes of carbon (GtC) as well as approximate movements of carbon about the carbon cycle in gigatonnes of carbon per annum (GtC year^{-1}). It is important to realise that there are a number of uncertainties that are the subject of current research (see text). The figure includes 2001 IPCC estimates for annual fluxes (1990–1999), marked *.

importance of both the new molecular biological sciences as well as the whole-organism approach to biology.

A first impression of the biology associated with the carbon cycle might focus on the plant activity sequestering carbon dioxide from the atmosphere. But this is just one aspect, albeit a key one. There is also plant and animal respiration, and the respiration of bacteria and fungi, returning carbon dioxide to the atmosphere. We might also think of biomes (climatically determined regional groups or assemblages of ecosystems) and how these might affect the global carbon cycle in terms of marine productivity or of the carbon in a biomes' biomass, be it in terrestrial tropical rainforests, temperate wetlands, etc. We might also consider the effect of climate change on such ecosystem assemblages as well as individual ecosystems; and indeed we will later in the book. But there is also the biomolecular dimension.

Included in the biomolecular perspective is the role enzymes play. Rubisco (ribulose-1,5-bisphosphate carboxylase oxygenase, which is sometimes portrayed in print as RuBisCO) is the most common enzyme on the planet and is fundamental to photosynthesis. It is therefore probably the most common protein on Earth; it constitutes about half of leaf proteins and is synthesised in chloroplasts. All carbon dioxide captured by photosynthesis – be it in algae or multicellular plants – is handled by this one enzyme. That is about 200 billion t (or 200 Gt) of carbon a year! Another important enzyme is carbonic anhydrase, which catalyses the hydration of about a third of the carbon dioxide in plants and in soil water. As we shall see, the way these enzymes handle carbon dioxide molecules with different isotopes of carbon, or oxygen with different isotopes of oxygen, help us in our understanding of carbon-cycle details.

It is important to note that there are some uncertainties in the estimates of the rate of flows between the various sources and sinks (Figure 1.3). This is the subject of on-going research. Part of the problem (other than the cycle's complexity) is that the carbon cycle is not static: there are varying transfers of carbon, so altering the amounts within reservoirs. For example, during cold glacial times (such as 50 000 years ago) the amount of atmospheric carbon (as both carbon dioxide and methane) was less than today. Conversely, currently the atmospheric carbon reservoir is increasing. This dynamism is not just because of human action; it also happens naturally, although human action is the current additional factor critical to what is called global warming.

In practical terms today, carbon-cycle uncertainties manifest themselves in a number of ways. Of particular concern is the mystery as to where roughly half the carbon dioxide released into the atmosphere by human action (from fuel burning and land-use change) ends up: from measuring the atmospheric concentration of carbon dioxide we know that half of what we burn does not

Table 1.3. *Estimated annual carbon emissions in 1980–9 and 1990–9 to the atmosphere and annual transfer from the atmosphere to sinks (IPCC, 2001a).*

	Carbon transfer (Gt of carbon year^{-1})	
	1980–9	1990–9
From fossil fuels to atmosphere	5.4 ± 0.3	6.3 ± 0.4
From deforestation and land-use change to atmosphere	0.2 ± 0.7	1.4 ± 0.7
Accumulation in atmosphere	3.3 ± 0.1	3.2 ± 0.1
Uptake by ocean	1.9 ± 0.6	1.7 ± 0.5
Net imbalance in estimates	0.4 ± 1.7	2.8 ± 1.7

remain for very long in the atmosphere. This imbalance is so significant that there is debate as to whether a major carbon-cycle process has been overlooked. Alternatively it could be that the current estimates as to the various flows have a sufficient degree of error that cumulatively manifests itself as this imbalance.

The broad (IPCC, 2001a) estimates (together with estimates of uncertainty) as to the contributions of burning fossil fuels and deforestation to atmospheric carbon dioxide as well as the entry to carbon sinks from the atmosphere for the decades 1980–9 and 1990–9 are listed in Table 1.3. The estimates are derived from a computer simulation.

As we can see from Table 1.3 there is this net imbalance between the estimates of carbon dioxide entering and of those accumulating and leaving the atmosphere. More carbon dioxide is being released into the atmosphere than is retained by the atmosphere or thought to be absorbed by the oceans. Where is this carbon dioxide going? This net imbalance is both large and of the same order of magnitude as existing global flows. This is why some think that a major route of carbon from the atmosphere has not been identified, or alternatively that one or more of the existing carbon-flow estimates (Figure 1.3) is considerably off the mark, or perhaps a bit of both. As regards the flow of fossil-fuel carbon to the atmosphere, this is quite well documented due to the economic attention paid to the fossil-fuel industry: we know how many barrels of oil are sold, tonnes of coal are mined, etc. So we can be fairly certain that this estimate is broadly accurate.

Looking at the other side of the equation, the accumulation of carbon in the atmosphere is also accurately charted as it can be, and has been, directly monitored over many years from many locations. It is because we know exactly how much extra carbon we release, and have released, from fossil fuels into the atmosphere as well as how much actually stays in the atmosphere, that we can be certain that there is a shortfall and so that some part of the carbon cycle has either not been properly quantified or even perhaps not properly identified.

It is uncertainties such as this, and that the global climate-warming signal had to be sufficient to be discernable from the background natural variation (noise), that has helped some argue that global warming is not taking place. As we shall see over the next few chapters, the climate has changed in the past (affecting biology and vice versa) and the atmospheric concentrations of greenhouse gases have played a major part in these changes.

Interestingly, the current year-on-year accumulation in atmospheric carbon dioxide, as measured in either the northern or southern hemisphere, is not smooth. Rather, there is an annual oscillation superimposed on the rising trend. The oscillation occurs because of seasonality outside of the tropics. During winters in the temperate zone there are no leaves on the trees and in the boreal zone too there is little photosynthesis on land or in the sea (in the main by algae). But in the summer there is considerable photosynthesis and so more carbon dioxide is drawn into plants and algae. In winter respiration continues, even though photosynthesis is reduced and so, on balance, more carbon dioxide is released into the atmosphere. So there is this annual cycle of waxing and waning of atmospheric carbon dioxide in the northern and southern hemispheres (see Figure 1.4). Indeed, because while in one hemisphere there is summer and the other winter, the carbon dioxide oscillations in the two hemispheres are opposite and complement each other. However, the seasonal variation of carbon dioxide in the southern hemisphere is not nearly so marked, as the southern hemisphere is dominated by ocean, which has a strong ability to buffer carbon dioxide. Oxygen, as the other gas concerned with photosynthetic and respiration reactions, also shows a seasonal variation in each of the hemispheres but one that is more marked than that for carbon dioxide (as oxygen does not buffer in the seas). Like carbon dioxide, the seasonal variation of the atmospheric concentration of oxygen is equal and opposite in the northern and southern hemispheres.

The changes in the atmospheric reservoir of carbon are quite well understood (even though our knowledge of other reservoirs is not so complete) because we can measure atmospheric carbon dioxide directly, and because atmospheric mixing within hemispheres is reasonably thorough. We can therefore be quite certain that our knowledge of this atmospheric part of the carbon cycle is fairly accurate. The problem of the missing carbon seems to be associated with one or more of the other carbon sinks: possibly the accumulation of carbon by terrestrial plants, or alternatively by absorption into the

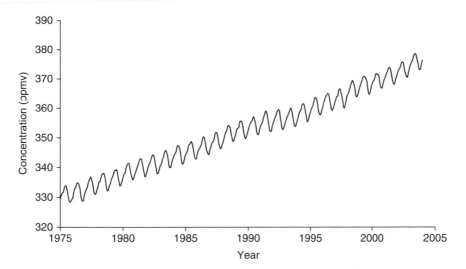

Figure 1.4 Outline of hemispheric seasonality effects of atmospheric carbon dioxide concentration of the late 1970s through to the early twenty-first century. Note that in addition to the annual waxing and waning, there is also an overall trend of growth in atmospheric concentration. ppmv, parts per million by volume.

oceans. Nonetheless, the annual waxing and waning of atmospheric carbon dioxide does illustrate the power of the photosynthetic carbon pump (the power of plant and algal photosynthesis to drive the carbon cycle).

Turning to human impacts on terrestrial reservoirs of carbon, estimates of deforestation do have greater error associated with them than changes in atmospheric gas concentrations, which can be directly measured. So it may well be that part of the missing carbon flow in the carbon cycle is associated with this component of the cycle. However, instances of more detailed scrutiny of deforestation data have revealed that official estimates are invariably underestimates, and that deforestation probably accounts for a greater contribution to atmospheric carbon dioxide, not less, and so the amount of missing carbon is greater. This means that, if anything, our understanding of the cycle's carbon imbalance is not even as good as we think. Of course, deforestation is but one dimension, of a number, to the changes of the terrestrial reservoirs of carbon.

The importance and power of the photosynthetic pump driving part of the carbon cycle is corroborated by the magnitude of the seasonal oscillation in carbon dioxide. Nonetheless land-use change, along with terrestrial-biome change, as well as ocean accumulation, are key areas of uncertainty (either singly or together) that might account for the missing carbon. It could be that oceans are accumulating more carbon than we think and/or that the increased atmospheric carbon dioxide along with global warming is encouraging terrestrial photosynthesis, drawing down carbon into plants over much of the globe. As stated above, because of the magnitude of seasonal variation in atmospheric carbon dioxide outside of the tropics we can see how powerful the photosynthetic and respiratory carbon pumps truly are. *If* this missing carbon sink is terrestrial (and given that much of the planet's land is in just one, the northern, hemisphere) it appears that the carbon may be being sequestered not in a temporary way by annual plants but in a longer-term way by perennials, and especially temperate and boreal trees. Then there is the carbon stored in soils and detritus. The total carbon store in soils, at 1500 GtC, is over twice that stored as biomass and is also more than the atmospheric carbon reservoir. Currently the reservoir of soil carbon is a net sink, although carbon flows to and from it are far less than those to and from either biomass carbon or atmospheric carbon reservoirs. It is important to note that *currently* soil acts as a global net carbon sink. 'Currently' because a world that is just a few degrees warmer could see the soil reservoir act as a net source of carbon as it would be released as carbon dioxide into the atmosphere. In such an instance soils would act to further global warming. (We will return to carbon in soils later in this chapter and again in Chapters 5 and 7.)

Again, as previously mentioned, alternatively (or in addition – we just do not know) the oceans could be a greater sink of carbon than we realise, so they could account for the carbon imbalance. Either way, it is almost certain that the driving force behind this missing sequestration of additional atmospheric carbon is photosynthesis, even if we are unable to say whether it is marine or terrestrial, let alone where on Earth it is taking place.

With regard to the scale of carbon flow between carbon-cycle reservoirs due to photosynthesis, the annual natural seasonal variation in atmospheric draw-down and replenishment we observe is considerable, and it is greater than the year-on-year increasing trend due to the human addition of fossil fuels and land-use change. So just as human action (the action of just one species) is responsible for the current growth in atmospheric carbon dioxide, so one of the most fundamental biological processes – photosynthesis – (through many species) can almost certainly be involved in its amelioration. Again this emphasises that biology and climatology are closely entwined.

Here the ways that photosynthetic and respiratory enzymes handle carbon and oxygen are beginning to illuminate the problem of missing carbon. As stated above, Rubisco is the enzyme globally responsible for fixing atmospheric carbon dioxide as part of the photosynthetic process. However, not all atmospheric carbon is in the form of the ^{12}C isotope. Around 1% is ^{13}C. Rubisco evolved to handle the almost universal ^{12}C and so discriminates against ^{13}C, leaving it behind in the atmosphere. If photosynthetic activity increases (as it does each summer in each hemisphere) then the increase in atmospheric ^{13}C left behind can be measured. This also works if photosynthesis increases due to global warming, because in a warmer world the thermal growing season (TGS) is longer. On the other hand, isotopes of ^{12}C and ^{13}C dissolve more or less equally well in sea water: in fact, if anything ^{13}C dissolves slightly more easily. Consequently if we can detect changes in atmospheric ^{13}C above and beyond the seasonal changes in a hemisphere that we would expect, we can see whether the photosynthetic pump is working harder or not. Similarly, carbonic anhydrase works on the ^{16}O isotope of oxygen and discriminates against ^{18}O. Isotopic studies are therefore an increasingly important tool in understanding how the carbon cycle works.

Ascertaining the carbon cycle's details is the subject of considerable on-going research. In terms of addressing the problem of increasing atmospheric carbon dioxide, the solutions will almost inevitably involve modifying carbon flows between reservoirs such that the atmosphere's carbon burden will be reduced. We will return to this towards the end of this section. Before doing so it is important to note that this on-going carbon-cycle research does not just provide extra detail but still turns up major surprises.

One such recent (2006) surprise was the possibility that plants in aerobic conditions (with oxygen available) produce methane (Keppler *et al.*, 2006). Indeed it was so surprising that *Nature* ran a small article in its news section entitled 'How could we have missed this?' The discovery was almost fortuitous in that it had been thought that all the principal sources of atmospheric methane had been identified even if their individual quantification needed to be refined. To ensure that it was methane from plants (and not microbes) the researchers attempted to kill off bacteria on the plants with radiation. They also removed methane from the air in the incubators in which the plants were to be grown. Although the amounts of methane detected from individual plants were small, globally it amounted to a significant source. The researchers who made the discovery could only make a very rough estimate (as work has yet to be done on an appropriately representative range of species and conditions) but they thought that the annual atmospheric contribution could be between 60–240 million t (Keppler *et al.*, 2006). This is between one-twelfth and one-third of the annual amount entering the atmosphere.

As for the year-on-year growth in atmospheric carbon dioxide since the beginning of the Industrial Revolution (as stated above) this has been carefully charted. It has been done so in two main ways. First, in recent times (since the middle of the twentieth century) there has been direct measurement of atmospheric carbon dioxide. Second, there has been historic measurement from bubbles of air trapped in ice caps (mainly from either Greenland or Antarctica)

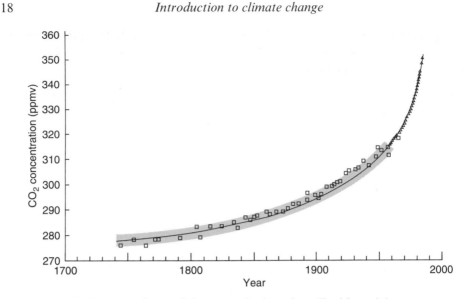

Figure 1.5 The annual growth in atmospheric carbon dioxide. □, Measurements on air trapped in ice from Antarctica; △, direct measurements taken in Hawaii.

as snow fell and then the snow turned to ice. Together these show a continuous growth in the atmospheric concentration from around the time of the Industrial Revolution through to the present (see Figure 1.5).

Much (but, remember, not all) of this growth in atmospheric carbon dioxide is due to the burning of fossil fuels. In essence this represents a short-circuiting of the aforementioned deep carbon cycle: the part of the carbon cycle that takes millions of years to complete. For while, as discussed above, much of the annual movement of carbon dioxide is due to respiratory and photosynthetic processes and contributes to the fast carbon cycle, there is also a slower accumulation of about 1 billion t of carbon a year in marine sediments and more being trapped terrestrially in soils and wetlands. After millions of years much of this ultimately ends up as coal and oil, the larger deposits of which we have only recently (in geological terms) mined and burned as fossil fuels. We are burning these fossil fuels at a far faster rate, thousands of times faster, than they are currently being formed, for the process of fossil fuel formation continues today. It is as if there are two carbon cycles within the overall carbon cycle. One is driven by photosynthesis, respiration and forest fires and operates over a short period of time. The other is the deep carbon cycle operating in so-called deep time of many millions of years and is driven by the geological formation and entrapment of fossil fuels and the tectonic subduction of carbon-rich sediments at the edge of some plates followed by the emission of carbon from volcanoes.

There is currently much research being done to elucidate the carbon cycle's operation, and to measure or infer carbon sinks (Houghton, 2002). This work might be grouped as follows.

1. Global budgets based on atmospheric data and models. The use of data from nearly 100 sites around the Earth of atmospheric carbon dioxide and isotopes.
2. Global budgets based on models of oceanic carbon uptake. The use of models of the oceans' carbon cycle and chemistry linked to those of terrestrial and atmospheric sources.
3. Regional carbon budgets from forest inventories. Many developed nations have national forest inventories and changes in volumes over time can indicate sources or sinks of carbon. However, we are a long way from making this accurate, either on a national basis or for global coverage, although progress is being made.
4. Direct measurements of carbon dioxide above ecosystems. Using stand towers in forests it is possible to measure changes in carbon dioxide being given off or absorbed. (This is quite different from measuring atmospheric carbon dioxide on top of a Hawaiian volcano (one of the key measurement sites) combined with scores of other direct atmospheric measurements to obtain a hemispheric average.)
5. Earth-system science modelling using ecosystem physiology. Using global models built up from global biome and ecosystem data. One of the big questions here (apart from the accuracy of the size of the ecosystem components) is whether all the important ecosystem processes have been included or properly quantified.
6. Carbon models based on changes in land use. This is related to item 5 and has similar constraints.

Having looked at the broad areas of research into the carbon cycle, this leads us on to the question of whether, because we are already altering one part of the carbon cycle so as to increase atmospheric carbon dioxide, we can alter another part to counteract this effect? Given that carbon dioxide is the principal anthropogenic greenhouse gas, and that the carbon cycle is the determining phenomenon in its atmospheric concentration, it would at least appear logical that we might alter the way carbon is currently cycled so as to offset atmospheric carbon increases. One way would be to increase terrestrial photosynthesis through planting new forests, so sequestering carbon, and we will return to forests and biofuel options later (see the end of Chapter 7). Another might be to increase marine photosynthesis.

Marine photosynthesis is mainly carried out by phytoplankton in the open oceans. Here the dominant species are principally the prokaryotes (organisms without internal structures surrounded by cell membranes) *Prochlorococcus* and *Synechococcus*. Both are cyanobacteria (also known as blue-green algae). In terms of crude numbers, *Prochlorococcus* is probably the most populous species on the planet. In addition to sunlight, carbon dioxide and water, these plankton species also require nitrates, phosphates and small amounts of metals. In the ocean, close to the surface, more than enough sunlight is present to drive photosynthesis, but it has been found (in parts of the Pacific at least) that raising the concentration of iron to about 4 nM (nano moles per litre) results in planktonic blooms and associated increased photosynthetic production. The most dramatic of these experiments have been the IronEx I (1993) and II (1996) experiments that covered an area of about 70 km^2, although an area larger than this (1000 km^2) had to be surveyed due to the blooms' drift. This has led to the speculation that it may be possible to use oceanic iron fertilisation to sequester atmospheric carbon. However, modifying the base of some of the planet's major ecosystems, such as in this way, may well carry with it unacceptable ecological risks. Furthermore, it is one thing for winds to carry a global load of minerals to fertilise the oceans and quite another for humans to do so. Indeed, it appears that the energy required to distribute the iron over the ocean surface would roughly equate in fossil-fuel terms with the carbon assimilated. Even so, it does appear that in the past natural changes in the carbon cycle, almost certainly involving the marine component, have had a major effect on the global climate (Coale *et al.*, 1996).

1.4 Natural changes in the carbon cycle

We know that atmospheric carbon dioxide plays a major role in contributing to the natural greenhouse effect and we also know that this natural greenhouse effect has varied in strength in the past. Perhaps the most pronounced evidence comes from Antarctic ice cores. Snow falls in Antarctica to form ice and in the process tiny bubbles of air from the atmosphere become trapped and sealed within it. As more snow falls, more ice with bubbles builds up. By drilling a core into the ice it is possible to retrieve atmospheric samples of times past. Indeed, cores at one spot, Vostok in eastern Antarctica, have provided an atmospheric record going back well over 100 000 years. This is a long-enough time to cover the last glacial–interglacial cycle (and more; a glacial being the cool part of the current ice age compared with the warmer interglacials, such as the one we are presently in; see Chapter 3). The ice at Vostok is well over 2 km thick and the cores retrieved in the mid-1980s through to the present have clearly showed that concentrations of carbon dioxide and methane were far lower during the cool glacials than they were during the warmer interglacials. These palaeo-concentrations can be directly compared to the estimated

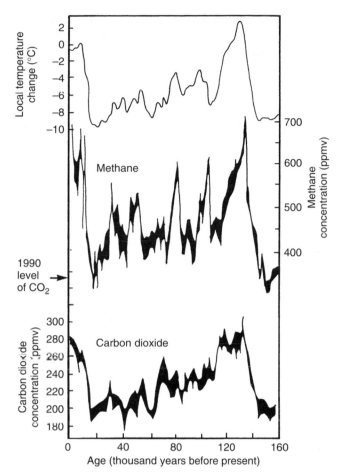

Figure 1.6 Atmospheric carbon dioxide and methane palaeo-record for the past glacial–interglacial cycle plotted with regional temperature change as indicated by the ice hydrogen isotope proxy. Adapted from Barnola *et al.* (1987) and Chappelaz *et al.* (1990), reproduced from the IPCC (1990) with permission.

difference – from the ice-water's deuterium concentration – in temperature between the oceans from which water at that time evaporated and when it fell as snow. This is because it takes more energy (heat) to evaporate water containing the heavier deuterium isotope of hydrogen (^{2}H) than water made up of the common isotope of hydrogen (^{1}H). Therefore, a plot of the ice-core's deuterium concentration gives an indication of regional temperature. Such temperature changes, it can be seen, closely correlate with carbon dioxide and methane concentrations (see Figure 1.6). Such ice-core evidence suggests that atmospheric carbon – be it carbon dioxide or methane – really is linked to climate. Because we know from laboratory analysis that these gases are greenhouse gases (absorbing long-wave infrared radiation), we can deduce that this is the mechanism linking them to climate. Equally importantly, because we know that the atmospheric concentrations of both these gases are affected, if not determined, by the carbon cycle, we have a direct link between the carbon cycle and climate. As one of the carbon cycle's key drivers is photosynthesis, we can see that life is clearly linked to the global climate.

1.5 Pacemaker of the glacial–interglacial cycles

There have been a score or so of these glacial–interglacial cycles over the past 2 million years and, at first sight, these have an apparent regularity. The question then arises as to whether there is anything causal driving this periodicity.

In the 1920s the Serbian mathematician Milutin Milankovitch (although owing a debt to the work of James Croll in 1864) suggested that minute variations in the Earth's orbit could affect the sunlight reaching its surface. This happens largely because the Earth's geography is currently asymmetric: at the present moment in geological time the northern hemisphere is largely land-dominated while the southern is ocean-dominated, so each hemisphere differs in the way it absorbs the Sun's heat. Also at the present time the positioning (through plate tectonics) of the North and South American and the Afro-Eurasian continents restricts oceanic currents, and hence heat transport, about the planet. Oceanic currents are further restricted at the poles, in the north by the North American and Asian continents and in the south by Antarctica. The way heat is accepted by and transported about the Earth under these constraints makes the planet prone to glacials. Milankovitch calculated a theoretical energy curve for changes over time based on the variations of three orbital parameters: angle of axial tilt, or obliquity, which varies between 22° and 24.6° over a 41 000-year cycle; orbital eccentricity (the degree of elliptical deviation from a true circle), which varies every 93 000–136 000 years (so, approximately 100 000 years) between a true circle (an eccentricity of 0) and an ellipse (of eccentricity 0.05) and back again (there is also a 400 000-year resonance); and precession of the equinoxes (the way the Earth's axis spins slowly like a gyroscope), which takes place with a cycle of roughly 19 000–23 000 years (effectively 22 000 years; see Figure 1.7). The climatic relevance of this last, put simply, means that at a specific time of year, say mid summer's day, the Earth is closer or not to the Sun than on other mid summer days. Pulling all these factors together Milankovitch could predict when the Earth was likely to experience a cool glacial or a warm interglacial (Milankovitch, 1920).

Milankovitch concluded that a drop in the amount of sunlight falling on the northern hemisphere at the end of both the precession and tilt cycles would make it more likely for there to be a glacial and that when the opposite happened it would coincide with the timing of an interglacial. For many years Milankovitch's theory did not have much currency. This was due to

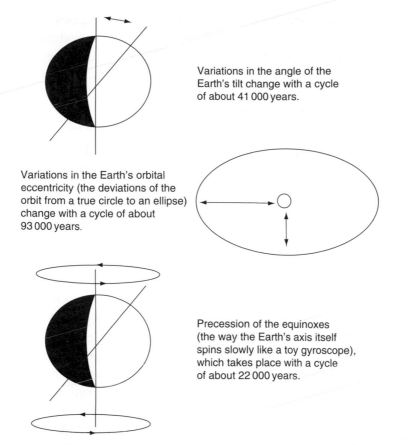

Variations in the angle of the Earth's tilt change with a cycle of about 41 000 years.

Variations in the Earth's orbital eccentricity (the deviations of the orbit from a true circle to an ellipse) change with a cycle of about 93 000 years.

Precession of the equinoxes (the way the Earth's axis itself spins slowly like a toy gyroscope), which takes place with a cycle of about 22 000 years.

Figure 1.7 Milankovitch orbital parameters of eccentricity, axial tilt and precession of the equinoxes. Each parameter varies through a cycle of different period. Each affects the way the Earth is warmed by the Sun.

two main things. First, for much of this time there was no real understanding of when in the past glacials and interglacials had actually taken place, so Milankovitch's theory could not be checked. Second, the variations in the solar energy falling on a square metre in the northern hemisphere that Milankovitch was talking about were of the order of $0.7\,\mathrm{W\,m^{-2}}$; in other words less than one-tenth of 1% of the sunlight (solar constant being $1400\,\mathrm{W\,m^{-2}}$) bathing the planet. What turned things around was related to the palaeo-evidence from Antarctica's ice cores in the 1970s and 1980s of when glacials and interglacials had taken place. This confirmed Milankovitch's timings and, as we shall see below, there is a considerable body of other biotic evidence corroborating the timing of past climates and climatic change.

If Milankovitch's theory simply provides a glacial–interglacial pacemaker but does not account for sufficient energy changes needed to instigate and terminate glacials, then what is amplifying this signal? The answer lies within the complexity of the global bio-geosphere system (from here on simply referred to as the biosphere system). There are numerous factors determining the global climate. Some, such as silicate erosion (see Chapter 3), affect the planet over long timescales. Others, such as the burning of fossil fuels (a major factor), stratospheric and tropospheric ozone (medium factors) and biomass burning, mineral-dust aerosols and variations in the Sun's energy output (very low factors), affect the climate on timescales of far less than a century. Other factors we still know little about and so their climatic effects are hard to quantify (such as aircraft con (condensation) trails; see Chapter 5). Complicating matters further still, there are many factors that conspire, or interact synergistically, to affect the climate with positive or negative feedback. These feedbacks either amplify climate change or have a stabilising effect. This text focuses on the biology of climate change but it is important for life scientists interested in climate change to have at least a basic appreciation that such feedbacks exist. Figure 1.8 illustrates three such feedbacks (there are many). Figure 1.8a and 1.8b are physical systems and might operate on a lifeless planet. These are both examples of positive feedback that add to, reinforce or amplify any forcing of the climate. That is to say, if something (be it a release of a greenhouse gas, either human-made or natural) forces the climate to warm up, then these feedbacks will serve to amplify the net warming.

Figure 1.8c represents a biophysical feedback system of a different kind. This is an example of a negative-feedback system that dampens any net change in the climate. We have already referred to iron that can fertilise the oceans, so encouraging algae that in turn draws down carbon dioxide from the atmosphere, so reducing the greenhouse effect and cooling the planet. Take this a step further. Consider a world slightly warmer than ours. Being slightly warmer there is more evaporation from the oceans; more evaporation means more rainfall (and/or snow), which in turn means more erosion. This increased geological erosion increases the amounts of iron (eroded from minerals) transported to the oceans that in turn encourages algae, which draws down atmospheric carbon dioxide. Of course, the timescale and magnitude of this natural effect may not be as large as some might wish, hence the discussion as to whether we should deliberately fertilise oceans with iron or take some other measure (although it would be unwise to tinker with the planet's biosphere mechanisms without a thorough understanding). Given that there are feedback processes that amplify change as well as those that stabilise the climate, it is not surprising that changes in the global climate are not always gentle. For example, we do not see a gentle segue from a glacial to an interglacial. Instead we see a sharp transition between the two (see Figure 1.6). It is as if positive feedback encourages sharp changes while negative feedback encourages stable

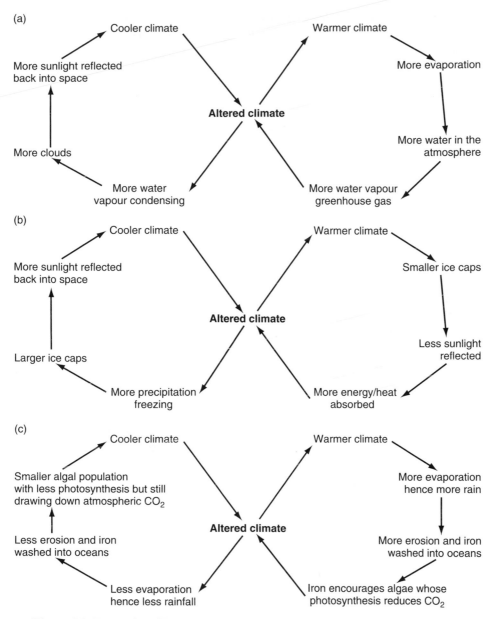

Figure 1.8 Examples of interacting positive- and negative-feedback mechanisms affecting climate change.

states and that either one or other of these two types of feedback dominates at any given time. The combined global climate picture is one of stable (or semi-stable) states between which there is occasional rapid flipping. Here one of the main pacemakers timing these flips is the combination of Milankovitch's orbital parameters.

   The question that arises from all of this is whether, with current global warming, the Earth is now shifting towards a new feedback system that may encourage further warming? One example of such a concern is that of carbon in soils, especially at high latitudes, and whether it may be released through warming into the atmosphere. Such soils include peatlands that are at such a high latitude that they are either frozen for part of the year or are permafrosts. Michelle Mack from the University of Florida and colleagues (Mack *et al.*, 2004), including those from the University of Alaska Fairbanks, have looked at carbon storage in Alaskan tundra. Such carbon storage in tundra and boreal soils is thought to be constrained by carbon–nutrient interactions because plant matter is the source of much (nearly all) soil carbon and plant growth is usually nitrogen-limited. Should soils warm in response to climate change, it is thought that nutrient mineralisation from soil organic matter will increase. This should increase plant growth. However, total-ecosystem carbon storage will depend on the balance between plant growth (primary productivity) and decomposition. Experiments at lower latitudes (in temperate and tropical zones) have given variable results (although we will cite a major (albeit rough) assessment of European soil carbon in Chapter 7), but high-latitude ecosystems, because of the large amount of carbon in their soil, show a clearer relationship between productivity and soil carbon storage. In 1981 Mack and colleagues began one of the longest-running nutrient-addition experiments in Alaska by adding $10\,g$ of nitrogen and $5\,g$ of phosphorus $m^{-2}\,year^{-1}$. This is about five to eight times the natural deposition rate in moist acidic tundra soils. Two decades later and poaceden or the graminoid (grass) tundra – which is dominated by the tussock-forming sheathed cottonsedge (*Eriophorum vaginatum*) – had changed to a shrub tundra dominated by the dwarf birch (*Betula nana*). The carbon above ground (in the form of plants and litter) had increased substantially; however, this was more than offset by a decrease in carbon below ground in the soil. This decrease in below-ground carbon was so great that the net result for the ecosystem was a loss of $2000\,g$ of carbon m^{-2} over 20 years. (The ecosystem's nitrogen did not change nearly so much, other than that a greater proportion of it at the end of the experiment was found above ground in the vegetation.) This loss of carbon is approximately 10% of the initial carbon in the ecosystem. The fear is that as globally soils hold $1500\,GtC$ (compared with some $750\,GtC$ in the atmosphere) and that as about a third of soil carbon is in arctic and boreal soils, that warming of these soils might make a significant contribution to atmospheric carbon. This carbon contribution would, should it occur, be beyond what is being added to the atmosphere through human actions such as fossil-fuel burning and land-use change. It would therefore exacerbate current global warming.

Returning to the aforementioned variable results of carbon-release change with temperature in temperate and tropical soils, it is beginning to look as if considering soil carbon to be in various forms with different turnover rates may explain matters. A three carbon-pool model, with each pool having differing turnover times, was proposed in 2005 (Knorr *et al.*, 2005). This was then applied to data from 13 previously published soil-warming experiments covering tropical and temperate soils that lasted over 100 days up to nearly 2 years. It gave somewhat varying results, but importantly this model was compatible with earlier work. What appears to be happening is that the faster-turnover pools of carbon mask the effect of pools with slower and larger turnover. This model also suggests that higher carbon release from warmed soils might continue over a number of decades. This last has yet to be tested, but we may get the chance to find out as the Earth continues to warm up.

To put the Alaskan experiment into context of current increases in atmospheric carbon dioxide and existing carbon pools, high-latitude warming could at some stage further increase the current rate of (largely fossil-fuel-driven) increase in atmospheric carbon dioxide by between half as much again and double, for a period of two decades or more. However, the exact global effect of climate change on soil carbon is uncertain. Nonetheless, the key point here is that warming would itself enable the release of more tundra soil carbon, which would fuel further warming, and so on. (This is another positive-feedback cycle.) The three-pool carbon soil model, applied to real experimental results, suggests that increases in soil carbon release with temperature rises also apply to temperate and tropical soils and not just to high-latitude (and high-carbon) boreal soils. As we shall see (Chapter 7), this could undermine the policy proposals of temperate nations to use soils as carbon sinks as a way of offsetting atmospheric carbon dioxide increases from fossil-fuel burning, but for now it is worth noting that soil carbon has a feedback to atmospheric carbon dioxide mediated by temperature change even if the precise strengths of this feedback have yet to be discerned.

So much for the broad picture of interacting feedback cycles and flips between semi-stable states. Not surprisingly, climatologists continually monitor research into these positive- and negative-feedback systems. Indeed, climate-modelling research has received multi-million pound investments in recent years, especially since the mid 1980s, when computer technology became sufficiently sophisticated to accommodate models of adequate complexity to provide tolerably useful (or at least interesting) output. The models are good in that they do broadly reflect the global climate, but they (currently) lack detail, both spatially and thermally, and there have been problems with some outputs that simply do not tie up with what we know (see Chapter 5). For example, in the 1990s global models were not particularly good at portraying climates at high latitudes (which are warming far faster than models predict), while the models of the early twenty-first century are including more developed biological components but still have a long way to go. However, the models are continually getting better as they are built with greater resolution (both vertically and horizontally) and incorporate more of the features and processes operating on Earth. Indeed, from the late 1990s onward, programmers turned their attention to including biological processes in their models, so continuing the trend of being able to increasingly match model outputs with expectations based on reality. Here biologists and geologists have much to contribute. For what has taken place climatically is frequently recorded biologically and preserved geologically. Not only do different species live under different climatic regimens but species are affected by climate and these influences can be laid down in ways that are long-lasting (for example, tree rings, to which we will return in Chapter 2).

Prior to the 1980s we had such a poor understanding of the way that the global climate operates that there was great uncertainty as to whether the Earth was warming or cooling. Indeed, when the first ice cores were analysed in the 1970s it was realised that the last glacial had lasted for roughly 100 000 years, whereas the previous interglacial was just 10 000 years. Furthermore, the change between the two was a sudden one and not gradual. Consequently, in the 1970s, given that our current (Holocene) interglacial has already lasted about 10 000 years, there was for a while a genuine concern that a new glacial was imminent and that this global cooling might even threaten civilisation. More research was required into the strengths of the various factors influencing (forcing) climate. It was realised in the 1980s that on balance the factors were warming our planet. Factors such as the current human generation of greenhouse gases were adding to natural warming processes and so were greater than the various cooling factors. The question that remained was how great a warming could we expect from our fossil-fuel generation of carbon dioxide and how would this compare against the current range of other climatic factors? This was the subject of the first IPCC report published in 1990.

1.6 Non-greenhouse influences on climate

Milankovitch variation and the changing sunlight-reflecting (albedo) properties of ice caps (and indeed the solar-reflecting differences of any surface) demonstrate that non-greenhouse considerations do play a part in climate change. They can, through feedback cycles, either increase or decrease the magnitude of change. They can also superimpose their own variability imprint on climate.

However, it is now accepted by nearly all within the climate community that the anthropogenic addition of greenhouse gases is the key factor determining current global warming. But this does not mean that these other, non-greenhouse, factors are not taking place and have no effect on climate.

One such factor has in particular caused some controversy, although it is generally accepted to have a minor effect compared to current anthropogenic climate change. This is the variation in solar output, or changes in the Sun's intensity and has had a major part to play over the three or so billion years of evolution of the Earth's biosphere. As we shall see in Chapter 3, the Sun, being a main-sequence star, is growing significantly warmer over a scale of hundreds of million to billions of years. This has considerable implications when it comes to elucidating the evolution of the biosphere. However, on shorter timescales of hundreds or thousands of years it is comparatively (but not entirely) stable. On even shorter timescales the 11-year sunspot cycle has an impact. But this last effect is small, causing a variation in irradiance of only 0.08%, which is too small to have much effect: changes here may affect the global climate by 0.02–0.4 °C. Changes in sunspot activity do seem to tie in with similar patterns of change in global temperature, but these are superimposed on larger climate changes determined by other factors (Foukal *et al.*, 2004).

The question as to whether the Sun is largely responsible for current global warming came about because the Sun has became slightly more active during the twentieth century. During much of this time the Earth's temperature rose and fell almost simultaneously with changes in solar activity. There is also the question of the so-called Little Ice Age and solar activity, and we will return to this in Chapters 4 and 5.

Since 1978 we have been able to take space-borne measurements of solar output and correlate them with sunspot activity, for which we have previously only had an observational record going back a few centuries. The relationship is not clear but there *is* a relationship. However, there is also another larger, longer-term (over a timescale of centuries) solar component whose exact historic magnitude is speculative, but roughly five times that of the solar variation reflected by sunspot activity. Furthermore, there does seem to be a correlation between global temperature and solar output (as we shall see in Chapter 4, which might account for the Little Ice Age) but is this relationship real, partial, or just a coincidence?

In 2004 a team of five European researchers from Germany, Finland and Switzerland used the carbon isotope ^{14}C from tree rings going back 11 000 years (Solanki *et al.*, 2004). It is usual to think of ^{14}C as a means of dating objects using radioactivity as ^{14}C is produced in the upper atmosphere at *roughly* a constant rate. ^{14}C is then absorbed into living things and begins to slowly decay at a known exponential rate. So the amount of ^{14}C left compared to the stable ^{12}C enables one to date objects, albeit with a certain amount of error. Yet ^{14}C is not produced at an *exactly* constant rate. ^{14}C is produced in the upper atmosphere due to the action of cosmic rays from the Sun on nitrogen and carbon atoms; the level of cosmic rays is an indication of the Sun's output. Counting tree rings from overlapping samples of wood enables each ring to be dated in a different, more accurate, way to carbon dating. It is therefore possible to deduce whether at any one time more or less ^{14}C was produced compared to what would have been produced if solar activity was constant. The research team's ^{14}C-determined calculation of solar output was corroborated by ^{10}Be (a beryllium isotope) from Antarctic and Greenland ice cores, as this isotope also relates to solar output. The researchers found that there was indeed unusually high solar activity at the end of the twentieth century and that this would have certainly contributed to some of the global warming experienced then. However, it could not account for it all and was 'unlikely to be the prime cause' (Solanki *et al.*, 2004).

What seems to be happening is this. There are many factors affecting the climate. Some 'force' it in a positive (warming) way and others in a negative way. Furthermore, some are strong climate forcers and some weak. Greenhouse gases are strong climate-forcing factors. Variations in the Sun's output over tens of thousands of years do occur but are comparatively small. Their effects may be superimposed on the climate change that is determined by the sum total of all other forcing agents and as such they may account for small changes in the climate (and possibly even the Little Ice Age). However, it is difficult for small climate forcing (such as the small increase in twentieth-century solar output) to account for the large temperature changes measured. Remember, the Milankovitch variations in energy reaching the northern hemisphere in the summer are small. As we shall see in Chapters 3 and 4, these small changes in solar radiation help trigger larger changes in carbon dioxide and methane greenhouse gases as the Earth moves between (glacial–interglacial) climate modes. One should not be surprised that greenhouse gases are not the only positive climate-forcing factors contributing to current warming. There are others (both positive and negative), and increases in solar output are but one. These others include volcanic activity, marine release of methane and (with regional effects on climate) oceanic and atmospheric circulation. We shall return to these later in the book. While these other factors can play an important part in some climate change, they are not the dominant factors of current global warming. Even so, they cannot be easily dismissed because, as we shall see, these factors could make a serious contribution to climate change in the future and in particular circulation changes can help flip climate regimens between semi-stable states.

1.7 The water cycle, climate change and biology

As something noted above, and to which we will occasionally return later in this book, one of the things one would reasonably expect in a warmer world is more evaporation from the ocean. Again, as previously noted, this expectation is at least in part corroborated by computer models of atmospheric water vapour: so, global warming affects the water cycle. This leads us to another expectation that we would reasonably anticipate from more evaporation, that there would be more precipitation (rain and snow) and so in turn also increased river flow. But how reasonable are these expectations?

The Earth's climatic system is complex. Furthermore, biology is not just affected by a changing climate; biology, as we shall see, itself plays a key part in affecting the nature of climate change itself and in affecting some of the consequences of that change. Having said this, not all these complications are biological. Nonetheless, these complexities need to be taken into account.

First of all, although a warmer world will lead to more evaporation (other factors, such as the complete solar spectrum, remaining constant), more evaporation does not in itself necessarily mean more precipitation. To take an extreme example, Venus is a far warmer planet than the Earth and its atmosphere imparts a far more powerful greenhouse effect. Indeed, Venus is so warm that water exists solely in the form of water vapour. There is no rain to soak the ground on Venus and there are no oceans. Second, although a warmer world results in more ocean evaporation, if, hypothetically speaking, part of this extra water vapour does return as a partial increase in precipitation this does not mean that river flow would necessarily increase. It could be, hypothetically, that the increased evaporation means that the excess precipitation evaporates before it reaches the rivers (we shall return to this later). Alternatively, it may be that rainfall would increase mainly over the oceans, so leaving river flow unaffected. Third, other factors and indeed biological processes may well play their part. So it is important to identify these factors (or as many as possible) and to try to quantify them.

Time for a quick reality check. First, as already noted, the global climate is an average made up of regional and annual variation. With a changing global climate not only will some climatic components change in some places, and at some times more than others, but it is possible that in some places (and/or times) there may be a change in the opposite direction to the overall trend. (This has been seized on by some, especially in the 1990s, to argue that global warming is not taking place. Such arguments are based on selective and atypical data, with the fallacious claim that the data is typical.) Second, there are other non-greenhouse-gas factors operating. Temperature is not the sole mechanism behind evaporation; direct sunlight (electromagnetic radiation of appropriate frequency) also plays a part. A photon of sunlight can excite a water molecule causing it to leave its liquid-state companions and become vapour. As we shall see later in this chapter changes in the amount of sunlight reaching the Earth's surface have been happening and have resulted in so-called global dimming due to pollution particles. Another factor affecting water becoming water vapour is biology.

The routes from liquid water back to water vapour are not restricted to straightforward evaporation but also plant transpiration as part of photosynthesis in terrestrial plants. For this reason evapotranspiration (the total water loss from an area through evaporation and vegetation transpiration) is important. Now let us return from theory to reality.

A warmer world due to increased greenhouse gases will, among other things, affect plant physiology. Plants exchange gas and water with the atmosphere through openings on the surface of leaves called stomata. Stomata open and close to regulate photosynthesis in the short term. (Also, as we shall shortly see, in the longer term, stomatal densities have been shown to vary on more geological timescales with atmospheric carbon dioxide concentrations – see Chapter 2.) In a warmer, more carbon dioxide-rich world with higher rainfall that serves to enhance photosynthesis, all other things being equal, we might expect plant homoeostatic processes (physiological mechanisms acting to keep functions stable) to dampen photosynthesis. If this were happening then we would reasonably expect plant transpiration to decrease. This in turn would lower a plant-covered water catchment's evapotranspiration and so more water would remain in the ground to percolate through to streams. River flow would increase. So much for theoretical expectations. The question then becomes one of how do the various factors of changed precipitation, warmth, plant physiology and river flow interact?

Here it is possible to model the individual processes. We do have climate records of temperature and precipitation covering many decades, measurements of solar radiation reaching the surface, as well as those of river flow. We also know about plant physiology and so can apply broad parameters to plant physiology over a region as well as factors such as deforestation and land use. In short, we know both actual river flow for principal catchments on each of the continents and, broadly, how the various factors that contribute to river flow have changed over the last century. A model of twentieth-century continental water runoff has been constructed that reflects actual river-flow measurements. It is then possible to examine the model conducting a 'sensitivity analysis' (or 'optimal fingerprinting' or 'detection and attribution', the nomenclature varies with research groups) to vary just one factor at a time and to see using statistics how this causes the predicted runoff to differ from reality. It is known that twentieth-century climate change alone is insufficient to account for runoff changes. However, such an analysis of a surface runoff model has indicated that including the suppression of carbon dioxide-induced stomatal closure makes the model's outcome consistent with actual runoff data (Gedney *et al.*, 2006; Matthews, 2006). In short, biology plays a key part in controlling the water cycle to such a degree that without it climate change and other non-biological factors cannot fully explain water-cycle trends.

As we shall see again and again throughout this book, climate and biology are connected: one affects the other and vice versa. Furthermore, humans, as a biological species that also rely on and who are in turn affected by other biological species, are caught in the middle of this climate–biology dynamic. That human action is also a significant driver of current climate change further complicates matters. We need to understand both how our species affects the global climate and the climate–biology dynamic if our growing population is to survive without a decline in either well-being or environmental quality.

1.8 From theory to reality

In the above review of the causes of climate change it can be seen that climate theories, be they that of greenhouse climatic forcing or that of Milankovitch's orbital effects of incoming solar energy, have to be validated. This is done by comparing theory with what actually happens, or has happened, in reality. What is happening can be measured today, although care is needed. For example, satellite remote-sensing devices need proper calibration and we need to know what their data actually mean. (I cite this example because in the 1990s incorrect assumptions were made about some satellite data.) The past 10 years has seen tremendous progress in understanding climate change on two fronts. First, there has been a steady improvement in computer models. This improvement continues but there is still a long way to go, both in terms of reducing uncertainty across the globe and in terms of spatial and temporal resolution (see Chapter 5). These inadequacies mean that even the best models can only present a broad-brush picture of a possible likelihood. The 2001 IPCC report provides some good illustrations of the limitations. For example, it presents (p. 10) two global models of projected changes in precipitation runoff from the 1960s to 2050. Both show decreases in runoff in the Amazon basin, much of the rim of Australia, and Central Europe, and runoff increases in south-east Asia, north-west Canada and southern Alaska. At the moment this does seem to tie in with *some* observations (IPCC, 2001b). This similarity of computer-model output with reality (even though we have yet to reach 2050) is not proof of the various models' accuracy but it does lend them added credibility: one corroborates the other. However, the same two models differ in that one shows a marked decrease in runoff in the eastern USA and the other an increase. There are also marked contrasting differences in runoff projected for much of the Indian subcontinent and much of north-west Europe. Such problems are not trivial even if progress is continually being made, with models becoming ever more sophisticated and now increasingly including biological dimensions.

Second, there has been a huge growth in understanding how the natural world reacts to climate change, be it in terms of a single organism or an ecosystem. And so a palaeorecord provides us with a proxy as to what might have really happened in one place at one time under a different climatic regimen. This is, if you will, just one pixel in a bigger picture of the Earth's climate history. But the past decade has seen much building up of records covering many periods of time and places. Such numerous records both help reduce uncertainty and inform. This is particularly useful because – although it is difficult to build up a picture, as if pixel by pixel, from biological and other indicators of past climates – such a picture reflects something far better than a computer model of the Earth: the Earth itself. Indeed, if at some time the Earth was warmer than today we can use climatic proxies (be they biological or chemical, laid down in the geological record) to determine what the planet was like and use that as an analogue in an attempt to predict what it would be like in a globally warmed future.

However, with this success the past decade has also seen problems. Policy-makers especially have held much faith in computer models. Some bioscientists on the other hand have found it difficult to relate their work on one (or a group of) biological palaeo-climate proxies with others. Finally communication between these modellers and bio- and geoscientists might arguably have been better. A 2001 editorial in *Nature* (v411, p1) summed it up thus:

To get the best return on [climate-change research] funding, there must be strong interdisciplinary collaboration between those taking data from sea sediments, ice and land, and also between those making measurements and those developing models. But above all, the community must deliver more by way of systematic deposition of its data.

In making the most of data, funding agencies, journals and researchers themselves all have a role to play. There seems to be too little awareness by researchers of what is admittedly something of a maze of publicly supported databases. And too often, as researchers will readily complain, trying to extract numbers in a large data set from the originators of published work is like pulling teeth.

This is not to decry much of the very good work that has been, and is being, carried out. The purpose of this text is simply to provide bioscience and other students with a broad-brush basic picture as to how climate affects biology (and vice versa). For others this book will serve as a background against which they can set their own more specialist work, with reference to human ecology and likely future climate change. However, before this can be done, it will be useful to review some of the key biological indicators of past climate change. This forms the subject of the next chapter.

1.9 References

Arrhenius, S. (1896) On the influence of carbonic acid in the air and upon the temperature of the ground of the absorption and radiation of heat by gases and vapours, and on the physical connection of radiation, absorption and conduction. *Philosophical Magazine*, **41**, 237–76.

Barnola, J. M., Raynaud, D., Korotkevitch, Y. S. and Lorius, C. (1987) Vostok ice-core provides 160,000-year record of atmospheric CO_2. *Nature*, **329**, 408–14.

Chappelaz, J., Barnola, J. M., Raynaud, D., Korotkevich, Y. S. and Lorius, C. (1990) Ice-core record of atmospheric methane over the past 160,000 years. *Nature*, **345**, 127–31.

Coale K. H., Johnson, K. S., Fitzwater, S. E. *et al.* (1996) A massive phytoplankton bloom induced by an ecosystem-scale iron fertilization experiment in the equatorial Pacific Ocean. *Nature*, **383**, 495–508.

Foukal, P., North, G. and Wigley, T. (2004) A stellar view on Solar variations and climate. *Science*, **306**, 68–9.

Fourier, J. (1824) Remarques générales sur la température du globe terrestre et des espaces planétaires. *Annales de chimie et de physique*, **27**, 136–67.

Fourier, J. (1827) Mémoire sur les températures du globe terrestre et des espace planétaires. *Mémoires de l'Académie Royal des Sciences*, **7**, 569 604.

Gedney, N., Cox, P. M., Betts, R. A., Boucher, O., Huntingford, C. and Stott, P. A. (2006) Detection of a direct carbon dioxide effect in continental runoff records. *Nature*, **439**, 835–8.

Houghton, R. A. (2002) Terrestrial carbon sinks – uncertain explanations. *Biologist*, **49(4)**, 155–60.

Intergovernmental Panel on Climate Change (1990) *Climate Change: the IPCC Scientific Assessment*. Cambridge: Cambridge University Press.

Intergovernmental Panel on Climate Change (1995) *Climate Change 1995: the Science of Climate Change – Summary for Policymakers and Technical Summary of the Working Group I Report*. Cambridge: Cambridge University Press.

Intergovernmental Panel on Climate Change (2001a) *Climate Change 2001: the Scientific Basis – Summary for Policymakers and Technical Summary of the Working Group I Report*. Cambridge: Cambridge University Press.

Intergovernmental Panel on Climate Change (2001b) *Climate Change 2001: Impacts, Adaptation, and Vulnerability – Summary for Policymakers and Technical Summary of the Working Group II Report*. Cambridge: Cambridge University Press.

Keppler, F., Hamilton, J. T. G., Brab, M. and Röckmann, T. (2006) Methane emissions from terrestrial plants under aerobic conditions. *Nature*, **439**, 187–91.

Knorr, W., Prentice, I. C., House, J. L. and Holland, E. A. (2005) Long-term sensitivity of soil carbon turnover to warming. *Nature*, **433**, 298–304.

Kuo, C., Lindberg, C. and Thomson, D. J. (1990) Coherence established between atmospheric carbon dioxide and global temperature. *Nature*, **343**, 709–13.

Mack, M. C., Schuur, E. A. G., Bret-Harte, M. S., Shaver, G. R. and Chapin, III, F. S. (2004) Ecosystem carbon storage in arctic tundra reduced by long-term nutrient fertilization. *Nature*, **431**, 440–3.

Matthews, D. (2006) The water cycle freshens up. *Nature*, **439**, 793–4.

Milankovitch, M. (1920) *Théorie Mathématique des Phénomènes Thermiques Produits par la Radiation Solaire*. Paris: Gauthier-Villars.

Soden, B. J., Jackson, D. L., Ramaswamy, V., Schwarzkopf, M. D. and Huang, X. (2005) The radiative signature of upper tropospheric moistening. *Science*, **310**, 841–4.

Solanki, S. K., Usoskin, I. G., Kromer, B., Schussier, M. and Beer, J. (2004) Unusual activity of the Sun during recent decades compared to the previous 11,000 years. *Nature*, **431**, 1084–7.

Tyndall, J. (1861) On the absorption and radiation of heat by gases and vapours, and on the physical connexion of radiation, absorption and conduction. *The London, Edinburgh, and Dublin Philosophical Magazine and Journal of Science, Series 4*, **22(146)**, 169–94 and **22(147)**, 273–85.

2

Principal indicators of past climates

If palaeoclimatologists had their wildest dreams come true then we would undoubtedly see them travelling back in time to make meteorological recordings before humans existed, let alone thermometers. Were that life be that obliging. Instead, preserved indications of past climate have to be identified, calibrated and then used to infer usually just one aspect of the climate, and frequently that one aspect at just one locality. Such a preserved indication provides us with a single proxy measurement.

Many measurements are needed from many proxies in order to build up a climatic picture of the past. The products of proxy analyses are often termed palaeoclimatic data. Palaeoclimatic data commonly come with considerable uncertainty in one or more of the following regards: the inferred temperature, the inferred time, and the space to which it applies. It is important to emphasise this and to recognise that most proxy data rarely ever give a direct measure of a single meteorological parameter. For example, tree-ring thickness is affected by numerous factors, not just climate. These include the aspect in which the tree is found, be it in a valley or on a hillside, and whether it is south or north facing; soils, which affect nutrients; and local geology, which affects water availability.

There are many palaeoclimatic proxies. Not only do they each come with their own vagaries as to their relationship with climate, but the passing of time also blurs information. For example, a sediment may contain an isotope or a biomolecule that relates to temperature, but the sedimentation process itself may not be uniform with time, or the sediment may itself be re-worked by biological and/or geological processes (such as burrowing organisms and/or strata movement, respectively). It is as if this book you are reading not only contains information, in the form of words on the page, but that the type of paper and even the page order had changed from when I wrote it to you reading it. An example of this sediment re-working is known as smearing.

Fossils being transported downwards into older sediments is termed backward smearing, and into younger sediments as forward smearing.

Another example of how external factors can affect the accuracy of proxy dating is exemplified in ^{14}C dating analysis. The radioactive ^{14}C isotope is formed in the upper atmosphere by the action of cosmic rays (mainly from the Sun) on nitrogen and carbon atoms in the atmosphere. Broadly, new ^{14}C atoms are produced at a rate of $2\,\mathrm{atoms\,cm^{-3}\,s^{-1}}$. This assumes that cosmic-ray (or solar) activity is constant with time. This is not so. However, there is sufficient homogeneity for some meaningful work, and steps have been taken to calibrate palaeo-^{14}C (see the previous chapter as well as the next section, on isotopic dendrochronology).

Because widespread direct instrumental climatic measurements have only been available for the past 100 years or so, we are dependent on proxy indicators for earlier information. To ensure their reliability, reconstructions based on these proxies must be validated by comparison with instrumental records during periods of overlap. Sometimes direct comparison is not possible and so this must be done indirectly by relating to another overlap with another proxy, and so commonly a number of proxies are used simultaneously.

Another problem is that all palaeoclimatic proxies only relate to a certain time period. Glacial terminal moraines largely tell us about the maximum extensions of ice, hence glacial maximums and not glacial beginnings. Furthermore, not only may there be a time window but the aspect of the climate being indicated may only relate to just part of the year. For instance, tree-ring data are only applicable to periods of time when trees grow and so largely relate to warm-season conditions and do not give an indication as to the severity of winter. (Of course, there are often exceptions and here, to continue with our current example, cold-season tree-ring information is available but largely limited to species from semi-arid or Mediterranean environments.) Such restrictions, in this case seasonal specificity, may give a misleading picture of large-scale temperature changes. For example, the climatic response to volcanic forcing can lead to opposite temperature changes in the summer (with cooling) and winter (warming) over continents.

It is for all these reasons that many proxies are required. Sets of these in turn are used to determine indices against which it may be possible to ascertain large-scale climatic changes. For example, one such set is used to determine what is called the North Atlantic Oscillation (NAO) index.

However, no matter how each proxy is subsequently used there is considerable need to add to our knowledge of existing proxies as well as to identify new ones to both establish further palaeoclimatic data and for corroboration.

The good news is that although as many climate proxies as possible are required to build up a picture of past climates, they do all begin to paint the same picture. So, on one hand there is uncertainty, but on the other, by looking at more and more proxies (in both number and type), there is increasing certainty.

Climatic proxies are many but can be divided into two principal groups, biotic proxies (terrestrial and marine) and abiotic proxies (physical and abiotic geological). We will shortly look at a few examples of the principal types of each of these, but because this text concerns the biology of climate change we will focus mainly on those that are in some way biological.

2.1 Terrestrial biotic climatic proxies

2.1.1 Tree-ring analysis (dendrochronology)

Dendrochronology was developed early in the twentieth century by, not a biologist but, an astronomer, Andrew Douglass. When he was 27, and working at the Lowell Observatory at Flagstaff, Arizona, Douglass discovered a possible relationship between climate and plant growth. He recorded the annual rings of near by pines and Douglas firs (*Pseudotsuga* spp.). In 1911 he made corresponding records among trees felled around 50 miles to the south-west of the observatory that prompted his study of sequoias. When viewing a cross section of a tree, it became clear in certain species that wide rings were produced during wet years, and narrow rings were produced during dry years (Webb, 1983).

Douglass coined the term dendrochronology, meaning tree-time-study, and worked on exploring the link between tree rings and climate. Between 1919 and 1936 he wrote a three-volumed work, *Climate Cycles and Tree Growth*. Collectively, the methods he pioneered are now invaluable for archaeologists to date prehistoric remains. It was first used this way in 1909 when Clark Wissler, of the American Museum of Natural History, put him in touch with the Archer M. Huntington Survey of the South-west. In 1918, Wissler (who recognized the value of tree rings for dating purposes) arranged for Douglass to receive nine beam sections from Aztec Ruin and Pueblo Bonito so he could begin cross-dating these two ruins. The final missing link to his studies came from an unstable beam that was extracted from the Whipple Ruin in Show Low, Arizona. That beam bridged the gap between living tree-ring chronology, and the archaeological tree-ring chronology that had been established.

The Laboratory of Tree-Ring Research, subsequently initiated by Douglass, has the largest accumulation in any research centre of tree-ring specimens from both living trees and age-old timbers.

Trees respond to seasonal and climatic fluctuations through their annual growth rings. The thickness of each ring reflects the climate of the principal growing season, primarily the summer, in which it is formed. The rings are made of xylem. Pith is found at the centre of the tree stem followed by the xylem, which makes up the majority of the tree's circumference. The outer cambium layer of the tree trunk keeps the xylem separated from the rough bark. Each spring or summer a new layer of xylem is formed, and so the rings are produced that may be counted as an annual record. More specifically, a tree ring is a layer of wood cells produced by a tree in 1 year, consisting of thin-walled cells formed in the early growing season (called earlywood), and thicker-walled cells that are produced later in the growing season (called latewood). The beginning of early-wood and the end of latewood forms one annual ring. In dendrochronology, these rings are then counted and their thickness compared. However, because (as discussed in the previous section) other local factors come into play – due to the tree's site, which determines the nutrients, water and sunlight it receives – statistical analysis of a number of tree-ring samples from the local area is required to discern the local climatic factor. Even so, trees are particularly useful climatic proxies when they are near their climatic limits of either rainfall or summer warmth as then it is possible to attribute variability in ring thickness with one, or other, of these climatic parameters. Conversely, trees in the tropics do not exhibit a pronounced annual cycle and so it is not possible to obtain climatic information from their rings.

Where possible, rings not only serve as climatic indicators but are also used in dating. Sections of tree trunks can be compared with others from either slightly younger or older trees in the region, and then the overlap in each tree record can be used to extend the overall ring record backwards or forward in time, as Andrew Douglass did. Indeed, starting from recent trees in the time of meteorological records, it is possible to make a calibration relating tree rings from appropriate samples directly to either rainfall and/or summer warmth back to centuries before meteorological records began. Starting from modern times and using overlapping sequences of ring it is possible to work back in time not just centuries but even, in some cases, a few thousand years. So successful has dendrochronology been that it has not just been of use to palaeoclimatologists but also to archaeologists wishing to date local wooden artefacts. Palaeoclimatologists who use dendrochronology have created a number of standard sequences against which new samples can be checked. Dendrochronology has proved particularly useful in understanding the palaeoclimate of part of our current interglacial, the Holocene (the time since the end of the last glacial to the present; see Chapter 4), which itself has so far lasted over 11 500 years (Briffa, 2000).

Further, more detailed information than just the nature of a particular season can be obtained by a closer examination of the wood structure within individual rings. Parameters such as maximum latewood density, minimum earlywood density, and width of early- and latewood growth can provide insight into weather fluctuations within the principal growing season.

There are a number of tenets underpinning dendrochronology. The main ones are described here.

The uniformitarian principle: variability evident in tree-ring growth can be linked to past environmental variability and used to identify environmental change.

The limiting-factor principle: the fastest rate that plant growth (and other processes such as seed production) can occur is equal to the greatest limiting factor.

The principle of ecological amplitude: tree species will be most sensitive to environmental flux at the latitudinal and altitudinal limits of their range.

The principle of aggregate tree growth: any individual tree-ring growth series can be broken down into an aggregate of environmental factors, both natural and human, that affected the patterns of tree growth over time (and these might include age-related growth trends, the climate that occurred over the course of the year, and various other factors that occur within and outside of the forest stand).

The principle of cross-dating: matching patterns in tree-ring width, density and other characteristics across several tree-ring series allows for the identification of the year in which the growth ring was formed.

The principle of site selection: sites can be identified based on the criteria that will produce growth-ring series that are most sensitive to the examining factor.

It is important to note these underpinning tenets (as it is for all assumptions made in science), for it is easy to assume that they will always hold. This is not so and invariably there are exceptions. Tree-ring series derived from high-latitude growth have provided a solid record for nearly all of the past two millennia. However ring growth in recent decades, principally since the 1960s, has deviated from the pronounced warming experienced (and instrumentally measured). Growth is less than is expected from temperature alone (see Figure 2.1). The reason for this is the failure of the above principles. There is some evidence, for example (Vaganov *et al.*, 1999), to suggest that what has happened is that our globally warmed world of recent decades has increased winter precipitation (snow) in these high-latitude regions. (One would expect in a warmer world more ocean evaporation hence, overall, more precipitation. More evidence from high latitudes is covered in Chapter 6.) In these permafrost environments this extra precipitation has delayed snow melt: a greater depth of snow takes longer to melt. As a result the initiation of cambial activity

Figure 2.1 Approximate, smoothed-average (with a *c*.70-year filter) northern high-latitude temperature changes over the past 2000 years, based originally on several dendrochronological series and deviating from the 1601–1974 average. Possible cometary debris in one instance, together with five volcanic eruptions, coincide with subsequent lower annual tree-ring growth. See Chapter 3: before 1000 the dendrochronology suggests it was warmer rather than cooler; also shown are the medieval warm period, 900–1200, the Little Ice Age, 1550–1750, and the twentieth-century warming. The decline in some boreal dendrochronological series at the end of the twentieth century (dashed line) is portrayed separately and is thought to perhaps be due to increased snowfall delaying spring greening or some other environmental factor impeding growth – see Chapter 5's section on the current biological symptoms of warming. Data in part from Briffa (1998).

necessary for the formation of wood cells is delayed. Fortunately, dendrochronological problems of such significance are rare.

Today there are a number of long, annual-resolution records of tree growth from temperate regions, determined by the aforementioned method of piecing together overlapping growth-ring patterns from successively older timbers. One of the lengthiest is the oak chronology in central Germany that runs continuously back to 8480 BC. This is even longer than the bristlecone pine chronology from the western USA, which has been available since the 1960s and which provides a record back to 6200 BC. There are some eight chronologies that span the last seven millennia; one from northern Russia, one from western North America and six European covering Ireland, England, Germany and Fennoscandia. The rate of development of dendrochronological databases in the latter half of the twentieth century is illustrated by the creation of the Irish oak chronology by researchers at Queen's University Belfast. Here thousands of samples of oak were collected from buildings and archaeological sites. Samples from living trees acted as a chronological anchor. By 1977 a continuous chronology ran back to 885 AD and by 1982 a continuous 7272-year Irish chronology had been determined. By 1984 sufficient correlation between Irish and German dendrochronological records had been ascertained for the announcement of the European oak chronology covering seven millennia.

It is important to remember that greenhouse gases are not the only factors forcing climate. For example, trees are sufficiently sensitive that dendrochronology can identify the years of cooling following major volcanic eruptions (see Figure 2.1). There are notable declines in tree-ring growth associated with large volcanic eruptions, such as Tambora (1815) and Krakatau (1883). The dust from these eruptions entered the stratosphere and resulted in spectacular sunsets for a few years as well as global cooling and disrupted harvests (see Chapter 5). Acid levels in ice cores (such as the Greenland cores discussed later in this chapter) suggest past supereruptions and these too have been connected to years of poor tree-ring growth. Such corroborative evidence is important not only to ensure that the climate inference is accurate but also to help provide further detail. For instance, several dendrochronological records suggest a twin climate event in the sixth century: one in the year 536 and something worse in the early 540s. The ice-core record is not nearly so clear; although there is evidence of some volcanic activity at that time there is nothing of the magnitude suggested by dendrochronology. However, one possibility is that the Earth may have accrued cometary debris around that time (Baille, 2003). Nonetheless, despite its limitations (such as its lack of effectiveness in the tropics) dendrochronology remains the only terrestrial biotic record detailing climatic change with annual detail covering the Holocene.

2.1.2 Isotopic dendrochronology

As will be discussed in more detail later, isotope analysis can be used to infer climate change. Neutrally charged isotopes of the same element have the same number of protons and electrons as each other but a different number of neutrons. Different isotopes of the same element therefore exhibit the same chemical properties but have different atomic weight and hence different physical properties. Most oxygen in the atmosphere (99.76%) consists of ^{16}O (with an atomic weight of approximately 16) but, of the remaining proportion, the next most common isotope of oxygen is ^{18}O. Because lighter ^{16}O evaporates preferentially to ^{18}O the ratio between the two varies, in atmospheric water vapour and in rain water, depending on climate. The oxygen in water is then incorporated into wood through photosynthesis and so a record of the isotopic ratio of the rain at the time the wood was growing can be determined.

This should give an indication of the climate. However, there are several problems with this form of analysis. These include rain-out effects: that the first precipitation from a cloud has a greater composition of the heavier ^{18}O. There are also altitude and ice/snow meltwater effects. Consequently isotopic dendrochronology has only met with limited success and not all such work has borne equivocal results. This form of palaeoclimatic analysis needs some further development. Yet, despite considerable limitations there have been some interesting possibilities. For example in 2006 the Swiss and German team led by Kerstin Treydte reported that a dendrochronological isotope record from the mountains in northern Pakistan correlated well with winter precipitation and that this *suggested* that the twentieth century was the wettest of the past 1000 years, with intensification coincident with the onset of industrialization and global warming (Treydte *et al.*, 2006).

Despite the extra caution required with isotopic dendrochronology, as we shall see using $^{16}O:^{18}O$ ratios in another biological context, Foraminifera (an order of marine plankton) can be employed as a kind of climate proxy but more exactly as an indication of the global amount of water trapped in ice caps.

Because dendrochronology gives precise annual rings, it has also been used as one of the means to calibrate radioactive ^{14}C dating (as discussed in Chapter 1). In 1947, the US chemist Willard Libby explained how the radioactive isotope ^{14}C was produced, for which he received the Nobel Prize for Chemistry in 1960. ^{14}C, with its half-life of 5730 years, is used for dating biological samples from several decades to thousands of years old beyond the last glacial maximum about 20 000 years ago: the limit of ^{14}C dating is usually considered to be about 45 000 years (and we will return to this when considering human migration during the last glacial in Chapter 4). ^{14}C is continually being generated in the atmosphere due to cosmic rays, mainly from the Sun, hitting nitrogen and carbon atoms. ^{14}C so formed is then incorporated into plants through photosynthesis. The common ^{12}C and the radioactive ^{14}C are in equilibrium with that in the atmosphere up until the plant dies, and at that point the various carbon isotopes become fixed. From then on the fixed reservoir of ^{14}C in the plant begins to decay and so the ratio between ^{12}C and ^{14}C changes. This ratio enables dating.

However, as mentioned, all this assumes that the rate of ^{14}C production in the atmosphere is constant. This is not precisely true as more ^{14}C is produced in the atmosphere at times of, for instance, exceptional solar activity, such as when there are many sunspots. Carbon dating has therefore had to be verified by comparison with other dating methods, including dendrochronology. (In recent times $^{12}C:^{14}C$ ratios have been upset by atmospheric tests of atomic bombs in the mid-twentieth century as well as by the Chernobyl nuclear power plant disaster.) Notwithstanding these difficulties, ^{14}C decays at an immutable rate of 1% every 83 years. If the rate of decay is measured to obtain an accuracy of, for instance, $\pm 0.5\%$, then the observation of around 40 000 decaying carbon atoms would be required. This would, in principle, allow the sample to be dated to an accuracy of ± 40 years and would require a sample of several grams to be measured in about a day. However, detecting radioactive decay is only one way of detecting (inferring) ^{14}C. A mass spectrometer enables direct detection of ^{14}C and ^{12}C and a similar accuracy can be obtained using a spectrometer with a few milligrams of sample in a few hours.

Isotopic dendrochronology has provided a lynchpin for both dendrochronology and ^{14}C dating. First, ^{14}C enables denrochronologists to date a sample of wood to within a few years or decades. Further accuracy may then be obtained by matching the sample to the corresponding regional dendrochronological record. Second, because ^{14}C production is not constant due to changes in solar radiation it is possible to use the established dendrochronological record as a base against which to determine variations in atmospheric ^{14}C production. Notably, two ^{14}C-calibration curves have been produced: one by researchers from the University of California, La Jolla, using the dendrochronological bristlecone pine data-set, and the other by those at Belfast University using oak. In particular the results show that ^{14}C production fluctuates with a peri odicity of around 200 years. Overall, for the past 6000 years both curves show that ^{14}C production has decreased a little from 6000 years ago to 2500 years ago before a marginal recovery. As ^{14}C production relies largely on solar radiation this helps inform as to alterations in the solar contribution to climate change over time. This can then be added into the mix of the various factors influencing (forcing) climate and so (since we know the atmospheric concentrations of the isotopes) help us estimate the greenhouse component over time.

2.1.3 Leaf shape (morphology)

In 1910 two explorers (Irving Bailey and Edmund Sinnott) noticed that the shapes of leaves appeared to relate to the climate of a given area. The most noticeable trait was that trees that grew in warm areas had large leaves with smooth edges, while trees that grew in cold places tended to have smaller leaves with very serrated edges (Bailey and Sinnott, 1916; Ravilious, 2000). This was largely forgotten until the 1970s, when a US researcher, Jack Wolfe, wondered whether fossilised leaves could reveal something of the climate if they were compared with their modern counterparts. He collected leaves from all over North America, recording aspects of their shape and relating these to regional weather records for the previous 30 years. Then, using statistical analysis he was able to compare fossil leaf assemblages with those of modern leaves and come up with past average temperature and rainfall values for the regions and times in which the fossil leaves once grew. For example, he predicted that the temperature in north-eastern Russia some 90 million years ago was roughly 9 °C warmer than it is today (Wolfe, 1995). Other early work following Bailey and Sinnott, such as by R. Chaney and E. Sanborn in 1933, showed that leaf size and apex shape were also related to climate. Apart from the Wolfe database standard for North America, other databases have been developed. For instance, in the late 1990s, Kate Ravilious worked on developing a similar database for the UK. These databases can be used in techniques such as leaf margin analysis (LMA) to infer temperature and leaf area analysis (LAA) to infer precipitation. LMA relates to the aforementioned changes in leaf shape with temperature, while LAA relates to changes in leaf area with precipitation level; for instance, plants with access to plenty of water can afford to lose it through transpiration over a large surface area (which also benefits sunlight capture, but which is not restricted by sunlight).

More recently still there has been a further twist in leaf morphology as an environmental indicator. At the turn of the millennium the UK team of Berling, Osborne and Chaloner developed a biophysical model that seemed to explain why leaves evolved in the first place. For the first 40 million years of their existence land plants were leafless or had only small, spine-like appendages. It would appear that the advent of megaphylls – leaves with broad lamina (leaf blades) – evolved from simple, leafless, photosynthetic branching systems in early land plants, to dissected and eventually laminate leaves. This took some 40 million years. It seems that these transformations might have been governed by falling concentrations of atmospheric carbon dioxide (some 90%) during the Devonian period 410–363 million years ago (see Figure 3.1). As noted earlier, atmospheric carbon dioxide has a bearing on climate so that the Devonian did represent a period of global climate change. (Although if we are to draw climatic analogies with the more recent (Quaternary) glacial–interglacial climatic cycles of the past 2 million years, we need to bear in mind that the Sun's energy output was slightly less at that time than it is today; see Chapter 3). This episode again demonstrates that climate change can influence evolution.

2.1.4 Leaf physiology

Whereas leaf morphology shows a general relationship with climate, one aspect of leaf physiology in particular has a close relationship with one of the key factors forcing global climate, that of atmospheric carbon dioxide concentration. A leaf's primary function is to enable photosynthesis, a case of form driven by function. Leaf stomata enable carbon dioxide to enter the leaf (stomata are small openings in plant leaves that facilitate gaseous exchange).

The theory underpinning leaf physiology as an indicator of atmospheric carbon dioxide, and hence global climate, is that the higher the carbon dioxide concentration the fewer stomata are needed to transfer the required amount of carbon dioxide the plant needs. So, there is an inverse relationship between stomatal densities and atmospheric carbon dioxide concentrations. This relationship was mainly deduced and established in the late 1980s.

Problems with this sort of analysis abound. First, there are all the dating problems commonly associated with terrestrial palaeontology (which we touch on elsewhere in this chapter). Second, the relationship with climate is less direct than other commonly used indicators. The indicator really is atmospheric carbon dioxide concentration and while carbon dioxide does force climate, climate change is known to occur without a major change in carbon dioxide. An example here would be the Little Ice Age (see Chapters 4 and 5), and, of course, weather (the day-to-day manifestation of climate) vagaries are independent of changes in carbon dioxide. Third, even the carbon dioxide/leaf stomata relationship produces only a coarse correlation with carbon dioxide concentrations.

Nonetheless, analysis of leaf physiology has on occasion provided some remarkable climatic insights. In 1999 Jenny McElwain and colleagues from Sheffield University (McElwain *et al.*, 1999) counted stomata densities in fossilised leaves from dozens of species from around the Triassic/Jurassic extinction some 205 million years ago (see Section 3.3.6). They found lower stomata densities after the extinction than before, which suggests that carbon dioxide levels were far higher. They calculated that carbon dioxide levels rose from around 600 parts per million (ppm) before, to between 2100 and 2400 ppm after the extinction. This would probably have forced the global climate to be some 4 °C warmer. The resulting climate would have been too warm for large leaves outside of high latitudes to photosynthesise properly and so could go a long way towards accounting for the extinction of 95% of all plant species at that time. The cause of this carbon dioxide pulse is thought to be volcanism during the split up of Pangaea. The area of volcanism required to generate the carbon dioxide pulse, the Sheffield team concluded, seems to fit with the likely area of the Pangaean split.

2.1.5 *Pollen and spore analysis*

Pollen and spores are critical parts of the life cycles of vascular plants. Because different species of plants thrive in different climatic conditions, it is possible to deduce past climates if one has evidence as to which species grew in those past times. Pollen and spore utility as palaeoclimatic indicators arises out of their resilience and durability with time. This is because they have very resistant cell walls, which in turn is due to the outer portion of their cells (the exine) that is composed of a waxy substance known as sporopollenin. Their cell-wall resistance and inert nature allows the preservation of pollen and spores in sediments under a variety of conditions. Pollen and spores typically are the most abundant, most easily identifiable and best preserved plant remains in sediments such as bogs and clays, as well as in the form of fossils in sedimentary rocks. Together it means that they can be used as a climatic indicator for both recent and geological timescales.

Spores, as referred to here, include the reproductive bodies of lower vascular plants such as club mosses, horsetails and ferns. The earliest occurrences of spores produced by land plants in the fossil record are in Lower Silurian rocks (around 430 million years ago; mya), slightly preceding the appearance of the first vascular plant fossils. Conversely, pollen grains are the gamete-carrying reproductive bodies of seed plants, including gymnosperms (such as conifers and cycads) and angiosperms (the flowering plants). Fossilised pollen first occurs in the Upper Devonian rocks (about 370 mya), corresponding to the occurrence of the earliest fossil seeds (*Archeosperma*). Pollen and spore walls of each plant species have a distinctive shape with characteristic apertures and these can be used to identify the types of plants represented by pollen in a sample from a given site. A microscope must be used for such identification because pollen and spores are small, typically between 10 and 200 μm. First though, pollen and spores must be isolated from sediments and rocks using both chemical and physical means, before they are mounted on microscope slides for examination and identification. For most geological and environmental applications of pollen and spore analysis, scientists count and identify grains from each sample using a microscope and generate pollen diagrams of the relative (per-cent) and absolute abundance of pollen in samples from a site's stratigraphic sequence (the geological age column): it is not just the presence or absence of a species used in this form of palaeoclimatology that is important but their abundance. Typically, the results of spores and pollen analysis from several species are pulled together to establish a picture of the changing plant ecology, and hence palaeoclimate, of a site.

Whereas dendrochronology provides an annual indicator of climate, pollen and spore analysis does not have such a high resolution. In some circumstances, specifically in well-dated sediment cores of high resolution, it is possible to obtain high-resolution records of botanical change at a decadal scale or less (in some cases to within a year) and also to document community changes over the last few centuries and even millennia. This, though, is rare. Typically the resolution achieved is of several years or even a few decades. Even so, it is important to bear in mind that the presence or absence of pollen or spores is not a hard-and-fast identifier of climate, as other ecological factors need to be considered.

For example, if the climate warms it will take a little while for a new species to establish itself, especially if the climate change is marked and sharp, so often necessitating species migration over a long distance. Indeed, there may also be other factors impeding migration. Suppose climatic cooling affects a northern boreal forest, even if the cooling is not marked enough to kill any trees, some species, such as black spruce, may stop growing cones and producing pollen. Black spruce pollen will then be absent from the record even though the species is as abundant as it was before, and it may even multiply by the growth of new roots where drooping branches touch the ground; so the trees will not die out even if they cannot seed. Yet, simply by being there they are impeding the arrival of new species whose pollen might be a better indication of the climate.

Aside from resolution and ecological problems, other difficulties with pollen and spore analysis as climate indicators include those of human land use. Humans tend to clear land for various purposes, from intensive urban land-scapes to semi-natural systems such as agriculture. Either way, the natural community is no longer there. Consequently spore and pollen analysis to elucidate past climates is best from sites in areas where there has been minimal human interference. (Although, of course, this particular problem can be turned on its head with pollen and spore analysis being used for archaeological purposes as an indicator of human activity.)

One also needs to recognise the problem of pollen decay. Pollen of some species are less resilient than others. For example, white pine pollen is resilient to breakdown whereas larch (*Larix* spp.), aspen (*Populus* spp.) and poplar (also *Populus* spp.) pollen decays quickly. So a sample of fossil pollen dominant in spruce (*Picea* spp.) could have come from a pure spruce forest or a mixed forest of spruce and poplar; there is no way of knowing. This means that pollen analysis rarely, if ever, provides a complete picture of a past ecosystem. Furthermore, some sites are better for conserving pollen and spores than others. The cold, acidic, anaerobic (oxygen-deficient) water of a peat bog is a better environment for preservation than warm, eutrophic (nutrient-rich) waters where bacteria thrive.

Considerable expertise by those who study pollen, palynologists, is needed to correctly identify which species particular pollen comes from. For example, the pollen grains of junipers (*Juniperus* spp.) and arborvitaes (or cedars, *Thuja* spp.) are indistinguishable. In other cases, although it may be possible to identify the genus of a pollen grain, it may be difficult to tell the species. Sometimes the grains from different species belonging to the same genus are identical except for a slight difference in size. Furthermore, pollen grains from

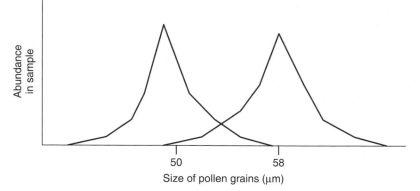

Figure 2.2 The variation size of pollen grains from two different species of the same genus may overlap, so impeding identification. Shown are data for lodgepole pine (*Pinus contorta*), with an average pollen-grain size of 50 μm, and western white pine (*Pinus montecda*), of average size 58 μm. Statistical analysis of difference of means of a sample against those from a known species may be required for identification purposes.

a single species do have natural variability in size and if the range from a single species overlaps with the range of another species in the same genus, say lodgepole pine (*Pinus contorta*) and western white pine (*Pinus montecola*), the identity of the grain becomes uncertain. In fact these two species have average pollen-grain lengths (excluding bladders) of 50 and 58 μm respectively (error, ±8 μm). In difficult instances where there is doubt a statistical analysis of difference of means is required (see Figure 2.2).

Consequently pollen and spores as palaeoclimatological indicators really only come into their own as a means of confirming (but not necessarily disproving) the picture portrayed by other palaeoclimatological proxies, be they local or hemispheric, and in this sense can assist in helping ascertain not only whether there was genuine climatic change of such-and-such a degree but whether it was a local variation or part of a more global climatic change. It is arguably better (as with species; see below) to think of pollen-and-spore evidence not in terms of climatic proxies but as evidence contributing to a picture of the effects of climatic change. We shall return to this later when discussing past ecosystems as being palaeoclimatic ecological analogues of possible future climatic change. (For an example of a typical summary pollen diagram see Figure 4.8.)

2.1.6 *Species as climate proxies*

It is well known that different fauna and flora live under different climatic conditions. So, as with pollen and spores, it might seem obvious to use species as climate proxies. Aside from the apparently obvious connection with climate, another attraction of considering species as palaeoclimatic indicators is that fossils are frequently all about us in unaccounted millions. Although large fossils are comparatively rare, small fossils, of a few millimetres or less in size, are frequently abundant. They come in two principal groups, macrofossils and microfossils, depending on whether a microscope is needed to identify them.

There is a considerable variety of small macrofossils. Usually only the hard parts of a plant or animal become fossilised. The soft parts decay quite rapidly after death. Animals are therefore often represented by the bones and teeth of vertebrates, the shells of molluscs (snails and clams), and the hard chitinous parts of arthropods (insects, spiders, etc.) These parts are rarely found intact but fragmentised to some degree. Among the commonest insect fossils are bits of beetles (Coleoptera), especially their upper wings (the elytra) and the shield covering most of their thorax (the pronotum).

Plants are commonly represented by fossil leaves, seeds, twigs and other wood fragments. Fossil charcoal gives evidence of past forest fires. As with animals, the soft parts of plants usually decay. However, fossil leaves are found if only because a single tree produces so many year after year and that they can be carried some limited distance. One therefore only needs a small fraction of these to be buried in a way that maximises the chance of fossilisation for there to be some samples surviving until today.

Microfossils are far more abundant. The most numerous are fossil spores and pollen, discussed in the previous section, but also there are diatoms (a class of eukaryotic algae of the phylum Bacillariophyta), Foraminifera (which belong to another class of eukaryotic algae), dinoflagellates, ostracods (a group of small bivalved crustaceans), chironomids (a family of midge) and stoneworts (part of the Charophyta algae). These are all important to micropalaeontologists. Microfossils can be found terrestrially and in the marine environment (see the next section).

As with pollen and spore analysis, there are many problems in using species as indicators of past climates and past climate change. Not least are problems of dating. Microfossils are capable of being washed from one place to lie in the sediment in another. Then there are ecological problems. Species do not migrate and become established instantaneously. Indeed, taking beetles as an example, a specialist plant-eating beetle needs for its food plants to be already established before it can migrate into an area and, if it is in turn preyed upon by a carnivorous beetle, then a population of the plant-eating beetle needs to established before the carnivorous predator can be sustained. All this takes time; meanwhile the climate has already changed. Consequently the remains of species are not ideally suited as palaeothermometers (except in a very broad-brush sense). Instead the remains of species are better viewed as indicators of the ecological impact of climate change. We shall return to this later when discussing the impact of climate change (Chapters 4, 5 and 6) and especially the likely effects of current global warming (Chapters 7 and 8) for which in some instances there are palaeo-analogues.

Of course, the above techniques only work if we have exact knowledge of species' climatic preferences: that is to say, that we can study live members of the species today. When trying to ascertain the climate many millions of years ago (as opposed to hundreds of thousands of years ago) we have to increasingly rely on fossils of species that are no longer with us. Consequently the next best thing that can be done is to look at the nearest living relatives of the fossil species. This does not give us as exact an idea of the past climate. The fundamental assumption behind analysis of nearest living relatives is that the climatic tolerances of living taxa can be extrapolated without modification to ancestral forms. The problem here is that this oxymoronically implies evolutionary stasis. However, confidence in nearest-living-relative species analysis can be increased if making a number of (hopefully reasonable) assumptions and adopting operational practices. Namely (1) that the systematic and evolutionary relationship between the extinct species and the living one is close; (2) being able to look at as large a number of fossil and nearest-living-relative pairs as possible, assuming that this will rule out individual species' idiosyncrasies; (3) using living relatives that belong to a diverse and widespread higher taxon as this increases the chance of finding a closer match with the fossil; and (4) using plant groups that have uniformly anatomical or physiological features that are constrained by climatic factors. Nearest-living-relative analysis can therefore provide us with a broad-brush picture of climate change many millions of years ago. (For examples see Allen *et al.*, 1993.) However, as this book is largely driven by the present need for science to underpin policy concerns over current global warming, much of this text's focus is on more recent change over the past 300 000 years or so (the past two glacial–interglacial cycles) and likely future change due to anthropogenic global warming. Over this recent and short time horizon many of the species that were around then are still with us. (Of course there are always exceptions and many large animal (mega fauna) species outside of Africa have become extinct over the last quarter of a million years due to human, as well as human combined with climate, pressures. African large animals were more resilient as they co-evolved with humans.)

Naturally, individual species are not alone in responding to climate change and it is often necessary to use more than one species to ascertain the timing and degree of the same climate change. For example Post and Forchhammer

(2002) published research on caribou and musk oxen on opposite sides of Greenland, demonstrating that populations of different species may respond synchronously to global climate change over large regions.

But terrestrial species are not the only ones to be used as palaeoclimatic indicators. Marine species are too (see the next section) and even those that can live both in the sea and on land. In 2004 it was reported that the migration of fish that can live for extended periods on land can be used as a palaeoclimate indicator, when Madelaine Böhme from the University of Munich used fossil snakeheads, a group of predatory fish (Channidae) native to Africa and south-east Asia, from the Miocene to Pliocene (23 million–5 million years ago; Böhme, 2004). The species provides a good indicator of summer precipitation. Its present-day distribution is limited to climates with at least 1 month of rainfall of 150 mm and a mean temperature of 20 °C. Böhme's study suggests that the most extended migration events (17.5 mya and between 8 and 4 mya) must have been linked to changes in northern-hemisphere air circulation, resulting in formerly dry summers having higher rainfall.

2.2 Marine biotic climatic proxies

Marine proxies for palaeoclimates have several advantages over many terrestrial indicators, especially biological ones. The sea surface is subject to fewer variables affecting biotic proxies compared to their terrestrial counterparts. Surface topography is constant and so there is no need to worry about whether the proxy originated on a long-since-gone slope. Vagaries in water availability for the proxy are also not a problem. Finally, the sea temperature (because of the sheer volume of water being affected by the climate) is more applicable to the regional climate (albeit affected by heat transport by ocean currents) than the local climatic vagaries that affect terrestrial biotic indicators. Marine biotic indicators are therefore particularly important when attempting to ascertain palaeoclimates. There are a number of marine biotic indicators of climatic proxies. These include terrestrial counterpart methods such as the use of species. However, of all the methods used, one in particular has generated much information as to climate change, both generally over several tens of millions of years and more recently in greater detail over hundreds of thousands of years. This method employs ^{18}O analysis.

2.2.1 ^{18}O isotope analysis of forams and corals

For decades now, the concentrations of the heavy isotope ^{18}O in the microshells of Foraminifera – an order of plankton – and also corals have been widely used to reconstruct the temperature profiles of ancient seas. Originally it was in 1947 that Harold Urey (who previously won a Nobel Prize for chemistry for his discovery of deuterium) pointed out that the oxygen-isotope composition of fossil seashells could serve as a palaeothermometer of past sea-surface temperatures. The stable heavy isotope ^{18}O accounts for about 0.2% of the oxygen in sea water's H_2O compared to the most common ^{16}O isotope. The greater proportion of ^{18}O that a shell incorporated, Urey showed, the colder the water in which it was formed. The slightly heavier nucleus of ^{18}O also affects evaporation and precipitation, but here it is crystal formation that is important. Urey argued, on thermodynamic grounds, that the formation of crystalline $CaCO_3$ (calcite) – with dissolved calcium in sea water and which is taken up by forams and corals – should involve some isotopic fractionation. The idea was that the fraction of ^{18}O incorporated into the calcite should decrease with increasing water temperature, essentially because the heavier nucleus slows down the vibrational modes of molecules with ^{18}O compared to ^{16}O-containing water molecules, which then have a thermodynamic advantage in the process of calcite formation. However, to exploit this isotopic effect for palaeothermometry, one has to correct for the global extent of ice caps and glaciers at the time in question. The reason for this is that water molecules with the normal, lighter ^{16}O preferentially evaporate compared to ^{18}O, which requires more energy that goes with higher temperatures; it is this last that underpins the theory behind ^{18}O dendrochronology mentioned above. So, using ^{18}O in shells for palaeothermometry is only really effective when there is little ice on the planet. When there is a lot of ice (such as now, with Antarctica and Greenland) ^{18}O in shells is a better indicator of the amount of ice worldwide than global temperature. (We shall return to this ^{18}O ice issue shortly.)

So, because ice caps take water molecules with a mix of oxygen isotopes out of the oceanic pool of water molecules, ice caps complicate things. However, from the late Cretaceous (65 mya) to the late Miocene (5 mya), the Earth had negligible ice caps and terrestrial ice (such as on mountain tops). Consequently ^{18}O palaeothermometry is an applicable palaeoclimatic tool for this geological window. There is a complication in that different plankton species, although they all float in the top few hundred metres of the ocean, live at different depths below the surface (hence possibly at different temperatures), and they often change habitat depth with the seasons. Some species, for example, keep to the upper layers because they live symbiotically with photosynthesising algae. Not knowing the detailed life cycles of all the long-extinct creatures means that some inferences need to be made, with some inadvertent erroneous assumptions. Consequently, in these palaeoclimatic studies researchers customarily deduce surface temperature for a given multi-species sample from the species that yields the lowest ^{18}O fraction, presumably corresponding to the habitat closest to the surface.

Since the mid 1980s, however, models of carbon dioxide greenhouse warming have confronted ^{18}O data from fossil planktonic (floating) Foraminifera with the so-called cool-tropics paradox. In stark contrast to many global climate models, the planktonic ^{18}O data seemed to suggest that 50 million years ago tropical ocean surfaces were about 10 °C *cooler* than they are now, even though this was a time when the carbon dioxide level was almost certainly much higher than it is today and the Arctic was warm enough for crocodiles and giant monitor lizards.

The controversy was resolved largely due to an analysis of planktonic Foraminifera from the late Cretaceous to the late Eocene (67–35 mya). This was conducted by Paul Pearson (University of Bristol) and co-workers in 2001, and does much to lay this troubling cool-tropics paradox to rest. Because planktonic Foraminifera, while they live, float at or near the surface, researchers had assumed that their ^{18}O concentration reflects the temperature of the sea surface. But Pearson and company, doing isotopic analyses of unusually well-preserved samples of pristine shells selected with the help of electron microscopy, conclude that the surprisingly high ^{18}O level of traditional samples is a misleading consequence of extensive recrystallization of the fossil shells in the much colder waters at the bottom of the sea (Pearson *et al.*, 2001). Daniel Schrag (Harvard University) reached much the same conclusion in 1999 by means of a mathematical model of the recrystallization of these very porous microshells buried in the sea floor (Schrag, 1999). At higher latitudes, where there is less temperature contrast between the surface and the bottom of the sea, this recrystallization issue is less of a problem for ^{18}O palaeothermometry.

There are other problems with this form of ^{18}O analysis, especially with regards to determining climates prior to the last glacial–interglacial cycle (of the order of a hundred thousand years) and going back millions of years. These relate to dating benthic (deep-sea) forams. First, more ancient deep-sea sediments are increasingly disturbed due to a variety of factors over millions of years, including plate-tectonic movement and the weight of fresher sediments being superimposed on top. Second, ^{14}C, as used in dendrochronology (discussed above), is not applicable over such long timescales and so radionucleotides with appropriately longer half-lives are required. However, this reduces the resolution of sediment age determination. Consequently, whereas we might *in theory* (see below) be able using various techniques to determine regional and global palaeoclimates down to virtually an annual level over the past glacial–interglacial cycle (of the order of 100 000 years), when attempting to go back tens of millions of years the palaeoclimatic resolution tends to

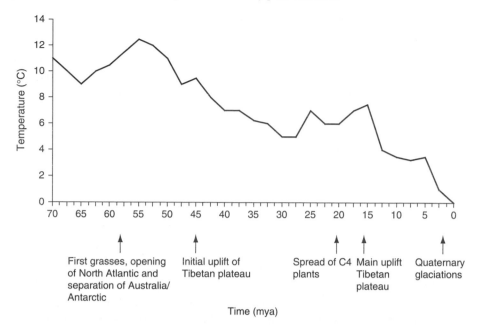

Figure 2.3 Cenozoic global temperature trends compared to 1990, as estimated from [18]O analysis of composite benthic foram records from the Atlantic Deep Sea Drilling Program sites. Based on data from Miller *et al.* (1997) and simplified from Willis and Niklas (2004). Indicated are some of the factors affecting the global climate. Grasses and C4 plants enabled enhanced photosynthetic sequestration of atmospheric carbon. The raising of the Tibetan plateau not only affected the continental Asian climate but also created a considerable body of mid latitude ice that affected planetary albedo. (Milankovitch effects occur on too short a timescale to be deduced in this way and resolved over this period of time.)

be of the order of several centuries, if not millennia, at best. However, it is possible to get a broad-brush (low-resolution) picture as to how the temperature of the planet has changed over such timescales (see Figure 2.3).

There are only a few long-half-lived isotopes, including those relating to the uranium-235 (^{235}U) decay series, that can be used to date remains further back in time than ^{14}C. And among these there are a number of radioisotopes that are not incorporated into biotic remains but with the surrounding geological strata instead. This obviously begs the assumption that the biotic proxy and the surrounding geological strata originated at the same time. Nonetheless, radioactive isotope dating methods can even be used over extremely long timescales that take us back to before life on Earth arose. Rubidium-87 (^{87}Rb), with its half-life of 4.8×10^{10} years, is used by geologists for dating over timescales back to the formation of the Earth.

Another problem is that of the resolution of change discernable from ^{18}O analysis. The problem is that the climate can sometimes change faster than ^{18}O concentrations. For ^{18}O concentration to properly reflect the climate there needs to be an appropriate amount of mixing of the oceans' waters and new balances struck with ^{18}O in its various guises in the atmosphere and ocean. This takes time. Broadly speaking, ^{18}O analyses are not particularly useful in resolving climatic episodes of much less than 1000 years in duration.

In short, ^{18}O Foraminifera and coral palaeothermometry is of value but (a) only for periods of time on the planet when there were virtually no ice caps, (b) if one concentrates on such proxies from high latitudes away from the tropics so as to minimise recrystallization problems, and (c) if the climate being elucidated is one that lasts for roughly 1000 years or more.

However, there is another value in ^{18}O analysis when there were or are large ice caps on the planet, such as now in Antarctica and Greenland, even in our warm (Holocene) interglacial, and especially in glacials with their substantive ice caps. This applies to all the Quaternary period of the past 2 million years. As noted above, ^{18}O-containing water molecules require more energy to evaporate from the oceans than ^{16}O water. The principal thing here is that as the lighter ^{16}O evaporates preferentially from the sea it tends to leave behind ^{18}O and so the sea becomes enriched in ^{18}O as the ice caps grow. This is reflected in ^{18}O in forams and corals. Consequently, whereas ^{18}O may be a reasonable palaeothermometer between the Eocene and Cretaceous (when there were no ice caps), when there are ice caps this isotopic partitioning between marine and terrestrial ice environments comes to the fore. At such times ^{18}O in marine calcite is an indicator of terrestrial ice and, hence, global sea levels (as opposed to temperature). The higher the concentration of marine calcite ^{18}O the greater the extent of terrestrial ice and the lower the sea level. (As terrestrial ice caps – rich in lighter isotopes – grow, accounting for more water in the biosphere, so the amount left for the oceans decreases.) Intuitively one might think that this indicator of ice volume is sort of a palaeothermometer in itself, as one might reasonably suspect that when the planet has large ice caps it must be cooler than when it does not. But this is not strictly true. Ice caps do not come and go overnight. We see substantial ice caps today in Antarctica and Greenland, on which snow falls even now in the middle of a warm interglacial (even if around the margins of these caps there is melting). So, at times when our planet (or in all probability any other Earth-like planet) does have ice caps, ^{18}O in marine calcite is an indicator of the extent of these terrestrial ice caps, and, commensurately, sea level, but not nearly so much of temperature.

Because ^{18}O concentration in forams and corals is determined primarily by two factors – temperature in the absence of terrestrial ice and sea level in the presence of terrestrial ice – a number of global palaeo-temperature graphs have two temperature scales, one relating to before 35 mya, when Antarctic glaciers began forming, and one after, when terrestrial ice became a major factor. This is because of the problems discussed above of this form of palaeoclimatic determination over periods of many million years. For this reason Figure 2.3 should actually be considered to be the merging of two ^{18}O graphs relating to temperature, one covering the period before 35 mya and one after. Alternatively, it could have been a single ^{18}O concentration curve but necessitating two temperature-related scales, again one before 35 mya and one after. Figure 2.3 was drawn the way it was to clarify matters simply.

But forams are not the only biological sources of ^{18}O that can be analysed. Coral analyses have certain advantages over foram analysis. Unlike forams, which are more subject to the vagaries of water movement, coral reefs are fixed. Consequently, problems with changes in temperature with depth are far less important.

Corals are animals in the cnidarian (coelenterate) phylum, order Scleractinia (class Anthozoa). Their basic body plan consists of a polyp containing unicellular endosymbiotic algae known as zooxanthellae overlaying a calcium carbonate exoskeleton and it is this calcium carbonate (calcite) that is used for ^{18}O analysis. Of particular value is that many corals outside of the tropics also have annual growth rings: actually they have daily rings that are further apart in the summer and narrower in the winter as well as during their monthly (tide-driven) breeding cycle. (This last has been used to determine the distance between the Earth and the Moon millions of years ago but that is a separate story.)

2.2.2 *Alkenone analysis*

Long-chain (C_{37}–C_{39}) alkenones are organic molecular 'fossils' from phytoplankton that represent about 8% of total phytoplankton carbon and which are cell-membrane lipids. These can be used to indicate the sea-surface temperature. The principle behind this is not exactly known, although the correlation between the unsaturated proportion (that with double and triple bonds in the chain) and temperature is undeniable. It is speculated that as temperature changes so does the rigidity, or some other physical property, of phytoplankton cell membranes. What is known is that, in culture, living algae respond to changing temperature conditions in a matter of days. Because of this they are particularly good for recording rapid temperature change subject to accurate dating. Also, alkenones are robust molecules that are capable of being preserved for many hundreds of thousands of years and so are useful in charting more than our current glacial–interglacial cycle.

One of the longest established alkenone-based proxies was devised at the University of Bristol in 1985. Known as the U^K_{37} index, it is based on the number of double bonds in alkenones with 37 carbon atoms that are found in many unicellular eukaryotic marine algae, including *Emiliania huxleyi*, one of the most common of such algae.

In 1986 another robust index was established, called the TEX 86 index, named after the Dutch island of Texel on which it was developed. TEX 86 is based on molecules that have 86 carbons. However, these molecules come from far less common single-celled Archaea, but because such Archaea have been found in sediments spanning the past 100 million years they represent a powerful complement to the U^K_{37} index.

2.3 Non-biotic indicators

Whereas (formerly) living organisms are most useful as an aid to estimating palaeoclimates because they either live solely within usefully narrow climatic restrictions, or because in the course of their lives they somehow incorporate a palaeobiotic indicator into their bodies, there are other indicators of past climates that operate completely independently of living creatures. Of these, one of the most useful in terms of helping determine climate change hemispherically (if not globally) over the past few hundred thousand years has been the isotopic analysis of water ice.

2.3.1 Isotopic analysis of water

Water is not just H_2O or HHO. As discussed in preceding sections, oxygen exists as two principal isotopes, ^{16}O and ^{18}O, together with another, ^{17}O. Further, there is also a second stable isotope of hydrogen in addition to the common ^1H and this is ^2H, or deuterium (sometimes portrayed as D). Only a tiny proportion of hydrogen and oxygen is not in their respective principal isotopic form, but nonetheless these rarer isotopes do exist. Consequently water is not just HHO, as in $^1H^1H^{16}O$, but has a variety of isotopic variations (see Table 2.1).

As has previously been noted the take-up by photosynthetic organisms of water containing ^{18}O for photosynthesis is affected by temperature and is complicated by preferential evaporation of water containing lighter oxygen isotopes. Yet such preferential evaporation of lighter isotopes with temperature can also be of use as a climate indicator. Deuterium-containing water, because of hydrogen's smaller size, is not so discriminated against by living things, so the preferential evaporation and condensing of water containing isotopes of deuterium is more solely temperature-dependent. More accurately, the deuterium concentration of ice in terrestrial ice caps such as Antarctica and

Table 2.1. *Isotopic species of water.*

Composition	Natural abundance (%)
$^1H^1H^{16}O$	99.732
$^1H^1H^{18}O$	0.200
$^1H^1H^{17}O$	0.038
$^1H^2H^{16}O$	0.030
$^2H^2H^{16}O$	0.020
$^1H^2H^{18}O$	0.006
$^1H^2H^{17}O$	0.001
$^2H^2H^{18}O$	0.0004
$^2H^2H^{17}O$	0.00008

Greenland reflects the temperature difference of the water (primarily the ocean) from which it evaporated and the air from which it froze out as snow. Deuterium concentration in such ice caps therefore reflects part-hemispheric more than global climate change. Yet, as we shall see in the next chapter, there is considerable agreement between the deuterium records of the northern hemisphere's Greenland cores and those from the southern hemisphere's Antarctic cores. However, there are also some interesting differences that will be discussed.

The problem with the isotopic analysis of water, or to be specific, ice, as a determinant of palaeoclimate is that it is only possible where ice has existed undisturbed as far back as the climate you wish to study. This restricts such palaeoclimatic studies, both in space and time; space because the places where ice exists undisturbed with longevity are rare, and time also because where it has lain undisturbed has not been long: even a stable ice cap will slowly build up so that the early lower layers come under increasing pressure, with an increasing likelihood of deformation. However, both the Greenland and Antarctic ice sheets have enabled us to piece together the global climate over the past couple of glacial–interglacial cycles. This has been a key validation of the Milankovitch theory of orbital climatic forcing: although we are aware from other evidence that Milankovitch climatic forcing has existed for many millions of years and so presumably for as long as the Earth has had a troposphere with discernable weather (as opposed to other factors affecting conditions on the planet's surface such as planetesimal accretion).

The great utility of the isotopic analysis of ice is that when it does lay undisturbed it is possible to discern annual layers of snowfall that subsequently have become compacted to form ice. But there are only a few places where ice is so undisturbed. As mentioned, one is at the top of the aforementioned

Greenland ice cap where there is minimal lateral movement of ice, and another is the ice cap over the lake at Vostok in Antarctica. The cores from these places enable us to discern isotopic proportions of ice with resolutions of a few years, yet the records cover hundreds of thousands of years. Although, even so, it just takes one layer to be less clear for the counting of subsequent layers to be out by one; there is a huge difference between much of a core having a high resolution and the entire core being so clear. Consequently, although a core of a thousand years of age may exhibit a high resolution, the precise dating of the layers through physical counting of the entire length of the core to the present may only be accurate to within a decade. Nonetheless, this means that it is possible to discern a particularly detailed picture of the palaeoclimate going back to more than two glacial–interglacial cycles and a highly detailed picture emerges from the end of the last glacial maximum some 20 000 years ago. As we shall see this has provided evidence that the hemispheric climate has 'flickered' in the past, changing rapidly over just a few years (within a decade) or a decade or two. As we shall see in the next chapter, climate change is not always as gradual as sometimes portrayed.

2.3.2 Boreholes

As the climate warms and cools, so varying amounts of heat are absorbed into the underlying geology. Consequently, it is theoretically possible to drill a deep borehole to measure the temperature at varying depths as an indication of past climate. Depths of borehole are typically of the order of a kilometre or more. Assumptions based on geological knowledge of the area, especially those related to geothermal heat, need to be made.

Problems with using boreholes as indicators of palaeoclimate include that the resolution of the thermal record is low and decreases with depth. However, in 2000 Shaopeng Huang and colleagues from Michigan University, USA, and the Western Ontario University, Canada, analysed temperatures from 616 boreholes from all continents, except Antarctica, to reconstruct century-long trends over the past 500 years (Huang *et al.*, 2000). The results confirmed the unusual warming of the twentieth century, which was greater than for the previous four centuries. Their results also corroborated the consensus as to past climatic change over this period from other palaeo indicators.

2.3.3 Carbon dioxide and methane records as palaeoclimatic forcing agents

In addition to indicators of past climates there are also records of palaeoclimatic forcing agents. Here the principal examples are the ice-core records of the past atmosphere. As snow falls on ice caps so bubbles of air become trapped and encased in ice. Bubbles in ice cores from Vostok (Antarctica), and Greenland provide a record of atmospheric composition going back hundreds of millennia. A history of both the key greenhouse gases, carbon dioxide (CO_2) and methane (CH_4; see Chapter 1), can be ascertained. These strongly reflect hemispheric climatic change as elucidated by the deuterium isotope record and so demonstrate the strong link these forcing agents have with climate. The methane record largely reflects the area of wetlands globally that produce the gas. Its concentration is higher in the warmer, hence wetter (due to increased ocean evaporation), interglacial times. Consequently, being a greenhouse gas, its atmospheric concentration change is another example of one of the many feedback systems operating in the biosphere. The warmer the global climate the more wetlands there are generating methane, which being a greenhouse gas further warms the climate, and so on. Conversely, the cooler the climate the less the wetland area, hence less methane is produced, and less climate warming from atmospheric methane.

Of course wetlands are not the only source of methane. Oceanic methane hydrates (clathrates) have been known to release pulses of methane into the atmosphere, providing a warming burst.

The problem with carbon dioxide and methane gases as a record of climate-forcing agents is that other forcing factors operate as well and so the record cannot be said to properly reflect climate. Indeed, one brief climatic return to glacial conditions (the Younger Dryas) prior to our current Holocene inter-glacial took place in the northern hemisphere but does not seem to have taken place to anything like the same extent in the southern hemisphere. It is also less prominently reflected in the Antarctic atmospheric carbon dioxide record. Furthermore, as we shall see (Chapters 4 and 5) the so-called Little Ice Age, in the fifteenth to seventeenth centuries, took place but was not reflected in changes in atmospheric carbon dioxide concentrations.

Another problem with relating atmospheric carbon dioxide to climate is that if the climate is changing and plants are absorbing more carbon dioxide, or rotting and producing more, or it is getting wetter and wetlands are expanding, thus pushing up methane concentrations, then it takes time for this to be reflected to any great degree in the atmospheric concentration. True, atmo-spheric carbon dioxide varies by 5 ppmv (parts per million by volume) every 6 months outside of the central tropics due to seasonal vegetation changes, but carbon dioxide concentration changes associated with glacial–interglacial transitions are of the order of 150 ppmv. If the hemispheric carbon pump can only cope (optimistically) annually with less than 10 ppmv then climate changes associated with greater carbon dioxide change than this fall outside of this window. In part related to this, there is the problem that the atmospheric residence time of a carbon dioxide molecule is of the order of a century or two. Consequently, climate change episodes not primarily driven by atmospheric carbon dioxide (such as those driven by changes in ocean circulation) and which occur quickly, cannot easily, if at all, be discerned from the atmospheric gas record. This is yet another reason why it is best to use as many as possible indicators of climate so as to understand what is really going on.

2.3.4 Dust as an indicator of dry–wet hemispheric climates

Dust blown many hundreds of miles is an indicator of how dry is the climate. In a wet climate rain is more likely to wash dust out of the air than in a dry climate. Indeed, we know that warm interglacials are globally far wetter than the cool glacials. As noted above, this is because in a warmer world there is more oceanic evaporation, hence precipitation, than in a cooler one. Of course, to discern this indicator one needs a recording location that is hundreds of miles from the nearest large source of dust. Such locations include the middle of ice caps and, as mentioned, one – Greenland – especially lends itself to the building up of ice layer by layer each year and has ice strata going back over hundreds of thousands of years. Measuring the dust trapped in the ice can be done by measuring the ice's electrical conductivity. (Conductivity being the reciprocal of resistance.) The more dust there is in the ice then the less conductive it becomes as the dust neutralises the acidic ions that facilitate conductivity. As with dendrochronology, providing dating can be accurately achieved, where there is clear layering it is similarly possible to discern conductivity levels on close to an annual basis. As we shall see in Chapter 4, ECM (electrical conductivity measurement) analysis of the Greenland ice core has revealed dry–wet (hence presumably cold–warm) climatic flickers of roughly a year or two, certainly far less than a decade (see Figure 4.12b).

2.4 Other indicators

The above sections deal with many of the key indicators (with a focus on the biologically related ones) used to determine past climates. However, it is important to note that there are others. Planktonic species abundance in oceanic cores (analogous to pollen analysis in peat bogs), lake-level reconstructions, lake sediments and glacial moraines are among a number of other palaeo indicators of climate. Together all indicators provide a valuable insight into past climates, over a variety of scales both spatially and temporally.

2.5 Interpreting indicators

Because proxy indicators are not *actual* instrumental measurements of climate but a *reflection* of climate it is important to recognise their limitations. One might consider the bringing together of several proxy data-sets to form a record of a region's climate as superimposing photocopies of different parts of a picture, with some pictures having parts that are different to others and others the same. Such superimpositions are perfectly readable but they are still blurred. Yet even statistical analyses of palaeorecords, even if they reflect when times were warmer or cooler, may not necessarily accurately reflect by exactly how much warmer or cooler past times actually were. Palaeo indicators such as tree rings are, as previously noted, themselves subject to a number of non-climatic factors such as whether the tree grew on the north or southern side of a hill, soil suitability and so forth. The data from such proxies are therefore said to be noisy.

In 2004 a European team, led by Hans von Storch, attempted to see whether past climate reconstructions from noisy proxies reflected climate by working backwards from a climate model that had already been shown to reasonably accurately reflect the northern hemisphere's climate for the past 1000 years. They used this model to generate data as if from climate proxies and then added noise. They then analysed the resulting proxy data to reconstruct the northern hemisphere's climate just as had already been done with real proxy data. They found that their simulated noisy proxy data did not accurately reflect the magnitude of the (simulated) past climate change. This suggests that some (actual) palaeoclimate reconstructions (including those used by the Intergovernmental Panel on Climate Change, IPCC) may only reflect half the actual change that has historically taken place since the Industrial Revolution (von Storch *et al.*, 2004). If this were so it means that the IPCC's future forecasts may not fully reflect the degree of climatic change we may actually experience (see Chapter 5 and the subsection on uncertainties and the IPCC's conclusions).

2.6 Conclusions

Each technique for ascertaining an aspect of past climates has its limitations in terms of: climatic dimension being indicated (or reflected), be it temperature or aridity; resolution in time and space; and finally the sensitivity of the proxy. These limitations have to be borne in mind when attempting to reconstruct past climates and any interpretation made with reference to other climatic proxies. However, palaeo indicators of climate have made it possible to begin to establish a picture of how the climate has changed in the past, even if this picture is more detailed in some places than others. Nonetheless, we now have a sufficiently good picture of climate change on the global scale going back the past few glacial–interglacial cycles at the decadal-to-century scale to begin to inform us about possible future climate change.

At the regional level in many places we also have a detailed understanding of the palaeoclimate with resolutions of a year or less going back many thousands of years. On timescales of back many millions of years we have a coarse understanding at the mid-latitude continental level and with resolutions of fractions of millions of years. Several palaeoclimatic proxies taken together can prove to be a powerful tool. One such multiple proxy index is the North Atlantic Oscillation index and such indices can be used to reconstruct large-scale hemispheric changes in temperature. Consequently not only is a picture of global climate change beginning to emerge but also – given that carbon dioxide is a principal forcer of climate – an appreciation of how the carbon cycle has broadly changed over the past few hundred years is beginning to emerge: although this, and palaeoclimatology itself, is still the focus of a considerable research effort in many countries.

2.7 References

Allen, J. R. L., Hoskins, B. J., Sellwood, B. W., Spicer, R. A. and Valdes, P. J. (eds) (1993) Palaeoclimates and their modelling with special reference to the Mesozoic Era. *Philosophical Transactions of the Royal Society of London Series B, Biological Sciences*, **341**, 203–343.

Bailey, I. W. and Sinnott, E. W. (1916) The climatic distribution of certain types of angiosperm leaves. *American Journal of Botany*, **3**, 24–39.

Baille, M. (2003) Dendrochronology and volcanoes. *Geoscientist*, **13**(3), 4–10.

Böhme, M. (2004) Migration history of air-breathing fishes reveals Neogene atmospheric circulation patterns. *Geology*, **32**, 393–6.

Briffa, K. R. (1998) Influence of volcanic eruptions on Northern hemisphere summer temperatures over the past 600 years. *Nature*, **393**, 450–8.

Briffa, K. R. (2000) Annual climate variability in the Holocene: interpreting the message of ancient trees. *Quaternary Science Reviews*, **19**, 87–105.

Huang, S., Pollack, H. N. and Shen., P. (2000) Temperature trends over the past five centuries reconstructed from borehole temperatures. *Nature*, **403**, 756–8.

McElwain, J. C., Beerling, D. J. and Woodward, F. I. (1999) Fossil plants and global warming at the Triassic-Jurassic boundary. *Science*, **285**, 1386–9.

Miller, K. G., Fairbanks R. G. and Mountain G. S. (1987) Tertiary oxygen isotope synthesis, sea-level history and continental margin erosion. *Paleoceanography*, **2**, 1–19.

Pearson, P. N., Ditchfield, P. W., Singano, J. *et al.* (2001) Warm tropical sea surface temperatures in the late Cretaceous and Eocene epochs. *Nature*, **413**, 481–7.

Post, E. and Forchhammer, M. C. (2002) Synchronization of animal population dynamics by large-scale climate. *Nature*, **420**, 168–71.

Ravilious, K. (2000) Leaf shapes: climate clues. *NERC News*, Spring, 6–7.

Schrag, D. (1999) Effects of diagenesis on the isotopic record of the late Paleogene tropical sea surface temperatures. *Chemical Geology*, **161**, 215.

Treydte, K. S., Schleser, G. H., Helle, G. *et al.* (2006) The twentieth century was the wettest period in northern Pakistan over the past millennium. *Nature*, **440**, 1179–82.

Vaganov, E. A., Hughes, M. K., Kirdyanov, A. V., Schweingruber, F. H. and Silkin, P. P. (1999) Influence of snowfall and melt timing on tree growth in subarctic Eurasia. *Nature*, **400**, 149–51.

von Storch, H., Zorita, E., Jones, J. M. *et al.* (2004) Reconstructing past climate from noisy data. *Science*, **306**, 679–82.

Webb, G. E. (1983) *Tree Rings and Telescopes*. Tucson, AZ: University of Arizona Press.

Willis, K. J. and Niklas, K. J. (2004) The role of the Ice Ages in plant macroevolution. *Philosophical Transactions of the Royal Society of London Series B, Biological Sciences*, **359**, 159–72.

Wolfe, J. A. (1995) Palaeoclimatic estimates of Tertiary leaf assemblages. *Annual Review of Earth and Planetary Sciences*, **23**, 119–42.

3

Past climate change

Post-industrial global warming is not the Earth's first major climate-change event. If we are to appreciate the significance and relevance to biology of current climate change, it is important to be aware of past climate events, at least the significant ones. This chapter summarises some of the major episodes between the Earth's formation and the beginning of the current, Quaternary, ice age.

3.1 Early biology and climate of the Hadean and Archeaen eons (4.6–2.5 billion years ago, bya)

3.1.1 The pre-biotic Earth (4.6 3.8 bya)

The Earth and the solar system formed some 4.6 bya, give or take a few hundred million years. It formed with the solar (Sol) system accreting out of a dust and gas cloud. The dust, ice (composed of various volatiles compounds and not just water) and rocks were not all small and particulate but themselves had accreted into small and large asteroid-sized bodies that ranged in size up to, and including, small planets. One of these, a Mars-sized planetoid (Thea) is thought to have had a glancing blow with the proto-Earth, exchanged material, and formed the Moon (Luna). This is not irrelevant to the nature of the Earth's climate. The lunar/Earth ratio of mass of 1:18.3 is much greater than any satellite/planet mass ratio for any other planet in Sol's system. Taking the Copernican principle, that there is nothing cosmologically special about the Earth as a life-bearing planet, this begs the question as to whether our large moon is a necessary factor facilitating a biosphere, or at least a biosphere with longevity. There is a suggestion that the Earth–lunar system is one that confers some axial stability to the Earth (affecting the variation in its angle of tilt), and hence climate stability, so enabling complex ecosystems to form. Indeed, we have already discussed (Chapter 1) axial tilt as a dimension of Milankovitch forcing of climate, but planets without large moons are prone to larger axial tilting and this means that good portions of such worlds spend half the year in sunlight and half the year in darkness. On the Earth this only takes place within the polar circles, which form a minor proportion of the planet.

Aside from the formation of the Moon and Earth's axial stability, the other likely result of the Thea impact is that it affected the Earth's atmosphere. Here the suggestion arises from the concentrations of inert gas in the Earth's atmosphere, and that of Venus (Genda and Abe, 2005). However, the notion has not become firmly established with any definitive link as to biosphere formation.

Following the creation of the Moon, on Earth oceans formed from water arising from the proto-Earth's rocks as well as from the comets, asteroids and planetesimals still bombarding it. At the time the Earth's atmosphere was very different to today, being largely oxygen-free and dominated not by nitrogen but by carbon dioxide and possibly methane. There is also some debate as to whether there was a significant amount of hydrogen in the very early mix. Either way, the early Earth's atmosphere must have conferred a powerful greenhouse effect. This was needed at that time as the early Sun (as a typical main-sequence star) was thought to be only 30% as bright as it is today. Studies of other main-sequence stars have made it clear that young main-sequence stars do not generate as much energy as when they are older. Given, from the previous chapter, that there is currently a natural greenhouse effect, the early Earth under a fainter Sun would have needed a far more powerful natural greenhouse effect if liquid water (that would greatly facilitate life) was common on the planet. So, as we shall see, greenhouse gases were vitally important to ensure that conditions on the Earth's surface were warm enough for life. Furthermore, as the Earth's atmosphere evolved so (a different mix of) greenhouse gases were required to maintain temperatures. Life (biology) was fundamental to this process.

3.1.2 The early biotic Earth (3.8–2.3 bya)

Life arose fairly quickly in planetary geological terms, if suggestive isotopic evidence (possible microfossils containing a high proportion of ^{12}C) is to be believed. It is thought to have begun somewhere between 3.5 and 4.2 bya, which is about when the heavy bombardment of the Earth ceased (somewhere around, or shortly after, 4 bya). However, both the Earth and the Moon were still subject to hits by large planetesimals, as the 1000-km-wide impact basins visible on the Moon testify, although the frequency of such hits gradually waned.

It is not known whether life based on RNA and DNA (and its RNA world precursor) was the very first bioclade; it may well have been that some other system did arise first (with a preference for ^{12}C that we may even eventually isotopically detect), but if it did it failed to become established. Equally, it may be (we simply do not know) that large impacts capable of sterilising large areas of the Earth's surface, and even the smaller impacts that continued up to 3.8 bya, made the surface too perilous for sustained life. Conversely, microbes sheltering deep in the Earth's crust would have survived: perhaps this is why the genes of living microbes suggest that the ancestor of all life had much in common with the hyperthermophiles that thrive in hot springs of 113 °C or more. It has even been contemplated that life arose on Mars, which had a far more benign environment at that time and presented a small target for asteroids, as well as having a lower gravitational well attracting them. Indeed, even if Mars had had an impact large enough to affect life (which it frequently did) the early Martian seas were so much shallower than Earth's that the damage would have been less, with the resulting steam condensing out in just decades. Life would have had a better chance of arising on Mars first and then hitching a ride to Earth; ejecta from an asteroid impact on Mars could have carried life to seed Earth. (Again, Mars' shallower gravitational well would have more easily facilitated material leaving it, compared with the Earth.) Conversely, others engaged in this scientific whimsy argue that as we know that life got off the mark quickly in geological terms, so the Martian hypothesis (albeit theoretically possible) is not necessary, and in any case would only have given a few hundred million years' (about half a billion years' at the most) advantage over any life arising on Earth. Furthermore, life needed to arise somewhere, be it Mars or Earth, and so the point is mute (unless one considers a non-planetary origin for life such as panspermia theories). Either way, as we shall see, consideration of the climate in which life arose remains central to discussion of its immediate subsequent evolution.

What is known is that our RNA/DNA bioclade arose in a biosphere very different from our own today. As stated above, the Sun was dimmer 3.8 bya than it is today: this is an important point, as atmospheric carbon dioxide has played a major role in helping stabilise the planet's surface temperature within a window enabling life to survive if not flourish. Second, also as noted above, the primordial Earth's atmosphere was markedly different from today's: it had little if any oxygen, and so concentrations were far from today's 21%. At that time carbon dioxide was possibly the major atmospheric constituent and a major greenhouse gas along with water vapour and methane. Consequently anaerobic prokaryotes (bacteria and algae) ruled the biosphere: aerobes came later (see below) followed by eukaryotic life (cells with internal membranes, including one around the genetic material) around 1.7 bya. The fossilised remains of some colonies of these anaerobic prokaryotes can be found today as stromatolites (although it is important to note that some stromatolites – clumped, fossilised, laminated sedimentary structures – are thought to have non-biological origins).

A key question, 3–3.5 bya on the anaerobic Earth, is whether carbon dioxide (and water vapour) alone would have been enough to stop the Earth freezing over under a far dimmer Sun. If carbon dioxide had been acting alone as the principal greenhouse gas then concentrations would be required of some 300–1000 times as much as the atmospheric levels today for greenhouse warming to counter the faint primordial Sun! But at such high concentrations the gas would have combined with iron to form iron carbonate (siderite, $FeCO_3$) in fossilised soil strata (palaeosols) and we simply have not found these today (although they have been found in ancient marine sediments). So, already primordial Earth gives us a greenhouse conundrum.

One view (from US cosmologists Carl Sagan and Christopher Chyba in the 1990s) is that ammonia (NH_3) could have contributed to anaerobic Earth's greenhouse effect. The problem with this suggestion is that ammonia is sensitive to sunlight and so would have required a high-altitude reflective methane haze for protection, but that haze itself would have reflected warming sunlight at that high altitude, so cooling the Earth's surface as much as the ammonia might have warmed it, and indeed if there was too warm an Earth then there would be little methane haze! (Note: methane as a haze would be reflective, as are water-vapour clouds, even though methane and water vapour as gases both have strong greenhouse properties.)

So, how was the early anaerobic life-bearing Earth kept warm? Possibly life itself could have provided a solution in the form of methanogenic bacteria. Methanogen ancestors of the bacteria now producing methane from marshes, river bottoms, landfills and sewage works, could have produced methane with levels 1000 times higher than today's concentration of 1.7 ppmv. Methane gas lower in the atmosphere is a greenhouse gas and has the opposite effect of high-altitude methane haze. Such a high methane concentration could only have happened in an anaerobic Earth, because if there had been any oxygen (as in an aerobic Earth) then it would have chemically mopped up the methane. Indeed, the average atmospheric residence time of a methane molecule today (in an aerobic atmosphere of about 21% oxygen) is about a decade, but back in the anaerobic Earth, without free oxygen in the atmosphere, a methane molecule could have survived the order of 20 000 years. With methane gas's greenhouse effect combined with that of carbon dioxide, the Earth could well have been warm enough not to necessitate a methane haze to protect photodegradable ammonia.

Nonetheless, the above synopsis needs to be treated with caution, as we simply do not yet have enough evidence for a firm view of primordial Earth's conditions. One contrary view is that there were trace amounts of oxygen in the anaerobic Earth's atmosphere and that this was enough to prevent siderite forming in ancient soils, although it would have done under water. Yet, if this was so then there could not have been appreciable quantities of methane in the atmosphere, although siderites have been found in a number of ancient marine sediments. Yet, if there was no methane then the level of atmospheric carbon dioxide would need to be at the very least roughly 100 times that of today to maintain a liquid ocean before 2.2 bya (Ohmoto *et al.*, 2004).

Nonetheless, despite which view ultimately prevails, we do know that a strong natural greenhouse effect (one far stronger than today) was required early on in the Earth's history and that carbon dioxide played a part.

Eventually, when Calvin-cycle photosynthetic metabolic pathways evolved in the primordial anaerobic Earth, a new energy source was tapped. This was to change the very nature of the biosphere with the introduction of copious quantities of oxygen.

However, there is much debate as to the early biotic Earth's chemistry and biochemistry. Some say that deep-sea hydrothermal vents were responsible for some early carbonaceous deposits and not life. Others say that anaerobic life was responsible. Indeed, recently there has been the suggestion that 3400-million-year-old carbonaceous strata in South Africa appear to have possibly been formed by an algal mat in a shallow sea, and that the algal mat was photosynthesising but in a way that did not involve oxygen production (Tice and Lowe, 2004). This debate continues.

3.2 Major bio-climatic events of the Proterozoic eon (2.5–0.542 bya)

3.2.1 Earth in the anaerobic–aerobic transition (2.6–1.7 bya)

Evolving robust, single-cell metabolic pathways takes time. Photosynthetic organisms evolved some 2 bya, around a billion years after life first arose. To put this into some temporal context, this means that for between one-third and one-half of the time of life on Earth it existed only in a simple unicellular form in an atmosphere that would be poisonous to nearly all of today's metazoans (multicellular animals). If some 3.8 billion years of life on Earth were represented as a single day then the anaerobic Earth would have lasted until close to midday.

These new photosynthetic organisms (prokaryotic algae) slowly began to change the atmosphere by consuming carbon dioxide and releasing vast amounts of oxygen. The oxygenation of the atmosphere was due to the photo-synthetic reaction (below) accompanied by the burial in marine sediments of some of the carbohydrate, so leaving behind an oxygen surplus (the same process that was to happen on a large scale in the Carboniferous and which is examined later in this chapter):

$$6CO_2 + 6H_2O \rightarrow C_6H_{12}O_6 + 6O_2.$$

The aerobic Earth (with an atmosphere that includes free oxygen) did not come into being overnight. Evidence exists for the first major rise in atmospheric oxygen some 2.23 bya (Bekker *et al.*, 2004). Over the next half a billion years or so (until 1 bya) the atmosphere's oxygen content continued to rise (notwithstanding the Carboniferous excess discussed below) until it was close to today's value (Holland, 1990). (The amount of oxygen in the atmosphere before 2.4 billion years is debatable. Although it is known that quantities were not significant compared to today's value, exactly how small they were, and when free oxygen first exactly could be found in the atmosphere, is still a subject of debate (Knauth, 2006).)

It is important to note that the generation of oxygen by primitive algal mats did not immediately result in a commensurate increase in the atmosphere's free oxygen. Just as today, there were buffering systems in the bio-geosphere (hereafter referred to as the biosphere). These buffers soaked up any free oxygen that was released and in the process the abilities of these buffers themselves would change. One of the first greenhouse casualties of the release of free oxygen would have been the primordial abundance of the greenhouse gas methane, irrespective of whether it was just high or extremely high. As previously noted, methane is quite easily oxidised by free oxygen and today has an atmospheric residence time of just 12 years (see Chapter 1). However, over 2 bya, when oxygen was beginning to be released into the atmosphere, initially it would be soaked up by the methane and any other agent (such as iron) capable of being oxidised. At some point, as oxygen continued to be released, much of the methane was used up, so lowering its atmospheric concentration, bringing it closer to today's low value. Also, iron oxide strata were laid down from the oxidation of iron in the primordial oceans to form characteristic banded formations.

We know that micro-organisms have flourished in the oceans since at least 3.8 bya, but there has been much debate as to when land was first colonised. Whereas some 2.7-billion-year-old stromatolite algal mats have been occasionally suggested as being of terrestrial origin, there has been doubt over this idea as such algal mats might more likely have been living in shallow marine or freshwater environments. However, in 2000 work by Yumiko Watanabe from Japan and colleagues in the USA and South Africa announced that elemental analysis of 2.6-billion-year-old palaeosols (fossilised soils) revealed ratios of carbon, hydrogen, nitrogen and phosphorus consistent with an organic origin. This is possible geochemical evidence for early terrestrial ecosystems (Watanabe *et al.*, 2000).

With regards to Earth's early, primordial climate, we simply do not know exactly what it was like between 1 and 2 bya during the very early part of the oxygenating phase. However, later, despite the Earth being subject to a far greater carbon dioxide greenhouse effect than today, in the absence of the additional methane greenhouse effect (that previously existed in the absence of oxygen) there appears to have been a major glaciation, roughly 2.2 bya; the so-called Snowball Earth I (from which you may deduce that there is at least one other Snowball Earth to come). There is some discussion that early life altering the atmosphere's greenhouse properties may have initiated (or at least contributed to) this major glaciation. Even so, obviously the Earth could not have been too cold for life to be completely wiped out, as you are reading this today, but there is much uncertainty regarding the details of Snowball Earth I (and II). It is likely that climatic feedback systems were either operating differently then and/or operating at a different magnitude to those in the present, due to the lower solar output. Nonetheless, Snowball Earth I was a cold place, for there is evidence of glaciation at what would have then been within tropical latitudes. What we do not know was whether the Earth was completely covered in ice right to the equator. The arguments against this would be that life in the sea's surface waters would surely have been wiped out. The arguments for complete ice coverage include that once ice had reached deep into the tropics then there would have been so much reflected sunlight back off the planet that the Earth would have cooled further still in a runaway feedback effect, until ice had reached the equator, with pockets of life only surviving near areas of geothermal activity. Although there is some debate as to the complete extent of ice coverage in Snowball Earth 1, we have evidence of glaciation as close as 6 and 16° latitude of the equator (Evans *et al.*, 1997). Either way we know that the Earth did not remain a snowball forever (after all we are here today). It is thought that volcanoes continued to steadily pump out geological carbon dioxide that boosted the atmosphere's greenhouse effect, which warmed the planet once more (Kirschvink, 1992).

Meanwhile, the Sun itself continued to warm. It is estimated that about 1.4 bya the Sun's luminosity was only 88% of its present value, and so its output had grown considerably since the Earth's pre-biotic days over 2.4 billion years previously (Clough, 1981). It has been shown that the solar 'constant', as a percentage of its current value of 1373 W m^{-2}, can be described as:

$$\text{Percentage change in solar constant} = [1 + 0.4(1 - t/4.7)]^{-1}.$$

t is the time in billions of years since the creation of the solar system (taken for this purpose as being 4.7 bya). So in another 4.7 billion years (when $t = 9.4$), the

Sun will be about 67% hotter, in terms of its outgoing radiation (as distinct from its core temperature).

3.2.2 The aerobic Earth (from 1.7 bya)

With the demise of an anaerobic atmosphere, the Earth had experienced its first pollution catastrophe: by comparison the current anthropogenic (human-generated) release of greenhouse gases is a mere hiccup in the biosphere's cycling processes. Referring to the creation of an oxygen atmosphere as 'the worst atmospheric pollution incident that this planet has ever known' (Lovelock, 1979) may seem somewhat of an odd statement but at the time it foreshadowed great change for the then living species. The anaerobic species that had dominated the Earth previously could not exist in the presence of oxygen and so became confined to anaerobic muds and other oxygen-free environments, where they are still found today.

The availability of oxygen, albeit at first in small concentrations, enabled life to exploit new thermodynamically advantageous metabolic pathways, so giving early oxygen-using species an evolutionary edge. This helped boost the populations of aerobic species and further enhanced the rate of oxygen production. It can be seen from much of this chapter already that even billions of years ago there was a link between early life (biology), the atmosphere and climate.

Indeed, very broadly speaking, the past billion years have seen an overall trend of a decreasing atmospheric concentration of carbon dioxide balanced by increased solar output. 'Broadly speaking' because there have been deviations from this trend and periods in time when the poles were ice-free and other times (glacials) where there was considerable ice reaching to today's temperate latitudes. (We will return to the question of the overall decline in atmospheric carbon over the past billion years or so shortly.)

Some 1.4 bya water vapour (which has been in the Earth's atmosphere virtually since the planet was formed) and carbon dioxide would have been two of the principal greenhouse gases. If there were not considerably more carbon dioxide in the atmosphere than today then, given the Sun's lower luminosity back then, there would have been a major glaciation. So given that 1.4 bya much of the primordial atmosphere's methane would have been soaked up by the rising oxygen, and that 'broadly speaking' water vapour was as it is today, then the question arises of just how much more carbon dioxide was in the Earth's early aerobic atmosphere.

In 2002 two US researchers, Alan Kaufman and Shuhai Xiao from Maryland University and Tulane University, respectively, ingeniously used biological processes to estimate the 1.4-billion-year-old atmosphere's carbon dioxide concentration (Kaufman and Xiao, 2003). Of carbon's five isotopes only the common ^{12}C isotope and the rare ^{13}C isotope are stable (we examined the utility of ^{14}C as a dating tool in palaeoclimatology in the previous chapter). Biological processes, including photosynthesis and the Calvin cycle, have evolved in an overwhelmingly ^{12}C-dominated environment. These processes therefore preferentially use this ^{12}C isotope over ^{13}C. (Hence the earlier reference to isotopic evidence for the possibility that some 'microfossils' were biotic.) The degree of this preferential biological fractionation of carbon isotopes depends on two things: the proportion of ^{13}C in carbon dioxide and the partial pressure of carbon dioxide (in other words, its atmospheric concentration). Kaufman and Xiao (2003) determined the level of fractionation 1.4 bya by isotopic analysis of microfossils of the acritiarch species *Dictyosphaera delicata* compared with carbonate produced by inorganic processes and this in turn enabled them to estimate the atmosphere's carbon dioxide concentration 1.4 bya. Their figures suggest elevated levels of between 10 and 200 times the present carbon dioxide concentrations. As a greenhouse gas, carbon dioxide has retained its importance ever since. But what brought us the oxygen in the first place? What were the early photosynthetic species?

Well, fossils have been found of one of the first aerobic species, and also an early eukaryote, dating from 2100 mya. It was similar to a species of alga living today, *Grypania spiralis* (Han and Runnegar, 1992). *G. spiralis* grows in long, spirally coiled, filaments about 1–2 mm wide. From over 2 bya the anaerobic species were to lose their domination over the planet to the aerobes that generated and/or used oxygen. Meanwhile, the huge carbon dioxide greenhouse blanket declined over the next billion years while the Sun continued to grow stronger. If it had not then the Earth would have risked a runaway greenhouse effect, with a warmer Earth causing more water evaporation from the ocean, and being a greenhouse gas water vapour would have generated more warming, so resulting in a positive-feedback loop (opposite to the albedo feedback loop that would have cooled the Snowball Earth).

But where did all this primordial carbon dioxide go? The carbon dioxide was chemically transformed by life into carbonates that were laid down as great beds of chalk as well as by abiotic, purely chemical, processes.

By 540–600 mya life had evolved to a multicellular level, including animals. This eon to the present day is known as the Phanerozoic (the age of visible life). The significance of multi-celled living things on the biosphere was profound. More than one cell means that it is possible to have a more advanced structure than simple cellular repetition seen in algal colonies. A more advanced structure in turn enabled specialization of cells for processing food, respiration and other metabolic functions. This meant that carbon could be processed in greater quantities more quickly. Greater carbon processing could have lead to a dip in atmospheric carbon dioxide. This would explain the global cooling thought to have taken place at that time. We know it took place because the cooling was so marked that large ice caps formed, leaving geological traces that can be seen today. The large ice caps reflected sunlight. This further cooled the Earth, so enlarging the ice caps, just as happened in Snowball Earth I, and this led to a second Snowball Earth. Snowball Earth II took place in the Precambrian's Proterozoic eon (to be precise, the Neoproterozoic era), and specifically in the aptly named Cryogenian period (meaning ice-generating) somewhere around 580–850 mya. The exact dates of Snowball Earth II have yet to be confirmed but a fairly convincing case has been put by two British and Australian geologists that the onset of Snowball Earth II took place after 592 ± 14 mya and lasted less than 40 million years (Schaefer and Burgess, 2003).

As with Snowball Earth I, there is considerable debate as to whether Snowball Earth II saw heavy glaciation across the equator. One recent idea (Pollard and Kasting, 2005) is that the Snowball Earth came about slowly and at the equator ice may have formed at rates as low as 7 mm a year and reached the equator, at least in part, by glaciers flowing down from higher latitudes. In this case, the researchers speculate, the sea glacier ice formed would have been far more transparent than we find in many of today's glaciers. Such transparent ice may have allowed photosynthesis to take place in water below. Whatever the case, there is little doubt that Snowball Earth I and II were extremely major events for the Earth's biosphere.

Again, as with Snowball Earth I, it is thought that volcanic carbon dioxide came to the rescue of Snowball Earth II by building up over several million years to atmospheric concentrations a few hundred times today's levels. As before this led to considerable global warming that ended the massive glaciation. With the ice caps gone and a super greenhouse effect in operation the average global temperature was far warmer than today and might even have been as much as 50 °C.

As with much palaeontology, especially relating to anything before the beginning of the Cambrian (542 mya), it is difficult if not impossible to derive a detailed picture of any exactness. So we do not know how long this warm period lasted. What we do know is that at some stage this super carbon dioxide greenhouse effect declined as carbon in the form of carbonate layers was deposited over the glacial strata from Snowball Earth II. We also know that once again, as followed Snowball Earth I, carbon-cycling feedback systems reorganised and stabilised matters and the climate.

Whereas global carbon cycling, through the evolution of photosynthesis and then photosynthetic plants with more structure, is one of the principal theories posited to explain Snowball Earth (which is presented here because this text deals with biology and climate) and subsequent ice ages, there are others. For example, ocean circulation plays an important role in transporting heat and it has been proposed that major global cooling only occurs when high-latitude oceanic gateways open and low-latitude ones close (Smith and Pickering, 2003). Geologists supportive of the gateway theory themselves have been quick to point out that even if gateways are important in regulating heat flow from the equator, they cannot be too critical a factor due to the lack of glacial deposits 1–2 bya: a time when oceanic gateways came and went.

Into this warm world came the surviving multi-celled plant and animal (metazoan) species (and of course the surviving older single-celled ones, both prokaryotic and eukaryotic) and beyond their Snowball refuge nearly a whole planet for them to colonise. From fossils it is likely (such ancient primitive fossils are scarce and their interpretation is difficult) that lichen-like symbiosis between coccoidal cyanobacteria (a group of spherical, photosynthetic, nitrogen-fixing bacteria) or algae and fungi took place some 600 mya before the evolution of vascular plants (Yuan *et al.*, 2005). The animal metazoans had all the basic body plans we see today, such as diploblastic (two multi-celled layers) and triploblastic coelomate (three layered with a body cavity) species. The time just prior to the Cambrian was a period of considerable speciation: which not surprisingly is known as the Precambrian boom. At first these animals had no backbones. These invertebrate species included trilobites, clams and snails. Between then and now terrestrial ecosystems have continued to evolve. But just as the time between an anaerobic Earth and an aerobic Earth (with its multi-celled creatures) saw periods of considerable biosphere reorganisation, the road between the Cambrian and now was not to be a steady one, as is evident from times of speciation as well as of extinctions both beyond and within background rates. It is not possible to review even a summary of the extensive palaeobiological literature here (for that the reader should seek out other texts such as Briggs and Crowther, 2001), but it is worth citing a few of the major climate-related biological events so as to put recent changes of the past million years or so, not to mention likely future changes over the next thousand (see Chapter 8), into a broader context.

3.3 Major bio-climatic events of the pre-Quaternary Phanerozoic (540–2 mya)

3.3.1 Late-Ordovician extinction (455–435 mya)

In the late Ordovician life was effectively marine. Life was flourishing in the seas and oceans but terrestrial life was restricted to algae in the splash zone by seas and possibly brackish environments. Yet in the late Ordovician life was threatened when globally about 22% of all families of species became extinct, which makes it one of the largest episodes of mass extinction in the Earth's history. The extinction is thought to have happened in two principal phases about 0.5–2 million years apart that wiped out nearly all major benthic and planktic groups. Among the molluscs, 125 genera of nautiloids were reduced to 25, the 88 gastropod genera went down to 55 and bivalve genera were down from 84 to 32. The two phases of the late Ordovician were associated with the growth and decay of a large Gondwanan ice cap when the southern continents drifted over the South Pole. Sea-water temperatures fell by as much as 8 °C, and a similar degree of cooling even took place in the tropics, which caused the loss of many tropical species. The sea level fell by between 50 and 100 m depending on regional eustatic conditions. Changes in ^{13}C of about 7‰ indicate a major change in carbon cycling. The rapid global cooling and growth of the ice caps contrasted with the greenhouse conditions prevalent before and after the event. The two extinction phases are thought to relate first to the cooling and then second the warming, so that those species that survived the first change were then faced with the second. What appears to be characteristic is that both these events were, geologically speaking, sudden. It is perhaps worth reflecting, as we shall see, that recent (Quaternary) glacials exhibit periods of rapid, rather than gradual, climatic change. As we shall see with some future major extinctions, there was a fungal spike associated with each, suggesting a world of rotting biomass. With the late-Ordovician extinction there was another spike, though not a fungal one but a microbialite spike. As noted above, microbialites (including stromatolites, thrombolites and dendrolites) were formed by microbial mats that were common in the late anaerobic Earth. They made a resurgence following the late-Ordovician extinction and in one sense this is analogous to the opportunistic fungi at the end-Permian and Cretaceous–Tertiary extinctions. However, in this case the opportunity came both in the form of a ready food supply – the seas with dead marine life would have provided nutrients for algae – and also a reduced pressure from grazing animals. The microbialite resurgence lasted some 5 million years (Sheehan and Harris, 2004). There is also some evidence for a similar stromatolite resurgence associated with the later end-Permian extinction (see below).

3.3.2 Late-Devonian extinction (365–363.5 mya)

The late-Devonian extinction appears to have been a succession of extinction pulses. Many cephalopods were decimated, as were armoured fishes that are associated with much of the Devonian. Many coral reef species and crinoids (sea lilies) were also lost. Both terrestrial and marine systems were affected and the tropics more so than the poles. There is much debate as to the cause of this cooling but one of the preferred theories is multiple (possibly five significant) asteroid strikes (there is evidence from Morocco). Another cause is likely to have been the spread (through changes in ocean circulation) of anoxic (oxygen-deficient) deep water. Of course, there could have been more than one factor operating.

3.3.3 *Vascular plants and the atmospheric depletion of carbon dioxide (350–275 mya)*

Vascular plants (the pteridophytes and seed plants) had a vascular system of xylem and phloem that enabled them to have a freestanding structure, and hence gain height on the land compared to more primitive plants. Equally importantly they had extensive roots that helped break up bedrock. Vascular plants evolved after some 400 mya and from around 350 mya larger vascular plants arose and started to spread across much of the terrestrial globe (as today, where conditions permit). These plants contained new types of biomass, such as lignin, which is important for imparting structural integrity to those plant species that have it. Some of these structural compounds were resistant to the then available pathways of organic degradation. This in turn increased their chance of burial. (Dead plants today tend to rot on the ground before they are buried.) Chances of burial were also further enhanced due to the Permo–Carboniferous glaciation, 330–250 mya (see below), with the repeated waxing and waning of its ice sheet causing the sea level to fluctuate with periods suggestive of Milankovitch influences. Burying this tremendous quantity of carbon led to a residual excess of atmospheric oxygen.

This oxygen excess can be easily understood when considering the biospheric photosynthetic–respiration equilibrium: carbon dioxide and water combining through photosynthesis to form carbohydrate and oxygen one way, and the respiratory reaction the other (see the equation below). With the biosphere in a steady state these reactions are equal and opposite (regional and seasonal variations notwithstanding). However, remove some of the carbohydrate by burial (shifting the equilibrium to the right) and an oxygen surplus results:

$$6CO_2 + 6H_2O \rightleftharpoons C_6H_{12}O_6 + 6O_2.$$

$\downarrow$ (Part burial leaves an
atmospheric oxygen surplus)
Geologically isolated organic
compounds $C_6H_{12}O_6$

Previously, 400 mya, the forests of the Earth were hugely different from those of today, or even those emerging 350 mya. These early forests of the Devonian were largely composed of moss-like species with primitive arthropods living among them (some of which were quite large, by comparison with their present counterparts, being a metre or more long). Such Devonian forests were only a few centimetres high, forming green mats on the land. However, the plants that followed had thick-walled cells to allow water to be conducted up their stems, but even so none of these late-Devonian plants was more than a metre or two in height.

It was the evolution of vascular plants in the Carboniferous period that allowed there to be forests of heights comparable with those today, even if the dominating species were different. Vascular plant roots were also more penetrating. These plants also allowed forests to have a higher biomass per unit area and, because of their roots, allowed a greater area of the land to be covered with plant life. Globally, therefore, the terrestrial biomass increased, but equally some of this biomass became buried, as described above. This buried material gave rise to the vast Carboniferous and Permian coalfields. With so much organic material taken out of the carbon cycle a significant oxygen surplus was left behind in the atmosphere. It is thought that oxygen concentrations started to rise after 400 mya and reached a peak 300 mya of 35–40% before returning to close to today's concentration of 21% over 200 mya. The high-oxygen environment would have facilitated forest fires and this would have helped raise carbon dioxide levels around 250 mya (see Figure 3.1), so helping end the Permo–Carboniferous glaciation (see below).

However, biological interactions with atmospheric carbon dioxide are a little more complex than this. They also depend on geological carbonate and geological silicate (summarily represented as SiO_2 below). Indeed, silicates previously trapped in the geology could be more easily accessed due to the root systems of the new vascular plants. Geological magnesium and calcium silicates have a chemical relationship with atmospheric carbon dioxide that may be succinctly summarised as follows:

$$CO_2 + CaSiO_3 \rightleftharpoons CaCO_3 + SiO_2.$$

Bringing all these factors together, the exploration of computer models (such as GEOCARB II) has found that the role of plants can greatly affect the level of atmospheric carbon dioxide through a feedback loop in a stabilising way. The evolution of plants with strong root systems enhanced the weathering of rocks, hence the release of calcium and magnesium silicates (driving the above equilibrium to the right) and so more atmospheric carbon dioxide became drawn down as calcium carbonate over the period around 350 mya: doubly so because the plants themselves have biomass so that carbon dioxide is also sequestered as organic plant material. With carbon dioxide drawn down the planet became

Figure 3.1 Broad outline (low resolution that excludes sharp spikes and short-lived fluctuations) of Phanerozoic atmospheric carbon dioxide. RCO_2 means the carbon dioxide relative to mid twentieth-century carbon dioxide levels of 300 ppmv. The best-estimate curve shown is summarised from Berner (1998). Berner's original work also includes upper and lower estimates and relates this curve to fossil stomata and isotope data which broadly relate to the key features, namely: that for the past 500 million years carbon dioxide has declined to present levels with a dip 300 mya due to the rise of vascular plants. Berner's x axis is reversed but the convention with this text is that palaeoclimate graphs read from the left to right into the past from the present, as do future projections into the future from the present.

cooler. This would have resulted in less evaporation from the oceans, so less rainfall. A cooler, lower-carbon dioxide and low-rainfall world is one in which water erosion and plant growth is suppressed. Less water erosion and less plant-root erosion reduces silicate-bearing rock weathering. Of course, all this assumes that other contributions to atmospheric carbon dioxide (such as volcanic activity) remained constant (Berner, 1998): as we shall see below, this was not always so. However, the lesson to be learned here is that life does have an effect on the biosphere and plays an important part in the long-term carbon cycle. Some argue – for example Britain's Andy Watson (2004) – that the effect of life on climate is largely a stabilising one other than in periods of evolutionary innovation, during which there is some reorganization of the carbon cycle. As discussed in Chapter 1, it is the short-circuiting of this long-term or deep carbon cycle that is causing much of the current, twenty-first-century, global warming.

3.3.4 Permo–Carboniferous glaciation (330–250 mya)

During the Carboniferous period the temperature dropped for two reasons. First, global climatic forcing from atmospheric carbon dioxide was low, for the reasons discussed above; vascular plants were becoming established, so increasing mineral weathering, and more biomass was being buried. Second, many of the Earth's landmasses came together to form the supercontinent Gondwana. This stretched from the equator to the South Pole and indeed a large part of Gondwana sat over the pole. With oceanic currents unable to transport warmth from the equator, a terrestrial ice sheet formed, covering much of Gondwana (as has one on Antarctica today). The rest of Gondwana would have been ice-free but somewhat arid, since a cool Earth sees less oceanic evaporation and hence low global precipitation levels.

Back then, the present-day North America and Eurasia lay together on the equator, and a steamy coal-forming belt stretched in a continuous band from Kentucky to the Urals that was then the warmest place on the planet. The coal formed due to the reasons, discussed in the previous subsection, of enhanced carbon cycling and carbon burial, especially in swampy delta-like areas. The burial rates were further increased by the Permo–Carboniferous glaciation due to the repeated flooding of continental margins arising out of the waxing and waning of Gondwana's South Polar ice sheet. It is now well established that many of these Carboniferous sea-level cycles were approximately over a range of 75 m. (This is only slightly less than the sea-level cycles associated with the current (Quaternary) series of glacial and interglacials, which are closer to 120 m.) It is also well established that each cycle saw the sea slowly decline over tens of thousands of years yet rapidly increase within a thousand years or so; again this is similar to sea-level change in our current glacial–interglacial series. There is therefore a strong suspicion that the waxing and waning of the Gondwana ice sheet in the late Carboniferous was driven by Milankovitch parameters (see Chapter 1).

The ecosystems that were largely to end up as coal-bearing strata were dominated by two tree-sized club mosses (Lycophyta): *Lepidodendron* and *Lepidophoois*. Meanwhile, inland and away from the poles, during the glacials much of what are today temperate latitudes were cooler and dryer. There is evidence that during the warm Permo–Carboniferous interglacials these inland dry zones became far wetter than during the glacials. Because Gondwana was so large, huge drainage systems were formed through erosion. Often these carried material that covered the organic matter in the continental margin swamplands and this again enhanced organic burial. Meanwhile the resulting atmosphere's oxygen excess enabled giant flying insects to evolve, as flying requires a fast metabolism and hence effective oxygen transition, which is greatly facilitated by the gas having a higher atmospheric partial pressure. Metabolic models of these insects have been constructed. We know their size from fossils and can reasonably assume a similar density to modern-day insects; hence we can deduce their mass. We can therefore estimate the energy they would have required for flight and so in turn their oxygen requirements. This gives us another way of estimating the oxygen excess 300 mya discussed above: such estimates seem to fit in with estimates derived from the carbon model.

3.3.5 End-Permian extinction (251 mya)

The largest of all the mass extinctions of the past 600 million years was the end-Permian or Permo–Triassic event 251 mya. The marked turnover of species, both fauna and flora, marine and terrestrial, is used to mark the boundary between the Palaeozoic and Mesozoic eras. At that time more than 60% of animal species, both marine and terrestrial, disappeared. Overall (for plants and animals) 90% of ocean species vanished, as did 70% of vertebrate families on land.

The tremendous volcanic activity of the Siberian Traps at the end of the Permian undoubtedly played a major part in this extinction. The Siberian Traps are a large igneous province that formed as a result of a mantle plume. They are centred around the present-day Siberian city of Tura and also encompass Yakutsk, Noril'sk and Irkutsk. This was a continental-flood basalt event. The Traps' area, including associated pyroclastics, is just under 2 million km^2 and greater than that of present-day Central and Western Europe combined. Estimates of the original volume of the Traps range from 1 up to 4 million km^3 (3 million km^3 is often cited). The Siberian Trap eruptions lasted at full intensity for about 1 million years, which coincides with the extinction. The most accurate dating method available at the moment is argon–argon radiometric dating which still contains too many uncertainties to conclusively prove the exact timing. The eruptions ejected volcanic ash and sulphur aerosols into the atmosphere. This would have caused a global cooling that probably lasted for decades or even centuries after the main eruptions ceased. To put this into some kind of present-day context, the largest eruption of the twentieth century, Mount Pinatubo, was tiny compared to the Siberian Traps but even that caused a 0.5 °C drop in global temperatures the following year. The largest eruption in historic memory occurred in Iceland, 1783–4, and generated some 12 km^3 of lava onto the island: the Siberian Traps by comparison released 1–4 million km^3. The poisonous gases given out by the Icelandic event are recorded as killing most of the island's crops and foliage and lowering global temperatures by about 1 °C. It is therefore most likely that the end-Permian event caused a short, sharp cooling which itself may well have killed off much life. In addition to ash and sulphur aerosols there would have been the release of considerable quantities of greenhouse gases such as

carbon dioxide, methane and water vapour. Carbon dioxide stays in the atmosphere much longer than dust and sulphur aerosols, especially if the Earth's
photosynthetic pump is disabled, so its climate-changing effects last longer. The
warming from any major carbon dioxide release would have become apparent
only once the cooling dust and aerosols had been removed from the atmosphere.

If the Siberian Trap volcanic activity were not enough, it has been suggested
that the resulting global warming from volcanic carbon dioxide may have been
enhanced by large releases of methane from the oceans. Methane hydrates (or
clathrates) exist in considerable quantities off the continental shelf (see the
subsection on methane hydrates in Chapter 6 and the Toarcian and Eocene
warming events below). Methane hydrate stability depends on pressure and
temperature: reduce the former or increase the latter and they quickly dissociate,
releasing methane gas, which (as noted in Chapter 1) is a powerful greenhouse
gas. If, at the end of the Permian volcanic carbon dioxide (or carbon dioxide or
methane from magma-warmed, organic-rich sediments) had warmed the planet
then some of this heat would have been conveyed to oceanic deep waters and
possibly destabilised methane hydrates. The methane would have further
enhanced any global warming, so further stressing ecosystems globally.

As if this methane enhancement were not enough there is evidence to suggest
that much of the extinction 'could have occurred in a single bad day' (Ward *et al.*,
2000). Geological strata, where the Karoo Basin subsequently formed in South
Africa, imply that the forest was literally stripped away in a very short time (Ward
et al., 2000). Could it be that in addition to the Siberian Trap vulcanism generating
global cooling and then warming, and the consequential oceanic methane release,
that there was an asteroidal, or cometary, impact (Jin *et al.*, 2000; Becker *et al.*,
2001)? In 2004 an Australian/US team of researchers reported their discovery of
an impact crater that may have been responsible for the end-Permian extinction.
Located off the north-west Australian coast (due west of Roebuck Bay and north
of De Grey) the crater may be about 100 km in diameter but this estimate is
speculative due to erosion (Becker *et al.*, 2004). Even so, it is the same order of
magnitude as the Chicxulub crater in Yucatan, Mexico, that is thought to have
played a part in the Cretaceous–Tertiary dinosaur extinction (see below).

Of course, even without evoking climate change, there were other, minor
(non-global), effects of the Siberian Trap vulcanism. Being in or around
Siberia at the time would not have been healthy simply due to the widespread
volcanic activity taking place. Another minor effect (comparatively speaking)
would have been the destruction of the ozone layer caused by gas emissions.
Chlorine and fluorine gases are erupted from almost all volcanic eruptions and
these destroy the ozone layer. Without the ozone layer, harmful ultraviolet rays
can kill exposed terrestrial organisms, so contributing to the mass extinction.

The cause of the end-Permian mass extinction appears to involve 'a tangled web rather than a single mechanism' (Erwin, 1994). Whatever the cause, what emerges is that much of life was simply wiped out. But not all life. Indeed, for a brief time it appears that fungi flourished. There is a fungal spike in the geological record around the time of the onset of the Siberian Trap eruption and it is supposed that this was due to fungi living off of dead and rotting vegetation. (A similar fungal spike has been identified in strata associated with the Cretaceous–Tertiary extinction – see below.) There is also some evidence of a stromatolite resurgence, similar to that of the late-Ordovician extinction. This is thought to be due to a decline in marine animals that would feed off of algal mats. Isotopic evidence using ^{13}C suggests disrupted global carbon-cycle activity following the extinction event for 3 million years, although the exact reasons for this remain unclear, but it suggests a relationship between biodiversity and Earth system (biosphere) function (Payne *et al.*, 2004). This ^{13}C evidence comes in the form of a carbon isotope excursion (CIE) and we will return to CIEs shortly.

3.3.6 End-Triassic extinction (205 mya)

The end-Triassic extinction has long been recognised by the loss of most ammonoids (a now-extinct group of single-shelled gastropod molluscs – the same group as present-day snails), many families of brachiopods (a shrimp subclass of Crustacea), Lamellibranchiata (bivalve molluscs), conodonts (a now-extinct group of fishes) and marine reptiles. There were also terrestrial extinctions. The cause for this extinction is unclear though asteroidal impacts have been suggested and there is some evidence that suggests a possible multiple strike, with a 40-km-diameter crater in Saint Martin, western Canada, a crater 100 km wide in Manicouagan in eastern Canada and a 25-km crater in Rochechouart, France. The arguments against this are that the craters have different magnetic fields. (As the crater cools so its magnetic field aligns with the Earth's, but the Earth's field takes a thousand years or more to re-set, which suggests that the craters were formed at least that far apart and not simultaneously.) The counter to this argument is that the largest crater had such a volume of molten rock that it took over a thousand years to cool to the Curie point (580 °C for pure magnetite) when the Earth's magnetic field would have become embedded in the rock (Vent and Spray, 1998).

An alternative explanation for the extinction is volcanic action due to the break up of the old continent of Pangea. This would have injected carbon dioxide (warming) and/or sulphates (cooling) into the atmosphere causing a climate blip that resulted in the extinction. Another explanation is that both these scenarios happened simultaneously. What is known is that the geological strata of the

Chon Aike (volcanic) province of South America and Antarctica were created by a super volcano 170 mya, as was the Karoo-Antarctica province 183 mya. These eruptions lasted millions of years and would have had an impact on the global climate. So, from the end of the Triassic through to the first half of the Jurassic there were a number of global climate-related extinction events of varying degrees.

3.3.7 Toarcian (early (late lower) Jurassic) extinction (183 mya)

As noted above, the first half of the Triassic saw a number of events but there has been particular recent interest in the Toarcian extinction due to both its magnitude and growing evidence for its similarity to that of the Eocene thermal maximum, or Palaeocene–Eocene thermal maximum (see below). For a brief time during the Toarcian, many of the Earth's ecosystems experienced severe disruption. Carbon laid down in Toarcian strata reveal a pronounced decline of approximately 5–7‰ in the proportion of ^{13}C. This suggests that a pulse of ^{12}C diluted the carbon pool. The source of this ^{12}C includes carbon from marine carbonate, marine organic matter and terrestrial plant material.

This decline in ^{13}C is sometimes referred to as a carbon isotope excursion (CIE). As previously discussed (Chapters 1 and 2) the majority (98.9%) of stable carbon exists as the ^{12}C isotope but some exists as other isotopes and radioactive (hence unstable) ^{14}C is used in carbon dating. However, around 1.1% of stable carbon exists as ^{13}C but the Rubisco enzyme in photosynthesis discriminates against ^{13}C. This means that organic matter (photosynthetic organisms and the animals that live off of them) has proportionally far more ^{12}C than ^{13}C compared to non-organic geological strata. This then relates to the isotopic balance in photosynthetically fixed carbon (even if the carbon is from animal remains further up the food chain or on a higher trophic level). However, there is also carbon in carbonate ions that is not photosynthetically fixed ^{12}C and here its isotopic balance is subject to the environmental sources (of which previously photosynthesised carbon is still one) and sinks of the various carbon isotopes. This carbonate is present in carbonate-shelled creatures such as Foraminifera. Such analysis of the carbon isotopes in such species provides an indication of the isotopic carbon balance in their environment. The implication for the Toarcian CIE (and for that matter the end-Permian, above, and the Eocene CIE, below) of a decrease in ^{13}C is that at the time the sediments were formed there must have been comparatively far more ^{12}C in the atmospheric carbon pool that was then fixed by living creatures into the carbonate in the strata being laid down. As organic matter is the major ^{12}C-rich source then it is likely that there was an injection of organically sourced carbon (be it fast or deep) into parts of the fast carbon cycle from which the deep carbon-cycle sediments were formed and which we analyse today. This carbon conceivably could have come from forests (although this is unlikely). (If it had all come from forests then woodlands planetwide would have needed to contribute the carbon, but we do not see such an extinction of tree species.) Alternatively, the ^{12}C could have come from organic carbon that had originally been buried in soils, or the combustion of carbon-rich sediments (such as coal) or methane hydrates in the ocean. This last is discussed further below with respect to the Eocene.

The Toarcian CIE (as we will also see with the Eocene CIE) also suggests that the time was one of elevated atmospheric methane and carbon dioxide. (Note: even if initially there was an increase in atmospheric methane, much would then become oxidised to carbon dioxide.) Of course, both methane and carbon dioxide are greenhouse gases and there is evidence of a simultaneous rise in global temperatures. The Toarcian CIE was also a time of a 400–800% increase in global levels of erosion (suggesting a marked increase in global precipitation) and elevated levels of carbon burial (which again one would expect in a warmer and wetter greenhouse world). There was also a terrestrial and marine mass extinction and strata from present-day Yorkshire (UK) reveal a decline in marine invertebrate species of greater than 50%. The event in total lasted some 200 000 years and is coincident with the geological formation of the Karoo-Ferrar large igneous province in South Africa and Antarctica. There is slowly mounting evidence suggesting that the relationship between these two is causal. It could be that magma from the Karoo-Ferrar province intruded into, so igniting or driving off methane from, Gondwana coal seams (McElwain *et al.*, 2005). It could be that the subsequent warming destabilised marine methane hydrates, so releasing more ^{12}C into the atmosphere. Much depends on the estimated size of the methane hydrate reservoir at the time and whether there was enough to account for the Toarcian CIE. There is also a notion that matters within the CIE event might have been further timed by Milankovitch astronomical pacing (Kemp *et al.*, 2005), but this last is fine detail (although it is likely to be important when considering critical thresholds in biosphere stability).

In short, what is known is that the Toarcian extinction was caused by a perturbation in the global carbon cycle that resulted in a period of greenhouse warming (above that of the average greenhouse warming of the age) and a perturbation in ocean chemistry.

3.3.8 Cretaceous–Tertiary extinction (65.5 mya)

Evidence abounds that the Earth during the mid-Cretaceous was warmer than it is today. For example, fossil remains from the period of the breadfruit tree

(*Artocarpus dicksoni*), which has present-day relatives that flourish in a tropical 15–38 °C, have been found in western Greenland, as have remains of large crocodile-like champosaurs in the Canadian Arctic archipelago. These suggest that in the Arctic there was a mean ambient temperature exceeding 14 °C. This has been supported by biomolecular climate proxy records based on the composition of membrane lipids that exist in the commonly found Crenarchaeota marine plankton. (A warm Arctic, however, contradicts computer climate models that to date do not appear to reflect high-latitude temperatures nearly as well as they do mid latitudes; see also Chapters 5 and 6.) Indeed, Arctic sea-surface temperatures are thought to have at times reached as high as 20 °C (Jenkyns *et al.*, 2004).

The plate-tectonic distribution of the continents was different in the Cretaceous than today and so, therefore, were the Earth's ocean and atmospheric circulation systems transporting heat and moisture about the planet. However, while warm for much of this time, a stage of the Cretaceous (known as the Aptian) was significantly cooler. Looking primarily at ^{18}O isotope levels in marine bivalve molluscs from latitudes between 8° and 31°N Thomas Steuber and colleagues (2005) have shown that when the Cretaceous was warm, high summer temperatures reached approximately 35–37 °C but with low seasonal variability (<12 °C) at latitudes 20°–30°N. But when cool (cooler by about 6–7 °C) seasonal sea-surface temperature variability increased to up to 18 °C. To put this last into context, it is comparable with the seasonal range found today. This raises the question as to whether a warm greenhouse world (of the kind we are moving towards in the twenty-first century and beyond) may be one of warmth but with low seasonality. Although one still has to remember that planetary circulation was different in the Cretaceous, current computer modelling (of both modern and palaeo Earth) is not sufficiently developed to accurately reflect palaeo-global climates that the results can be viewed with confidence.

A warm Cretaceous Earth was ideal for reptiles but they were to find that the Earth was too cool during the latter half of the Cretaceous period. The Arctic appears to have gradually cooled by 10 °C over 20 million years (Jenkyns *et al.*, 2004). Nonetheless it was not this cooling (at least not by itself) that caused the extinction event.

The Cretaceous–Tertiary, or end-Cretaceous, extinction is probably the most famous mass extinction event of all as it was the one that saw the demise of the dinosaurs. Between 75 and 80% of terrestrial species became extinct and possibly somewhat fewer at higher latitudes. It is now well established, due to an iridium layer, that this was due to an asteroidal impact.

An iridium layer is a reasonably reliable indicator of an asteroidal impact due to the way the Earth's crust was formed. When, in the beginning, the Earth was molten, heavy metals like iridium sank to its core, leaving the Earth's mantle depleted in such metals (in comparison with asteroids). In the late 1970s Alveraz and his team in the USA detected significant levels of iridium in the Cretaceous/Tertiary layer and it was this that pointed to an asteroid impact (Alveraz *et al.*, 1980). It was this discovery that spawned investigations into possible extraterrestrial causes of other mass extinctions.

The Cretaceous–Tertiary impact is thought by many to have taken place in central America at Chicxulub, Yucatan, Mexico. (There is some debate about this.) Whether or not this was the impact that caused the extinction there was an impact in middle America somewhere around the extinction time and this caused ejecta that reached up well into the USA. Ejecta dust undoubtedly entered the stratosphere that would have resulted in diminished photosynthesis as well as cooling on a global scale for several years (not to mention spectacular sunsets). As with the end-Permian extinction, a fungal and algal spike has been identified (Vajda and McLoughlin, 2004). This was followed by a recovery led by pteridophytes and gymnosperms (ferns and seed-bearing/woody plants). The Cretaceous–Tertiary and end-Permian vegetation recoveries represent similar responses to terrestrial ecosystem collapse, although the end-Permian biodiversity crisis was much more prolonged. Whether or not there may have been a subsequent warming depends on how much carbon dioxide was also released. Either way, a change in the climate, albeit for a few years, would have been the main cause of the dinosaur extinction worldwide, although on a continental basis there would have been shock events and raining ejecta wiping out life. Here endeth the summary story ... well, not quite.

There are still some who say that an asteroid did not wipe out the dinosaurs. As a point of pedantry this view has much merit. To begin with, only some genera of dinosaurs became truly extinct, other genera only suffered a pseudo-extinction: they did not die out as such but new species evolved. For example, birds evolved. Second, the dinosaurs were already in decline prior to the asteroid impact. The 60-million-year-long Cretaceous period was the third period (the Triassic and Jurassic being the other two) in the Mesozoic era of the dinosaurs, beginning 230 mya. The Mesozoic saw considerable evolution in the dinosaurs and some dinosaur species that were around in the Triassic (at the beginning of the Mesozoic) were not around in the Cretaceous. Conversely, some dinosaurs did not evolve until the Cretaceous. Indeed, the *Tyrannosaurus rex* star of the book (and film) *Jurassic Park* is associated with the Cretaceous, not the Jurassic. Overall, many dinosaur genera were in decline well before the end of the Cretaceous and its asteroid impact, and so the asteroid cannot have been responsible. The possible culprit behind the – if you will – pre-asteroid dinosaur decline may well be the considerable volcanic activity associated with the Deccan Traps in north-west India. The Traps represent the remains of a continental flood basalt event (like the end-Permian Siberian Traps) which is thought to have been active (albeit to varying degrees) for 8 or 9 million years somewhere around 60–65 mya, although there is some associated geology that is earlier (70 mya) than this. The continental flood basalt event would have released cooling dust and ash as well as warming carbon dioxide and methane. Either way, a global impact of the continental flood basalt event was virtually unavoidable. As such, this combination of volcanic traps and an asteroidal impact is similar to the end-Permian extinction.

Then again, the dinosaurs were not the only life on Earth at that time: indeed, there were other newcomers. The late Cretaceous and the early Tertiary were periods of vegetation change due to the appearance, and then increasing dominance, of flowering plants (angiosperms). There is also evidence of a fungal spike similar to that found associated with the end-Permian extinction. In all likelihood fungi would have flourished in a world of much dead and rotting biomass.

It is well established that in the Cretaceous the large Pangean supercontinent continued to break up and there was considerable volcanic activity (see the Deccan Traps in north-west India, above). New oceanic currents were established; old ones diminished or ceased. The late Cretaceous was a time of major sea-level transgressions and regressions and in particular no less than six such events have been identified. Globally, oceanic isotopic data (such as forams – see the previous chapter) from the tropics and also belemnites (a small, extinct cephalopod or octopus-type mollusc) from higher latitudes suggest a consistent picture of at least two periods of warming and cooling (Frakes *et al.*, 1992). Add to this the regional climatic changes due to the aforementioned circulation and times were hard, and especially so for those creatures more wedded to a specific biome. Considered this way, it is not surprising that the dinosaurs were in decline. Whether or not an asteroidal *coup de grâce* was required to terminate the Mesozoic age of the dinosaurs we will never know. That such was necessary for so complete a termination is, however, likely. For some the debate will continue (as it did with a lucid exchange of articles and letters over many issues of *Geoscientist* in 2003/4).

3.3.9 *Eocene climatic maximum (55–54.8 mya)*

Following the Cretaceous–Tertiary dinosaur extinction (65 mya) the archaic mammals began to diversify, but further change was to come. Prior to 55 mya, although the temperature globally was warmer than today, there is a case that the terrestrial climate was more seasonal than it had been. Many of the major landmasses had already tectonically moved towards the northern high latitudes and so globally there may have been a more deciduous type of vegetation. Certainly folivorous (leaf-eating) mammals were not apparent until the Eocene.

The climatic background prior to the Eocene climatic maximum was one of an already warm Earth and the result of a number of factors. The Palaeocene (the first epoch in the Tertiary period, beginning 65.5 mya) was a time of higher atmospheric carbon dioxide concentration than the present (although it had been higher beforehand; see Figure 3.1). So, there was more carbon dioxide greenhouse forcing than now, but not as much as in earlier times (66–240 mya). At the time of the Eocene, the ocean circulation transporting heat about the planet was substantially different than today, and more efficient, although the distribution of the continents on the Earth's surface was beginning to become recognisable as the present layout. Back then, a number of tropical oceanic gateways were open, such as the seaway between Indonesia and Papua New Guinea as well as that between North and South America. The high-latitude southern-hemisphere gateways of the Tasmanian Straits and the Drake Passage between South America and Antarctica were more constrained than today. These differences together helped transport heat away from the tropics. More heat then went to higher latitudes and less heat was re-radiated back into space from the tropics. With less heat leaving the planet the result was an overall warmer Earth, but it was to get even warmer, with the Eocene climatic maximum.

The Eocene climatic maximum is sometimes known as the initial Eocene thermal maximum (IETM), or alternatively the Palaeocene–Eocene thermal maximum (PETM). Also, it has sometimes been referred to as the Eocene climatic optimum; however, the former terms are preferable as, as far as existing ecosystems were concerned, there was little that was optimal about this event, such was the resulting ecosystem disruption. (Indeed, as we shall see in the next chapter there is a similar nomenclature problem with the Holocene climatic maximum, which used sometimes be known as a climatic optimum.) The very beginning of the IETM for geologists actually defines the start of the Eocene epoch and technically the start of the Lower Eocene, or Ypresian as it is sometimes known.

As for the event itself, at the beginning of the Eocene (54.97 mya) the sea-surface temperatures quickly (geologically speaking) rose by some 7 °C in high latitudes, with polar temperatures rising by as much as 8 °C. This lasted for tens of thousands of years, but probably far less than 100 000 years as an extreme event (although less significant warming may have lasted longer and for a further 100 000 years). Evidence also suggests that temperatures in the mid latitudes of 45° were some 18 °C warmer than they had been at the end of the Cretaceous (the extinction event and aftermath notwithstanding), and about 30 °C warmer than the same latitude is today. In short, the IETM was the most pronounced transient warming of the last 100 million years.

In 1990, marine scientists James Kennett and Lowell Stott, both then at the University of California in Santa Barbara, reported an analysis of marine sediments showing that, not only had the surface of the Eocene Antarctic ocean heated up, but that the entirety of the ocean column had warmed to some degree and that its chemistry changed markedly (Kennett and Stott, 1991). Such complete ocean warming would be expected with a warming event lasting at least tens of thousands of years. Subsequent work, such as by Tripati and Elderfield (2005), showed that Atlantic and Pacific tropical and subtropical bottom waters warmed by some 4–5 °C. Again, this was indicative that the warming had lasted for a number of thousand years; indeed, it would have taken this length of time for the oceans to warm this much from the climate, such are ocean mixing times. Further, the volume of the Earth's oceans having been warmed, the ocean mass provided an additional thermal buffer, prolonging the warming. The warming caused profound environmental change and ecosystem dislocation, comparable with, if not greater than, that associated with more recent (Quaternary) glacial–interglacial transitions. Indeed, above the Arctic Circle the very warm climates of this period permitted the growth of moderately diverse and productive deciduous forests. With regards to the ocean's chemical changes there was severely reduced oxygen in deep-sea waters (warmer waters tend not to hold dissolved gases so easily), and 30–40% of deep-sea Foraminifera abruptly became extinct, this last in no small part due to increased ocean acidity.

But what caused this abrupt warming and, indeed, the increased ocean acidity? The picture of what probably happened has been built up over a number of decades of research and even now there is considerable debate as to both some of the various factors involved and the detail as to how they interacted. Yet, as we shall see, a broad picture is emerging and one that increasingly appears to be relevant in the twenty-first century.

Analyses of both marine carbonate fossils and terrestrial fossils reveal a sharp decline in the normal (low) ratio of ^{13}C to the common ^{12}C. The logical explanation for this carbon isotope excursion (CIE) is that a reservoir of ^{12}C flooded the active biosphere (i.e. its living components). As noted previously (with the Toarcian event, above), because photosynthesis prefers ^{12}C to ^{13}C it is likely that the flooding carbon was ultimately biogenic (generated from life). However, such was the amount of ^{12}C flooding the biosphere it must have come from a ^{12}C-dominated reservoir not usually associated with the short-term carbon cycle, but from the deep carbon cycle. ^{12}C reservoirs include: the living part of the biosphere itself (short-term or fast carbon cycle); fossil fuels of the deep carbon cycle (or fossil-fuel-type sources, since no one was refining fuel back then); methane hydrates (clathrates) in submarine continental slopes and some sea floor sediments (both deep carbon cycle); and volcanic gases (which ultimately owe much to the tectonic recycling of organic sediments and which are again part of the deep carbon cycle and so can be ^{12}C rich).

With regard to the first of these options, if all the carbon associated with the living terrestrial component of the biosphere was turned to carbon dioxide (for example, by some sort of cataclysm causing, say, global forest fires) then this still would not have been enough to provide the amount of ^{12}C released. More than three times the amount found planetwide in terrestrial plants and animals would have been required. Soils and detritus globally contain sufficient carbon (see Chapter 1) but releasing *all* of this would have necessitated an extinction event of unprecedented severity and this is also not seen. Alternatively, the burning of a massive peat deposit could have caused the CIE. This last could be a possibility but the source of this deposit has not been found. Deep (abyssal) marine sediments could have provided more than enough carbon but disruption of these on a planetary scale would have provided too much and in any case would have also disrupted continental shelf methane hydrates. Other proposals include a volatile-rich asteroidal impact, but again there is no evidence for this nor is there the iridium layer that is associated with other large, global climate-changing asteroid-impact events. This leaves fossil-fuel-type sources, marine methane hydrates (clathrates) or volcanic gases as source candidates.

In 1995 Gerry Dickens and colleagues from the University of Michigan proposed that only geological methane gas had enough light carbon (^{12}C) to produce the early Eocene CIE. They proposed that a belch of methane escaped from methane hydrates in continental slope sediments as the Earth warmed during the late Paleocene (Dickens *et al.*, 1995). But if this occurred, what triggered the release? As we shall see in the next chapter, methane 'burps' from marine landslides on the continental slopes have affected the climate during the last glacial, but not on the IETM scale.

Nonetheless, in 1999 US geologists Miriam Katz and Dorothy Pak, together with Gerald Dickens working in Australia (Katz *et al.*, 1999), reported the discovery of a submarine landslide off the Florida coast dating from the time of the Eocene CIE and, from the benthic forams found, ^{18}O analysis suggested that bottom waters rose in temperature by at least some 6 °C. The landslide was just the sort one would expect with methane hydrate dissociation. The problem was that some 1000–2000 Gt of ^{12}C would need to be released to generate such a pronounced CIE and this was not possible from this single submarine avalanche alone. The Katz team therefore predicted that if their 'smoking gun' hypothesis was true then there must have been other Eocene submarine landslides, the remains of some of which should still be around to discover today. Certainly globally there should have been enough methane hydrate to cause the CIE.

Then in 2004, Henrik Svensen and fellow Norwegian colleagues announced they had found evidence of literally thousands of hydrothermal vents in a complex off the More and Voring basins in the Norwegian sea, dating from the Paleocene–Eocene boundary. Could the intrusion of mantle-derived melts in the carbon-rich sedimentary strata in the north-east Atlantic have caused an explosive release of carbon? Such a structure on land is associated with the Siberian Traps that resulted in the end-Permian extinction approximately 250 mya (discussed earlier in this chapter). Meanwhile, the complex of the More and Voring basins are part of a greater North Atlantic volcanic province.

Another option has recently been proposed by a European and American team of geologists led by the Dutch (Lourens *et al.*, 2005). They have found another smaller warming event and the timing of this and the IETM are possibly suggestive that long-term Earth orbital factors may have played a part, just as Milankovitch orbital variations pace Quaternary glacials (see Chapters 1 and 4). (There is a similar suggestion concerning the Toarcian event described earlier in this chapter.) However, whether or not this orbital role was causal in (rather than tweaked the timing of) the dissociation of methane hydrates remains to be seen: there is a lot to discover about the IETM. Nonetheless, orbital factors could well have played a part in the pacing of climate detail within the event. Anything that increased the temperatures of intermediate sea waters would increase the chance (and hence the pacing) of methane hydrate dissociation.

Carbon dioxide, be it from part of the volcanic event, which itself might have been methane-dominated, or as a result of the oxidation of released methane, would easily dissolve in the oceans, so increasing their acidity. This in turn would increase the dissolution of calcite shells of microplankton which are the dominant component of sea-floor sediments, leaving behind only non-soluble clays; hence the foram-extinction event. Other species with calcareous shells are also likely to have been affected (see the ocean acidity subsection in Chapter 6). As it happens, it is well documented that a change in the colours of sediments, from bright white carbonate to deep red clays, marks the Paleocene–Eocene event. Normal deposition of microscopic carbonate foram shells on the deeper reaches of the sea floor did not resume for at least 50 000 years, and the total recovery time to a so-called normal state took as long as 100 000 years. Finally, ocean sediments also reveal a boom in the number of planktic Foraminifera species a few million years after the IETM (Emilliani, 1992). This might be expected if the ocean contained an abundance of material from which to make carbonate. However, whereas this peak begins immediately after the IETM event, a more gradual rise in foram biodiversity began prior to the IETM, so making it difficult to claim with certainty that the foram biodiversity peak and IETM were connected, even if this is arguably likely.

It is worth noting that Svensen's team estimate that the *rate* of carbon released as methane during this Norwegian sea event – over a period of between 30 and 360 years – is comparable to an average year's worth of carbon release from the burning of fossil fuels as carbon dioxide in the 1990s. The suggestion is that our current burning of fossil fuels might eventually warm the planet and hence, as in the Eocene, eventually the oceans, enough to disrupt methane hydrates and so trigger further greenhouse warming above and beyond that which we are currently doing through burning of fossil fuels and clearing forests.

In addition to Svensen's team's estimate as to the *rate* of carbon release from these vents, they also estimate that the total carbon release over the period of the CIE might be 1500–15 000 GtC. Given that the CIE is thought to involve around or less than 2000 GtC, the vents themselves may have been responsible for the CIE on purely carbon volume terms (Svensen *et al.*, 2004). However, this alone is unlikely. We know the CIE happened and it resulted in global warming that affected the oceans' deep-water temperatures and so thermal dissociation of methane hydrates would be virtually inevitable. Also, with warming carbon was also likely to have been released from soils. The question then becomes one of whether the North Atlantic volcanic province was the trigger for a separate release of methane from marine clathrates and also (perhaps subsequently) other terrestrial carbon pools.

As for comparisons with current carbon releases driving present-day global warming, it is worth noting a few figures. The release of at least some 1000–2000 GtC was needed to generate the Eocene CIE. This compares with, in 1990, some 7 GtC released that year alone into the atmosphere through human fossil-fuel burning and industry (such as cement works). Meanwhile, the Intergovernmental Panel on Climate Change (IPCC, 2001a) forecast that the total twenty-first-century fossil carbon release is likely to be between 1000 and 2100 GtC. In short we could be on our way to creating an analogous climatic event to the Eocene thermal maximum.

One key question that follows is what was the atmospheric concentration of carbon dioxide during the Eocene peak? There are a number of ways to calculate this, notwithstanding the volume of carbon needed to cause the

^{12}C isotope excursion. Another is to use (non-biological) chemical estimates such as that by UK geologists Paul Pearson and Martin Palmer in 2000. They looked at boron-11 (^{11}B) concentrations of marine calcium carbonate sediments as boron in solution occurs as $B(OH)_3$ and $B(OH)_4^-$, the equilibrium between which is pH-dependent, and marine pH is largely governed by atmospheric carbon dioxide concentration. They give an IETM atmospheric carbon dioxide concentration peak of some 3800 ppm (with a lower error limit of 2800 and a higher of 4600 ppm). On one hand this is far higher than the IPCC consider in their stabilisation scenarios of current carbon emissions: the highest being 1000 ppm (IPCC, 2001b). On the other it needs to be remembered (from Figure 3.1) that this peak was superimposed on the existing higher-end Paleocene and early Eocene atmospheric carbon dioxide concentrations. Consequently the early Eocene *increase* in atmospheric carbon dioxide above pre-Eocene levels was around 1000 ppm or more. So again this nearly chimes with the current carbon pulse generating the global warming the IPCC are examining.

In contemporary greenhouse terms, assuming a similar CIE today (which is neither an insignificant nor an unrealistic working assumption), then this carbon would need to be added to the pre-industrial atmospheric concentration of 280 ppm, so making a total of 1280 ppm or more. This is clearly higher than the highest stabilisation scenario the IPCC considered in their 2001 report (IPCC, 2001b).

Another matter that springs to mind, from Chapter 1, is that if carbon dioxide has an atmospheric residence time of just a century or two, how come the approximately 100 000-year Eocene thermal maximum lasted so long?

Matters are even more complex than a simple carbon pulse. In 2004 an Anglo-American research team led by Gabriel Bowen and David Beerling suggested that terrestrial and, importantly separately, oceanic carbon cycling may have changed significantly as a result of the Eocene carbon release and that this in turn would have a consequential regional climate-change effect (as opposed to the average global increase in temperature expected). Soil organic-matter turnover therefore probably increased dramatically, suggesting vigorous plant growth in a number of terrestrial ecosystems (due to the extra warmth, precipitation as well as increased carbon dioxide from the slow oxidation of methane), so doubling terrestrial carbon cycling. The picture they construct is that there would need to be an increase in soil moisture and humidity of at least 20%. Water vapour is itself a greenhouse gas and this would have helped further the Eocene's thermal maximum during which the Earth appears to have switched into a new climatic mode. However, water-vapour concentration is a response to warming and would not itself maintain it, as we perceive from the geological record, but reinforce something else that was going on. The Bowen–Beerling team could not state with confidence what this factor (or factors) might be but did make a number of suggestions. These included: changes in atmospheric and/or ocean circulation and heat transport; higher atmospheric carbon dioxide due to circulation and ocean chemistry changes; or higher methane due to increased wetlands arising from the higher precipitation, humidity and temperature as well as some of the carbon release from soils once global temperatures started to rise (Bowen *et al.*, 2004).

All the above notions are likely factors but in themselves do not fully explain the duration of the Eocene CIE. Yet, pulling the evidence published over the past decade together, the following scenario is one well worth considering.

This is what may have happened. An event such as magma intrusion into biogenic carbon rich sediments as reported by Svenson *et al.* (2004) could (if the conclusions of the analysis are to be believed) easily have resulted in the release of more than enough ^{12}C to generate the Eocene CIE. Yet, even if this were so, the raising of deep-ocean water temperature by 4–5 °C was very likely to have destabilised those marine methane hydrate deposits that were already close to dissociation. After all, the degree of methane hydrate stability is a function of both temperature and pressure so that those deposits that were closer to the surface would be more likely to dissociate first than those deeper. These latter deposits would only dissociate with further abyssal warming and that would take time. Furthermore, there is evidence suggesting that there was an abrupt change in ocean circulation during the Eocene warming event, from a more temperature-driven thermohaline circulation (see Figure 4.1 and associated text) to a more salinity-driven halothermal circulation. Foram benthic (deep-water) carbon-isotope evidence from four ocean basins dating from the time of the Eocene thermal maximum event also suggests a switching of ocean circulation during the climate event itself (Nunes and Norris, 2006). This would have exposed other methane hydrate strata to thermal destabilisation that had previously been protected by cooler currents. In short, methane hydrate deposits did not all dissociate simultaneously but gradually, as the oceans slowly warmed and as the ocean circulation changed. Further, once the planet begun to warm up, other carbon reservoirs may have seen changes in carbon flow – soil carbon being one likely example. In short, a pulse of light carbon (^{12}C) could have cascaded from one biospheric carbon pool to another, to another and so on, largely via the atmosphere which was greenhouse-warmed. One might speculate that fossil carbon triggered warming that initiated the release of ocean hydrate carbon that lengthened, and deepened, the CIE and warming that in turn released high-latitude soil carbon, and that atmospheric carbon was in part absorbed by the oceans as well as by terrestrial ecosystems that slowly released it via rivers and so forth (in addition to the atmospheric route). Then there would be a recovery period during which the carbon pools were (largely) restored. Unfortunately at the moment computer models are rather limited in being able to reproduce the Eocene CIE and warming. This is because ecosystem and marine hydrate climate and detailed carbon responses to warming are not included in models. (Although biome cover in the albedo sense has been from the late 1990s and biome carbon tentatively in the first years of the twenty-first century, I say tentatively because the biology of carbon flows has not been elucidated in nearly sufficient detail; this is a current focus of research (see Chapter 7).)

In terms of modern anthropogenic global warming, the suggestion that lends itself from the whole IETM episode is that if humanity continues to release carbon, as it has recently, then the resulting warming might eventually disrupt present-day methane hydrate sediments, so *further* increasing the atmospheric carbon burden. However, this extra burden would not be as carbon dioxide but (initially at least) as the more greenhouse-potent methane. This would be an analogous event to the IETM itself. Broadly this means that should we experience a repeat of this Paleocene–Eocene methane event then the current Intergovernmental Panel on Climate Change (IPCC) warming forecasts of 1990–2001 (see Chapter 5) would be greatly exceeded. There may also be a switch to a new climatic mode, or at least a very warm interval, lasting many tens of thousands of years. Either way, if current fossil-fuel emissions of carbon do trigger a methane hydrate release, then the total greenhouse burden from carbon may well exceed that of the Eocene. This is because we are already set to release broadly 2000 GtC this century from fossil fuel (similar to the carbon involved in the Eocene CIE), so that any methane hydrate release from the oceans would be extra, and this does not even include likely carbon releases from boreal soils or further carbon-pool cascade effects.

The aforementioned notion, if it came to pass, should not be considered an undue surprise. The IPCC clearly warn that their forecasts do not include surprise events and emphasise that we should be aware of climate surprises (see the section on possible surprises in the latter half of Chapter 6). As such, the IETM might represent a possible quasi palaeoanalogue (perhaps even only a modest one) to future anthropogenic warming should we continue to pump fossil carbon into the atmosphere (see Chapter 8). Quasi because if the previous paragraph's scenario did take place then it would be a more severe climatic event than the Eocene thermal maximum and one that would have a far faster onset: the Eocene CIE took some 30 000–40 000 years from its beginning to the start of its peak, whereas we are set to release 2000 GtC in just 100 years.

The impact on the marine environment, especially some calcareous species, is noted above, but what of terrestrial biological impacts? It is generally accepted today that the warmer (and wetter) the environment the greater the terrestrial species diversity. So what was species diversity like during the IETM/PETM? Tropical South America has the highest plant diversity of any region today. In 2003 a team of US palaeobiologists led by Peter Wilf reported an analysis of fossil flora in sediments from a caldera lake in Patagonia (which today is well south of the tropical zone). They recorded 102 leaf species of dicotyledons (the most diverse group with 88 leaf species found at the site), monocotyledons, conifers, ginkgophytes (one of the five main divisions of seed-bearing plants but with only one living genus, *Ginkgo*), cycads (another of the five main seed-bearing plant divisions) as well as ferns (Wilf *et al.*, 2003). This, adjusted for sample size, correlates with a diversity that exceeds that of any other (non-thermal maximum) Eocene leaf diversity in that area. So parts of the Earth during the IETM had sufficient rainfall to maintain phototranspiration and plant diversity at levels seen today in parts of Brazil's further north (and other) tropical rainforests but back in the early Eocene at palaeolatitude (to allow for a little continental drift) approximately 47°S in the south of the South American continent. What we do not (yet) know is exactly how much of the planet's land area sustained plant diversity and how much did not. Nor do we know the detail of how plant–animal interactions fared across the planet.

We do know a little, however, and indeed some of Peter Wilf's earlier work has shown that back in the Eocene, where today's south-western Wyoming is located, that fossil leaves exhibit greater signs of insect herbivory (both in terms of frequency and types of damage) than just 1 or 2 million years earlier in the late Paleocene (Wilf and Labandeira, 1999). Today we see greater such insect herbivory in warmer tropical forests than in temperate ones so while this discovery is not unexpected it is an important corroboration of the broader ecological changes that one might expect with an episode of global warming.

If tropical species migrated to what are today temperate latitudes, what happened nearer the poles? In 2006 a clutch of papers was published in *Nature* of an international investigation of the palaeo-Arctic environment during the IETM and summarised the same issue by Heather Stoll, a geologist at the Universidad de Oviedo, Spain (Stoll, 2006). The teams, mainly European and US but including some Japanese, analysed marine sediment cores beneath the Arctic. This was no mean feat as in the Arctic sea ice is constantly on the move, so a new technique was tried to ensure that drilling could continue for 9 days over one spot. While the drill ship maintained a fixed position, it was protected by two other ships that broke up the ice around it and pushed the larger pieces away. The cores reveal that the surface of Arctic waters during the IETM rose to around 18 °C in the summer, which is comparable to that in Brittany, France, today. (Although it needs to be remembered that seasonality was probably higher, with a greater difference between Eocene summer and winter temperatures compared to lower latitudes.)

It also appears that Arctic waters then were fresh and not saline as the remains of freshwater ferns, *Azolla*, were found. *Azolla* are a genus of aquatic floating ferns that get their nitrogen from symbiotic nitrogen-fixing blue-green algae. As today *Azolla* flourishes only in waters with less than 0.2% salt (for comparison, average sea-water salinity today is 3.47%), why was the Arctic so fresh? Two factors played their role. The first is that the Arctic back in the Eocene was (due to continental drift) smaller and far more enclosed by land, hence more cut off from the rest of the Earth's oceans. Indeed, it was the beginnings of the spread of the Atlantic that was related to the North Atlantic volcanic province that, as noted above, in turn could have triggered the release of greenhouse gases that caused the IETM warming. This more isolated Arctic water was still fed by fresh water from rivers off of what would become the north of North America, Greenland, Europe and Asia. Second, the warmer Eocene and its greater rainfall would have further contributed to this fresh-water supply.

Even though we do not know much of the detail of ecological change that took place during the IETM, what we do know is that many ecosystems saw great disruption. For example, a study of plant fossils in present-day USA showed that Eocene maximum floras were a mixture of native and migrant lineages and that plant shifts were large and rapid (occurring within 10 000 years). Leaf margin analysis (LMA) of 23 dicotyledon species suggested that temperatures in the present-day USA area rose by approximately 5 °C during the Eocene maximum compared to LMA of fossils 250 000 years beforehand and is consistent with isotope-derived estimates. Leaf area analysis (LAA) suggests increased precipitation but some local reductions and a reduction (approximately 40%) early on in the event. This overall increase from LAA is again in line with isotopic evidence elsewhere on the planet that suggests an increase in precipitation during the maximum (Wing *et al.*, 2005).

It is likely that a similar event today, with a global human population of several billion fragmenting the natural landscape, would not see such successful species migration as happened in the Eocene. Furthermore, if the event were faster (due to our twenty-first-century pulse of a similar amount of carbon being more rapid) there would be considerable dislocation of largely human-managed ecosystems (such as agricultural land). Consequently, it is likely that there would be much human suffering (and economic dislocation). Whether it would have significant long-term evolutionary implications is simply not known; even without climate change, humans are affecting many ecosystems globally and causing biodiversity loss, which itself is significant.

In terms of the IETM's long-term biological impact, as a result of the early Eocene climatic-maximum ecosystem disruption the then existing mammals began to decline and were replaced by a diversification of the modern orders of mammals. This initial diversification favoured smaller creatures. Smaller creatures have a higher surface-area-to-volume ratio and so tend to lose heat faster than larger animals. This would have been an advantage in a hot climate. Carnivorous mammals also diversified but at first were more generalist and they did not assume the role of larger predators until the late Eocene (Janis, 2001). The Eocene also saw the order Primates begin to diversify, having first appeared in basal (early) Eocene deposits of North America and Europe and slightly later in Asia.

Meanwhile, leaving the IETM/PETM behind, another major extinction was to follow.

3.3.10 Eocene–Oligocene extinction (approximately 35 mya; or 33.9 mya?)

The Eocene Oligocene boundary has recently been re-dated (hence the two dates given in this subsection's title, with the latter younger date being the one appearing in some recent literature). It was a time noted for being a cooler period, with palaeotemperatures at a latitude of 45° dipping to within a few degrees of today. Antarctica, which was moving to cover the South Pole, also saw the beginnings of an ice cap. It was a time noted for having one of the biggest extinctions since the Cretaceous–Tertiary extinction. In Europe this extinction was known as the Grand Coupure although extinctions have been noted on the other continents and the full explanation is not entirely clear. Nonetheless, approximately half the mammal genera became extinct within about a million years (Hallam, 2004).

But, was this extinction due to cooling due to indigenous planetary climate forcing (be it greenhouse or atmosphere–ocean circulation changes) or some other factor? Evidence emerged in the mid 1990s of an asteroid impact aged 35.5–35.2 mya at Chesapeake Bay off the Virginia Shore, USA. This is a structure of 85 km in diameter and the probable source of the huge field of tektites (characteristic glassy nodules) covering mainly Georgia some 1100 km away and microtektites in Barbados 2250 km away. This was clearly a significant impact, albeit one that was smaller than that associated with the end-Cretaceous event.

If this were not enough, soon after the Chesapeake discovery there was the dating of a second crater, the 100-km-diameter structure in Popigai, northern

Siberia, which is therefore possibly the fifth largest impact crater currently on the planet. (The Cretaceous–Tertiary structure at Chicxulub, Yucatan, being 180–310 km in diameter.) The Popigai structure contains some 2000 km^3 of impact melt rocks and an iridium anomaly has been detected in Italy (over 3200 km away) exactly matching the Popigai dates. The structure had been known for many years but its date of origin was ambiguous, with estimates ranging from between 5 and 65 mya. But in 1997 Canadian and Russian researchers announced that they had managed to refine the dating to 35.7 ± 0.2 mya (Bottomley *et al.*, 1997). This suggests two likelihoods. Either the Earth was subject to two major hits within a few hundred thousand years, or it was subject to two major ones simultaneously. Both are possible. Given that major hits are rare events a few tens of millions of years apart one might at first think that the two were independent and not contemporaneous: that sheer chance meant that two just happened in fairly quick succession, in geological terms. However, many asteroids are known to exist as binaries orbiting each other. Also, it is well known that an asteroid passing within the Roche (tidal stress) limit of a large gravity well (say, that of the gas giant Jupiter) could be disrupted, so ending up as two or more asteroids on a similar trajectory, which in this case intercepted the Earth's orbit.

Either way, both the impacts would have been profound enough to have resulted in a mass extinction of the magnitude we see at the Eocene–Oligocene boundary with an initial physical disruption of large regions of two continents, climatic cooling (and certainly photosynthetic disruption) for a number of years due to stratospheric dust, and possibly longer-term warming if there was a carbon dioxide pulse. Further, there is evidence to support a short, sharp cooling event and Antarctic ice first appeared as it did in some other parts of the planet in a 400 000-year glacial (Zachos *et al.*, 2001).

Having said this, it is important to re-emphasise that there is some debate surrounding the Grand Coupure. Tony Hallam of the University of Birmingham, whose career has in no small part had a focus on mass extinctions, does recognise the cooling taking place at the time. But he also notes that land bridges formed around that time between North America and Asia as well as North and South America and that immigrant species contributed to a number of post-Grand Coupure regional biomes (Hallam, 2004). Whether or not species migration causally contributed to the extinction event is an interesting point.

The Earth emerged from the Eocene–Oligocene extinction warmer than today but the global temperature was 2–3 °C cooler than it had been prior to the extinction event (i.e. 3 or 4 °C warmer than today) and possibly there was also greater seasonality with cooler winters and warmer summers. This relatively cool period lasted for roughly 10 million years until about 24 mya, before it warmed up, back close to end-Eocene temperatures. Although, once Antarctic ice appeared it never went away, even if it did reduce at the end of the afore-mentioned cool period. After this the planetary climate slowly cooled 3 or 4 °C from then to the late Miocene, 5.3 mya, after which this overall slow cooling trend continued.

3.3.11 Late Miocene expansion of C_4 grasses (14–9 mya)

Conventional wisdom has it that the general decline in atmospheric carbon dioxide (as seen in Figure 3.1) was enhanced from the Miocene (25 mya) due to the photosynthetically efficient C_4 plants and this efficiency enabled the C_4 photosynthetic grasses to expand considerably in the late Miocene (approxi-mately 8–4 mya). While C_4 grassland biomes had a significant presence on the Miocene Earth, the grasses did evolve earlier and grass seeds have been found in dinosaur coprolites (fossilised excreta). The idea is that not only did the C_4 grasses help drive atmospheric carbon dioxide concentrations down, at lower carbon dioxide levels they had more of an evolutionary advantage (plants' other evolutionary factors notwithstanding) over those plants with a lower photosyn-thetic efficiency.

Modern grasses use two different photosynthetic pathways, the more ancient C_3 pathways and the more recent C_4 pathway, where C_3 and C_4 refer to the number of carbon atoms in the first product during carbon dioxide assimilation. The C_4 pathway includes a carbon dioxide-enrichment mechanism that enables it to photosynthesise more efficiently than C_3 plants at lower carbon dioxide partial pressures (which is effectively similar to atmospheric concentration). Indeed, it is worth noting now, for we will only briefly return to this point later, that the photosynthetically efficient C_4 grasses – being ideally suited to a low-carbon dioxide atmosphere – will undoubtedly play a significant role in the biosphere's continued evolution on a scale of hundreds of millions of years.

Other environmental factors facilitating the rise of grasses was that the late Miocene was also the time of orogenesis (mountain building) that resulted in the Rocky Mountains, Andes and Himalayas and in turn in regional climatic effects from rain shadows and rainy seasons, and of course the uplift areas themselves were cooler.

While it is true that carbon dioxide levels were considerably lower in the Miocene than compared with the previous couple of hundred million years, and that this gave C_4 plants and grasses a competitive advantage, there is some evidence to suggest that carbon dioxide levels had largely already stabilised (recent Quaternary glacials notwithstanding) by around 9 million years ago. As we shall see in the next chapter, although the evolution of C_4 plants was earlier, their actual global spread did not take place until from about 8 mya and there was not a decisive lowering of carbon dioxide at that time to act as a trigger for the grassland expansion. It is likely that C_4 expansion was driven by additional factors such as tectonically related episodes, such as orogenesis (mountain building) or vulcanism (Pagani *et al.*, 1999), which caused cooling climate change that disrupted ecosystems, so enabling the grasses to capitalise on their advantage in the already lower carbon dioxide levels.

This distinction may seem trivial and the point a subtle one, but it is none-theless worth noting. As we have seen with regards to climate change itself, and as we shall see in later chapters with regards to ecological change in response to climate change, the resulting biological change is frequently a result of a necessary interplay of a number of factors.

The importance of such evolutionary subtlety is also relevant to the broader change in mix of global plant species. For example, it has long been recognised that ferns are an old group of species and that one type, the leptosporanginates (which feature the majority of contemporary or extant ferns), is more than 250 million years old and today number over 10 000 extant species. However, it used to be thought that the golden age of a high number of fern species was prior to the Cretaceous (135 mya). It was thought that this time before the Cretaceous, when fern-dominated biomes were far more common, was when the great diversity of ferns arose. It was similarly thought that there were fewer species of fern evolving actually in the Cretaceous due to the rise and increase in angiosperms (flowering plants). Supposedly, the rise and diversification of one offset the diversification of the other, as they were competitors. Of course, some fern species – indeed many – have died out over the past couple of hundred million years, but there is evidence that the richness of fern species we see today does not appear to have stemmed from before the Cretaceous. Genetic and systematic analysis (of species' evolutionary relatedness) sug-gests a later development of the range of modern ferns that coincided with – not preceded – the rise of angiosperms in the Cretaceous.

This may at first seem counterintuitive. It would seem likely that there were more of today's species early on when the ferns dominated many terrestrial ecosystems and not later, with only a few surviving extinction episodes and competition from the rise of more advanced plants. However, what actually appears to have happened is that the new angiosperm communities being established in the Cretaceous also created new environmental niches, and of course the ferns were already there, poised to exploit them (Schneider *et al.*, 2004). The rise of angiosperms was therefore coincident with, and not subse-quent to, the diversification of the fern species we see today and the fern diversification was in fact a response to the rise in angiosperms (Figure 3.2).

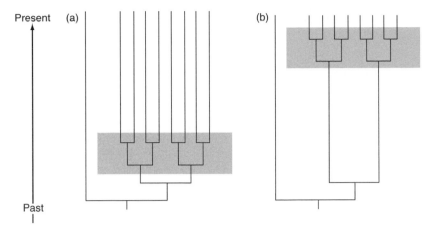

Figure 3.2 The wide diversification of ferns did not solely take place back when they dominated many terrestrial biomes (as represented in panel a) but extant ferns diversified later (more as represented in panel b), filling the new environmental niches created by the angiosperms.

The relevance of this to the late-Miocene expansion of C_4 grasses (14–9 mya) is that the grass expansion too provided new ecological niches, hence the evolutionary opportunity for speciation (the formation of new species). Arguably the speciation that began at this time that was ultimately of greatest impact on the planet was of the Hominoidea (apes). This branch of speciation continued ultimately to include modern humans, *Homo sapiens*, but began back in the late Miocene. One of the early Hominoidea speciations was that of the sivapithecines, whose remains have been found in Turkey, India and Pakistan. They are thought to have survived up to about 8 mya and are evolutionary cousins (as opposed to grandparents) of *H. sapiens*.

There is other, more tangential, relevance of the drivers behind the earlier Cretaceous diversity of ferns to the Miocene expansion of grasses. We have more recent evidence that other factors were at work than the decline in atmospheric carbon dioxide, with which C_4 plants are better able to cope than C_3 flora. Isotopic studies of microfossils from the end of the last glacial 13 000–21 000 years ago from two Central-American lake beds show that C_4 plants began to flourish around one in Guatemala but not the other, in Northern Mexico, even though the atmospheric carbon dioxide would have been the same for both. The reason was a climatic change at the end of the last glacial and local environmental conditions pertaining to each lake. Consequently, while earlier in the Miocene, low carbon dioxide may have facilitated C_4 grass expansion, other factors such as regional climate were critical to determine where this happened. Where the conditions were right, C_4 species flourished (Huang *et al.*, 2001). This is exactly analogous to the

Cretaceous fern diversification that took place where the conditions were right (then in niches created by the rise of angiosperms). In short, major evolutionary drivers are rarely down to single factors and a holistic approach to understanding the development of ecosystems is required.

We will return to this theme of a holistic approach in later chapters when looking at the biological consequences of future climate change, as this will not be just dependent on the climate change itself but also the nature and distribution of ecosystems in the landscape.

Other evolutionary developments associated with the rise of grasses and grasslands are frequently tied to the observation that equids (the Equidae, the family of horses, asses and zebras) and other mammals developed hypsodont (high-crowned) teeth in the early Miocene (beginning 25 mya) in North America. This perspective is supported in that fossilised grasses and grass pollen appear in abundance in the Miocene and become increasingly common to the present. Grasses are far tougher to eat (see below) compared to the berries and leaves of tropical forests and woodlands that equids used to inhabit. However, since the late 1990s it has been noted that the increase in hypsodonty in mammals occurred in the Oligocene (beginning 35 mya) in South America. It is now thought that grasses were significant a little earlier in South America and maybe also in Africa. Unfortunately for palaeontology, grasses are most abundant in oxidising and well-drained environments that do not encourage fossilisation. So, field evidence is literally thin on the ground.

The Miocene increase in open grasslands instead of the closed forest meant that equids faced new pressures. In forests, being short helped movement through the undergrowth, where manoeuvrability was the key to escaping from predators. Conversely, in open grasslands it was speed that was essential. Those individuals with the longest feet, the ability to use their toes to give them a little extra speed, tended to survive while slower individuals did not. Equids became taller and evolved into (hooved) ungulates.

Grasses belong to the family Poaceae (formerly known as Graminae) of the order Poales (which in turn is sometimes referred to as Glumiflorae) and all belong to the monocotyledon class (Monocotyledonae) of angiosperms (flowering plants). They include the C_4 species of maize, wheat and rice and so are of central importance to agrarian and technological humans. (C_4 plants form carbohydrates through photosynthesis, using a different metabolic carbon pathway from C_3 plants.)

Grasses, particularly C_4 grasses, have several properties marking them as different from C_3 dicotyledons. (Dicotyledons have two cotyledons – leaves emerging from the seed – as opposed to the grass monocotyledons' one.) First, grasses usually have a lower nitrogen content, which makes them intrinsically less nutritious. (Note: as food plants humans use the seed part of monocotyledons.) Second, C_4 and C_3 grasses rarely have secondary toxins that conversely are common to many C_3 dicotyledons. Third, protein in C_4 plants is protected by a bundle of sheath cells making them harder to digest (hence the agricultural pre-processing of such crops). And finally, grasses have a higher silica content than dicotyledons. This makes them more abrasive and this abrasiveness during mastication provides the link with hypsodonty that enabled mammals to take advantage of the new food source.

Herbivorous animals' food can be determined one way through carbon-13 (^{13}C) analysis. C_3 plants (including some grasses) have ^{13}C values of 26‰ whereas C_4 plants (which are mostly grasses) have ^{13}C values of 12‰; consequently fossil teeth provide a record of the proportion of C_4 and C_3 plants in herbivore diets and palaeosol (fossilised soil) carbonates. However, there is a problem: C_3 (monocotyledonous) grasses have the same ^{13}C content as C_4 dicotyledonous plants. This means that ^{13}C analysis only reveals the expansion in C_4 grasses. Conversely, most C_4 plants are grasses and the C_4 grasses are the dominant grasses at low latitudes (around the tropics). C_4 plants can have a carbon dioxide photosynthetic advantage over their C_3 counterparts. Carbon dioxide (on an annual basis) is effectively the same the planet over and so at low latitudes C_4 plants are better placed to make use of the higher-intensity insolation (the solar radiation received at the Earth's surface) and temperature. C_4 grasses are therefore the dominant grasses in the tropics whereas C_3 grasses are found at higher latitudes (or altitudes, if in the tropics).

The ^{13}C record of North American equids shows an abrupt increase starting about 7 mya, although Pacific coast and Canadian equids' ^{13}C values suggest that this increase occurred later. A similar change was noted at the end of the Miocene (approximately 7–5 mya) in Pakistan, Kenya and South America.

Just as ferns did in the Cretaceous, over the Tertiary period (65.5–1.8 mya) mammals diversified. Back in the Mesozoic era (the age of the dinosaurs) the largest mammal was the size of a house cat today. In the Tertiary this was to change. Here again, equids are illustrative of mammal diversification generally. The early horses (such as Eohippus) were small and well adapted to the global tropical-like forests of the Tertiary period's early Eocene epoch (55.8–33.9 mya). Their later diversification into increasingly larger forms reflects their being more suited to a cooler climate (increased size has a lower surface-area-to-volume ratio, and hence reduced heat loss) and to eating nutritionally poor grasses (which necessitates gastrointestinal specialisation that takes up room). And, as previously noted, these larger forms were more suited to grassland plains than the smaller forms which were more suited to moving among the tangle of forests. (Once again, note the interplay of factors.)

During this time North America was geographically separated from South America (there was no connecting isthmus). Consequently the smaller early horses could not migrate south towards the equator. If they had been able to then the modern horses may have evolved allopatrically (in a population in a different geographic area) with the more ancient forms surviving today, somewhat in the way the mouse deer (tragulids) did in the Old World tropics.

And of course, of perhaps special interest to readers, as noted above, the precursors to the Hominoidea began diversifying 14–9 mya and this continued with the evolution of the australopithecines that lived 5–1.2 mya.

From the end of the Miocene, and throughout the Pliocene (2–5 mya), the Earth cooled. This trend culminated in our current (Quaternary) ice age consisting of a series of glacials interspersed by warmer interglacials such as the one we presently experience. This will be examined next.

3.4 Summary

A key feature of the biosphere's evolution (including its abiotic components of geology and atmosphere) up to the current (Quaternary) ice age has been the interrelation of biology and climate. Key evolutionary steps in biological evolution – such as the developments of photosynthesis, multicellular organisms, vascular plants and new photosynthetic processes – resulted in changes in both the atmosphere and climate. A number of these steps have had consequences for the reorganisation of elemental cycling, especially carbon, its various reservoirs and in relations between them. This implicitly affected atmospheric carbon (both carbon dioxide and methane) and so changes in the greenhouse forcing of climate, hence climate itself. Taking all these factors together provided further evolutionary opportunities that included those that (much later ultimately) allowed for the rise of *H. sapiens*.

In addition, the biosphere's evolution has been punctuated by a number of extinction events imposed independently of biological evolution but which have affected species' evolution (or ended it). These events have principally been vulcanism-related or asteroidal. Of the former some have been associated with (or at least helped to trigger) carbon releases and both have had global climate effects. However, even asteroidal biological extinctions, though not directly related to the biosphere's evolution per se, are necessary components of a cosmology that enables Earth-like planets in general, and the Earth specifically, to exist. Asteroids and planetesimals are necessary in a proto-planetary system for planets to form. In contrast, vulcanism is a manifestation of the energy that drives plate tectonics, hence the deep carbon cycle, and so is a major factor enabling allopatric speciation and biosphere stability (see the discussion on ice-world recovery earlier in this chapter). Some of the main causes of mass extinctions therefore relate to factors that enable life. Others, such as methane releases from submarine methane hydrate (clathrates), ultimately arise out of life itself. Life and mass-extinction events are intricately connected well beyond the obvious symmetry of one virtually being an antithesis of the other and that you cannot have an extinction without life.

With regard to our current twenty-first-century anthropogenic global warming and climate change, there is much we can learn from the biosphere's history. There are perhaps two key points of particular relevance. First, reorganisation of the carbon cycle, as we are now doing by liberating formerly geologically isolated carbon into the atmosphere, results in major environmental change. (A point that may seem obvious to many now, but which was not a decade or two ago – see the IPCC's first assessment, 1990.) Second, such environmental change affects the number and range of species extant on the planet both positively and negatively. This should be of considerable human concern, for not only are we an animal species in our own right, but we rely on many species of plants and animals for our daily survival. We rely on an even greater population of species through what ecologists call ecosystem function beyond the daily time frame to maintain environmental quality. These impli cations will be explored in later chapters.

3.5 References

Alveraz, L. W., Alveraz, W., Asaro, F. and Michel, H. V. (1980) Extraterrestrial cause for the Cretaceous-Tertiary extinction – experimental results and theoretical implications. *Science*, **208**, 1095–1108.

Becker, L., Poreda, R. J., Hunt, A. G., Bunch, T. E. and Rampino, M. (2001) Impact event at the Permian-Triassic boundary: evidence from extraterrestrial noble gases in fullerines. *Science*, **291**, 1530–3.

Becker, L., Poreda, R. J., Basu, A. R. *et al.* (2004) Bedout: a possible end-Permian impact crater offshore of northwestern Australia. *Science*, **304**, 1469–75.

Bekker, A., Holland, H. D., Wang, P.-L. *et al.* (2004) Dating the rise of atmospheric oxygen. *Nature*, **427**, 117–20.

Berner, R. A. (1998) The carbon cycle and CO_2 over Phanerzoic time: the role of land plants. *Philosophical Transactions of the Royal Society of London Series B Biological Sciences*, **353**, 75–82.

Bottomley, R., Grieve, R., York, D. and Masaitis, V. (1997) The age of the Popigai impact event and its relation to events at the Eocene/Oligocene boundary. *Nature*, **388**, 365–8.

Bowen, G. J., Beerling, D. J., Koch, P. L., Zachos, J. C. and Quattlebaum, T. (2004) A humid climate state during the Palaeocene/Eocene thermal maximum. *Nature*, **432**, 495–9.

Briggs, D. E. G. and Crowther, P. R. (eds.) (2001) *Palaeobiology II*. Oxford: Blackwell Science.

Clough, D. O. (1981) Solar interior structure and luminosity variations. *Solar Physics*, **74**, 21–34.

Dickens, G. R., O'Neil, J. R., Rea, D. K. and Owen, R. M. (1995) Dissociation of methane hydrate as a cause of isotope excursion at the end of the Paleocene. *Paleoceanography*, **10**, 965–71.

Emilliani, C. (1992) *Planet Earth: Cosmology, Geology and the Evolution of Life and the Environment*. Cambridge: Cambridge University Press.

Erwin, D. H. (1994) The Permo-Triassic extinction. *Nature*, **367**, 231–6.

Evans, D. A., Beukes, N. J. and Kirschvinck, J. L. (1997) Low-latitude glaciation in the Palaeoproterozoic era. *Nature*, **386**, 262–6.

Frakes, L. A., Francis, J. E. and Syktus, J. I. (1992) *Climate Modes of the Phanerozoic*. Cambridge: Cambridge University Press.

Genda, H. and Abe, Y. (2005) Enhanced atmospheric loss on protoplanets at the giant impact phase in the presence of oceans. *Nature*, **433**, 842–4.

Hallam, T. (2004) *Catastrophes and Lesser Calamities: the Causes of Mass Extinctions*. Oxford: Oxford University Press.

Han, T. M. and Runnegar, B. (1992) Megascopic eukaryotic algae from the 2.1 billion year-old Neguanne iron formation. *Science*, **257**, 232–5.

Holland, H. D. (1990) Origins of breathable air. *Nature*, **347**, 17.

Huang, Y., Street-Perrott, F. A., Metcalfe, S. E. *et al.* (2001) Climate change as the dominant control on glacial-interglacial variations in C3 and C4 plant abundance. *Science*, **293**, 1647–51.

Intergovernmental Panel on Climate Change (1990) *Climate Change: the IPCC Scientific Assessment*. Cambridge: Cambridge University Press.

Intergovernmental Panel on Climate Change (2001a) *Climate Change 2001: Mitigation – a Report of Working Group III*. Cambridge: Cambridge University Press.

Intergovernmental Panel on Climate Change (2001b) *Climate Change 2001: Impacts, Adaptation, and Vulnerability – Summary for Policymakers and Technical Summary of the Working Group II: Report*. Cambridge: Cambridge University Press.

Janis, C. M. (2001) Radiation of Tertiary mammals. In Briggs, D. E. G. and Crowther, P. R., eds. *Palaeobiology II*. Oxford: Blackwell Science.

Jenkyns, H. C., Forster, A., Schouten, S. and Damsté, S. (2004) High temperatures in the late Cretaceous Arctic ocean. *Nature*, **432**, 888–92.

Jin, Y. G., Wang, Y., Wang, W. *et al.* (2000) Pattern of marine mass extinction near the Permian-Triassic boundary in South China. *Science*, **289**, 432–6.

Katz, M. E., Pak, D. K., Dickens, G. R. and Miller, K. G. (1999) The source and fate of massive carbon input during the Late Palaeocene Thermal Maximum. *Science*, **286**, 1531–3.

Kaufman, A. J. and Xiao, S. (2003) High CO_2 levels in the Proterozoic atmosphere estimated from the analyses of individual microfossils. *Nature*, **425**, 279–81.

Kemp, D. A., Coe, A. L., Cohen, A. S. and Lorenz, S. (2005) Astronomical pacing of methane release in the early Jurassic period. *Nature*, **437**, 396–9.

Kennett, J. P. and Stott, L. D. (1991) Abrupt deep sea warming, palaeoceanographic changes and benthic extinctions at the end of the Palaeocene. *Nature*, **353**, 225–9.

Kirschvink, J. L. (1992) A palaeogeographic model for vendian and Cambrian time, Chapter 12. In Schopf, J. W. and Klein, C., eds. *The Proterozoic Biosphere*. Cambridge: Cambridge University Press.

Knauth, L. P. (2006) Signature required. *Nature*, **442**, 873–4.

Lourens, L. J., Sluijs, A., Kroon, D. *et al.* (2005) Astronomical pacing of late Palaeocene to early Eocene global warming events. *Nature*, **435**, 1083–7.

Lovelock, J. E. (1979) *Gaia: A New Look at Life on Earth*. Oxford: Oxford University Press.

McElwain, J. C., Wade-Murphy, J. and Hesselbo, S. P. (2005) Changes in carbon dioxide during an oceanic anoxic event linked to intrusion into Godwana coals. *Nature*, **435**, 479–82.

Nunes, F. and Norris, R. D. (2006) Abrupt reversal in ocean overturning during the Palaeocene/Eocene warm period. *Nature*, **439**, 60–3.

Ohmoto, H., Watanabe, Y. and Kumazawa, K. (2004) Evidence from massive siderite beds for a CO2-rich atmosphere before ~1.8 billion years ago. *Nature*, **429**, 395–9.

Pagani, M., Freeman, K. H. and Arthur, M. A. (1999) Late Miocene atmospheric CO2 concentrations and the expansion of C4 grasses. *Science*, **285**, 876–9.

Payne, J. L., Lehrmann, D. J., Wei, J. *et al.* (2004) Large perturbations of the carbon cycle during recovery from the end-Permian extinction. *Science*, **305**, 506–9.

Pearson, P. N. and Palmer, M. R. (2000) Atmospheric carbon dioxide concentrations over the past 60 million years. *Nature*, **406**, 695–9.

Pollard, D. and Kasting, J. F. (2005) Snowball Earth: a thin-ice solution with flowing glaciers. *Journal of Geophysical Research* **110**, C07010.

Schaefer, B. F. and Burgess, J. M. (2003) Re – Os isotopic age constraints on deposition in the Neoproterozoic Amadeus Basin: implications for the 'Snowball Earth'. *Journal of the Geological Society London*, **160**, 825–8.

Schneider, H., Schuettpelz, E., Pryer, K. M. *et al.* (2004) Ferns diversified in the shadow of angiosperms. *Nature*, **428**, 553–7.

Sheehan, P. M. and Harris, M. T. (2004) Microbolite resurgence after the late Ordovician extinction. *Nature*, **430**, 75–7.

Smith, A. G. and Pickering, K. T. (2003) Oceanic gateways as a critical factor to initiate icehouse Earth. *Journal of the Geological Society London*, **160**, 337–40.

Steuber, T., Rauch, M., Masse, J.-P., Graaf, J. and Malkoč, M. (2005) Low-latitude seasonality of Cretaceous temperatures in warm and cold episodes. *Nature*, **437**, 1341–4.

Stoll, H. M. (2006) The Arctic tells its story. *Nature*, **441**, 579–80.

Svensen H., Planke, S., Malthe-Sørenssen, A. *et al.* (2004) Release of methane from a volcanic basin as a mechanism for initial Eocene global warming. *Nature*, **429**, 542–5.

Tice, M. M. and Lowe, D. R. (2004) Photosynthetic microbial mats in 3,416-Myr-old ocean. *Nature*, **431**, 549–52.

Tripati, A. and Elderfield, H. (2005) Deep-sea temperature and circulation changes at the Paleocene-Eocene Thermal Maximum. *Science*, **308**, 1894–8.

Vajda, V. and McLoughlin, S. (2004) Fungal proliferation at the Cretaceous-Tertiary boundary. *Science*, **303**, 1489.

Vent, D. V. and Spray, J. G. (1998) Impacts on the Earth in the Late Triassic (separate letters). *Nature*, **395**, 126.

Ward, P. D., Montgomery, D. R. and Smith, R. (2000) Altered river morphology in South Africa related to the Permian-Triassic extinction. *Science*, **289**, 1740–3.

Watanabe, Y., Martini, J. E. J. and Ohmoto, H. (2000) Geochemical evidence for terrestrial ecosystems 2.6 billion years ago. *Nature*, **408**, 574–7.

Watson, A. (2004) Gaia and the observer. In Schneider, S. H., Miller, J. R., Crist, E. and Boston, P. J., eds. *Scientists Debate Gaia, The Next Century*. Cambridge, MA: MIT Press.

Wilf, P. and Labandeira, C. C. (1999) Response of plant-insect associations to Paleocene-Eocene warming. *Science*, **284**, 2153–6.

Wilf, P., Cuneo, N. R., Johnson, K. R. *et al.* (2003) High plant diversity in Eocene South America: Evidence from Patagonia. *Science*, **300**, 122–5.

Wing, S. L., Harrington, G. J., Smith, F. A. *et al.* (2005) Transient floral change and rapid global warming at the Palaeocene-Eocene boundary. *Science*, **310**, 993–5.

Yuan, X., Xiao, S. and Taylor, T. N. (2005) Lichen-like symbiosis 600 million years ago. *Science*, **308**, 1017–34.

Zachos, J., Pagani, M., Sloan, L., Thomas, E. and Billups, K. (2001) Trends, rhythms and aberrations in global climate 65 MA to present. *Science*, **292**, 686–93.

4

The Oligocene to the Quaternary: climate and biology

We are currently in the middle of an ice age! This ice age is known as the Quaternary ice age, and it began roughly about 2 million years ago (2 mya). ('Roughly' because how much ice do you need on the planet to say that it is an ice age?) We might not think we are in an ice age and this is because we are in a warm part, called an interglacial. As we shall see, there have been a number of glacials and interglacials in our Quaternary ice age. However, this ice age did not just start by itself but arose out of a number of factors that became relevant earlier, in the Oligocene (34–23 mya) and Miocene (23–5.3 mya) eras, well before the beginning of our ice age and its Pliocene and Pleistocene glaciations (Zachos *et al.*, 2001). These glaciations actually had their beginnings some 5.3 mya. To understand how our Quaternary ice age came about we will need to briefly re-cap part of the previous chapter and note some other material to provide a more biological perspective while leaving out the extinction events.

4.1 The Oligocene (33.9–23.03 mya)

Between 35 and 15 mya the Earth's temperature was roughly 3–4 °C warmer than today and atmospheric carbon dioxide concentrations were twice as high. However, climate-forcing factors were coming into play that were to cool the planet. Carbon dioxide levels were falling and, as noted in the previous chapter, this fall could only have been furthered by the new C_4 plants (even though their period of major expansion was not to take place until 8 mya: see below).

A major forcer of global climate came about by the collision of the Indian and Asian plates. These plates first made contact with each other around 55 mya (coincidentally roughly around the time of the Eocene climatic event) and they continued to compact. This resulted in the subsequent uplift of what was to become the Tibetan plateau that gradually rose, and is still rising today.

The plateau increasingly affected global climate in three ways. First, it forced the re-organisation of atmospheric circulation, with the creation of the East Asian monsoon and associated climatic patterns. Second, the uplift caused ice to form in the mountains and notwithstanding this there were other albedo effects (due to altitude and changing nature of the land's surface) so reflecting more of the Sun's energy. This is one of the main reasons why Milankovitch curves are frequently calculated on the solar energy received in northern latitudes between 28° and 40° north. Third, it increased erosion so increasing the amounts of silicate, hence calcium silicates that could react with carbon dioxide, precipitating it in solid form as calcium carbonate and so encouraging the further drawdown of atmospheric carbon dioxide (see Chapter 3). These factors helped to switch the bio-geosphere from being more stable, in a greenhouse mode (the Eocene and earlier), to a cooler mode (the Oligocene and earlier).

$\delta^{18}O$ Analysis of forams and ocean sediments to ascertain past calcite compensation depths (CCDs) – the water depth at which the calcium carbonate rain is balanced by the dissolution rate due to ocean acidity – suggests that the first indications of the aforementioned shift in stability took place around 42 mya in the Eocene. A more definite shift then took place somewhere around 34 mya at the Eocene–Oligocene boundary (Tripati *et al.*, 2005). (This of course is distinct from the comparatively transient (in geological terms) change in ocean chemistry associated with the Eocene thermal maximum event; see Chapter 3.)

Earlier, in 2003, US palaeoclimatologists, Robert DeConto and David Pollard, demonstrated through a general-circulation computer model coupled with atmosphere and ice-sheet components, and incorporating palaeogeography of the time, that a combination of declining Cenozoic carbon dioxide (from four times to just double the pre-industrial level) and the thermal isolation of Antarctica enabled an ice sheet in the east of the continent to begin to form 34 mya. This is supported by geological sediment evidence. The thermal isolation was caused by the opening of oceanic passages between South America and Antarctica (the Drake Passage) and Australia and Antarctica (the Tasmanian Passage). This enabled the Antarctic Circumpolar Current (or ACC as it is sometimes summarised) to form. However, in this instance of long-term global cooling, ocean circulation change does not appear to have been as important early on as the decline in atmospheric carbon dioxide. This may be because a threshold of depth to allow a critical volume of water to flow needed to be reached. But aside from carbon dioxide, it appears that an ice albedo positive-feedback effect (see Chapter 1) also played a major role. Here the expanding ice sheet reflected more sunlight and so cooled the continent, allowing the ice sheet to expand further still, which reflected even more sunlight, and so on.

The first Antarctic ice sheets were sensitive to Milankovitch forcing (see Chapter 1) much as the northern-hemisphere ice sheets were during the Quaternary glacial–interglacial series of waxing and waning ice sheets (which we will examine shortly). It has been suggested that orbital frequencies of 40 000 years (obliquity) and 125 000 years (eccentricity) dominated the way the ice margins behaved at that time.

4.2 The end Miocene (9–5.3 mya)

As mentioned in the previous chapter the new C_4 photosynthetic plants were better able to take advantage of the atmosphere's lower carbon dioxide compared to those earlier in the Tertiary (prior to 20 mya). This gave them an advantage over C_3 plants where other conditions (such as soil, temperature and moisture) were favourable. As noted in the last chapter, C_4 plants originated well before their spreading to become established globally: global expansion takes time. Indeed, it is thought that atmospheric carbon dioxide had already lowered to the equivalent of Quaternary levels (of the past 2 million years or so) by 15 mya, and if anything atmospheric carbon dioxide is thought to have been on the increase between 15 and 5 mya. So, although the lowering of atmospheric carbon dioxide over much of the Tertiary facilitated speciation and the rise of C_4 plants, lowering atmospheric carbon dioxide by itself cannot explain their global spread at the end of the Miocene. It may be that tectonic activity, changes in seasonal precipitation, and some other factor(s) or combinations thereof facilitated matters (Pagani *et al.*, 1999). Indeed, as mentioned in the previous section there was the opening of the high-latitude southern-hemisphere oceanic passages and of course uplift, forming the Tibetan Plateau. By 10 mya the east-Antarctic ice sheet had grown (but was still smaller than today) from its beginnings 35 mya, and one 10 mya was already on Greenland and itself slowly growing. Superimposed on this overall trend of a cooling Earth and increasing ice, was Milankovitch-driven ice-sheet waxing and waning. So there was climate change taking place.

As noted in the last chapter, the spread of C_4 plants during this time has been charted in a number of ways. The US geophysicist Thure Cerling and colleagues published in 1993 the results of isotope analysis of soil and tooth remains: C_3 and C_4 plants' metabolic pathways have different preferential selections (or fractionation) of the ^{13}C and ^{12}C carbon isotopes. Initially there was some discussion that this interpretation might be undermined if animals migrated from where they fed to where they died. However, in 1997 Cerling and colleagues published the results of carbon analysis of tooth remains from more than 500 hypsodont equids from Asia, Africa, Europe and North and South America. In addition, they analysed the teeth of fossil proboscideans (elephants and allied species) and notoungulates (an extinct order of South American mammals). The advantages of looking at teeth, as opposed to soils, are that animals' selective feeding enhances the isotope signal, and tooth identification and dating is comparatively easy, whereas soil contains a mix of ^{13}C and ^{12}C originating from both C_3 and C_4 remains. The results of Cerling's team show that large mammals from before 8 mya from a global population all had diets either dominated by C_3 plants or which were purely C_3. But by 6 mya equids and some other large mammals from many regions in latitudes below 37° (the sub-tropics and tropics) had C_4 diets.

During this time the Drake and Tasmanian Passages widened further, so increasing Antarctica's thermal isolation and facilitating growth of its ice sheets. By 6 mya ice-sheet growth in west Antarctica had become significant. This ice sheet, and that already existing in east Antarctica, were subject to positive albedo/temperature-feedback effects and so ice slowly grew to fill the Antarctic continent.

4.3 The Pliocene (5.3–1.8 mya)

The Pliocene was the last epoch in the Tertiary period. It was a time of global transition from warm conditions with global surface temperatures about 3 °C warmer than today, smaller ice sheets and higher sea levels compared with current conditions. In 2004 a team, led by Ana Ravelo, from the University of California and Boise State University, published an analysis of several palaeoclimate proxy records. These included an ^{18}O isotope record from benthic Foraminifera, an alkenone record, a biogenic carbonate, an organic carbon record, and a biogenic isotopic carbon-sediment record. Together these records and those of other researchers provide a reasonably clear picture as to how the climate changed, although of course the detail is limited.

Relative to today (the dawn of the twenty-first century), the approximately 3 °C warmer early Pliocene Earth had a sea level some 10–20 m higher than now and an enhanced thermohaline circulation (to which we will return later). Atmospheric and ocean-circulation patterns were in places significantly different from today. For example, it appears that the Pacific was subject to permanent El Niño-like conditions and different ocean- and atmosphere-circulation patterns (Wara *et al.*, 2005). (Today El Niño is a semi-regular event; see Chapter 5 and the Holocene summary as well Chapter 6.) That the early Pliocene can be considered a palaeoclimatic analogue of a possible late twenty-first-century global-warmed Earth means that Pliocene research is of interest to those concerned with present-day climate change.

Antarctica's ice sheets were growing throughout the Pliocene, but were still smaller than today. Similarly there was ice in the northern hemisphere, and again this was greatly reduced compared to today. Atmospheric carbon dioxide, although lowering was also slightly (approximately 30%) higher later in the Pliocene than recent pre-industrial levels, at around approximately 360 ppm. This compares with carbon dioxide levels in the 1990s, but the reason that the Pliocene Earth was warmer than the 1990s is because its oceans had already been warmed, while its ice caps were smaller than today (reflecting less sunlight), and there were also some atmosphere- and ocean-circulation differences.

Associated with this last, tectonic movement of the continents also continued through the Pliocene that impacted on the global climate. The aforementioned Drake and Tasmanian Passages continued to widen but also there was the closing of a key tropical gateway linking the Pacific and Atlantic Oceans, with the joining of North and South America. This affected heat transportation from the tropics as global ocean circulation reorganised.

In addition to ocean circulation, which is important for heat distribution about the planet, there were the Milankovitch orbital factors affecting the amount of energy the Earth receives between hemispheres (before albedo reflection considerations) and at different times of the year. As stated, the rise of the Tibetan Plateau was central to this. Evidence from late-Pliocene temperate pollen analysis suggests a temperate-zone climate periodicity of 124 000 years, with a non-linear response of the climate system to Milankovitch forcing.

Not only did some temperate ecosystems wax and wane, in extent and location due to climatic changes, but plant evolution was also taking place. The rise of C_4 plants was a major evolutionary step, but there were also more modest, but still important, speciations to exploit the cooler Earth. Much Arctic flora is thought to have originated some 3 mya towards the end of the Pliocene. They evolved to survive both the cold and the climatic waxing and waning. The likely pattern, as exemplified by Richard Abbot *et al.* (2000), is that an Arctic species evolved in one location (different locations for one or more different species) and then spread in a ring around the pole. As the Pliocene Earth warmed and cooled these species extended from, and returned to, various ecological refuges. Abbott and his colleagues demonstrated this by looking at the chloroplast DNA of species of saxifrage (an order of dicotyledon trees and shrubs), and specifically *Saxifraga oppositifolia*. As time passes mutations arise in the DNA and these increase additively with each generation. Furthermore, as populations become increasingly isolated, when the plant is confined due to adverse climate regimes to small refugia, the population retains its characteristic set of mutations. It is therefore possible, through genetic analysis, to ascertain to which refuge population a plant belongs, and indeed the primary location where a plant originally evolved. The evidence of Abbott *et al.* suggests that in the case of *S. oppositifolia*, the species originated in Beringia, part of which was the now-submerged land bridge between present-day Siberia and Alaska. It then spread around the pole to Greenland, Iceland, northern British Isles, Scandinavia, Siberia and Canada, and then relied on some 14 refugia when times changed and were harsh (during glacials). Refugia are ecologically important in climate-change terms, as we shall see later.

It is worth noting now that in terms of polar ecology, Beringia was of hemispheric significance – and we will mention it again later. Beringia, when an exposed land area, was covered with vegetation even in the coolest of times (including more recently in the depth of the last glacial some 24 000 years ago, as then the land-bridge area was substantive due to the sea level being 100 m lower than today). The Pliocene Beringia vegetation supported populations of large mammals including mammoths, horses and bison. This Pliocene vegetation was unlike that found in modern Arctic tundra. Both the Pliocene and modern vegetation communities can sustain a relatively low population of mammals, but the Pliocene Beringian community was a productive dry grassland ecosystem that resembled modern sub-Arctic steppe (dry and treeless) communities. It is likely that Beringia ecologically consisted of a mosaic patchwork with wet tundra in poorly drained lowlands as well as marked ecological differences between the central and its eastern-most regions (Zazula *et al.*, 2003). This added to the habitat diversity: it was of an appropriate quality for an ecological refugium, enabling it to provide habitats for a range of species.

For millions of years up to somewhere between 4.8 and 7.4 mya, Beringia was important in terms of both global climate and ecology. Ecologically it formed a bridge between the North American and East Asian landmasses (enabling animals to cross between the two) and, in terms of marine ecology and hemispheric climate, it was a barrier between the Pacific and Arctic oceans (preventing the exchange of marine life). The tectonic separation of North America and East Asia, between 4.8 and 7.4 mya, split Beringia allowing water – at times of high sea level – to flow between the Arctic and the Pacific while hindering terrestrial fauna and flora travelling between the two continents. This dating of this significant opening was estimated due to the presence of a formerly Arctic and North Atlantic bivalve mollusc, *Ascarte* (Marincovich and Gladenkov, 1999). The separation facilitated the high-latitude reorganization of ocean circulation including the way heat entered and left the North Polar region. It was yet another factor forcing global climate. As we shall see, this land bridge continued to form and close as the sea level rose and fell with the major glacial cycles throughout the Quaternary to the present. One consequence of this sea-level change was that Beringia's role as a refuge was insecure. Because it did not flood uniformly in the warm interglacials of the past million years, its highest hills tended to survive, and do even today, as islands.

This flooding also happened more recently between the last glacial maximum and the early parts of our current (Holocene) interglacial. Indeed, it is known that some now-extinct Pleistocene fauna survived well into our current Holocene interglacial. For example, mammoths (*Mammuthus* spp.) survived on St. Paul Island as late as 7900 years ago and Wrangel Island 3700 years ago, which is well after the end of the last glacial. Conversely, mammoths that were less isolated on the Californian Channel Islands became extinct far earlier, around the time of human arrival *c*.11 000 years ago (Guthrie, 2004). The relationship between this mammoth extinction and human arrival is thought to be causal. Had the Beringian islands been much larger, so as to enable larger populations (and hence reduced inbreeding), then it is possible that mammoths might have survived for longer, perhaps even to the present day, in the absence of humans. Consequently Beringia should not be considered as just a land bridge but, for reasons alluded to above, also a land area where many characteristic Pleistocene species and communities evolved.

Meanwhile, returning to between 5.3 and 2.7 mya, the Earth gradually cooled due to the aforesaid combination of factors of tectonic movement and changes in air and ocean circulation. At about 2.75 mya a sharp change in this global cooling took place, in all likelihood due to the interaction of climate-influencing factors that prompted new ocean- and/or atmospheric-circulation patterns, and not a single threshold event. Although notably North and South America had joined by this time which would have had a pronounced effect on ocean circulation. What is known, from alkenone analysis and oxygen isotopes from diatoms, is that at this time late-summer sea-surface temperatures in the sub-Arctic Pacific rose. At first it may seem paradoxical that a sea-surface temperature rise is associated with a glaciation. What seems to have happened is that at this time, with the reorganisation of ocean circulation, there was increased water-column stratification (temperature difference), hence the surface water failed to sink. Meanwhile, winter sub-Arctic Pacific sea-surface temperatures did cool. Consequently, the late-summer warm water increased precipitation in the autumn so encouraging the build-up of snow and ice over neighbouring Canada which, with the cooler winters, was preserved for longer into the following year (Haug *et al.*, 2005). The northern-hemisphere ice sheets grew: ice sheets can only grow if the preceding year's snow and ice survives the summer into the subsequent winter. In this way the North American Laurentide ice cap began to expand to complement the Fennoscandian ice cap over northern Europe and Russia. The Fennoscandian ice sheet obtained much of its water from prevailing winds running above the North Atlantic Drift (commonly called the Gulf Stream). The overall result was a step-like progression in the Earth's cooling that marked the onset of northern-hemisphere glaciation (NHG). This is reflected in a number of palaeoclimatic indicators including high-latitude benthic ^{18}O, the mass accumulation rate (MAR) of some northern-hemisphere oceanic biogenic opal (this almost ceased), the magnetic susceptibility record of the sub-Arctic Pacific (an indicator of ice-rafted debris which increased sharply having previously been at a low level) and other isotopic MAR evidence of foram species (Ravelo *et al.*, 2004).

   With the onset of NHG approximately 2.75 mya, which complemented the already existing Antarctic ice, the current Quaternary (and Pliocene) ice age could be said to have begun. Having said this, the key ice deposits that, according to the International Commission on Stratigraphy, officially mark the start of the Quaternary (and Pleistocene) date from 1.806 mya. As is often the case, the trigger for climate change came first and the key geological strata (used for key dating) came later. Similarly, today we already have had significantly elevated levels of atmospheric carbon dioxide for the best part of a century (depending on how you quantify a significant elevation) but only recently have we begun to discern the climate-change fingerprint both biologically and above meteorological records (see Chapter 6). This biological fingerprint is likely (almost certainly) to become more pronounced and then leave behind remains that may be discerned in the future as indicators of environmental change.

4.4 The current ice age

Before reviewing the current ice age it is worth noting the nomenclature appearing in the academic literature as variations in usage can confuse. The depth of the current Quaternary ice age is defined broadly in the nomenclature of geological timescales in terms of two parts or epochs. First, there was a series of glacial and interglacial cycles that comprise the Pleistocene. Second, there is the Holocene: this is just the current interglacial in which we find ourselves today. Confusingly the Pleistocene (1.806–0.0115 mya) is sometimes referred to as an ice age in its own right, which conjures up a picture of a single cold spell for the planet, as opposed to oscillations between glacial and interglacial times. (Indeed, some parts of past interglacials were slightly warmer than today.) Similarly, the end of the Pleistocene is sometimes described as the end of the ice age when in fact it is just the end of the last glacial. Notwithstanding human greenhouse considerations, our Holocene interglacial will eventually end with a return to glacial conditions. This confusion in terminology is simply a result of the historical way we have uncovered what has happened in the past. It reflects the way our understanding has changed since the term ice age was first coined by the Swiss naturalist Louis Agassiz in 1837 (after whom the large last glacial maximum (LGM) lake was named; see later). This ambiguity of nomenclature is just something of which the student of palaeoclimatology simply has to be aware. (Such anomalies are also common in other areas of science, such as the use of the term fertility; see Chapter 7.)

Recently, aside from the Holocene and Pleistocene, a new term, the Anthropocene, has begun to be used. This denotes the time from which humans had a discernable impact on the (global) biosphere. Even here there is some confusion. Some say that this has only happened since the Industrial Revolution, but others, notably Bill Ruddiman of the University of Virginia, say that humans have been affecting the global climate since the spread of agriculture some 8000 years ago (Ruddiman, 2003). Irrespective of the exact time when humans began to have a discernable effect on the biosphere – specifically global cycling systems (such as for carbon, nitrogen and phosphorus) and physically (such as climate) – humans today are certainly imparting such an impact. Today, globally some 15% of ice-free land is dominated by human use (such as urban areas and roads) whereas 55% is part-dominated by human use (such as agriculturally dominated landscapes and canalised rivers). This is so much so that it is estimated that just under a third of terrestrial primary productivity (the amount of energy, carbon or biomass) fixed by photosynthesis is influenced by humans (Imhoff *et al.*, 2004), so that saying that we are today in the Anthropocene is not inappropriate. However, notwithstanding the vagaries of terminology, the onset of 'major' glaciation in both hemispheres can be said to mark the onset of the current Quaternary ice age of the past 2 million years.

The previous gradual cooling that had been taking place over a few million years before the onset of significant NHG was not entirely smooth but consisted of temperature oscillations with a period of a few tens of thousands of years. After the onset of the NHG these oscillations quickly (within a hundred thousand years) became more marked and at that time had a period approximating 41 000 years. Although this cannot be attributed to a single event (again rather a combination of circumstances) it is perhaps useful to re-emphasise that the Isthmus of Panama formed before some 2.5 mya, physically separating the Pacific and Atlantic Oceans. Even though the isthmus formed then, as the two continents drew closer, the water flow between the two oceans would have gradually declined over previous millennia and this factor might have, in association with others, become more climatically significant at some point prior to the continents' physical joining. The 41 000-year periodicity subsequent to the onset of NHG reflects Milankovitch obliquity and it was to dominate global climate variations 1.75–1 mya (see Chapter 1); all the while, the Earth was cooling with ice volumes increasing with each glacial cycle.

In addition to the changes in surface ocean circulation, which helped cool the higher latitudes, there were amplifying feedbacks. One of these came to play a significant part once the Earth had undergone some initial cooling and so encouraged further cooling.

Carbon resides in a number of reservoirs within the bio-geosphere (see Chapter 1). This includes reservoirs in the atmosphere (of climate significance) and both in the surface and deep waters of the oceans. Both these latter reservoirs are larger than the atmospheric reservoir and so flows to and from the oceans have a major impact on the atmospheric reservoir; indeed, the size of the ocean deep-water carbon reservoir is over 45 times that of the atmosphere's (see Figure 1.3). In fact, the flow of carbon between these three reservoirs exceeds that of the flow to the atmosphere by human fossil-fuel burning by at least an order of magnitude and so these natural flows are important determinants of the global climate. For carbon dioxide to return to the atmosphere from the deep ocean there needs to be mixing with surface waters. It has been suggested (Sigman *et al.*, 2004) before about 2.7 mya that at high latitudes the oceans were sufficiently warm that, by contrast, winters would have seen a cool surface with a temperature not that far off the then deep-water temperature (which was warmer than that today). This would allow surface water to sink and so elsewhere for deep water to surface and return carbon to the atmosphere. However, after 2.7 mya the Earth's deep oceans became so cold that vertical mixing was reduced. This is because warm water when cooled a degree becomes proportionally more heavy than cool water cooled a degree. In other words, the relationship between sea-water density and temperature is not linear (see Figure 4.1). It is an idea that seems to fit in with the aforementioned sub-Arctic North Pacific scenario. Meanwhile, in today's world there is less deep water returning to the surface bringing with it carbon dioxide. More carbon dioxide remains trapped within ocean deep water. This tendency for lower atmospheric carbon dioxide after 2.7 mya added an additional cooling forcing factor to the already cooling global climate.

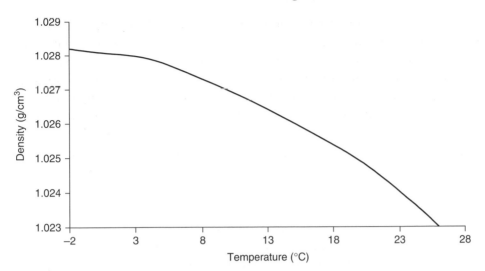

Figure 4.1 Sea-water density and temperature. As sea water cools it gets more dense but as the temperature approaches freezing point it gets denser at a slower rate. Therefore, a degree-cooler water in cool seas has less negative buoyancy than a degree-cooler water in warm seas, which will have a greater tendency to sink.

As we shall see later this chapter, when discussing the Broecker circulation, today there is some climatically important mixing of surface and deep waters but this takes place only in special places where the sea is both more salty (and hence more dense) as well as cool. It is this combination of salt (halinity) and temperature (thermality) affecting density that drives the Broecker thermohaline circulation, and not temperature and density by themselves.

The idea that changes in sea-water temperature/density relationships changed 2.7 mya is substantiated by biological evidence other than that from the sub-Arctic North Pacific (Sigman *et al.*, 2004). This is exemplified by two examples that relate to both the northern and southern hemispheres. First, opal is a mineral composed of amorphous silica and forms a constituent of the cell wall of a group of phytoplankton. It is a useful palaeo indicator because it is comparatively resistant to decomposition. Second, living organisms prefer to take up the nitrogen isotope ^{14}N to ^{15}N from the fixed biosphere pool of nitrogen. So, when there are more living organisms consuming nitrogen the ratio between the two isotopes changes. Looking at sediments around Antarctica and the Arctic, Sigman *et al.* found that both the rate of opal production and the nitrogen isotope proportions changed around 2.7 mya. This signifies a change in nutrient mixing, hence surface- and deep-water mixing necessary to support a large phytoplankton population. Finally, this scenario ties in with ocean sediments that suggest that ice rafting also commenced about this time. Therefore it would appear that ice rafting began around a time of changing surface- and deep-water mixing and a possible causal relationship between climate and surface-/deep-water mixing.

The high-latitude changes in ocean circulation and the onset of significant NHG were not only inter-related; the high-latitude cooling (hence the polar ice) also resulted in the sub-tropical regions beginning to cool. There is a clear and obvious synchronicity between the two and so we know that the cooling experienced then was a planet-wide event. The tropics themselves, though, did not see any fundamental changes, although undoubtedly there were changes in tropical biome extent: there was sufficient environmental similarity to warmer times in some parts of the tropics to provide a refuge for many species. This lends support to the view that the rapid climate transition about 2.75 mya must have primarily involved extra-tropical processes such as ocean (thermohaline) circulation, ice albedo, bio-geochemical processes and so forth that came together in an ice-sheet determined and/or tectonic and greenhouse (some combination thereof) threshold.

As the planet continued to cool a new threshold was reached some 2 mya with new modern tropical circulations. These included the development of what is known as the Walker Circulation and cool sub-tropical temperatures. The term Walker Circulation was first introduced in 1969 by Professor Jacob Bjerknes for two circulation cells in the equatorial atmosphere, one over the Pacific and one over the Indian Ocean. Schematically these are longitudinal cells where, on one side of the ocean, convection and the associated release of latent heat in the air above lifts isobaric surfaces upward within the upper troposphere and creates a high-pressure region there (Gilbert Walker was an early twentieth-century British climatologist who studied air circulations over the Pacific Ocean). This modern atmospheric circulation, combined with the previous reorganisation of the northern hemisphere's ocean circulation, again served to cool the Earth further.

Another threshold was reached about 0.75 mya when the interglacial–glacial cycles became more pronounced and took place with an increase in periodicity from 41 000 years (reflecting the Milankovitch obliquity cycle) to approximately 100 000 years (reflecting 93 000-years eccentricity). This 100 000-year periodicity and degree of glacial–interglacial temperature swing has persisted to the present day (Ravelo *et al.*, 2004): our present day, of course, being a warm interglacial.

The switch between the 41 000-year global climate cycle to 100 000 years appears to be more driven by a change in ocean circulation than the gradual overall decline in atmospheric carbon dioxide across glacial–interglacial cycles. Oxygen-isotope analysis (see Chapter 2) of planktonic Foraminifera reveals changes in Pacific circulation at this time (de Garidel-Thoron *et al.*, 2005). This is not to dismiss atmospheric carbon dioxide as a driver of climate change but other factors also play their part and may be the dominant (as opposed to sole) factors initiating a change in the Earth's climate system at some threshold point.

One might summarise the Quaternary and Pliocene trend in global climate as one that had three characteristics. First, overall long-term cooling. Second, Milankovitch-driven oscillations between warm and cold. Third, an increase at a transition point in both the period of climate oscillations and their degree.

In addition to isotopic analysis of biological remains in sediments, the fossil record and other techniques (involving biology and/or geology) used to discern global climate change over the past few million years, ice cores from Antarctica have been useful in ascertaining regional climate change in the southern hemisphere over the past few glacial–interglacial cycles. Greenland ice cores have provided similar information, particularly relating to the last glacial–interglacial cycle. (Ice cores can be analysed for variations in deuterium, dust and greenhouse gases as discussed in Chapter 2.) The longest ice core to date comes from Dome C in Antarctica. It provides a continuous record of chemical climate proxy data covering 740 000 years and several glacial–interglacial cycles. In terms of the core's physical length it is some 3 km long. We will shortly return to ice cores, but for now note that the Dome C record does strongly suggest that iron-dust deposition was higher in the cool parts of glacials (when globally there was less rain, and hence it was drier on land) and that subsequently the glacials reached their coldest phase. It is thought that iron fertilisation of the oceans encouraged algal blooms that further drew down carbon dioxide into the oceans, so causing the coldest periods of glacials (Wolff *et al.*, 2006).

One question to arise is the nature of glacial pacemaking. It is clear from so much evidence that the Milankovitch orbital parameters drive the glacial–interglacial cycles. However, do the Milankovitch orbital variations *lead* the driving, or forcing, of the climate? Or in other words does global climatic change *lag* some time behind the parameters as opposed to the parameters having a more or less instantaneous effect? Finally, does the waxing and waning of the ice sheets that amplify such forcing do so in time with either global climatic change and/or Milankovitch forcing? As noted in Chapter 1, Milankovitch factors provide the pacemaker for glacial–interglacial cycles, but are the two (orbital and glacial timing) and the changes in atmospheric greenhouse gases perfectly synchronised?

Nicholas Shackleton of Cambridge University's Godwin Laboratory addressed these questions in 2000. After all, Milankovitch forcing is weak. Although we are talking about changes of about 15% on 440 W m^{-2} on the Earth's surface at 65°, this is countered by opposite changes in the southern hemisphere. The actual difference taken globally is less than $1\,W\,m^{-2}$ or less than 0.1% of the Solar constant (the energy received per square metre perpendicular to the Sun in the Earth's orbit). This is due to the solar energy the entire planet receives and absorbs which in turn arises from the summation of a variety of factors: its albedo at various locales about the planetary sphere and the time of year these locales receive the energy as well as the distance from the Sun again integrated across the planetary sphere. So how does this weak signal magnify into a glacial–interglacial transition? Shackelton shed some light on this when comparing the timing of changes in benthic ^{18}O foram palaeorecords with the ice core and trapped gas Antarctica records as well as June insolation at 65°N (another often-used latitude for Milankovitch forcing curves that crosses both the former Laurentide and Fennoscandian ice sheets). He found that while the *extremes* in Milankovitch forcing are in time with changes both in global climate and atmospheric carbon dioxide, lagging behind by a few thousand years is ice volume. This might suggest that northern-hemisphere summer sunlight (or perhaps the northern-hemisphere thermal growing season?) affects the carbon cycle such as to drive atmospheric carbon dioxide concentrations and that this in turn has an immediate affect on temperature. This, while initially plausible, is unlikely because the annual variation in atmospheric carbon dioxide is too small to have a significant climate effect. On the other hand, whereas ice albedo effects may possibly magnify temperature changes (see Figure 1.8) the thermal inertia in the vast ice caps of the Antarctic and Greenland means that they take longer to respond and so only respond to sustained year-on-year climate forcing factors as opposed to factors that take place within a year.

As we shall see in subsequent chapters, this is somewhat relevant to our current anthropogenic release of carbon dioxide from fossil-fuel combustion. The anthropogenic future will not so much be determined by the weak Milankovitch forcing but our effects on the carbon cycle on a year-on-year (as opposed to within-a-year) basis, as well as albedo changes as glaciers and ice caps retreat. This makes policies, such as energy, that determine human carbon dioxide releases of fundamental importance should we be concerned about future climate change.

4.5 The last glacial

4.5.1 Overview of temperature, carbon dioxide and timing

The last glacial began about 115 000 years ago with temperatures that oscillated towards ever-colder conditions. It reached its climax (the last glacial

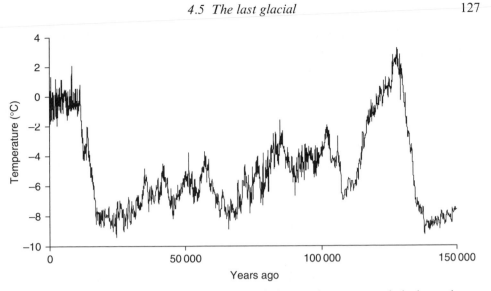

Figure 4.2 Temperature changes around Antarctica as revealed through deuterium analysis of ice cores. Based on data from the National Oceanic and Atmospheric Administration (see Acknowledgements).

maximum; LGM) some 18 000–22 000 years ago and ended (after a false start to the current interglacial 14 000 years ago) approximately 12 000 years ago (see Figure 4.2).

The importance of understanding the last glacial–interglacial cycle is that it illuminates the planet's current climatic mode. It is also important because, being closest to us in time, it has been the easiest era for which to gather data. Nonetheless, although the last glacial–interglacial cycle is fairly typical of the series for the past 700 000 years (in the 100 000-year Milankovitch-dominated Earth; see the preceding section), the fourth cycle ago features an interglacial that may be the best analogy to the current interglacial in its natural state (without anthropogenic carbon dioxide). Milankovitch forcing changes four cycles ago are similar to those now. However, as the last cycle is the easiest to study, and indeed in the history of palaeoclimatic research it was the one studied closely first, it is appropriate to look at it.

The big difference in global climate between glacial and interglacial times relates to global precipitation and, obviously, temperature. In glacial times there is less rainfall globally simply due to the fact that there is less evaporation from the oceans. With regards to temperature, the glacial Earth is broadly about 4 °C cooler than its interglacial counterpart and possibly at least 5 °C cooler during the LGM, 18 000 years ago. This, though, represents a global average which, as we shall see, is the same order of magnitude (i.e. single-figure degrees Centigrade) as is anticipated for global average anthropogenic warming over the next century. However, at high latitudes, near and on the ice sheets, temperatures during glacials were far cooler and more like 12–14 °C cooler than now.

The tropics were less affected by glacials but they did not entirely escape the cooling effects. Analysis of forams from the Pacific, and other palaeoclimatic indicators, suggests that the tropical Pacific sea surface was about 3 °C cooler during the last glacial maximum (LGM) than today. The Pacific, being such a vast reservoir of heat, is an important component of the Earth's climatic system. If the high latitudes were to experience a greater glacial–interglacial temperature difference than the 5 °C global average then one might expect that major parts of the tropical zone would see a smaller deviation to maintain the average. However, this is less true on land than in the sea. On land the vagaries of mountain ranges on climate and other factors come into play. Some parts of the tropics were clearly less affected by the glacial as tropical species survived not just the last glacial but the previous several that took place in the Quaternary. Yet this was not so everywhere. For example, some parts of Brazil were over 5 °C cooler during the LGM than today (Stute *et al.*, 1995).

Furthermore, climate-related events in high latitudes near the North Polar circle did affect the tropics. Biological productivity in the tropics off of Venezuela appears to have increased and decreased, reflecting periods of iceberg discharge from the large glacial Laurentide ice sheet over northern North America, as represented in the climate change recorded in the Greenland ice sheet. Chemical change in the sediments off of Venezuela bears an even closer correlation with the Greenland ice-core record for the ^{18}O isotope, which more than adequately reflects regional climate change within the glacial (Peterson *et al.*, 2000). Having said this, the Venezuela sediment record may relate more to deep-ocean ventilation than the tropical climate per se because, as we shall shortly see, global ocean circulation helps transport heat hemispherically as well as link the oceans globally and transport heat away from the tropics. Changes in ocean circulation also manifested themselves elsewhere in the tropics. For example, the African Kalahari dessert saw significant arid events approximately 41 000–46 000, 20 000–26 000 and 9 000–16 000 years ago, which are thought to be times of circulation change (Stokes *et al.*, 1997).

The detail required to understand and climatically model the planet is far greater than we have so far acquired, as we shall see in Chapters 5 and 6. So there is a clear need to obtain as much of such palaeoclimatic information as possible from wherever possible if we are to accurately determine how the palaeoclimatic systems worked. Much research is still required today. As there remains some uncertainty as to the exact conditions in the past, the validation of climate computer models cannot be undertaken with the exactness that policy-makers require or in a number of instances realise (see also Chapter 5, section 5.3.1).

If some tropical palaeoclimatic records correlate with polar records, it should not be surprising to find that many current temperate-zone palaeorecords exhibit similar correlations. For example, rapid environmental changes similar to those indicated by Greenland ice cores are discernable in southern European (Italian) palaeorecords covering the last glacial (Allen *et al.*, 1999).

4.5.2 Ice and sea level

During the LGM ice sheets of up to 3 km thick covered much of northern North America down to the Great Lakes (the Laurentide sheet; Figure 4.3) and also much of northern Europe, Scandinavia and the northern British

Figure 4.3 Approximate southern extent of ice over northern North America during the LGM. Note that the present-day coastline is portrayed whereas during the LGM the lower sea levels would have meant a larger land area.

Figure 4.4 Approximate glacial ice extent over Europe during the LGM. Extending over all of northern Europe, this ice sheet fixed a considerable body of water and was in places over a mile thick. Note that the present-day coastline is portrayed whereas during the LGM the lower sea levels would have meant a larger land area.

Isles (the Fennoscandian sheet; Figure 4.4). Meanwhile, in the southern hemisphere, much of New Zealand, Argentina and Chile sported ice caps, as did mountains in South Africa and southern Australia. With so much of the Earth's water trapped on land as ice sheets, the sea level fell by about 120 m, so exposing much of the continental shelves.

4.5.3 Temperature changes within the glacial

The temperature change in the last glacial in the regions surrounding Antarctica is depicted in detail in Figure 4.2, which is based on a deuterium isotopic analysis (see Chapter 2) of the Vostok core. Note that the current (up to the present) and previous interglacials are short, of the order of 10 000 years or

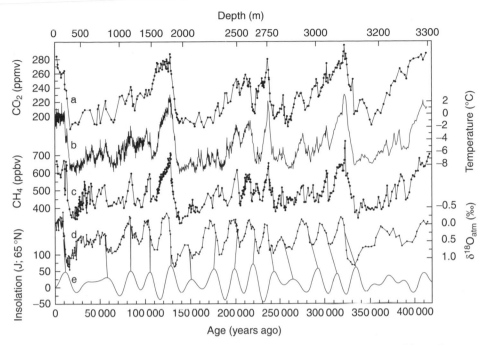

Figure 4.5 Carbon dioxide and methane relate to temperature. (a) carbon dioxide, (b) temperature (as derived from isotopic composition of the ice) and (c) methane. The patterns for all three are broadly similar. (d) Track changes in the ice's ^{18}O isotope. (e) Milankovitch insolation (or solar energy received) as calculated for June at 65 °N. This shows how peaks and troughs in Milankovitch climate forcing act as a pacemaker of glacial–interglacial cycles. Reproduced with permission from Pétit *et al.* (1999).

less, whereas the glacial itself was about 90 000 years in duration. Also note that the beginning and end of the glacial were comparatively sharp, as were some of the less dramatic changes in temperature within the glacial itself. What the glacial–interglacial cycles are not are gently sinusoidal temperature oscillations.

The temperature graph from Figure 4.2 is shown on Figure 4.5 and added to it is the ice-core record of past atmospheric carbon dioxide and methane concentrations covering additional glacial–interglacial cycles. Here it can be seen that carbon dioxide and temperature effectively parallel each other, again further evidence as to the importance of greenhouse gases in determining global climate. Methane too, as we shall shortly see, shows a broadly similar pattern of high concentration during interglacials and low during glacials, with the lowest concentration during the LGM. Atmospheric methane reflects the global extent of wetlands (including, in modern times, rice paddies). As previously mentioned, when the Earth is warmer the greater ocean evaporation leads to greater rainfall which in turn adds to the extent of wetlands. Conversely, when the planet is cooler and drier, there is less methane produced. So methane concentration can also be loosely said to track temperature and indeed, being a powerful greenhouse gas (see Chapter 1), amplify temperature changes.

Importantly the temperature and gaseous changes recorded in the Vostok cores in Antarctica are reflected in their northern-hemisphere counterparts. The ^{18}O isotope records of Vostok and Greenland cores broadly, though not nearly exactly, follow each other. Figure 4.6a shows a low-resolution graph (with 25 000-year filter) of how ^{18}O has changed over the past 90 000 years from the present to early in the last glacial. The LGM transition to the current Holocene interglacial can be clearly seen.

The reason for using a filter on the above data is that a high-resolution analysis of northern- and southern-hemisphere ocean temperatures during the last glacial reveals a time-lagged exchange of heat between the hemispheres. The temperatures around Greenland and Antarctica vary by about 4–6 °C with a period of about 2500 years, each being out of phase with the other (Knutti *et al.*, 2004). Heat is thought to be transferred from one hemisphere to the other through the Broecker thermohaline conveyor (see Figure 4.7; and see below) of surface and submarine currents that connect all the oceans. The result is that there is a time delay during a glacial between when one hemisphere feels the effects of warmth and when the other hemisphere does.

However, were that life was this simple. Whereas there is the aforementioned ocean temperature evidence, there is other evidence suggesting that terrestrial temperatures at mid latitudes within glacials differ differently between hemispheres to the ocean time lag suggested above. Local insolation (the level of solar energy locally received) in mid latitudes of the southern hemisphere is almost completely out of phase with that of high northern latitudes. Mid latitude long-term terrestrial records going back one or more complete glacial–interglacial cycles are rare but one such exists in the form of a 10-m-deep peat bog in Okarito Pakihi outside the limits of ice during the LGM in southern New Zealand (see Figure 4.8). Palaeo-vegetation analysis of bog peat cores has provided an indication of climate change going back two glacial–interglacial cycles. This revealed an earlier onset and longer duration of the LGM. The explanation for this is likely to relate to the Milankovitch-driven levels of insolation at that latitude (Vandergoes *et al.*, 2005).

The atmospheric gas composition of bubbles of northern- and southern-hemisphere ice cores track more exactly, which is not surprising given that global mixing of the planet's atmosphere takes a little over a year or two. Figure 4.6b is a low-resolution portrayal of methane from both hemispheres over the entirety of the past glacial–interglacial cycle and demonstrates synchrony.

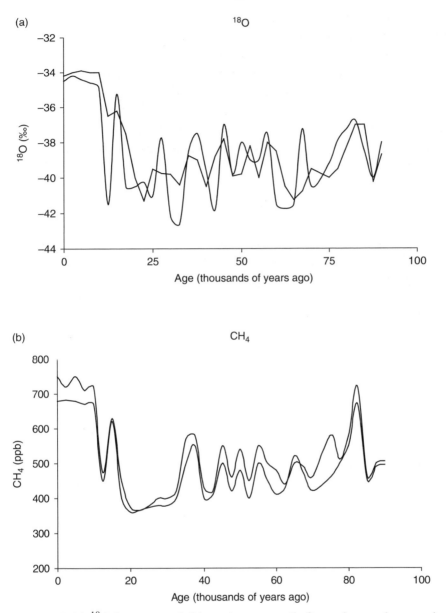

Figure 4.6 (a) ^{18}O isotope and (b) methane records from the northern and southern hemispheres (Greenland and Antarctica) smoothed (with a 5000-year filter).

The cores clearly demonstrate that glacials and interglacials are not just hemispheric events but planetwide in nature. Further, they show that some trends and features within the glacial are also global (even if hemispherically they are not exactly in phase).

Cold, deeper
salty current

Oceanic conveyor belt

Warm, less salty
current at surface

Figure 4.7 A simplified representation of the Broecker conveyor (reproduced from Cowie, 1998).

The core records also demonstrate the broad structure of the glacial and interglacial cycle. For example, the LGM occurred at the end of the glacial as opposed to the middle, which would be likely if glacial–interglacial temperature cycles were uniformly sinusoidal. However, cores have shown that other features that are pronounced in one hemisphere are only weakly reflected in the other. The end-glacial hiccup return to glacial conditions, as the Earth was beginning to emerge into the current interglacial (Holocene), is strongly represented in the Greenland ice cores but only weakly in its Antarctic counterparts. This suggests that some climatic-forcing events were driven by factors located in one hemisphere rather than the other, even though glacials and interglacials overall are planetary in nature.

Looking at ice cores in more detail reveals that the temperature changes are not smooth. There are a number of factors causing this. Two of the main ones are changes in ocean circulation and also, not surprisingly, greenhouse-gas concentrations.

First, as previously noted, ocean circulation is a very important mechanism for transporting heat away from the tropics and so it is worth examining aspects of this in a little more detail. In recent geological times, in the land-dominated northern hemisphere (which loses heat quickly), one of the major

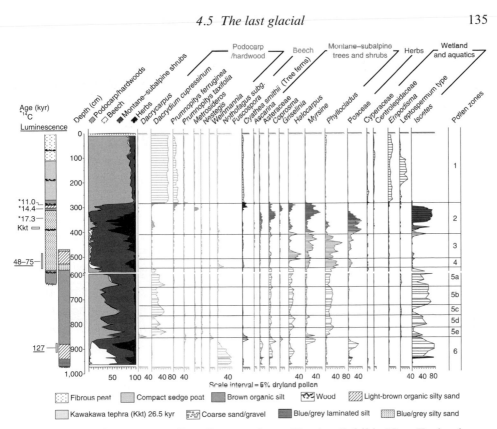

Figure 4.8 Summary pollen diagram from Okarito Pakihi, New Zealand. Note that the timescale gets more condensed the further into the past. Also, the current interglacial (Holocene), LGM and early glacial vegetation types show up clearly. The last interglacial vegetation does not and local insolation was low at that time. Reproduced with permission from Vandergoes *et al.* (2005). Kyr, thousand years.

currents transferring heat from the tropics has been the North Atlantic Drift (or Gulf Stream). This runs from east of the Gulf of Mexico, up the south-east coast of the USA and north east across the Atlantic up to Iceland as well as the British Isles and the western coasts of Scandinavia. In winter months this warms the western edge of northern Europe so that it is 10–15 °C warmer there at latitudes of 40–60°N compared with Russia and much of Canada: the heat transported from the tropics is considerable. On its journey north evaporation takes place so that by the time the water has reached Icelandic latitudes the current is more concentrated in salts than other parts of the ocean. All this time, as well as evaporating, the current is shedding heat (and hence moderating winters in north-west Europe) and cooling. At Icelandic latitudes the North Atlantic Drift is cooler and saltier and so it sinks. But the current does not end there. This action of dense (cool and saltier) water sinking is one of the main drivers of a global current that extends from the surface to beneath the oceans. It forms a global conveyor driven by thermohaline gradients (*thermo* being Greek for heat and *haline* meaning salt.) The conveyor runs either on the surface or beneath the major oceans and importantly connects the North and South Atlantic with the Pacific (see Figure 4.7).

In 1992 Wallace Broecker of Columbia University argued that the global thermohaline conveyor might be disrupted if one or more of its driving components were disabled. One way to disrupt the conveyor would be to dilute the saltier sea water at points where it sank. Adding fresh water at these points could do this. If this were done in the North Atlantic then the water would not sink and indeed block further water coming up from the south. Of course, the North Atlantic gyre would not cease, as Coriolis forces from the Earth's rotation would remain. Instead, the northern part of the North Atlantic Drift would be displaced south and so cease to warm north-western Europe in winter but further warm Spain and North Africa. Broecker suggested that the switching off of the conveyor would result in a sudden cooling of north-western European in winters.

The following year, in 1993, fellow Columbia University researchers, with Gerard Bond and colleagues, presented evidence and a picture of what might have been happening during the last glacial. It had been known since the late 1980s that at several distinct times during the last glacial icebergs had carried moraine (rocks and geological detritus) into the North Atlantic. There the icebergs melted, releasing the moraine; this can be seen in ocean sediment cores. These events were summarised in 1988 by Hartmut Heinrich. What Bond and colleagues did was to tie these Heinrich events into a class of glacial temperature changes within the last glacial (known as Dansgaard–Oeschger cycles) as revealed by Greenland ice cores. They used a palaeoclimatic bioindicator, the abundance of planktic foram *Neogloboquadrina pachyderma*, in North Atlantic sediment cores. The theory that was to emerge was that in the northern hemisphere, as the glacial progressed, the ice sheets grew. As the one over North America (the Laurentide sheet) became thicker it pressed down on the underlying geology as well as providing increasing thermal insulation. Geothermal heat and the ice's weight caused basal melting and the sheet slipped, fracturing to release icebergs into the North Atlantic. This iceberg armada freshened the Atlantic sea water, so shutting down the Broecker conveyor and in turn cooling the North Atlantic. This then facilitated ice-sheet build-up until the process repeated while the thermohaline circulation slowly recovered. There is also evidence to suggest that the presence of a large ice sheet over northern North America affected air circulation but that ice-sheet collapse may have allowed the air circulation to return to a pattern similar to today's, which would have helped the restoration of the conveyor. Meanwhile, with each cycle the ice sheet grew bigger and although its oceanic margins could see iceberg loss, its terrestrial margins would not. Similarly the Fennoscandian ice sheet bordering the sea over northern Europe also discharged icebergs while its terrestrial-bounded edges did not.

Whereas the Heinrich events and associated climate changes are largely instigated by northern-hemisphere factors, only the larger of such events of the last glacial are reflected in the southern hemisphere's Antarctic ice-core record; the weaker ones are not.

Nonetheless, whereas Heinrich events are a major part of northern-hemisphere climatic events and only weakly reflected at the South Pole, they do have a pronounced effect in the tropics in terms of both climate and biology. The Intertropical Convergence Zone (ITCZ) is a band of rainfall that spans the Atlantic. It migrates north–south with the seasons. The effect of the ITCZ in our current interglacial is that it makes a band across Brazil wet and a parallel area to the south dry. This arid zone of eastern South America currently separates two tropical rainforests, the Brazilian and the Atlantic forests. In 2004 researchers, led by Xianfeng Wang, reported on a geological climate record of speleothem and travertine deposits – calcium carbonate carried by water – in a cave. Layers in the deposits were accurately dated using uranium and thorium isotopes and this showed periods of time when there was considerable rain (pluvials) and other periods when it was dry. They found that these pluvials coincided with Heinrich events. During these pluvials the two tropical rainforests, Brazilian and Atlantic, were able to merge. This merging enabled species (and gene) transfer between the two zones (Figure 4.9).

The Broecker thermohaline conveyor is not just important in that it helps explain the drunken walk of global climate change during the last glacial but, as we shall see later (Chapter 6), it is of relevance with present-day concerns about anthropogenic climate change.

The second major factor affecting temperatures within a glacial arises from the concentration of greenhouse gases. As noted methane effectively tracks glacial–interglacial temperature largely due to the global extent of wetlands. These decline as the planet cools and the atmosphere becomes less moist. Atmospheric moisture (water vapour) itself is a greenhouse gas and so a cooler, hence drier, atmosphere holds less moisture and also has a smaller greenhouse forcing of climate. Both atmospheric moisture and the extent of wetlands, hence methane release into the atmosphere, are further examples of positive-feedback phenomena as noted in Chapter 1.

Another source of methane comes from methane hydrates (or methane clathrates as they are sometimes known) that exist in considerable quantities in permafrosts as well as in the seabed on the slopes of the continental margins (see the Eocene climatic maximum section in the previous chapter). Estimates

Figure 4.9 The Intertropical Convergence Zone (ITCZ) and key areas of South American tropical forest. Wet periods coincident with Heinrich events enabled the Amazon and Atlantic forests to periodically join up in the depth of glacials. Reproduced with permission from Wang *et al.* (2004).

of the total quantity of methane so stored vary mainly between 1×10^{19} and 2×10^{19} g. This is some 4000 times the amount in the atmosphere today. Because methane hydrates have a smaller proportion of the ^{13}C isotope compared to atmospheric carbon dioxide – you will recall that the enzymes driving photosynthesis prefer ^{12}C to ^{13}C – they are biogenic in origin. As noted in the last chapter when discussing the end-Permian extinction, the physico-chemical properties of methane hydrates are that they are meta-stable and will easily dissociate into hydrogen and water with an increase in temperature or a decrease in pressure.

In 2000 James Kennett and his team, from the Universities of California and California State, discovered oscillation in carbon isotopes in forams that could be explained by methane releases from hydrates. The Kennett team discovered four major releases including one just before the beginning of the Holocene interglacial, and these took place at the same time as four major Heinrich events at the beginning of Dansgaard–Oeschger cycles.

Guy Rothwell of Britain's Southampton Oceanography Centre in 1998 suggested the biggest climate change over the last glacial–interglacial cycle possibly attributable to a methane release from marine methane hydrates. Seismic studies of the Balearic abyssal plain to the west of Sardinia reveal an extensive ancient submarine landslide three times the size of Wales involving 500 km^3 of mud. Rothwell's team radiocarbon-dated sediments above and below the landslide to get an estimated age of 22 000 years. This date ties in with the LGM. Could it be that the landslide was the result of a massive methane release well in excess of half a billion tonnes? This would have represented a very significant input of gas to the Earth's atmosphere at a time when it contained just 0.8–1.0 billion t. Methane is far more powerful a greenhouse gas, molecule for molecule, than carbon dioxide, but lasts for less time in the atmosphere (see Chapter 1). Even so, such a methane release could easily have given the Earth's climate system a jolt. A decade or so of warming would trigger other changes. For example, continuing with methane, warming could have released more of the gas (and other forms of carbon) from permafrosts, and there would be implications for atmospheric and ocean circulation. What is known for certain is that there was a short period of warming about this time, in the depth of the last glacial, even though temperatures did not return to anywhere near interglacial levels.

But what triggered the massive landslide? It could have been coincidence as such landslides happen by chance. While chance is certainly a factor in marine landslides, in that they are fairly common, what made such a large area collapse all at once? The suggestion is that as the glacial progressed and ice caps grew, sea level dropped. As the sea level dropped the pressure of water stabilising the meta-stable hydrates reduced. Hydrates in the region risked dissociating and all it took was one trigger event (an initial slide) to trigger others. Of course, sea levels did not just decline in the Mediterranean, so if this climate-change mechanism has validity one would expect other major slides worldwide but perhaps spread out over a few centuries. Interestingly, Antarctica and Greenland ice cores reveal that there was a major methane overshoot prior to the beginning of the previous interglacial and this might signify a number of major landslides occurring more or less simultaneously, geologically speaking (say within a century or two).

Before we leave methane it is worth mentioning here that the Intergovernmental Panel on Climate Change (IPCC) has warned policy-makers to be wary of surprises and not to rely solely on climatic forecasts (we shall return to this in Chapter 6). However, one such surprise we may encounter as the Earth continues to warm might be a major release of methane from marine hydrates as temperature destabilises them. Of course, since in the twentieth century sea levels globally have been rising marine methane hydrates are under more pressure rather than less and this will confer greater stability. Yet if the temperature of the water around the hydrates were to increase by about 5 °C then there may be a massive release of methane, as was likely back in the Eocene. One recent estimate of the methane that could be released this way globally was provided by a geological team lead by Matthew Hornbach (2004). They suggest that a release of some 2000 Gt of methane is possible. To place this in the context of present-day anthropogenic warming from carbon dioxide releases, this is roughly three times the mass of carbon as was released as carbon dioxide in the final decade of the twentieth century. Furthermore, from Chapter 1, bearing in mind methane's higher global-warming potential (GWP) then the warming possible over two decades of such methane release would be broadly equal to nearly 2000 years of late-twentieth-century carbon dioxide from fossil-fuel burning. However, methane in the atmosphere over several years gets oxidised to carbon dioxide, which is a comparatively weaker greenhouse gas but which stays in the atmosphere for a century or two. Fortunately, a methane-hydrate-destabilising 5 °C rise in abyssal waters is highly unlikely this century (although we should heed the IPCC's warning of surprises and other destabilising mechanisms). Nonetheless, such a risk becomes increasingly relevant should anthropogenic warming and fossil-fuel consumption continue to grow and especially if it does into the next century.

It is also worth emphasising that wetlands are not the only terrestrial methane source. Peatlands represent a huge store of carbon. High-latitude peatlands, with between 180 and 455 Gt of carbon, represent up to about a third of the global soil carbon pool. Most have been formed since the last glacial maximum (LGM) and so represent a major terrestrial carbon sink with 70 Gt of carbon being sequestered since that time. Much of this initial peatland formation took place in Siberia's western lowland before their counterparts in North America: all told, Russia's peatlands contain perhaps half of the planet's peat. Peatlands are, in interglacial times, an atmospheric methane source but much is still unknown as to how exactly they fit into the dynamic carbon cycle over glacial–interglacial cycles (Smith *et al.*, 2004). Indeed another carbon-cycle unknown is the degree to which these high-latitude peatlands and permafrost soils might release their tremendous store of carbon if warmed (see Chapters 1 and 7). These are just two more reasons why climate computer models are not complete.

Another reason given for the suddenness of the warming after the LGM is that the Broecker thermohaline conveyor played a significant part. But was the change in the planet's surface and deep-ocean circulation initiated in the northern hemisphere? There has been considerable attention given to the circulation's North Atlantic pump, but there other possibilities. One, according to a Canadian team led by Andrew Weaver (2003), is that a pulse of comparatively warm water could have originated from Antarctica that would have warmed due to Milankovitch and methane factors. Had a southern hemisphere event affected the global conveyor then – because the conveyor is global – this would have affected matters in the northern hemisphere. The jury is still out, but we should not dismiss the possibility that factors (note the plural) in both hemispheres played their part.

4.5.4 Biological and environmental impacts of the last glacial

The depth of the last glacial saw increased aridity over much of the planet. With ice caps extended and the sea level lowered, there was a considerable migration of species. The former induced migration to lower latitudes and the latter facilitated migration by exposing land bridges. Because the Pleistocene has seen a number of glacial–interglacial cycles, flora and fauna have had evolutionary selection pressures favouring those that are capable of migrating from one place to another, often in suboptimal conditions. Many species (but not all) that survived from one climatic cycle to another did so in ecological refuges. Such refuges might be latitudinal, that is, closer or further away from the equator, or alternatively altitudinal in mountain ranges with vertical migration upwards in response to warm interglacials and downwards in cooler glacial times.

Genetic analyses provide insights into the dynamic ecology of refugia over time. The southern part of South America provides a good illustration of how such insights can be gained. There the Andean mountain chain, separating Chile and Argentina, was glaciated during the last glacial maximum while the lowlands on either side were ice-free. Pollen analysis reveals that there were trees there throughout the LGM, suggesting that these lowlands were refugia for temperate forest species. The question then arises as to whether there was just one refuge during the LGM on one side of the Andes or two distinct refuges, one on each side. If the former then an analysis would reveal that the two populations were genetically similar. Conversely, if there were two separate refuges then populations would exhibit significant genetic differences. In 2000 Andrea Premoli from Patagonia and researchers from the University of Colorado looked at a tree species, the Patagonian cypress (*Fitzroya cupressoides*). Today it can be found substantially spread out in lowland Chile but only in scattered populations on the Argentinean side of the Andes. They examined 11 enzyme systems associated with 21 genetic loci in this species and found that spatially distinct populations on both sides of the Andes did have distinct genetic differences, indicating that there were two separate refugia.

Were life as simple as the above suggests. It would be facile to conclude that just because one species reached out from, and retreated to, two refugia that temperate forest species all behaved in a similar way. *Fitzroya* is a species that exhibits considerable longevity, with individuals living in excess of 3000 years, but it has poor regenerative powers. This indicates that it is not best suited to migrating. However, other temperate forest species are and they may be able to relocate further afield. In short, component species of the same ecosystem do not necessarily respond uniformly to climate change; indeed, most do not.

There are many examples of genetic analysis illuminating how a population outrode the last glacial. One European study, by the French botanist Rémy Pétit, examined the chloroplast DNA of 22 widespread species of European trees and shrubs (all woody angiosperm species), from the same 25 European forests. The advantage of using chloroplast DNA is that it accompanies seeds only (not pollen, which can travel considerably further). The forests were located in Sweden, Scotland, Germany, England, Austria, Hungary, Italy, France, Greece, Slovakia, Croatia, Romania and Spain. During the last glacial European forests were considerably more restricted than those in North and South America, or for that matter South Africa, due to the Mediterranean Sea forming a barrier to latitude migration. Nonetheless, one might expect the greatest genetic diversity to be found in those trees and shrubs in the south, as the glacial refugia would be more likely to be located there. The reality is more complicated. True, the southern European latitudes were important for providing refugia during the LGM and the Iberian, Italian and Balkan peninsulas were all important locations. But pollen and fossil analysis also indicates that several tree species were localised in small, favourable spots at the southern edge of the then cold, dry steppe-tundra in eastern, central and south-western Europe. With post-glacial warming some of the surviving populations expanded (horizontally), some migrated altitudinally (vertically up mountains) and some unable to migrate in time became extinct. One would expect that those populations that had been able to migrate and persist would show the greatest genetic diversity. One might also anticipate a gradual decrease in diversity away from source populations. Again the reality was more complex.

Petit and colleagues (2003) analysed their data in two ways. First, the genetic diversity of species within populations. Second, the genetic divergence of populations in forests from all remaining populations in other forests. This last is an indication of the genetic melting pot as to how the separate populations are related to each other.

It transpires that for the most part there is a greater genetic diversity in southern forests which suggests that their population persisted within the region throughout the last glacial. There was, though, one exception and that was in the Iberian peninsula, whose populations exhibited low genetic diversity. This may well be because that peninsula seems to have been exposed to particularly severe climatic episodes of cold and aridity, and because the Pyrenees formed a particularly formidable barrier. (The lower LGM sea level did not provide the extensive land bridges there, as there were between the British Isles and mainland Europe or Italy and the Balkan peninsula.)

The second thing discovered was that the genetic divergence of populations from all remaining populations was highest not in the low-latitude refugia but across France, Germany, Austria and Hungary.

What appears to have happened is that species were constrained to southern refugia during the glacial, as well as a few fortunate higher-latitude refugia. However, after the glacial each refuge made its own contribution to the genetic diversity of the current interglacial higher-latitude forests on land in between the refugia. Interglacial populations of trees close to glacial refugia retained the genetic make-up of the refuge populations. Populations further away and equidistant from refugia tended to share source-refugia population genes. However, with the rapid transition from glacial to interglacial conditions and back it was the populations of trees that were closest to refugia that were able to migrate back while the mixed-gene populations furthest away tended to die out. Consequently, the refuge populations tended to retain their genetic identity.

Of course, as with any study one has to be aware of limitations and this analysis only looked at diversity within species (intra specific diversity) and not inter specific biodiversity. Furthermore, whereas this scenario applies in this circumstance and with these species, it does not necessarily apply to other species or circumstances, such as with the refuge (or refugia) being further away.

The above examples of refugia are all terrestrial but they can be aquatic too. Furthermore, the above examples using genetic analysis tend to suggest that refugia, where species ride out times of climatic change, only have a minimal changing effect on species populations within the refugia. However, as with climate change per se, the comings and goings from refugia can result in significant evolution. In 2005 results of genetic, morphological (shape) and geographic analyses by a European and South African team of researchers showed how climate change in association with a transitional refuge had a lasting effect on biodiversity on a near-continental scale. In this case, which we look at shortly, the example is aquatic (Joyce *et al.*, 2005).

Of course, in one sense all refugia are transitional in that at a certain latitude an area of land mass may at one time favour a species, enabling it to survive, while at another be inhospitable to it. In both hospitable and inhospitable times the land area (be it a mountain or whatever) physically remains but in inhospitable times the refuge literally disappears. An example is the Lake palaeo-Makgadikgadi that existed in the lowest part of the Kalahari Plain (present-day northern Botswana) during the last glacial which was for part of the mid-southern African watershed a pluvial (a time of rain). During this time the lake was large and comparable in size to present-day Lake Victoria further north in eastern Africa. Today little remains other than a mix of salt pan and swamp: primarily the Makarikari Salt Pan and Okavango Swamp.

The importance of Lake palaeo-Makgadikgadi was discovered because of the rich biodiversity of cichlid fish species in sub-watersheds 800 km or so apart that would have once fed into this lake. Conversely the rivers associated with Lake Victoria, and the rivers flowing into other large African lakes, have only two widespread riverine lineages of haplochromine cichlids. This contrasts with the lakes themselves, which have a high cichlid biodiversity. The reason for this, it is supposed, is that these large lakes provide many opportunities for speciation whereas the rivers do not. So why are the rivers that flow into the Kalahari Plain species rich?

The analyses carried out suggest that Lake palaeo-Makgadikgadi when it existed did provide many opportunities for speciation. This biodiversity was physically forced to move into the various surrounding river systems as the lake gradually dried out over the period from when the Holocene began to about 2000 years ago. Had the lake dried out extremely quickly instead, then this movement of populations would have been unlikely to take place. Indeed, had the lake been very small then this might well have happened. The palaeo-lake's large size also enabled there to be a variety of ecological niches that facilitated speciation. Finally, it is likely that this speciation happened over more than one glacial–interglacial cycle.

Although important, not all species rely on refugia and with some species the entire population migrates as the climate changes. So the biological response to climate change is complex and we will return to this later when considering the response to possible future climatic change.

Biological change did not just occur at the time of glacial-to-interglacial transition, it also took place with climatic change within glacials themselves. For example, the aforementioned Italian palaeorecords show that the Lago Grande di Monticchio region saw switches between three broad biome types. Oak (*Quercus* spp.) and elm (*Ulmus* spp.) temperate woodland dominated both the current and previous Eemian (or Sangamon as it is sometimes known) interglacial. Conversely, pine (*Pinus* spp.) and juniper (*Juniperis* spp.) dominated the LGM and other cooler parts of the last glacial, whereas warmer interstadial climatic events during the glacial saw wooded steppe dominate.

The genetic consequences of all these biome transitions both within and between glacial and interglacial conditions were profound on many plant species. For tropical forests, they were not only reduced during glacials but their present distribution areas were dissected. Then again aridity increased in glacials and so there was the expansion of desert areas to five times their present extent by the LGM. Also, as mentioned, in present-day Europe and North America, high-latitude boreal species moved south ahead of the

European (Fennoscandian) and North American (Laurentide) ice caps as they expanded. There was a considerable dynamic as ecological communities reorganised in terms of composition, latitude and altitude.

Typically, when a plant species migrates it does so through seed dispersal. The range seeds travel from a source plant is largely determined by probability with a curved distribution: the majority of seeds travel a short distance and a few travel longer distances. But largely is not the same as always. Circumstances, such as a storm or a seed getting stuck in an animal's fur, can see seeds transported considerable distances beyond the norm. So, if one were to plot a graph of the number of seeds against distance one would not only see a probability curve but also a small leptokurtic kink representing a small distribution further afield. (This leptokurtic phenomenon is the same as those bedevilling concerns over the separation of genetically modified crops from their non-modified counterparts, as well as concerns over invading species. For while one may plant a genetically modified crop so far apart from its non-genetically modified counterpart as to be assured that 99.9% of pollen from one crop will not reach the other, there is still a small chance of contamination.)

That a small proportion of seeds will travel exceptionally long distances has profound genetic consequences during climate-change events. Plants migrating out of refugia due to changing conditions will be moving into lands that up to the change had been occupied by other species, many of which will now be ill-suited to the new conditions. So those few seeds travelling the farthest first grow up with little competition and so subsequent generations can grow with comparative impunity. Northern-hemisphere interglacial populations at the edge of their range would, with glacials, experience southern migration into large areas of suitable territory with leaders establishing new colonies long before the rest can arrive. Indeed, those behind now have to compete to migrate. Since only a small number of plants make up this leading wave, the genetic diversity of the new colonies covering a large area is reduced compared with those found in the refugia, which cover smaller areas. Indeed, this also happens with some animal species, for example, as illustrated by an analysis of 41 species of North American fish (Bernatchez and Wilson, 1998).

In addition to expanding populations having restricted genetic diversity, so a retreating population will suffer dissection and extinctions so that the last survivors at the edge of the population's range will also tend to have restricted genetic diversity. And it is these individuals that will lead the charge with the next climatic cycle (Hewitt, 1996).

With migration due to climate change, animals and plants will inevitably encounter some environmental parameters different to those found in their refuge. For example, changing latitude, so as to broadly maintain a constant thermal regimen, also sees a change in day length with season. This is not entirely insignificant. The reproduction of many plants is determined by temperature and day length, as are the migration dates of some birds, and day length even affects the reproductive cycle of some insects. So while genetic diversity is restricted, there is some genetic variation, which is necessary for survival.

It is also important to note that whereas one place may be a climatic refuge for some species, it may not be for another. A consequence of this is that a single region in an interglacial may receive different species from different refugia. For example, at the end of the last glacial Britain received oaks (*Quercus* spp.), shrews (*Sorex* spp.), hedgehogs (*Erinaceus* spp.) and bears (*Ursus* spp.) from Spain, but grasshoppers (*Chorthippus* spp.), alder (*Alnus* spp.), beech (*Fagus* spp.) and newts (*Triturus* spp.) from the Balkans. There are also co-evolutionary implications if a population is to spend some of its time with one community and other times with a quite different community in terms of genetic make-up, if not species make-up.

Matters in Europe and North America differed. In Europe the Alps, Pyrenees and Carpathians provided latitudinal barriers, as indeed did the Mediterranean. In North America the mountain ranges run more north–south and the area of uninterrupted land is more extensive than in Europe. The species and the genetic diversity of the south-eastern USA suggest that the area was a refuge for many species during glacials (Hewitt, 2000).

However, Europe and North America were not totally isolated. As previously mentioned (Chapter 3), during glacials the sea level lowered and there was a Beringial land bridge between present-day Russia and Alaska. Circumpolar species saw repeated expansions and dissections with the glacial–interglacial cycles. This repeated fragmentation has allowed speciation to produce subspecies and sometimes new species. Animals so affected include guillemots (*Uria* spp.), dunlins (*Calidris* spp.), reindeer (*Rangifer* spp.) and bears (*Ursus* spp.; Hewitt, 2000).

Another instance of such population change and fragmentation was provided late in 2004 by Beth Shapiro from Oxford University, who was the lead author of a paper arising out of an international collaboration of more than 15 institutions. The team looked at bison (*Bison* spp.) and sequenced mitochondrial DNA from some 450 individuals from the present to more than 60 000 years ago. By ascertaining the geographical distribution of genes as well as which genes survived to the present, and when others died out, they were able to deduce how the bison population changed. They showed that the bison population across Beringia declined catastrophically from 42 000 to

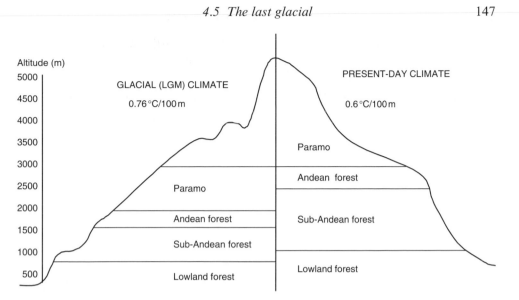

Figure 4.10 Altitudinal distribution of main vegetation belts across a Colombian Andean mountain. After Hooghiemstra and Van der Hammen (2004).

35 000 years ago. This was the time in the run-up to the LGM. Unlike with mammoths, mentioned above, this was before humans had established a significant presence in the area. Given that this was a time of lowering sea levels (hence an expanding Beringian land area) the most likely explanation is that the collapse was climatic. The research also revealed the fragmentation that the bison population underwent during the LGM. They found that today's North American bison (*Bison bison*) was separated from the rest of the bison population (including those that ultimately became the European bison; *Bison bonasus*). The North American bison arose from a subpopulation separated from the rest by the Laurentide ice sheet. These subpopulations survived in ecological refuges to the south of the major northern-hemisphere ice sheets.

As mentioned, mountains can also provide refugia vertically away from areas that have become inhospitable. Some tropical mountains especially have been ecologically important in providing stable refugia where there is vertical migration with climatic change but, importantly, comparatively little horizontal migration. Species can move up and down the mountain as the vertical ecological zonation changes with climate change (see Figure 4.10). Having said this, do not assume that the vegetation zones all change uniformly. Often there are fundamental differences in altitudinal ecological zonation between glacial and interglacial times. For example, a study of a Colombian Andean mountain (Hooghiemstra and Van der Hammen, 2004) has shown that at the LGM the rate of temperature decrease with altitude was 0.76 °C/100 m compared with 0.6 °C/100 m today. This is due to the drier conditions at the LGM, (cool air holds less water than warm air, which is a different, but related, phenomenon to cooler seas evaporating less; the moisture content of the air affects its heat-carrying properties). Furthermore the sub-Andean forest occupied a narrower band that may be due to the lower altitudinal limit of glacial night frost, again due to the drier conditions. The message here is that proportional and linear relationships of climate and ecological parameters between different climate modes cannot be assumed.

Nonetheless, some tropical mountains, according to DNA divergence, have provided a home for populations for over 6 million years. This is a similar duration to that associated with some regions of African and Amazonian forest that were refugia for many species. These tropical mountain and lowland refugia are so stable that they also exhibit great genetic diversity that in some instances facilitates speciation itself (Hewitt, 2000). Because of the short migration distances (short vertical rather than long horizontal) and because mountain ranges sport so many environmental niches, tropical mountains may *partially* explain greater species richness in the tropics.

4.6 Interglacials and the present climate

4.6.1 Previous interglacials

We are currently in an interglacial, the Holocene interglacial, but there have been several other Quaternary interglacials. The last one before ours was the Eemian (sometimes known as the Sangamon) interglacial some 125 000–115 000 years ago. It began just like our current interglacial with a geologically short period (around a century or two) of rapid warming. Indeed, the warming may well have been faster due to changes in ocean circulation but tempered due to the massive ice sheets that took time to melt following the previous glacial's maximum. What is known is that the temperature early on in the Eemian interglacial was a degree or so warmer than at present in our own Holocene interglacial. This lasted for a few centuries before cooling to present-day temperatures. The warming gave a boost to sea levels during the Eemian, which were at least 3 m, but possibly 5–8 m, higher than today. Regional crust movement in northern North America and northern Europe, due to the weight of the glacial Laurentide and Fennoscandian ice sheets, lowered the land in those areas and these only slowly recovered once the ice melted: indeed, following the LGM these regions are still slowly rising today. This vertical crust movement on top of the mantle is known as isostasy, and the resulting apparent change in sea level is known as 'isostatic' change. It complicates attempts to identify past sea levels. (As we discussed earlier and will develop in Chapter 6, sea levels can also change due to terrestrial ice formation and melting as well as thermal expansion.)

One important question is where did this extra water early in the last interglacial come from? The reason why this question is so important is because if we are currently warming the planet, then we might see sea levels similar to those back in the Eemian. If we knew where the extra sea water came from then we would know where to monitor matters today. At first the preferred theory was that the melting of the west Antarctic ice sheet contributed much of this rise. However, recent computer modelling tied in with the ice-core records suggest that Greenland may have made a significant, if not a major, contribution (Cuffey and Marshall, 2000).

Previous interglacials were often considered in the 1990s to be a palaeo-analogue of our current interglacial. However, there are notable differences between the various glacial and interglacial cycles, both as predicted from Milankovitch theory and seen practically from the geological record. Fortunately the latter tend to confirm the former. Over the years this understanding has grown as analyses of the geological record have extended further back from one interglacial cycle in the 1970s through to eight in 2005 from the European Project for Ice Coring in Antarctica (EPICA, 2005) and associated analyses (Siegenthaler *et al.*, 2005; Spahni *et al.*, 2005). Glacial–interglacial cycles, we now know, are not exactly the same, and other evidence corroborates this.

Even before the 2005 EPICA cores we knew that not all interglacials were the same, especially the current Holocene and previous Eemian interglacials. First, the present Holocene interglacial appears to be more climatically stable; although climatic stability during the previous Eemian interglacial has been found in France (Rioul *et al.*, 2001) and Greece (Frogley *et al.*, 1999) according to a number of biological palaeoclimatic indicators, but less so elsewhere. Second, ocean circulation during the Eemian appears to be different (Hillaire-Marcel *et al.*, 2001), which may explain the stability contradiction. Third, the early Eemian saw a period of warmth significantly warmer than the early Holocene. Finally, Milankovitch theory suggests that there was greater seasonality in the Eemian, with warmer summers and cooler winters and there is evidence for this, such as from coral oxygen-isotope climatic indicators from the Middle East (Fells *et al.*, 2004). One of the main reasons for the differences between the Eemian and Holocene interglacials is that although the Milankovitch curve insolation climatically forces interglacials, its various contributing orbital factors can combine in a number of ways.

Figure 4.11 The regional temperature of the present Holocene interglacial (broken line) superimposed on the one following Termination V (solid line) as represented by the change in Antarctic ice-core ²H. The top horizontal timescale represents that of the current interglacial to the present (0) and the bottom timescale relates to the interglacial following Termination V. The positioning of the superimposition is arbitrary but an alignment has been made between the respective glacial terminations. If our interglacial and that of the one following Termination V can be considered similar (in Milankovitch terms) then the superimposition suggests that we are roughly half way through our current interglacial from EPICA (2004). Reproduced with permission.

Therefore, early in the twenty-first century when the EPICA ice-core data from previous cycles became available the fourth interglacial before ours – beginning 420 000 years ago – was preferred for comparison with our Holocene (EPICA, 2004; Figure 4.11). This is because of the aforementioned complexity of the Milankovitch curve that incorporates three orbital variables (see Chapter 1). At times (allowing for climatic-system inertia) of low Milankovitch forcing we see a glacial, and conversely with high forcing we see an interglacial. There are a number of ways in which the three orbital components to Milankovitch can come together to form these peaks and troughs. Unlike the previous Eemian interglacial, the fourth past interglacial (following what geologists call Termination V) was borne out of Milankovitch circumstances similar to those of our current Holocene interglacial. That the Milankovitch circumstances were different for the more recent Eemian compared to our Holocene is the likely explanation for the differences in ocean circulation (the continental tectonic arrangement in the last interglacial being barely different from today's). The Milankovitch complexities in turn possibly explain why high-latitude palaeorecords indicate a different temperature profile for the Eemian interglacial compared with the present Holocene, while European palaeorecords suggest a similar climatic stability.

Importantly, the previous (Eemian) interglacial was short, of the order of 10 000 years or so. Were our Holocene interglacial truly similar, then this would suggest (in the absence of present-day anthropogenic global warming) that we are due to enter another glacial in the geologically very near future, as the Holocene is already 11 500 years old. (Indeed, this worried climatologists in the 1970s.) However, it transpires that the fourth interglacial ago was longer, lasting at about 25 000 years. It also differed in its structure compared to the Eemian. The Eemian began with an extremely warm period of a thousand years, a time far warmer than our Holocene has been to date. Our Holocene's initial millennium was also warmer but only marginally compared to the rest of the interglacial. The fourth (Termination V) interglacial ago similarly lacked the Eemian's notably warmer initial spell and in that sense was similar to the Holocene. However, whereas our Holocene is comparatively stable to within a degree or so, the temperature of the fourth interglacial ago seems to have risen to a peak and then fallen more gently. Superimposing Antarctic deuterium data from both interglacials, one over the other one, shows that the deuterium figures between the Holocene and the 4th interglacial ago (see Figure 4.11) are closer than that between the Eemian and the Holocene, but still are not the same. The way temperature changes within the current interglacial is different from how it changed in the fourth-past interglacial. Could this be due to other factors such as the gradual evolution of carbon reservoirs over successive glacial–interglacial cycles? Or perhaps solar forcing, changes in the Sun's output, affected the climate differently between the two interglacials? After all, approximately 10% of changes in positive radiative forcing of our current climate since 1850 are estimated as being due to changes in the Sun's energy output, although there is considerable uncertainty associated with this (IPCC, 2001).

If our Holocene interglacial is, in Milankovitch terms, similar to the one 420 000 years ago following Termination V, then it could be 25 000 years long. This would mean that we are only half way through our present interglacial. This contrasts with palaeoclimatogists' views back in 1972, when comparisons were made with the Eemian and they thought that all interglacials were as short, suggesting that the next glacial was imminent (Kukla *et al.*, 1972). However, more recent analyses (Berger and Loutre, 2002) of future Milankovitch curves and climate models suggest that we are indeed at least only half way through our current interglacial, just as the latest ice-core evidence suggests. Furthermore, continuing current carbon dioxide emissions throughout this century could conceivably mean extending our current interglacial possibly as much as another 50 000 years, but almost certainly another 30 000 years (assuming depletion over the coming centuries of all economically recoverable fossil fuels without any carbon abatement). Either way, the next glacial is not now thought to be imminent.

Why the temperature curves of the Milankovitch-similar Holocene and Termination V interglacial are different is a good question. The aforementioned reasons are all likely in that the cause lies in the other variables in the system. But there are other possibilities too. For example, the Earth's plate-tectonic arrangement, while only slightly different in the last interglacial, was more different still four glacial–interglacial cycles ago. For instance, the Himalayas were lower 420 000 years ago and this would have affected albedo and atmospheric circulation. Even so, these differences over hundreds of thousands of years are paltry compared to the difference in plate tectonics over tens of millions of years mentioned in Chapter 3. Another difference between the Termination V interglacial and ours is that our interglacial is four more along the series of Quaternary interglacials that have seen a broad trend of deepening intervening glacials. So, the Termination V interglacial may not be analogous in all ways to our Holocene, but both Milankovitch theory and ice-core evidence suggests that their durations may be similar.

4.6.2 *The Allerød, Bølling and Younger Dryas (14 600–11 600 years ago)*

The end of the last glacial in the northern hemisphere was marked by a series of major events of sufficient strength to be reflected in the southern hemisphere, albeit more weakly. This is reflected in Figure 4.12, which portrays ice-core records of the past 15 000 years. Figure 4.12a is a detail of the deuterium record from Vostok, Antarctica, which reflects regional temperature change. Figure 4.12b is a detail of the northern hemisphere's Greenland ice core showing its electrical conductivity measurements (ECMs). Here conductivity of ice is improved with acid ions and decreases when these acids are neutralised by dust. Ammonia – such as from nearby biomass burning (water-soluble ammonia is easily absorbed by water vapour and then washed out of the air) – will also neutralise acid, but the ammonia

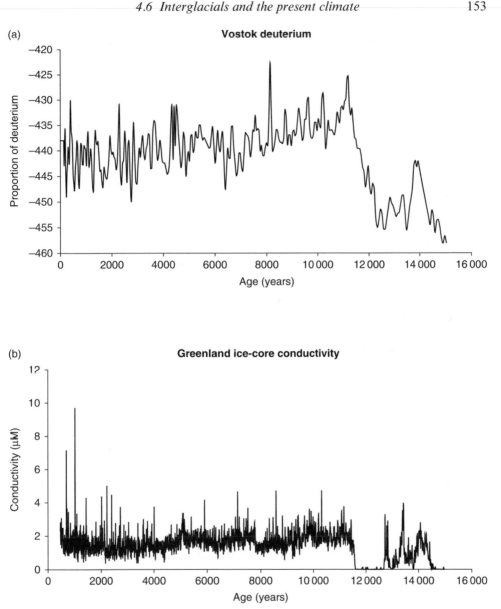

Figure 4.12 (a) Vostok deuterium record over the past 15 000 years, reflecting Antarctic regional temperature change. Data from Pétit *et al.* (2001), *Vostok Ice Core Data for 420,000 Years*, IGBP PAGES/World Data Center for Paleoclimatology Data Contribution Series #2001-076. NOAA/NGDC Paleoclimatology Program, Boulder, CO, USA. (b) Greenland ice-core electrical conductivity over the past 15 000 years reflecting North Atlantic Arctic precipitation, and hence indirectly temperature. Data source: IGBP PAGES/World Data Center for Palaeoclimatology; as used in Taylor *et al.* (1993).

record in Greenland's cores (fairly distant from biomass burning) is reasonably constant from 11 000 to 15 000 years ago so the ECM record reflects acid-neutralising calcium dust, which comes from the erosion of calcium carbonate rocks such as limestone or chalk. However, there is another factor: the presence of dust reflects the amount of precipitation. The more rain, the more acid-neutralising dust is washed out of the atmosphere. Dust is less sensitive than ammonia to washing out of the air and so, in the absence of nearby extensive vegetation, dust is a good indicator of regional precipitation as water vapour will tend not to take it out of the air, although rain will. Put the aforementioned data all together and the Greenland ice-core ECM record reflects regional precipitation. Furthermore, this is one more example of the climatic relevance of dust.

As we have already frequently noted, the warmer the planet the more ocean evaporation hence rain: increased temperature and rain go together. Consequently dust, which would only be present when it is dry, is in turn a secondary indicator of hemispheric temperature (Taylor *et al.*, 1993). More exactly, as temperature is not being directly reflected, the level of dust indicates whether the Earth is in a glacial (low rainfall/high dust) as opposed to an interglacial (high rainfall/low dust) mode. Finally, from Chapter 2 we know that deuterium is an indicator of regional climate. Given this, it possible to see how the panels of Figure 4.12 relate to climate.

Some 15 000 years ago the Earth was warming after the LGM. However, this warming was not permanent and cooler conditions returned, mainly in the northern hemisphere. The temporary warm period itself was comprised of two shorter warm periods, or interstadials, called the Allerød and Bølling, which in turn were followed by a cool period, the Younger Dryas, from 12 850 to 11 650 years ago before there was a warming again to the present Holocene interglacial. Figure 4.12a is an enlargement of part of Figure 4.2 from an Antarctic Vostok ice core and last shows that the Younger Dryas was not so strongly reflected in the southern hemisphere: in other words, the Younger Dryas was predominantly a northern-hemisphere event. Looking closely at Figures 4.12a and b it can be seen that the Antarctic cooling in fact began roughly 1000 years before the Younger Dryas in the northern hemisphere. So, though the two events are linked, they are different and the cooling in Antarctica is known as the Antarctic Cold Reversal. There is other evidence that the northern hemisphere's Younger Dryas cooling had less impact south of the equator, such as in the pollen-grain record from lake sediments in southern Chile (Bennett *et al.*, 2000). However, not all the evidence is so minimal and little impact compared to the northern hemisphere is not the same as no impact. Other pollen evidence from Chilean lakes

(Moreno *et al.*, 2001) demonstrates that there was enough alteration of the climate for some significant ecosystem change in the southern hemisphere. Coral evidence from the south-western Pacific suggests a sea-surface cooling somewhere between 3.2 and 5.8 °C. Furthermore, the Younger Dryas cool period was of sufficient strength for both Swiss and New Zealand alpine snowlines to drop to about 300 m below the lowest levels reached in the subsequent Holocene. In short, there was linkage of *effects* to the southern hemisphere from the northern, but was there hemispheric linkage regarding the Younger Dryas' *cause*?

So what did trigger the Younger Dryas? It has been suggested that the early melting of the North American Laurentide ice sheet added fresh water to the Gulf Stream in the Gulf of Mexico as the presence of the remaining ice sheet to the north prevented the meltwater discharging into the North Atlantic. The latter would, as we shall see, most likely disrupt the Broecker thermohaline circulation. Conversely, the addition of fresh water far further south in the Gulf of Mexico could have enhanced the Broecker circulation and so helped the transportation of heat from the tropic via the Gulf Stream. This could perhaps have been a significant additional factor in pulling the planet so rapidly out of the last glacial. The consensus today is that the thermohaline circulation was disrupted, so causing the cooling of the Younger Dryas. Although there is still debate as to exactly where the fresh water came from that did the disrupting, geological evidence clearly shows that there was flooding south of the Laurentide ice sheet. However, there is some computer-model evidence to suggest that about half of the northern-hemisphere meltwater came from the Laurentide sheet and that here the largest pulse (and even if the models are wrong, certainly a significant one) came, not via the Mississippi or Hudson routes, but via the Fram Straits off the east of northern Greenland (Tarasov and Peltier, 2005).

But are changes in the North Atlantic circulation the only possible causes for the end of the short Younger Dryas cool period? Could a marine methane release, such as happened in the depth of the last glacial, be a cause? The answer is probably not. Terrestrial and marine biological sources of methane both have the same amount of hydrogen (four atoms per methane molecule) but some of these hydrogen atoms are in the form of the deuterium isotope of hydrogen. Here the amount of deuterium from marine and terrestrial sources of methane markedly differs: marine methane has more deuterium. Analysis of the $^2H/^1H$ ratio of air trapped in the Greenland ice sheet suggests that the marine methane clathrates were stable during these particular warming periods and did not contribute to this episode of climate change (Sowers, 2006). (This may perhaps be because methane releases from clathrates earlier in the glacial took place when the sea level (hence stabilising pressure on the clathrates) was lower. At the time of the Younger Dryas sea levels were already rising and so even though the planet was warming clathrate stability would have benefited from the extra pressure of water.)

If, as it seems, the Younger Dryas was caused by some North Atlantic disruption of the thermohaline circulation then the part of the Earth most significantly affected would be north-west Europe and the British Isles, which would no longer be warmed by the Gulf Stream. Here we do find that in England's Lake District the glaciers returned (or started growing again) so quickly that living trees were *caught in* their advance.

The Antarctic Cold Reversal, due to it starting earlier, cannot be entirely explained by any disruption in the North Atlantic of the thermohaline circulation that led to the Younger Dryas. The Antarctic Cold Reversal could be explained by Antarctic melt as the Earth left the LGM behind. This could disrupt part of the Broecker thermohaline circulation in the southern hemisphere. If this happened it would weaken the global thermohaline circulation. Because the northern hemisphere had large land-based ice sheets outside of the Arctic Circle (unlike the Antarctic Circle) these would have taken longer to react to coming out of the LGM. If the global thermohaline circulation had already been part-disrupted by the Antarctic Cold Reversal then it would be more susceptible to subsequent northern-hemisphere disruption. Again the lesson to learn is that changes in ocean circulation (and atmospheric circulation for that matter) have a profound effect on the climate.

As discussed in the previous section, while the pacemaker of the glacial–interglacial cycles is the change in Milankovitch insolation, the vagaries of warming and cooling within glacials were driven either by changes in greenhouse gases and or changes in ocean or atmospheric circulation and in particular the Broecker thermohaline circulation. Indeed Milankovitch insolation in July at 60°N had already peaked following the LGM by the time of the Younger Dryas and remained at a high value throughout it, as well as for a time into the early Holocene. We have already noted how greenhouse gases parallel each other in both northern and southern hemispheres, which means that the likely candidate *primarily* causing the northern hemisphere's Younger Dryas return to glacial conditions was to do with atmospheric and/or ocean circulation and not greenhouse climate forcing. Having said this, there was a noticeable dip in both carbon dioxide and methane concentrations at the time of the Younger Dryas, but it is likely that this was the result of a biotic response to the cooling rather than a cause of it.

Yet if it was the melting of the northern-hemisphere ice caps that caused the Younger Dryas temporary cooling, why did this happen suddenly and not gradually as the ice caps melted? Furthermore, why was there not cooling of a similar degree in the southern hemisphere? Here, the geography of the northern and southern hemispheres is markedly different. A key factor is that in the northern hemisphere the landmasses of North America, Europe and Siberia enabled large ice sheets to form during the LGM. Conversely the southern hemisphere's ice sheet was centred on Antarctica and the Antarctic continental shelf and not on land at lower latitudes as in the north. Unlike their marine counterparts, terrestrial ice sheets push (terminal) moraine in front of them and as the ice retreats these can act as dams. Local terrestrial geography can also serve to constrain meltwater. So it is likely that as the Earth warmed, leaving behind it the depth of the last glacial, the melting Laurentide ice sheet created a volume of inland fresh water. It is this water that ultimately discharged into the salty northern Atlantic, probably in a pulse, thus switching off the Broecker thermohaline circulation.

This possibility is worth further consideration because, as we shall see, there was at least one other pulse of meltwater likely to have disrupted the Broecker circulation. So is there evidence of such a reservoir of fresh water? Indeed there is. The geological evidence for large lakes off of the Laurentide ice sheet at the end of the last glacial is clear and there were also numerous smaller bodies of water. Two lakes in particular were bigger than Canada's present Great Lakes and so they represented a considerable volume. Indeed, it is likely that the majority of freshwater fish in Canada and the northern USA today have ancestors that lived for some time in these two lakes. Both lakes were bounded on their north-eastern side by the Laurentide ice sheet. On their western side were the upland plains and hills that lie in the shadow of the Rocky Mountains. The largest meltwater lake, when at its greatest, was Lake Agassiz, at some 350 000 km^2, and it is named after the naturalist that coined the term ice age. These lakes were only temporary and just existed at the end of the glacial because of the ice dams. They were also dynamic. When the Laurentide ice sheet began to melt sufficiently, more water was fed into the lakes and part of the Laurentide ice retreated. The lakes therefore changed shape and size. Many of the lakes were from time to time connected to each other. The variability of ice aided this. Not only was there melting but also re-freezing. In addition, not only would there be terminal moraines and local geography but also ice dams were possible. Geologists think that occasionally an ice dam that had held back water would, once there was sufficient water behind it, float and so allow a pulse of water to escape. If the ice dam was at a key location then the volume of water released would be considerable. In North America sometimes the drainage would be to the north through the Mackenzie River and sometimes south through the Missouri and Mississippi Rivers:

throughout these regions there is ample geological evidence of brief torrential flooding dating from the end of the last glacial. It could well be that it was one of the larger of these pulses that precipitated the Younger Dryas with the suddenness we see on the Greenland ECM record. Indeed, this is the preferred theory, although it should be noted that there are others.

The other thing to note about the Greenland ice-core ECM is that it appears that the regional climatic changes associated with the brief interstadials after the LGM, and also the beginning and end of the Younger Dryas itself, took place extremely quickly: perhaps as fast as 5–50 years. Furthermore, a detailed examination of the ECM record reveals climatic flickers as short as 1 year: these could be changes in atmospheric circulation (Taylor *et al.*, 1993) and possibly represent flickers between two semi-steady states. Such rapid change serves as a reminder with regards to future warming by the Intergovernmental Panel on Climate Change of being wary of surprises. We shall return to this later (see Chapter 6).

Before leaving the topic of Greenland's palaeo-dust record, it is worth remembering the iron effect on plankton blooms (see section 1.3). The Greenland ice-core ECM markedly reflects the transition from a dusty glacial to a less-dusty interglacial environment. There is no reason to suppose that other types of dust than the calcium acid-neutralising dust were not similarly affected. The transition from glacial to interglacial would have also affected iron-containing dust. In a glacial mode more iron dust would be blown on to the oceans so encouraging algal blooms that draw down atmospheric carbon dioxide from the atmosphere and so helping to keep the planet cool (see Chapter 1). Similarly, under wetter, warmer interglacial conditions some iron-containing dust would be washed out of the atmosphere before reaching the oceans, this would reduce algal blooms, reduce carbon dioxide drawdown and so help maintain the atmospheric concentration and keep the planet warmer. These are other examples of feedback systems (again, see Chapter 1) that regulate the biosphere much in the way engineers describe a cybernetic system (Figure 4.13).

Although the climate warmed quickly after the last glacial, some other environmental parameters took longer. Sea-level rise took thousands of years. The Baltic Sea filled and Britain became an island only as recently as about 7000 years ago, some 4000 years after the end of the last glacial. From this we can see that the approximate average rate of sea-level rise over this time was a little more than half a metre a century. Having said this it is important to note that this is an average rate and there were times between the LGM and 7000 years ago when sea-level rise was faster (see Chapter 6), and others when it was slower.

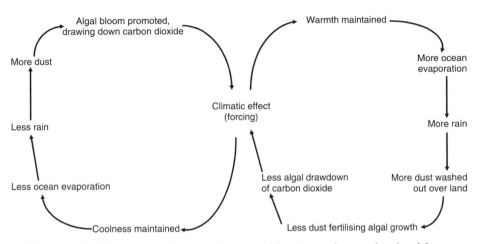

Figure 4.13 Climate–dust interactions combine to tend to maintain either a cool (glacial) or a warm (interglacial) mode. Just one of the processes that tend to keep the Earth in one of a number of climate states.

4.6.3 The Holocene (11 500 years ago – the Industrial Revolution)

The Earth finally emerged somewhere around 11 600 years ago from the last glacial into a warmer climate mode that has lasted until the present. (Although the official date, based on key – Global Standard Section and Point – geological strata, for the commencement of the Holocene is 11 500 years ago.) Average snowlines around the planet (where latitudinally climate allowed this) rose around 900 m after the LGM. On the basis of today's temperatures this suggests an atmospheric warming of about 5 °C and of the tropical sea surface of some 3 °C. Of course, the huge amount of ice left over from the last glacial did not disappear as quickly as the Milankovitch-forced climate had changed. It took some 3000 years for the Fennoscandian sheet over northern Europe and Russia to largely disappear and over 5000 years for the North American Laurentide ice sheet to collapse. This slower collapse may have been in part due to the climatic vagaries due to atmospheric circulation but was probably more connected with the proximity of the higher latitude Greenland ice sheet, which is still with us today, and of course which provides us with valuable palaeoclimatic records from its ice cores.

The melting of the ice sheets themselves continued to have an effect on the northern hemisphere's climate. It is thought (Barber *et al.*, 1999) that a pulse of meltwater around 8470 years ago (Rohling and Pälike, 2005) from the North American glacial lakes of Agassiz and Ojibway, until then held back by the ice, disrupted the Broecker thermohaline circulation by freshening the cold, salty water in the North Atlantic. Again, as with the Younger Dryas, this prevented the water there from sinking, one of the processes driving the global conveyor. It cooled the temperature in central Greenland by 4–8 °C and regionally in the surrounding North Atlantic sea and land by 1.5–3 °C, again a similar climate response to the Younger Dryas in the previous section.

Some 10 000 years ago Milankovitch forcing (as determined by the 60°N July insolation) peaked. However, inertia within the planet's climatic system, and other forcing factors combined, delayed the peak in the planet's climate until 6000 years ago – the peak of the Holocene climatic maximum. While the climate below latitudes of about 40° was only a little warmer (mostly about 0.5 °C or less) than today (the early twenty-first century), the climate at higher latitudes was roughly 1.5 °C above the Holocene average. Further, at extremely high latitudes over much of the North Pole and Greenland, as well as a pocket in Russia north of the Caspian Sea, the climate was as much as 3 °C warmer than at the end of the twentieth century. At lower latitudes this warm period manifested itself not so much with a warmer climate but in other ways as more energy entered parts of the climatic system such as the monsoon's summer circulation. So, there was more precipitation in many parts of the sub-tropics. Consequently, the tropics during the maximum were warmer and wetter than today. However, due to current global warming, from the various IPCC scenarios (see section 5.3.1), it is likely that we will reach temperatures equivalent to the Holocene climatic maximum well before the mid-twenty-first century and possibly as soon the end of its first quarter.

It is worth noting that in the past the Holocene climatic maximum has been known as the Holocene climatic optimum. Many palaeoclimatologists have ceased to use this term as the use of the word optimum does not mean that the climate was in any sense climatologically or biologically optimal, but that the time represented a thermal maximum for the pre-twenty-first-century Holocene. Having said this, we are still in the Holocene and present-day global warming is taking us beyond what was the Holocene climatic maximum and so it may be that terminology will again change in a decade or two.

In addition to a warmer climate, and resulting sea-level rise, the Holocene saw humans contribute their own ecological impact. The degree these various factors of climate, sea-level rise and humans together impacted on species varied with location and, of course, the species in question. So the resulting picture was complex. Yet, just one of these factors was unique to the end of the last glacial and the current interglacial and that was the spread of modern humans (*Homo sapiens sapiens*). There is no single mechanism responsible for the last-glacial and current interglacial extinctions of every species in every location. As noted earlier in this chapter (see section 4.3) some Pliocene large animals (megafauna) survived to quite recently in geological terms. Mammoths (*Mammuthus* spp.) survived up to as recently as 4000 years ago due to isolation on Wrangel Island. Another species, the Irish elk (*Megaloceros giganteus*), survived to about 7000 years ago. Before 20 000 years ago the Irish elk was widespread from Ireland to Russia (west–east) and Scandinavia to the Mediterranean (north–south) but at the time of the LGM (15 000 years ago) they appear to have become more restricted, probably to refuges in the shrub steppes of central Asia. From there the Irish elk recolonised much of their lost ground as the Alpine and Scandinavian ice sheets retreated. It is possible that the productive shrub and open grass community necessary for the Irish elk persisted in the eastern Urals (Stuart *et al.*, 2004). The picture is further complicated in that some large herbivore species (possibly including the Irish elk) not only respond to changes in their food-plant communities but themselves affected plant communities and plant productivity through nutrient redistribution and the grazing selection of food species. Further, by trampling tundra mosses and stimulating grasses through grazing the woolly mammoth might have helped maintain and even expand its own habitat (Pastor and Moen, 2004). Just as a prolonged period of climate change (albeit vulcanism-induced) put pressure on the dinosaurs (whose extinction was then sealed with an asteroid impact) so climate change appears to have put pressure on Pliocene megafauna whose extinctions were finally sealed due to human hunting; after all, the megafauna had survived a number of previous glacial–interglacial climatic shifts.

Evidence from Alaska and the Yukon territory (Guthrie, 2006) suggests that there mammoths (*Mammuthus primigenius*) and a regional species of horse (*Equus ferus*) died out around 12 000 years ago, around the time when humans arrived, with very little overlap. Again the picture is more complex than first appears for along with the arrival of humans there is also the appearance of moose (*Alces alces*). Pollen analysis shows that the area saw a transition through three principal types of biome. The first was a cold, dry steppe up to around 13 500 years ago in which mammoths thrived. The second was a transitional moister biome that existed from 12 000 to 13 500 years ago. Finally there was the rise after 12 000 years ago of mesic taiga and tundra with dwarf birch (*Betula*). The suggestion is that plant species in this ecosystem did not support many large herbivores. That humans arrived around that time may have added critical pressure on top of the radical environmental change taking place both at the time of their arrival and the episode of ecological change 1500 years earlier. This environmental change was driven by changes in the global climate as the planet moved out of the last glacial. In this instance, humans themselves were not (solely) responsible for the extinction.

Table 4.1. *The numbers of mammalian megafaunal genera (groups of closely related species) in the Pleistocene extinction (from Barnosky et al., 2004).*

Continent	No. of genera extinct	% extinct on each continent
Africa	5	18%
Australia*	14	88%
Eurasia**	5	36%
North America	28	72%
South America	49	83%

*Australia also has seven extinct (and no surviving) genera of megafaunal bird.

**Only northern Asia due to insufficient data on southern Asia.

It is interesting to note that the late Pleistocene and early Holocene extinctions were largely driven by a combination of both climate change and human pressure, for today's current extinctions are similarly driven. The only difference between these two extinction periods is that the late Pleistocene and early Holocene extinction was part driven by *natural* climate change whereas the current extinction involves *anthropogenic* climate change. It is also interesting to note that the degree of extinctions on each continent varied depending on the length of human existence (see Table 4.1). Because humans evolved in Africa, and then spread to Eurasia, animals here had a chance to co-evolve. (We shall shortly return to this in the next section on biological response.)

The time from 8000 to 5000 years ago marks the aforementioned period known as the Holocene climatic maximum. In Britain trees grew on land 180–300 m higher than the present-day tree line, and species such as lime (*Tilia* spp.) and elm (*Ulmus* spp.) were more common than today. Trees also grew in the Orkney Islands, Faroe Islands and Iceland. Winters then were probably far milder than today (Lamb, 1965).

However, there is some ambiguity in the climate record as some parts of the Earth appear to have cooled for part of the climate maximum period, especially in high latitudes (Antarctica, Greenland and eastern Canada). One suggestion is that this high-latitude cooling decreased precipitation and reduced terrestrial biomass. The carbon from this biomass then entered the atmosphere and indeed a small increase of 10 ppm in atmospheric carbon dioxide is detected between 7000 and 5000 years ago. (However, this is small by late-twentieth-century increases of more than 60 ppm over the century.) The Holocene climatic maximum increase in greenhouse forcing may in part account for the maximum and this period of time may represent a redistribution of carbon between various pools (Steig, 1999). Such problems are the stuff of some current research.

Palaeoclimatic records for the last 4500 years generally indicate that temperatures were lower than the Holocene climatic maximum. There was a general cooling of 1–2 °C known as the Iron Age neoglaciation that took place between 4500 and 2500 years ago.

There was then a return to warmer conditions, but not nearly as warm as the Holocene climatic maximum, around the beginnings of the Roman Empire. Changes of just a degree, or at most around two, mark the nature of much of the climatic changes seen within the Holocene interglacial. (Remember that regional warming and cooling can be greater than the global average change. Furthermore, a comparatively small global change in temperature might become manifest in greater or less seasonality.)

A return to a cooler climate took place 1500–1000 years ago (AD 500–1000), which coincided with the historical period in Europe known as the Dark Ages. This was followed by the Medieval climatic anomaly (MCA) or Medieval warm period (MWP; broadly around AD 1100–1300). The MCA is also sometimes called the medieval climatic optimum but human ecologists and environmental scientists are wary of this term as it implies that this time was somehow optimal. During the MCA European temperatures reached some of the warmest levels for the last 4000 years since the Holocene climatic maximum. Although not as warm as the Holocene maximum was for Europe, the MCA anomaly was roughly a degree warmer than in the last quarter of the twentieth century. Estimates for continental Europe's climate are mainly 1.0–1.4 °C, and for Greenland 2–4 °C, warmer during the MCA compared to the end of the twentieth century.

There is evidence that the MCA was not regional (confined to Europe) and was to some degree at least hemispheric. Preserved tree stumps have been found in marshes and lakes in California's Sierra Nevada that have been radio-carbon dated and dendrochronology used to determine the climate. This has revealed that the Sierra experienced severe drought for over two centuries before AD 1112 and for more than 140 years before AD 1350 (Steine, 1994). Diatom (common members of algal flora) analysis of sediment cores from Moon Lake, in the Upper Colorado River, infer salinity levels. These suggest that during the MCA, further to the north west, the northern great plains that cover much of Montana, Wyoming and North and South Dakota, saw increased drought frequency compared to more recent, hemispherically if not globally, cooler times. Today the area is one of considerable agricultural productivity that was famously greatly undermined by droughts in the 1890s and 1930s (Laird *et al.*, 1996). The implications for today are that should twenty-first-century emissions of greenhouse gases continue then that area could again suffer from severe and prolonged drought. (It is interesting to note that since the late 1980s part of western North America has experience a marked lowering of rainfall.)

Having said this there is a lack of evidence for the MCA from some parts of the northern hemisphere (such as dendrochronological evidence from the Urals). Further, it may be that the warming was confined to certain seasons. The picture is unclear. It *might* be that there was a general warming but accompanied with some circulation changes that counteracted the warming in some areas. This is not too unsurprising as climate change is never completely uniform all over the globe. However, to date (2007) the MCA picture emerging is not yet as coherent as climate researchers would like.

One of the more significant of the 'minor' (±0.5 °C) climate fluctuations in comparatively recent historic times was the Little Ice Age, notably between 1550 and 1850 but divided into two parts, the latter half of the sixteenth to close to the end of the eighteenth century (about 1550–1770) and then a return in the nineteenth century (about 1800–50). It appears that during the Little Ice Age a tongue of Arctic ice reached down as far as the Faroe Islands while glaciers advanced in Europe, Asia and North America. The Little Ice Age even affected ocean circulation and the sea-surface temperatures off west Africa were markedly reduced, by 3–4 °C (deMenocal *et al.*, 2000). The Little Ice Age brought extremely cold winters to many parts of the Earth, but is most thoroughly documented in Europe and North America. In the mid-seventeenth century, glaciers in the Swiss Alps advanced, gradually engulfing farms and crushing entire villages. The River Thames in England, and the canals of the Netherlands, froze over during the winter, and people skated and even held fairs on the ice. In the winter of 1780, New York Harbour froze, allowing people to walk from Manhattan to Staten Island. Sea ice surrounding Iceland extended for miles in every direction, so closing the nation's harbours.

That the Little Ice Age is not so well documented outside of Europe and North America suggests that it may not have been a completely global event, or if it was global it may have had a focus in one hemisphere. Although the temperature evidence is less clear there are data (Verschuren *et al.*, 2000) of climatic resonance with the Little Ice Age derived from lake-sediment strata and fossil diatom and midge assemblages in equatorial east Africa that suggest a drier climate there during the MCA (1000–1300) and wetter conditions broadly corresponding with the Little Ice Age (AD 1270–1850). In South Africa the time of the Little Ice Age was more arid compared to the earlier MCA when it was wet, and in Kenya it was dry (Tyson *et al.*, 2002). If the Little Ice Age were global we may not have picked up such a shallow climate change from natural proxies and it may simply be that it was the rise of meteorological instrumentation in Europe and North

America, which naturally led to much of the meteorological recording taking place there and so better documentation of the Little Ice Age. Interestingly, ice cores reveal that atmospheric carbon dioxide did not show a marked decline during this time. (There was a dip of around 10 ppm in concentration from 284 ppm in the middle of the sixteenth century, which is minor compared to the subsequent increase from the eighteenth century onwards of around 80 ppm.) So, at the time of the Little Ice Age there was no significant decline in the forcing of climate from one of the major greenhouse gases. This leaves other factors.

Of the non-greenhouse contenders that could have generated the Little Ice Age there are two more likely: volcanic eruptions (see Chapter 2 and Figure 2.1) and variations in solar output, hence insolation on the Earth. Indeed, there were a number of major volcanic eruptions during the Little Ice Age including Billy Mitchell in the south-west Pacific (approximately 1580), Huaynaputina in Peru (1600), Parker in the Philippines (1641) and Long Island in New Guinea (approximately 1660). Then in the second part of the Little Ice Age there was Laki in Iceland (1783), Tambora in the Lesser Sunda Islands (1815) and Krakatau west of Java (1883). These can only have served to cool, or further cool, the planet away from the late Holocene average. Of these Huaynaputina was one of the worst and the acid spike in the Greenland ice cap is higher for 1600 than for 1883 (Krakatau). After Huaynaputina the following few years saw very low temperatures. In Europe the summer of 1601 was cold with freezing weather in northern Italy extending into July and overcast skies for much of the year. In parts of England there were frosts every morning throughout June. Less violent and more recent was the eruption of Laki in Iceland in 1783. It is estimated that 80 Mt of sulphuric acid aerosol was released by that eruption. To put it into a modern context, this is four times more than El Chichon (Mexico, 1982) and 80 times more than Mount St. Helens (USA, 1980). In terms of a vulcanologists' perception of climate interactions, the Laki eruption illustrates that low-energy, large-volume (14 km^3 basaltic lava and over 100 million t sulphur dioxide for Laki) and long-duration (8 months) basaltic eruptions can have climatic impacts greater than quick, large-volume, explosive, silica-rich eruptions. The sulphur contents of basaltic magmas are 10–100 times higher than silica-rich magmas. Immediately following the Laki eruption the winter in the eastern USA was about 4.8 °C below the 225-year average, while the very broad estimate (to be treated with caution) across the northern hemisphere was of a 1 °C drop. What is clearer is that it took under 2 years to recover from half the temperature drop and somewhere between 5 and 6 years for complete recovery. Another reason for viewing the effects of the Laki eruption with a little caution is that the same year saw the eruption of a smaller volcano, Asama in Japan. Even so, whereas high-sulphur volcanic eruptions may cool the planet for a few years, they by themselves cannot explain all of the Little Ice Age's climate.

With regards to solar output, during the period 1645–1715, solar activity as inferred by sunspot activity, was low with some years having no sunspots at all. This period is known as the Maunder Minimum after the astronomer E. W. Maunder who noticed, from records, that there were few sunspots during that period. For example, during one 30-year period within the Maunder Minimum, astronomers observed only about 50 sunspots, as opposed to a more typical 40 000–50 000 spots. What the precise link is between low sunspot activity and cooling temperatures has not been firmly established, but the coincidence of the Maunder Minimum with the deepest trough of the Little Ice Age is highly suggestive.

One might suspect that if solar output, hence Earth's insolation, during the Little Ice Age was less that this would have a uniformly global effect. This is not necessarily so, given (as noted in Chapter 1) that the northern hemisphere is predominantly land and the southern ocean and that these react differently to incoming energy. Second, in terms of biology, reduced insolation through-out the year may only affect part (and/or the length) of the thermal growing season. This may explain why Himalayan tree-ring series show little evidence of the Little Ice Age as these dendrochronological series are mostly affected by pre-monsoon temperatures and rainfall (Boragaonkar *et al.*, 2002).

Of course, in terms of nomenclature, the Little Ice Age was neither an ice age nor was it a glacial. That such a palaeoclimatologically powerful term has come to describe what was a short and modest shift in climate is simply due to it being the major climatic event taking place within modern recorded history prior to the end of the twentieth century. In terms of psychology and the social-science dimensions of climate change, the use of the term Little Ice Age demonstrates that hemispheric and global climate change is hugely relevant to human society. We will return to the biological and human ecological effects of Holocene climate change shortly, but for now it might be interesting to note that our cultural heritage of a Dickensian white Christmas owes itself to the Little Ice Age.

The recovery from the end of the Little Ice Age is well documented by over a century of direct observation in Europe and North America. Globally snow-line retreat since 1860 to the end of the twentieth century suggests a global warming of 0.6–0.7 °C. And between 1880 and 1930 records at Oxford, England, show that there was a 10% increase in the thermal growing season.

In terms of present-day questions of global warming, this period is critical. It was during this time (after inception of the Industrial Revolution) that significant amounts (although small at first) of greenhouse gases were added to the atmosphere from human activity. Whereas before the twentieth century the global climate was largely (but not solely) determined by non-human factors, during the twentieth century anthropogenic warming increasingly became manifest.

This point is an important one because it is possible to mistake (and indeed some vocal people have mistaken) anthropogenic global warming for some other factor. That the early twentieth century saw a warming out of the Little Ice Age does not mean that there was just one factor at work. Some (such as the popular thriller author Michael Crichton) put this climate change down to 'natural' climatic vagaries, whereas others (such as some astronomers) put it down to changes in solar output. However, the IPCC, the considered scientific authority on climate change, is clear that many factors affect climate, including changes in the Sun's output, but that this does not prevent human-generated greenhouse gases becoming an ever increasingly powerful factor, from small beginnings with the Industrial Revolution to a dominant (if not the dominant) climate-change factor by the end of the twentieth century.

Before leaving our climate review of the pre-industrial Holocene, it is worth noting that there has been some discussion with regards to human influences on the global climate system over this time (see Anthropocene earlier in this chapter) Analysis of methane concentrations from Antarctic ice cores covering the past 2000 years reveal an interesting ^{13}C isotope pattern embedded within the record of change in overall methane concentrations, containing both ^{13}C and ^{12}C.

As noted previously, plant photosynthesis sequesters more ^{12}C from the atmosphere than ^{13}C but this preference not only differs between C_3 and C_4 plants, the proportions of ^{13}C in methane given off by microbial decomposition of organic material in wetlands as well as of ^{13}C in fossil fuels are also different (this last itself has only been a significant factor for the past 200 years and is reasonably well documented). Therefore, in theory it should be possible to begin to disentangle the way various methane sources contributed to atmospheric concentrations. Here analysis of methane trapped in ice-core bubbles has showed that changes in $^{13}CH_4$ (represented as $\delta^{13}CH_4$) were such that the $\delta^{13}CH_4$ was about 2‰ above expected values between the years AD 0 and 1000. Conversely, $\delta^{13}CH_4$ values were about 2‰ below expected values in the subsequent 700 years, up to AD 1700. The MCA also shows up as expected in the methane pattern. However, interpreting the various isotopic $\delta^{13}CH_4$ is difficult and so as yet (without other corroborating evidence) conclusions should be viewed with caution. Yet the suggestion is that both human activities (such as land clearance and use of wood as a fuel) and natural climate change have influenced the pre-industrial carbon cycle and that, up to the early twenty-first century, researchers have understated the human influence on late pre-industrial Holocene methane budget research (Ferretti *et al.*, 2005).

4.6.4 *Biological response to the last glacial, LGM and Holocene transition*

The biological response to glacial and Holocene interglacial climate change is in the main manifested in the form of species migration. However, fixing one's self in the landscape, an observer would see any one location undergo considerable ecological change as species come and go. For example, the Beringian land bridge, the aforementioned area that was ecologically important before the Quaternary, re-appeared during glacials with their lower sea levels. Due to marine geological cores containing pollen, plant macrofossils and insect remains, it is possible to determine how the Beringian environment changed over a glacial–interglacial cycle. Close to the LGM Beringia's ecology was one of a dominated heathland with some shrubs and the occasional birch tree (*Betula* spp.). During the Younger Dryas birch increased, as did grasses, and there was a patchwork of ponds filled with weeds. By the time of the early Holocene, Beringia's open ground – which was being reduced by sea-level rise – was that of mesic tundra, half way between polar tundra and a wetter temperate biome, where summer temperatures were warmer than present-day Alaskan northern slopes (Elias *et al.*, 1996).

The transformation between glacial and interglacial conditions was arguably the greatest in North America as, at the LGM, its glacial ice sheets were at least 50% greater than those in Europe and Asia combined. As such, the ice sheet was as large as that of Antarctica, covering all of Canada and the north of the USA (see Figures 4.3 and 4.4). The Canadian ecologist Chris Pielou (1991) has reviewed many of the biological changes and some of these are summarised here along with other research observations.

A key problem facing species riding out climatic change is where can they live? The concept of 'refugia' has been introduced in previous sections and so, just as there were refugia for present-day temperate species during the last glacial, so there must have been refugia for peri-Arctic species when the ice sheets were at their greatest extent. It is likely that during the LGM exposed mountain peaks rising above the glacial ice sheet (nunataks) would have provided a possible refuge for Arctic stony species such as the woolly lousewort (*Pedicularis lanata*) and Ross's sandwort (*Minuartia rossii*) although Beringia would have undoubtedly been an important refuge for such species (and this would explain their presence in both present-day Siberia as well as Canada and Greenland). Nunataks nearer the coast would have seen a comparatively milder climate (though still very cold) and so possibly be refugia for species such as the green spleenwort (*Asplenium viride*) and the mountain holly fern (*Polystichum lonchitis*), whose present-day ranges cover much of the Rockies at, or near, the tree line as well as parts of Greenland.

Long-distance horizontal, as opposed to short-distance vertical, migration was an important mechanism for many species to survive the interglacial–glacial climatic changes. However, the migrations southwards (at the beginning of glacials) were different from those northwards (at a glacial's end) as the environment around expanding ice sheets is different to that around contracting ones. At a time of a cooling climate, the cooling takes place before ice sheets expand. This would leading to freezing ground (permafrost) before the ice sheet, or an arctic tundra-like environment, and if there was such an environment already there then this zone would have widened. Either way plants cannot live in this zone and any that previously had would die. As noted earlier in this chapter, plants themselves do not, as such, migrate; rather, the descendents from the seeds of those furthest away from the ice distribute even further still, and so it is the respective plants' ranges that shift. The reverse process tends to take place with contracting ice sheets, except that contracting sheets do not have soil present underneath them and so colonisation is delayed until soil forms. Conversely, contracting sheets expose clear ground devoid of trees and this void can facilitate seed dispersal.

Horizontal species range migration as a result of major glacial–interglacial climate change often results in species migrating over considerable distances. As has been mentioned, in Europe latitudinal migration was restricted by the Mediterranean, Pyrenees and Swiss Alps. However, in both North and South America it was possible for many species to migrate 1700–2400 km as did the lodgepole pine (*Pinus contorta*) from Washington State to the Yukon, or the maple (*Acer* spp.) and chestnut (*Castanea* spp.) from Louisiana to Ohio, Pennsylvania and New York State. Migration began at the end of the Younger Dryas, although there is evidence of some earlier migration during the Allerød–Bølling interstadial, especially along north–south mountain ranges that also allowed for vertical migration. (The interstadial prompted both vertical and horizontal migration. Vertical migration, where possible, especially enabled species to outride the cool Younger Dryas.) However, some species reacted quicker than others. For example, the eastern white pine (*Pinus strobes*) began its migration about 1000 years before the eastern hemlock (*Tsuga canadensis*).

While this migration broadly began with the onset of climate change, species ranges do not all shift at the same rate. In North America following the last glacial, the range of pine trees (*Pinus* spp.) shifted at an average speed of just under 200 m/year. Travelling at such speeds it can take millennia to cover the 3000 km that many species did. Indeed, in North America the lodgepole pine (*P. contorta*) reached its northern limit probably only less than a century ago, and of course with current anthropogenic warming its range can move still further north. Conversely, the white spruce (*Picea glauca*) migrated some 50% faster, at an average of around 300 m/year.

One of the key factors determining the speed with which a species' range can change is the nature of the plant's seeds. Maple (*Acer* spp.), with its winged seeds, changes its range faster on average than the chestnut, with its heavy seeds (*Castanea* spp.): 200 and 100 m a year respectively. It should be noted that not all winged seeds enable their parent species to move faster than their non-winged competitors. Oaks (*Quercus* spp.) were sufficiently large to be attractive to animals but small enough for them to be carried more easily and faster than a number of species with winged seeds. In North America, oak ranges moved by an average of 350 m/year. Indeed, climate change exerts an evolutionary pressure favouring migration. It has been noted that the lodgepole pine's seeds at the northern end of their range had smaller wing loadings (lower weight/wing area ratio) than those at the southern end (Cwynar and Macdonald, 1987).

From intra-species genetic variation it can be seen that the LGM–Holocene transition also facilitated the speciation of animals. Indeed, many species would have to change and sometimes in ways fundamental to their life cycle. For example, in a warmer world species' migration patterns would change. The three North American bird species – the common flicker (*Colaptes auratus*), yellow-rumped warbler (*Dendroica coronata*) and the dark-eyed junco (*Junco hyemalis*) – each have their own eastern and western North American race or subspecies. These races have their present-day geographical distribution either side of a line that runs from Canada's north-western British Columbia to New Orleans, Louisiana, USA. That each of these species has two subspecies with the same geographical range is not accidental but reflects that during the glacial the two populations migrated south to two different parts of the USA separated by a large dry desert, so enabling allopatric speciation, before migrating back at the glacial's end.

Other pairs of bird species are more distinct (but still owe their evolution to the last glacial–interglacial cycle). These include the northern shrike (*Lanius excubitor*) and loggerhead shrike (*Lanius ludovicianus*); the Bohemian wax-wing (*Bombycilla garrulous*) and the cedar wax-wing (*Bombycilla cedrorum*); and the northern three-toed woodpecker (*Picoides tridactylus*) and the black-backed three-toed woodpecker (*Picoides arcticus*). It is thought that for each of these pairs there was a common ancestor but that during the LGM one population rode out the worst of the glacial in Eurasia and the other in America. Then after the glacial the two species moved north to overlap their distribution in North America.

In plants one of the adaptations many would have to make would be to adjust to the different day lengths throughout the year at different latitudes.

Photoperiodism (the phenomenon by which plants flower at a specific time of the year and day) needs to change when species migrate latitudinally. Many high-latitude plants are long-day plants; they cannot flower until spring has become summer. Conversely, many temperate-zone plants are short-day plants. Change in photoperiodism manifests itself in species at the level of different plant races. So, the glacial–interglacial climate change was the impetus for the creation of numerous new races of plants. But it was not all new races and speciation. The end-glacial climate change brought extinctions too.

We briefly looked at the Earth's major extinction events in the previous chapter and saw that many of these were climate-related: even if the immediate cause was not climatic (such as an asteroidal or bolide impact), climate change often followed. So, not surprisingly the Quaternary series of glacials and interglacials have been associated with extinctions. In the main these extinctions happened at the end of the glacials. One of the biggest of these took place at the end of the last glacial.

True, North America had the largest non-polar ice sheet during the last glacial, but the extinction at the end of the last glacial was not confined to that continent: there were extinction events in South America, Africa, Asia, Australia and Europe. Not all species extinctions take place at the glacial's end, but there are more species threatened during this time if only because that is the part of the glacial–interglacial cycle that sees the greatest rapid climatic change. However, there are other times during glacials where sudden, albeit smaller, climatic change takes place and extinctions (especially locally) can result.

One key question is why was the end of the last glacial associated with a particularly major extinction event compared to those associated with other glacial cycles? One answer, mentioned previously, concerns humans. As noted (Chapter 3) the late Cenozoic spurred hominid evolution. Indeed, the repeated series of glacials and interglacials would favour hominids that were generalists and could adapt to different environments, and especially those that could artificially modify their own personal environment such as through using fire. Although there is evidence of an earlier minor human presence in North America before 13 000 years ago, when there was the hunting of megafauna in the south-west of the present USA (perhaps as early as 18 000 years ago), the first major incursion of humans towards the glacial's end had crossed the Beringia land bridge from Asia into North America. To establish themselves in that continent they probably kept close to the coast and away from the Laurentide ice sheet that still dominated inland North America. Prominent among these were people belonging to what is termed the Clovis culture, although as noted there are clear signs of pre-Clovis presence. The Clovis were big game hunters, which is typical of nomadic people as opposed to agrarian cultures that necessitate fixed abodes.

Among the pre-Clovis signs are those at Monte Verde in Chile that date from around 12 500 years ago. Further, human footprints have been found fossilised in volcanic ash in the Valsequillo Basin, Mexico, which have been dated to before 40 000 years ago. So, while the Clovis migration was an important event, they were clearly not the first on that continent.

The timing of the Clovis migration itself into North America is confined by the melting of the Laurentide ice cap. The only likely terrestrial route they could have taken would have been a narrow gap between the ice sheet and the glaciers of the coastal Pacific mountains in the Yukon. From carbon dating we know that this passage was not available much before 13 000 years ago as the Laurentide ice sheet would have been too big. Conversely, the problem with considering the pre-Clovis cultures' route on to the continent is how did they overcome the glacial obstacles? One answer may be because the sea was lower and they could have skirted the coast. The coastal termination of glaciers would have made this difficult but, especially with primitive boats, not impossible. Unfortunately, evidence for such a passage is now under water (Marshall, 2001).

In the Old World, humans had moved through Africa and into Eurasia. There are signs of tool making going back as far as 125 000 years and earlier, back to the last interglacial. In short, humans had the means and opportunity to provide added pressure on species already stressed by climate change.

Yet this situation is at first perplexing. If humans were in Africa well before the American continents, why has the New World seen a greater extinction of large animals compared to Africa? (see Table 4.1). The answer is likely to be co-evolution. African large animals evolved alongside hominids. Those animals that learned to fear and avoid humans survived, while those that did not perished. Conversely in the New World, where human co-evolution had not taken place, so large species were less able to cope with the arrival of humans.

At this point it is perhaps worth digressing from the late Quaternary and briefly return to our previous mention of early humans a couple of million years ago. As noted (Chapter 3), human evolution owes much to climatic change. There is debate as to the exact nature of our human precursors' evolution but considerable evidence points to an African origin for hablines (*Homo habilis*). (Earlier sivapithecine remains have been found in Turkey and India 14–8 mya: sivapithecines are closer related to orang-utans than chimpanzees, gorillas and humans.) The relationship (and there is little doubt that at least in part there is one) between climate and human evolution has yet to be clearly elucidated. This is because terrestrial bio-geological and geological records of east African environmental change are rare, geographically dispersed and incomplete, Atlantic Ocean and (perhaps more importantly) Indian Ocean sediment records have been used to reconstruct climatic changes in the region. In 2005 Martin Trauth and colleagues reconstructed the east African climate 0.5–3 mya from 10 lake-basin records in the area. They identified three possible hominid periods: 2.7–2.5 mya, 1.9–1.7 mya and 1.1–0.9 mya. They note that these could well have had an important impact on the speciation and dispersal of mammals and hominids at that time.

The Quaternary (from 1.8 mya) and its series of glacial–interglacial cycles was the period of time when much of modern human evolution took place. The earliest direct evidence of hominid stone-age technology dates from 2.5 mya in the Ethiopian Rift Valley. Conversely, the bones and tools of *Homo erectus* are found both in Africa and Eurasia from about 1.5 mya. One evolutionary offshoot (quite probably from the *H. erectus* line) is that of *Homo floresiensis* (named after the island of Flores east of Java where the remains of the first specimens were found in September 2003). Analysis (^{14}C, luminescence, uranium series and electron-spin resonance) suggests that *H. floresiensis* existed from before 38 000 years ago (probably well before and perhaps as much as 800 000 years ago) to as recently as 18 000 years ago (Morwood *et al.*, 2004) although there has been some speculation of continued survival to more recent times. Either way, they therefore evolved either in the last glacial or over several glacials, but whichever it is, it is likely that a glacial maximum, with its lower sea levels and extensive land bridges, facilitated the island of Flores' initial habitation, even if some sea crossing was required. *H. floresiensis* were short hominids of about 1 m in height with a chimpanzee-sized brain (approximately 380 cm^3), but which did use crude stone tools and fire. Whereas climate change may very well have facilitated their getting to Flores, it was the limited size of Flores itself, together with its semi-isolation, which facilitated the evolution of *H. floresiensis*' short stature.

So, from Africa early human (*Homo* spp.) species then migrated to other lands. Note: modern humans (*Homo sapiens sapiens*) left Africa much later, sometime between 85 000 and 55 000 years ago.

In Europe Neanderthals (*Homo sapiens neanderthalensis*) evolved from 300 000 years ago (or about three glacials ago), although DNA analysis suggests their genetic divergence from a common ancestral stock some 600 000 years ago. Neanderthals have sequences not found in modern humans (*H. sapiens sapiens*). Previous glacials appear to have caused a *H. sapiens* population bottleneck with a population in Africa of about 10 000 in the Pleistocene. (Having said this, there is some debate as to what caused this bottleneck and speculation ranges from global environmental change due to a super volcano; see Chapter 6. It may be that it was the combination of a glacial and these factors, or even another factor entirely.) Subsequently this African population expanded some 100–50 000 years ago, during which some left Africa and subsequently, somewhere around 23 000 years ago (in the depth of the last glacial), there was expansion in Europe. However *H. sapiens sapiens* began replacing *H. sapiens neanderthalensis* earlier from about 41 000 years ago (that coincides with a colder Europe due to a short-term Heinrich event that made the North Atlantic cooler than it was at that time in the already cold glacial). Neanderthals subsequently vanished about 30 000 years ago. At the height of the LGM, 24 kya, humans in Europe would have been forced south. Mitochondrial DNA (mtDNA) analyses of humans today show six principal lineages from this Upper Palaeolithic age. (It is important to stress, though, that these dates are not exact. We will return to this shortly.)

Consequently much of the Pleistocene (1.8 mya–11 500 years ago) and its series of deepening glacial–interglacial cycles saw considerable human evolution and speciation as well as migration. *H. sapiens neanderthalensis* and *H. sapiens sapiens* were the most similar to each other, whereas *H. floresiensis* was the most different. The shorter *H. floresiensis* appears to have evolved from a more encephalised (larger-brained) ancestor so its discovery markedly added to the biological spectrum of humanity. At one time, about 40 000 years ago, all three species of hominid must have flourished simultaneously, albeit separately. *H. floresiensis'* small size is possibly due to their island location, which also was the habitat for dwarf elephants (a *Stegodon*) and the large Komodo dragon (*Varanus komodensis*). (The island of Flores was home to both the smallest elephant species and human but a large predator. Small prey tends to escape notice in a confined environment compared to large prey but large predators outcompete rivals where resources are clearly finite.)

Genetic analysis has yet to reveal the full picture of modern human's expansion across the globe. This is partly because the genetic techniques only began to become practically available to researchers around the turn of the millennium. It is also partly because some nations, especially in the Middle East, are resistant to western researchers conducting a genetic analysis of their populations. As said, modern humans arose around 150 000 years ago in east Africa during the glacial before last. We can now see that east African human genetic diversity is to this day particularly high and early colonization within Africa can be seen by branches in mtDNA – which is transferred down the female line – and Y-chromosome branches carried down the male line. These branches are known as haplogroups and can be found as Kalihari Bushmen (Khoisan) and certain pygmy populations. Haplogroups also contribute to our understanding of the aforementioned modern-human expansion into Europe. Haplogroups therefore help illuminate understanding of modern human exodus from Africa and migration to Australia. Peter Forster and Shuichi Matsumura of Cambridge University usefully summarised this research in *Science* (2005) by way of introducing two quite independent papers in the same issue by Vincent Macaulay *et al.* and K. Thangaraj *et al.* Up to 2005 there was some debate as to how modern humans originally left Africa for Asia and the initial colonization of India and the Malay peninsula and beyond. Perhaps the most obvious route would be up the Nile and across into Sinai (the route likely used for the aforementioned European expansion) or, alternatively, some 2400 km south, across the narrow Bab-el-Mandeb Strait at the southern end of the Red Sea and into what is present-day south Yemen.

Both research teams looked at small, long-established, indigenous populations (that are today dwindling fast and becoming diluted genetically). They both found evidence that corroborates a common picture. A population containing some 600 women (possibly slightly fewer and maybe as many as 2000 or more) left east Africa somewhere around 60 000 years ago. This was in the middle of the last glacial but a good 40 000 years before the glacial maximum. However, it was a cooler part of the glacial and although sea levels were not as low as during the later glacial maximum it is possible (if not pragmatically likely) that conditions were then favourable for migration across the Middle East. This early colonising population appears to have taken a coastal 'express' route arriving in southern Australia by 55 000–46 000 years ago (see below). This suggests a migration speed of somewhere between 0.7 km and 4 km/year and is of the same order of magnitude as other genetically dated inland journeys of migrant populations during the last glacial.

It should be noted, as discussed in Chapter 2, that (as with all dating techniques) ^{14}C dating is prone to error. In the case of ^{14}C dating some error comes from the rate of ^{14}C production in the atmosphere not being constant (it varies with solar activity) and variations in the Earth's magnetic field. Other error stems from sample contamination by more recent carbon (say through carbon carried by percolating water), which affects potential sample ^{12}C:^{14}C ratios. This particularly affects the dating of older samples (with the least ^{14}C). For example, if a 40 000-year-old sample was contaminated by just 1% with modern carbon then the dating error would be around 7000 years. With regards to the dating of human remains there have been two recent innovations. In 2004 a technique was developed for extracting collagen (a family of fibrous proteins found throughout vertebrates) from *within* bone, so helping to eliminate contamination. Second, since 1988 there has been a continually improved understanding of variations in ^{14}C production over the past

50 000 years. Although this does not substantially affect the picture portrayed here, it has been argued that modern humans migrated through Europe faster than had been thought, and that their period of co-existence with Neanderthals was shorter (Mellars, 2006).

Early humans, and allied hominids, were hunter-gatherers. The remains of plant grinders or pounders (such as grinding slabs, mortars and pestles) make their first appearance in the Upper Palaeolithic from 45 000 years ago and suggest simple food processing. The beginning of baking was an important step in human nutrition that facilitates carbohydrate entering the bloodstream being metabolised to glucose. While such evidence is compelling, there are only a few examples where food remains have survived to permit identification in association with such preparatory artefacts. (After all, pestles and mortars could be used to make dye not food.) The earliest direct evidence of simply-processed food remains comes from Galilee about 23 000 years ago with remains of wild barley (*Hordeum spontaneum*) and emmer wheat (*Triticum dicoccoides*). This was close to the LGM and the plant species were wild. Domestication of wheat and barely subsequently took place around 10 000–9 000 years ago, during the beginning of our Holocene interglacial. It therefore appears that a series of key developments in human food security took place under different climatic conditions and across glacial–interglacial change. (We will return to food security when addressing current and future food-security concerns in Chapter 7.)

The Holocene also saw human migration across the Pacific. The nature of this migration has been hotly debated. However, as new genetic-analytical tools become available it has been possible to obtain a clearer picture. mtDNA analysis indicates that the ancestors of the Polynesians originated from somewhere in eastern Indonesia and then migrated east and south, through and around Australasia approximately 11 000 years ago before moving on to Polynesia, reaching Fiji 3000 years ago and then on to Samoa and Tonga before 2500 years ago. There is a genetic cohesiveness among the Polynesians, from Hawaii in the north to New Zealand in the south and including Melanesia, the peoples of which Captain Cook thought were different (Gibbons, 2001).

This wave of migration was subsequent to the earlier colonising 'express' route down to Australia by 55 000 and 46 000 years. The Polynesians clearly had to be accustomed to substantial sea voyages even if they did as much island-hopping as possible. However, the earlier coastal express route would have benefited from the lower sea levels, and the resulting land bridges, in the latter third of the last glacial, which would have greatly reduced the need for extensive sea voyages.

The arrival of humans in Australia marked the onset of a megafaunal extinction event. There has been much debate as to whether this was due to climate change or the presence of humans (although the coincidental timing makes the latter difficult to ignore). It is true that there were climate fluctuations during the last glacial. However, there were similar fluctuations earlier in the glacial and indeed previous glacial and interglacial climatic cycles. It is therefore unlikely that climate alone was the cause of this extinction.

Some 60 Australian taxa are known to have become extinct, including all large browsers, whereas large grazing forms, such as the red (*Macropus rufus*) and great grey (*Macropus giganteus*) kangaroos, survived. The selective loss of browser-dependent species is suggestive of environmental change but, because of the aforementioned argument, climate change alone cannot account for this. So what may have happened?

As discussed in the previous chapter when looking at the late-Miocene expansion of C_4 grasses, herbivorous animals' food can be determined through ^{13}C analysis. C_3 plants (including some grasses) have ^{13}C values of $-26\permil$ while C4 plants (which are mostly grasses) have ^{13}C of $-12\permil$; consequently, things like fossil teeth or eggshells provide a record of the proportion of C_4 and C_3 plants in past herbivore diets. Research by Gifford Miller and colleagues (2005) looked at the proportion of ^{13}C in the eggshells of the now-extinct flightless bird *Genyornis newtoni* and the Australian emu *Dromaius novaehollandiae* to provide a record covering the past 140 000 years in South Australia's semi-arid zone. They found that a marked shift in the proportion of ^{13}C took place 45 000–50 000 years ago. They also tested the ^{13}C record against other glacial–interglacial times (in 15 000-year groupings) to see if climate change was affecting the ^{13}C record. It was not. The suggestion therefore is that the human newcomers were changing the environment through fire. The drought-adapted mosaic of trees, shrubs and grasslands became more like the modern fire-adapted desert scrub. Animals that could adapt survived; those that could not became extinct (Miller *et al.*, 2005).

Whereas this affirms the role of human activity in causing environmental change and so being an extinction driver, it does not, as it may at first seem, discount the impact of climate change on the extinction event. The above testing of climate change against the ^{13}C record only demonstrates that previous climate did not affect the ecology sufficiently to affect the animals' diets. It could be that human-induced (anthropogenic) ecological change did a certain amount of environmental modification and that this could have been furthered by climate change at the time. An alternative variation could be that anthropogenic ecological change put a certain level of pressure on the species and that climate change raised this to extinction level. Both these scenarios are relevant to aspects of twenty-first-century anthropogenic climate change in that human activity has modified a significant part of the modern global terrestrial landscape and that this, coupled with twenty-first-century climate change, is synergistically detrimental to many species as well as local ecosystems and regional biomes.

Meanwhile, the human migration down North America to South America continued. One of the earliest major South American agricultural civilizations that depended on a range of domesticated plants arose in Peru 5000–3800 years ago. The plants used included cotton (*Gossypium barbadense*), squash (*Cucurbita* spp.), chilli (*Capsium* spp.), beans (*Phaseolus vulgaris* and *Phaseolus lunatus*), lucuma (*Pouteria lucuma*; formerly known as *Lucuma obovata*), guava (*Psidium guajava*), pacay (*Inga feuillei*), camote (*Ipomoea batatas*), avocado (*Persea Americana*), maize (*Zea mays*) and achira (*Canna edulis*). This civilisation provided a foundation for many of the features of subsequent Andean civilisations through to early historic times (Haas *et al.*, 2004; Perry *et al.*, 2006).

Once the climate had settled in the current Holocene interglacial, the Neolithic agricultural revolution began to spread from the Near East around 9000–5000 years ago (Hewitt, 2000). Agricultural practice grew and became an increasing part of the landscape. Since the Industrial Revolution in the eighteenth century human population has increased markedly. The global population has been estimated to have been around 425 million in AD 1500 and 600 million in AD 1700. But from then it grew to 4430 million by 1980, an increase of nearly 640% (see Chapter 7). Commensurately, the 1700–1980 increases in cropland were over 466%, or 1.2 billion ha, whereas the larger area of forest and woodland worldwide declined by 18.7%. As we shall see, this will impede natural ecosystem adaptation to climate change. During this time agricultural efficiency increased, which is why the growth in agricultural lands was somewhat less than the growth in population. (As we shall see in Chapter 7, increasing direct and indirect energy inputs largely drove the improved agricultural productivity.) But it was the Industrial Revolution that was to spark the major human impacts on climate change that concern us today. The Industrial Revolution began in the latter half of the Little Ice Age and both these events begin our examination of the present climate and biological change in the next chapter.

4.7 Summary

The past glacial and the current interglacial have seen much biological change, which again clearly demonstrates that biology and climate are interrelated. The past glacial–interglacial cycle also saw the spread and expansion in numbers of modern humans. That humans have had a profound effect on ecology in various parts of the planet is most germane to present climate change that has an increasing anthropogenic component to it. That the climate has been warmer – marginally at times during this and previous interglacials and markedly before the Quaternary – provides palaeoanalogues from which lessons can be learned as the Earth warms again. To do this we need to see how the climate is changing today and this is the focus of the next chapter.

Past glacials and interglacials have also had a genetic impact on species as they have expanded from, and retreated to, their respective refugia. In a world without humans this broad pattern of genetic forcing and glacial–interglacial ecological change would have continued with future glacials and interglacials. However, our planet does have humans (and hence has a changed and more ecologically fragmented landscape) and we do have global warming. Once this warming takes us beyond past interglacial analogues (not unlikely in the twenty-first century), then species' respective genetic mix (hence characteristics) may not always enable them to flourish as they have during the past 2 million years of the Quaternary or even earlier still in the Pliocene. Further, given that component species may not flourish, there is the possibility in some circumstances that global temperatures greater than past interglacials, let alone human land-use impacts, could result in ecosystem dysfunction.

Finally, the growth of human civilisations during the current (Holocene) interglacial has largely been dependent on domesticating species for food. This domestication has taken place over the past several thousand years during which the global climate has been comparatively stable compared to the glacial–interglacial range. As we shall see, twenty-first-century warming is likely to take us beyond this range and so is likely to impact on global food security. The future of food security and climate change will be discussed in Chapter 7.

4.8 References

Abbott, R. J., Smith, L. C., Milne, R. J., Crawford, R. M. M., Wolff, K. and Balfour, J. (2000) Molecular analysis of plant migration and refugia in the Arctic. *Science*, **289**, 1343–6.

Allen, J. R. M., Bradnt, U., Brauer, A. *et al.* (1999) Rapid environmental changes in southern Europe during the last glacial period. *Nature*, **400**, 740–3.

Barber, D. C., Dyke, A., Hillaire-Marcel, C. *et al.* (1999) Forcing of the cold event of 8,200 years ago by catastrophic drainage of Laurentide lake. *Nature*, **400**, 344–8.

Barnosky, A. D., Koch, P. L., Fernac, R. S., Wing S. L. and Shabel, A. B. (2004) Assessing the causes of late Pleistocene extinctions on continents. *Science*, **306**, 70–4.

Bennett, K. D., Haberle, S. G. and Lumley S. H. (2000) The last glacial-Holocene transition in southern Chile. *Science*, **290**, 325–8.

Berger, A. and Loutre, M. F. (2002) An exceptionally long interglacial ahead. *Science*, **297**, 1287–8.

Bernatchez, L. and Wilson, C. C. (1998) Comparative phylogeography of nearctic and palearctic fishs. *Molecular Ecology*, **7**, 431–52.

Bond, G., Broecker, W., Johnsen, S. *et al.* (1993) Correlations between climate records from North Atlantic sediments and Greenland ice. *Nature*, **365**, 143–7.

Borgaonkar, H. P., Kolli, R. K. and Pant, G. B. (2002) Tree-ring variations over the Western Himalaya: little evidence of the Little Ice Age. *Pages News*, **10**(1), 5–6.

Broecker, W. S., Bond, G. C., Clark, K. G. and McManus, J. F. (1992) Origin of the northern Atlantic Heinrich events. *Climate Dynamics*, **6**, 265–73.

Cerling, T. E., Wang, Y. and Quade, J. (1993) Expansion of C4 ecosystems as an indicator of global ecological change and palaeoecologic indicators. *Nature*, **361**, 344–5.

Cerling, T. E., Harris, J. M., MacFadden, B. J. *et al.* (1997) Global vegetation change through the Miocene/Pliocene boundary. *Nature*, **389**, 153–8.

Cowie, J. (1998) *Climate of Human Change: Disaster or Opportunity?* London: Parthenon Publishing.

Cuffey, K. M. and Marshall, S. J. (2000) Substantial contribution to sea-level rise during the last interglacial from the Greenland ice sheet. *Nature*, **404**, 591–4.

Cwynar, L. C. and MacDonald, G. M. (1987) Geographical variations of lodgepole pine in relation to population history. *American Naturalist*, **129**, 463–9.

DeConto, R. M. and Pollard, D. (2003) Rapid Cenozoic glaciation of Antarctica induced by declining atmospheric CO_2. *Nature*, **421**, 245–9.

de Garidel-Thoron, T., Rosenthal, Y., Bassinet, F. and Beaufort, L. (2005) Stable sea surface temperatures in the western Pacific warm pool over the past 1.75 million years. *Nature*, **433**, 294–8.

deMenocal, P., Ortiz, J., Guilderson, T. and Sarnthein, M. (2000) Coherent high- and low-latitude climate during the Holocene. *Science*, **288**, 2198–202.

Elias, S. A., Short, S. K., Nelson, C. H. and Birks, H. H. (1996) Life and times of the Bering land bridge. *Nature*, **382**, 60–3.

EPICA (2004) Eight glacial cycles form an Antarctic ice core. *Nature*, **429**, 623–7.

Fells, T., Lohmann, G., Kuhnert, H. *et al.* (2004) Increased seasonality in Middle East temperatures during the last interglacial. *Nature*, **429**, 164–8.

Ferretti, D. F., Miller, J. B., White, J. W. C. *et al.* (2005) Unexpected changes to the global methane budget over the past 2000 years. *Science*, **309**, 1714–16.

Forster, P. and Matsumura, S. (2005) Did early humans go north or south? *Science*, **308**, 965–6.

Frogley, M. R., Tzedakis, P. C. and Heaton, T. H. E. (1999) Climate variability in northwest Greece during the last interglacial. *Science*, **285**, 1886–8.

Gibbons, A. (2001) The peopling of the Pacific. *Science*, **291**, 1735–7.

Guthrie, R. D. (2004) Radiocarbon evidence of mid-Holocene mammoths stranded on an Alaskan Bering Sea island. *Nature*, **429**, 746–9.

Guthrie, R. D. (2006) New carbon dates link climatic change with human colonization and Pleistocene extinctions. *Nature*, **441**, 207–9.

Haas, J., Creamer, W. and Ruiz, A. (2004) Dating the late Archaic occupation of the Norte Chico region in Peru. *Nature*, **432**, 1020–3.

Haug, G. H., Ganopolski, A., Sigman, D. M. *et al.* (2005) North Pacific seasonality and the glaciation of North America 2.7 million years ago. *Nature*, **433**, 821–5.

Heinrich, H. (1998) Origin and consequences of cyclic ice rafting in the Northeast Atlantic Ocean during the past 130 000 years. *Quaternary Research*, **29**, 142–52.

Hewitt, G. (1996) Some genetic consequences of ice ages, and their role in divergence and speciation. *Biological Journal of the Linnean Society*, **58**, 247–76.

Hewitt, G. (2000) The genetic legacy of the Quaternary ice ages. *Nature*, **405**, 907–13.

Hillaire-Marcel, C., de Vernal, A., Bilodeau, G. and Weaver, A. J. (2001) Absence of deep-water formation in the Labrador Sea during the last interglacial period. *Nature*, **410**, 1073–7.

Hooghiemstra, H. and Van der Hammen, T. (2004) Quaternary ice-age dynamics in the Colombian Andes: developing an understanding of our legacy. *Philosophical Transactions of the Royal Society of London Series B Biological Sciences*, **359**, 173–81.

Hornbach, M. J., Saffer, D. M. and Holbrook, W. S. (2004) Critically pressured free-gas reservoirs below gas hydrate provinces. *Nature*, **427**, 142–4.

Imhoff, M. L., Bounaoua, L., Ricketts, T., Loucks, C., Harriss, R. and Lawrence, W. T. (2004) Global patterns in human consumption of net primary production. *Nature*, **429**, 870–3.

Intergovernmental Panel on Climate Change (2001) *Climate Change 2001: Technical Summary of the Working Group II Report*. Cambridge: Cambridge University Press.

Joyce, D. A., Lunt, D. H., Bills, R. *et al.* (2005) An extant cichlid fish radiation in an extinct Pleistocene lake. *Nature*, **435**, 90–5.

Kennett, J. P., Cannariato, K. G., Hendy, I. L. and Behl, R. J. (2000) Carbon isotopic evidence for methane hydrate instability during Quaternary interstadials. *Science*, **288**, 128–33.

Knutti, R., Flückiger, J., Stocker, T. F. and Timmermann, A. (2004) Strong hemispheric coupling of glacial climate through freshwater discharge and ocean circulation. *Nature*, **430**, 851–6.

Kukla, G. J., Matthews, J. M. and Mitchell, Jr, J. M. (1972) The end of the present interglacial. *Quaternary Research*, **2**, 261.

Laird, K. R., Fritz, S. C., Maasch, K. A. and Cumming, B. F. (1996) Greater drought intensity and frequency before AD 1200 in the Northern Great Plains, USA. *Nature*, **384**, 552–4.

Lamb, H. H. (1965) Outline of Britain's climatic history. In Johnson, C. G. and Smith, L. P., eds. *The Biological Significance of Climate Changes in Britain*, pp. 3–31. London: Academic Press and The Institute of Biology.

Macauley, V., Hill, C., Achilli, A. *et al.* (2005) Single, rapid coastal settlement of Asia revealed by analysis of complete mitochondrial genomes. *Science*, **308**, 1034–6.

Marincovich, Jr, L. and Gladenkov, A. Y. (1999) Evidence for an early opening of the Bering Strait. *Nature*, **397**, 149–51.

Marshall, E. (2001) Pre-Clovis sites fight for acceptance. *Science*, **291**, 1730–2.

Mellars, P. (2006) A new radiocarbon revolution and the dispersal of modern humans in Eurasia. *Nature*, **439**, 931–5.

Miller, G. H., Fogel, M. L., Magee, J. W., Gagan, M. K., Clarke, S. J. and Johnson, B. J. (2005) Ecosystem collapse in Pleistocene Australia and a human role in megafaunal extinction. *Science*, **309**, 287–90.

Moreno, P. I., Jacobson, Jr, G. L., Lowell, T. V. and Denton. G. H. (2001) Interhemispheric climate links revealed by a late-glacial cooling episode in southern Chile. *Nature*, **409**, 804–8.

Morwood, M. J., Soejono, R. P., Roberts, R. G. *et al.* (2004) Archaeology and age of a new hominin from Flores in eastern Indonesia. *Nature*, **431**, 1087–91.

Pagani, M., Freeman, K. H. and Arthur, M. A. (1999) Late Miocene atmospheric CO_2 concentrations and the expansion of C4 grasses. *Science*, **285**, 876–9.

Pastor, J. and Moen, R. A. (2004) Ecology of ice-age extinctions. *Nature*, **431**, 639–40.

Perry, L., Sandweiss, D. H., Piperno, D. R. *et al.* (2006) Early maize agriculture and interzonal interaction in southern Peru. *Nature*, **440**, 76–9.

Peterson, L. C., Haug, G. H., Hughen, K. A. and Röhl, U. (2000) Rapid changes in the hydrological cycle of the tropical Atlantic during the last glacial. *Science*, **290**, 1947–51.

Pétit, J. R., Jouzel, J., Raynaud, D. *et al.* (1999) Climate and atmospheric history of the past 420 000 years from the Vostok ice core, Antarctica. *Nature*, **399**, 429–36.

Petit, R. J., Aguinagalde, I., de Beaulieu, J.-L. *et al.* (2003) Glacial refugia: hotspots but not melting pots of genetic diversity. *Science*, **300**, 1563–5.

Pielou, E. C. (1991) *After the Ice: the Return of Life to Glaciated North America.* Chicago: University of Chicago Press.

Premoli, A. C., Kitzberger, T. and Veblen, T. T. (2000) Isozyme variation and recent biogeographical history of the long-lived conifer *Fitzroya cupressoides*. *Journal of Biogeography*, **27**, 251–60.

Ravelo, A. C., Andreasen, D. H., Lyle, M., Lyle, A. O. and Wara, M. W. (2004) Regional climatic shifts caused by gradual global cooling in the Pliocene epoch. *Nature*, **429**, 263–7.

Rioul, P., Andrieu-Ponel, V., Rietti-Shati, M. *et al.* (2001) High-resolution record of climate stability in France during the last interglacial period. *Nature*, **413**, 293–6.

Rohling, E. J. and Pälike, H. (2005) Centennial-scale climate cooling with a sudden cold event around 8,200 years ago. *Nature*, **434**, 975–9.

Rothwell, R. G., Thomson, J. and Kähler, G. (1998) Low-sea-level emplacement of a very large late Pleistocene megaturbide in the western Mediterranean Sea. *Nature*, **392**, 377–80.

Ruddiman, W. F. (2003) The anthropogenic greenhouse era began thousands of years ago. *Climate Change*, **61**(3), 261–93.

Shackleton, N. J. (2000) The 100,000 year ice-age cycle identified and found to lag temperature, carbon dioxide and orbital eccentricity. *Science*, **289**, 1897–1901.

Shapiro, B., Drummond, A. J., Rambaut, A. *et al.* (2004) Rise and fall of the Beringian steppe bison. *Science*, **306**, 1561–5.

Siegenthaler, U., Stocker, T. F., Monnin, E. *et al.* (2005) Stable carbon cycle-climate relationship during the late Pleistocene. *Science*, **310**, 1313–17.

Sigman, D. M., Jaccard, S. L. and Haug, G. H. (2004) Polar ocean stratification in a cold climate. *Nature*, **428**, 59–63.

Smith, L. C., MacDonald, G. M., Velichko, A. A. *et al.* (2004) Siberian peatlands a net carbon sink and a global methane source since the early Holocene. *Science*, **303**, 353–7.

Sowers, T. (2006) Late Quaternary atmospheric CH_4 isotope record suggests marine clathrates are stable. *Science*, **311**, 838–40.

Spahni, R., Chappellaz, J., Stocker, T. F. *et al.* (2005) Atmospheric methane and nitrous oxide of the late Pleistocene from Antarctic ice cores. *Science*, **310**, 1317–21.

Steig, E. J. (1999) Mid-Holocene climate change. *Science*, **286**, 1485–6.

Steine, S. (1994) Extreme and persistent drought in California and Patagonia during the Mediaeval time. *Nature*, **396**, 546–9.

Stokes, S., Thomas, D. S. G. and Washington R. (1997) Multiple episodes of aridity in southern Africa since the last interglacial period. *Nature*, **388**, 154–8.

Stuart, A. J., Kosintsev, P. A., Higham, T. F. G. and Lister, A. M. (2004) Pleistocene to Holocene extinction dynamics in giant deer and woolly mammoth. *Nature*, **431**, 684–9.

Stute, M., Forster, M., Frischorn, H. *et al.* (1995) Cooling of tropical Brazil (5°C) during the last glacial maximum. *Science*, **269**, 379–83.

Tarasov, L. and Peltier, W. R. (2005) Arctic freshwater forcing of the Younger Dryas cold reversal. *Nature*, **435**, 662–5.

Taylor, K. C., Lamorey, G. W., Doyle, G. A. *et al.* (1993) The 'flickering switch' of late Pleistocene climate change. *Nature*, **361**, 432–5.

Thangaraj, K., Chaubey, G., Kivisild, T. *et al.* (2005) Reconstructing the origin of Andaman islanders. *Science*, **308**, 996.

Trauth, M. H., Maslin, M. A., Deino, A. and Strecker, M. R. (2005) Late Cenzoic moisture history of East Africa. *Science*, **309**, 2051–3.

Tripati, A., Backman, J., Elderfield, H. and Ferretti, P. (2005) Eocene bipolar glaciation associated with carbon cycle changes. *Nature*, **436**, 341–6.

Tyson, P., Odada, E., Schulze, R. and Vogel, C., (2002) Regional-global change linkages: Southern Africa. In Tyson, P., Fuchs, R., Fu, C. *et al.*, eds. *Global-Regional Linkages in the Earth System*. Berlin: Springer.

Vandergoes, M. J., Newham, R. M., Preusser, F. *et al.* (2005) Regional insolation forcing of late Quaternary climate change in the Southern Hemisphere. *Nature*, **436**, 242–5.

Verschuren, D., Laird, K. R. and Cumming, B. F. (2000) Rainfall and drought in equatorial east Africa during the past 1,100 years. *Nature*, **403**, 410–13.

Wang, X., Auler, A. S., Edwards, R. L. *et al.* (2004) Wet periods in northeastern Brazil over the past 210 kyr linked to distant climate anomalies. *Nature*, **432**, 740–3.

Wara, M. W., Ravelo, A. C. and Delaney, M. L. (2005) Permanent El Niño-like conditions during the Pliocene warm period. *Science*, **309**, 758–61.

Weaver, A. J., Saenko, O. A., Clark, P. U. and Mitrovica, J. X. (2003) Meltwater Pulse 1 A from Antarctica as a trigger for the Bølling-Allerød warm interval. *Science*, **299**, 1709–12.

Wolff, E. W., Fischer, H., Fundel, F. *et al.* (2006) Southern ocean sea-ice extent, productivity and iron flux over the past eight glacial cycles. *Nature*, **440**, 491–6.

Zachos, J., Pagani, M., Sloan, L., Thomas, E. and Billups, K. (2001) Trends, rhythms and aberrations in global climate 65 MA to present. *Science*, **292**, 686–93.

Zazula, G. D., Froese, D. G., Schweger, C. E. *et al.* (2003) Ice-age steppe vegetation in east Beringia. *Nature*, **423**, 603.

5

Present climate and biological change

5.1 Recent climate change

5.1.1 The latter half of the Little Ice Age

The seventeenth century was not just in the Little Ice Age, it is also noted (and for some better noted) for the Renaissance in the gathering of scientific understanding which in turn was to drive the Industrial Revolution of the eighteenth and nineteenth centuries. In Britain in the 1640s and 1650s scientists sought what they termed 'a great insaturation', which drew on the philosophies exposed by the likes of Francis Bacon. Among these, Bacon's principles of exact observation, measurement and inductive reasoning provided the intellectual tools for scientific advance. These advances had yet to percolate through to day-to-day application in technology, and life, society and its economy were still largely powered by humans and animals together with wood burning. Major global impacts from human activity were not yet manifest (although of course trace global signatures such as metals in Greenland ice cores can be found dating to thousands of years earlier).

In terms of climate and weather 1659, within the Little Ice Age, is an important date. Before that date we rely solely on the proxy indicators (see Chapter 2) for climatic information. After 1659 there began a source of new information: direct meteorological measurement. The first significant series of measurements began in 1659 and (much later) were compiled into a monthly series of temperature readings for rural sites in central England by Gordon Manley (1974): it is the longest homogeneous record and is still kept up to date by the UK Meteorological Office. The earlier measurements were varied but increasingly included, and were soon dominated by, instrumental measurements. The Central England Temperature (CET) records were soon accompanied by others to ultimately be built into a series such as those for De Bilt in The Netherlands from 1705. Such records are fundamentally important. As we have

Figure 5.1 Seven centuries of Holocene Greenland deuterium measurements. The proportion of deuterium (^{2}H) in the ice, compared to the common isotope of hydrogen in water, is proxy for regional temperature (see Chapter 2). Data from the National Oceanic and Atmospheric Administration (see Acknowledgements).

seen, whereas we can use a variety of proxies to build up, piece by piece, quite a good picture of past climate, deep-time climatic proxy indicators simply are either not sensitive or representative enough to tell us what was going on. This is especially true with regard to finer changes. For example, ice-core isotopic records are very fine for charting regional glacial and interglacial transitions of a few degrees but are less useful for discerning trends in changes of fractions of a degree within our current interglacial. For instance, Figure 5.1 is a portrayal of seven centuries of Greenland deuterium up to the late twentieth century. That it only represents the deuterium evapo-fractionation from the ocean surrounding Greenland restricts its usefulness. This is not just because of regionality but because different times of the year see different amounts of water transportation ending up as Greenland ice, so whereas some parts of the year may be getting warmer, others may be getting cooler, and what ends up in the ice may not be representative. Also, the climate is not one-dimensional. Temperature alone does not determine climate but a range of factors, such as humidity, precipitation, sunshine, seasonality and so forth. Instrumentation provides standard, more sensitive, forms of measurements independent of the constraints of many individual physicochemical, and biological, palaeoclimatic proxies. There is a somewhat ironic symmetry that we have relied on proxies, especially biological ones, to determine past climates before the Renaissance, and then

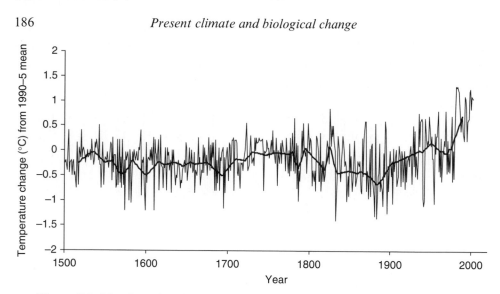

Figure 5.2 Northern-hemisphere climate trends over 500 years as discerned from instrumental, dendrochronological and Greenland ice-core records. Based on Luterbacher *et al.* (2004).

instrumentation afterwards. The Renaissance enabled there to be an Industrial Revolution that in turn saw the beginnings of extensive fossil-fuel use which is now beginning to perturb biological systems. The post-Renaissance period enabled calibration of the bioproxies used in the former. Since the Renaissance, instrumentation has enabled the discipline of meteorology. Its forecasts have been useful primarily because the weather affects our biologically based lives and, more recently, because our anthropogenic perturbations of the climate are affecting many biological systems globally. Nonetheless, at first (in the Little Ice Age) instrumental records were neither widespread nor as rigorously compiled as they became later. Combining early records with a variety of palaeo-climatic indicators does provide a useful picture, such as that in Figure 5.2.

Against the context of the Quaternary glacial–interglacial cycles that helped shape human evolution, the Little Ice Age is a comparatively minor blip in our current Holocene interglacial. There are of course even smaller blips and, regionally there are extreme meteorological events of even shorter duration. Yet these can have an acute impact on human activities, as we shall see in the next section. For now, one twentieth-century example will serve to illustrate the point: the dust-bowl years of the 1930s, especially 1934 and 1936, in the midwest USA. This extreme arose because temperate continental interiors are subject to great seasonality with extremely cold winters and hot summers. For farmers in the midwest this last was not too much of a problem as the summer heat enabled vast quantities of water to be transported to fall as rain. However, the years 1934 and 1936 were both extremely hot *and* extremely dry. The results, combined with poor agricultural practice, was not only crop failure but severe soil erosion. The thing to note here is that just as in more recent times there have been extreme regional events, so there were also in the Little Ice Age.

Notwithstanding isolated extreme events, the Little Ice Age itself was not a sustained period of cold. This is why perhaps there is not as much clear evidence from across the planet as one might like to define such a period as one of great global significance. As with the dust-bowl years, the Little Ice Age mainly shows up in specific climatic dimensions. The 1590s were cold years but frequent subsequent years were more notable specifically for their colder winters. But one of the first climatic events (albeit it a mini one) picked up by the CET record is that after the exceptionally cold 1690s there was a warming around the turn of the century. The 1720s and 1730s were as warm as the late twentieth century before suddenly returning to cooler conditions in 1740.

The climate, as far as Europe was concerned, showed no discernable trend over the next 180 years: there was nothing warmer than the 1730s and nothing cooler than the 1690s. This shows up in not only the CET and the De Bilt series but also records from other European cities. The summers were particularly consistent. What signals the emergence from the Little Ice Age is really reflected in less-harsh winters. So, just as the twentieth-century midwestern-USA dust-bowl years are defined climatologically through their hot dry summers, so the Little Ice Age is defined by harsher winters and shorter summers that may have been a little wetter and only marginally cooler than those in the twentieth century. This resulted in a shorter thermal growing season with atypical conditions at its beginning (sowing time) and end (harvest time). It is estimated that in England during the coldest decades of the Little Ice Age the thermal growing season was shortened by 3–4 weeks. As we shall see later in the chapter this is the same order of magnitude of biological effect we are seeing early in the twenty-first century compared with the mid-twentieth century due to global warming, although the effect is obviously the other way, with an extension of the growing season (see Chapter 6). As we shall also see in the next section, this shortening of the thermal growing season in the Little Ice Age would result in considerable human suffering.

Added to this change in the climate year, there were also a few volcanic eruptions that seem to have had a global effect, with years following the eruption being even colder than the other years of the Little Ice Age. These included Laki (1783), Tambora (1815) and Krakatau (1883; see Figure 2.1).

Interestingly, as in Europe, in eastern Asia the seventeenth century was also a cold period, especially around 1800. However, nineteenth-century eastern Asia did not see years that were as cold as those in western Europe or North America.

The Little Ice Age should therefore not be seen as a period of consistently uniform cold across the planet. There were, though, a few short cold episodes, either across the northern hemisphere or globally. These were the decades 1590s to 1610s, 1690s to 1710s and 1800s to 1810s. There were also a few slightly warm periods, including the 1730s and 1820s, before reaching the twentieth-century warming, which commenced in the 1930s (Burroughs, 1997).

The third quarter of the nineteenth century saw a brief easing of conditions, as signalled by glacier retreat in Europe. This did not last and the century's final quarter saw glacial expansions in Europe and Canada. In terms of high-latitude northern-hemisphere climate this quarter represented the harshest period of the Little Ice Age. As time moved on into the early twentieth century the Little Ice Age clearly came to an end.

Although we do not know the exact nature of the Little Ice Age (if it were truly global and not just restricted mainly to the northern hemisphere) it may still have been the harshest climate the Earth has seen since the Holocene's beginnings. Certainly, in terms of European glaciers, these were at their most advanced during the Little Ice Age compared with their decline early on in the Holocene. If the Earth were slowly sinking back into a glacial (and it would be slow, given that Milankovitch forcing will tend to promote an interglacial over the next 10 000 years), then the Little Ice Age might have been representative of a longer-term trend. If this is really the case then recent greenhouse climatic forcing due to human emissions of greenhouse gases has had a tremendous impact, for recent greenhouse warming would not only have to be compared against, say, an early twentieth-century standard, but also against a theoretical, cooler early twentieth century had the Earth's slow slide into a future glacial continued unaffected.

Looking at the Little Ice Age from a greenhouse perspective raises a question. Carbon dioxide levels during that time had not deviated particularly. So, what caused it? There are arguments that other factors played a part, such as the solar Maunder minimum (see Chapter 4) and also the number of significant volcanic eruptions is suggestive. What is known (even if the exact extent is still unclear) is that climate forcing from anthropogenic greenhouse emissions has increased markedly since the end of the nineteenth century.

5.1.2 Twentieth-century climate

Greenhouse-gas emissions from burning of fossil fuels and land-use change (mainly deforestation) increased dramatically in the twentieth century. More carbon dioxide was emitted in the second half of the twentieth century compared to the 100 years leading up to 1950. Further, emissions from burning of fossil fuels were four times higher by the end of the century (an increase of 300%) compared to 1950 (Cowie, 1998; we will return to the pattern of emissions later). As a result, the atmospheric concentration of carbon dioxide rose from its pre-industrial (mid-eighteenth century) level of 279 ppmv to 369.4 ppmv in the year 2000 (Worldwatch Institute, 2003).

Apart from some wavering (mainly natural variability), the twentieth century was notable for being one of warming to such an extent that the 1990s saw a concentration of record-breaking hot years.

5.1.3 Twenty-first-century climate

Such has been the anthropogenic injection of greenhouse gases into the atmosphere since the Industrial Revolution, and especially during the twentieth century, that there is little doubt that global warming will continue throughout the twenty-first century. The international scientific consensus as to twenty-first-century warming has been determined by the Intergovernmental Panel on Climate Change (IPCC). This will be covered in the section on the IPCC business-as-usual scenario (section 5.3.1).

5.1.4 The Holocene interglacial beyond the twenty-first century

Since the 1970s, when ice-core analysis revealed the broad nature of the glacial–interglacial cycle, there has been a general assumption that the length that glacials last is the order of 100 000 years; conversely, interglacials are an order of magnitude smaller and are 10 000 years long. Furthermore, as noted in Chapter 4, ice cores have showed that the transition between past glacial and interglacial states was sudden and not gradual. Early on (in the 1970s) there was a concern that as the Holocene was already about 10 000 years old the Earth might be about to enter a glacial period. As noted in Chapter 1, global warming and human-driven greenhouse concerns had long abounded and so the question for the late 1970s and early 1980s focused on the balance between these two warming and cooling forcing agents. The IPCC's first report in 1990 (see below) provided an international scientific consensus, concluding clearly that anthropogenic global-warming factors superimposed on the natural climate-forcing agents would dominate through the twenty-first century and, by implication, at least decades beyond, if not more.

However, in the 1990s two things became clear. First, in the northern hemisphere at least there had been a marginal cooling trend of 0.2 °C between AD 1000 (especially since AD 1350) and AD 1900. This cooling trend was both slight and only just above the background noise of climatic variability.

Nonetheless, it might possibly suggest that, as hinted above, in the absence of twentieth-century global warming the beginning of the end of the current Holocene interglacial would be upon us soon. (The fear in the 1970s had been that there would be global cooling into a glacial, and not global warming.) However, as noted above, this long-term cooling trend ended with the twentieth-century warming. The second thing that became clear in the 1990s was that both glacials and interglacials are complex, with vagaries of their own (such as Bond cycles and associated Heinrich events within a glacial). Furthermore, glacials and interglacials have their own individual characteristics. Just because the Eemian interglacial (which in some texts is referred to as the Ipswichian or Sangamon) 130 000 years ago was roughly similar in length to its predecessor interglacial 220 000 years ago (variously referred to in texts as the Hoxnian, Holstein or Yarmouth interglacial), this does not necessarily mean that our Holocene interglacial will be a similar length.

As noted in Chapter 1, the pacemaker of glacials and interglacials is the Milankovitch solar radiation curve. Again (see Chapter 1) this curve is made up of three cycles of varying lengths and so it is likely that, unless the Milankovitch circumstances are similar, each glacial and interglacial will have its own characteristics. Indeed, the closest of the interglacials similar to our own in a Milankovitch sense was the one following Termination V (five glacials ago), called the Waalian (sometimes known as the Pastonian) interglacial, which began 425 000 years ago. It is this interglacial, and not the Eemian or the Hoxnian, that should be used as an analogue for our current interglacial (Augustin *et al.*, 2004).

Using Milankovitch curves projected into the future (Berger and Loutre, 2002) it can be seen that, unlike the Eemian, the Holocene interglacial will be exceptionally long. Looking back at the Waalian interglacial following Termination V, we can also see from the Antarctic ice-core record that that interglacial lasted some 30 000 years, from 425 000 to 395 000 years ago (Augustin *et al.*, 2004). In other words, *without* global warming it is *unlikely* that we would soon see a return to full glacial conditions. Without global warming it is thought that we might have an overall cooling over the next 25 000 years, only after which would there then follow a sharper return to glacial conditions. However, assuming that we will continue to add substantial amounts of carbon dioxide to the atmosphere over the next couple of centuries, we will be forcing the climate well above what would have been expected for our current interglacial.

In short, the future sees us entering a climate mode that bucks the natural trend and which has not been seen during the past 2 million years of glacials and interglacials. Either way, the Holocene interglacial is not likely to end soon. Conversely, not only is it likely to continue for well over another 10 000 or even 25 000 years, but with human-induced warming it *possibly* (in the context of the last chapter's discussion of the early-Eocene carbon event) may last for well over 100 000 years: here at the very least the next Milankovitch-paced glacial would be skipped. Meanwhile, the next few centuries are likely to see global warming with temperatures unprecedented during our previous inter-glacials, let alone our current one, which saw the rise of early human civilisation.

5.1.5 Holocene summary

Whereas the Holocene represents a period of warmth as well as of climatic stability, the past 10 000 years have not seen a static climate. The Holocene climatic maximum saw considerable shifts in biomes, both before and after. More recently there was a second medieval (warm) anomaly and then the Little Ice Age cooling. This last in particular serves to demonstrate that there are many factors involved in the forcing of the global climate and not just green-house gases, be they natural or anthropogenic. Indeed other factors, such as volcanic and solar activity, play their parts.

So far we have only considered climate change in the broadest of contexts and with little detail. There is not the space in a book that introduces the interface between climatology, biology and human ecology to dwell at length on the intricacies of climate. For instance, one could spend much time debating where weather variability ends and climate change begins. Nonetheless, such intricacies do exist and the reader needs to be aware of this. For example, there are several oscillations taking place in the oceans and atmosphere that determine dry or wet, warm or cool, or combinations thereof in the seasons in any year.

For instance, in the south Pacific there is an oscillation, the El Niño Southern Oscillation (ENSO) that results in El Niño years. The El Niño is a current that flows along the coasts of Ecuador and Peru between January and March and which terminates the local fishing season. In some years it is warmer and lasts longer, so preventing a nutrient-rich upwelling. It is part of the bigger Southern Oscillation that sees the Pacific waters build up due to the prevailing winds in the western South Pacific. This build-up takes time and is released both when the build-up is sufficient and meteorology allows. There are a number of other key oscillations. One affecting North America and Europe is the Arctic Oscillation. Another, affecting the Antarctic, is the Antarctic Oscillation, which is some-times known as the Southern Annular Mode.

A question researchers have posed in recent years is whether these oscil-lations are themselves modulated over a longer time frame. For example, one recent analysis of layers in cores of 13 lakes in north-eastern USA has determined how the degree of storminess has varied over the Holocene. Here their results show a characteristic period of about 3000 years and that this pattern appears consistent with long-term changes in the average sign of the Arctic Oscillation (Noren *et al.*, 2002). Such oscillations are but one grouping of many climatic intricacies that shape the complex detail of climate change.

Notwithstanding this complexity, or the natural vagaries and climatic-related oscillations, the Holocene's most recent hundred years (the twentieth century) have largely been a period of warming. Be it based on palaeoclimatic proxies (such as the high-latitude dendrochronological records in Figure 2.1), or instrumental (be it satellite- or ground-based), or a combination of both (Figure 5.2), the late twentieth-century warming has been unprecedented since instrumentation began. Further, using palaeo proxies, the late twentieth century has seen the greatest warming for 1000 years. Indeed multi-proxy reconstructions of monthly seasonal surface temperatures for Europe back to 1500 show that the late twentieth and early twenty-first centuries were warmer than any other time during the past 500 years. This agrees with findings over the entire northern hemisphere. European winters were 0.5 °C warmer in the twentieth century compared to the previous four centuries. Conversely, summers have not been that much warmer, although individual ones have had warmer spells. Finally, in the half millennium up to 2000, the coldest European winter was that of 1708/9; so, not in (present human-generation) contemporary times but in the Little Ice Age (Luterbacher *et al.*, 2004).

That the twentieth century's warming trend is almost certainly going to continue is of tremendous concern. As noted above it looks as if the Earth is entering a climatic mode unprecedented in human history. Indeed, it is likely that in the twenty-first century the Earth will achieve temperatures not seen since the height of the last interglacial (140 000 years ago) and just possibly temperatures not seen during the Quaternary (the past 2 million years). Beyond this, further warming in the next century may result in temperatures not seen since before the Quaternary well over 2 million years ago. If so – and it is a real possibility – it will transform our planet in ways that many have not contemplated today. It is because we are biological creatures that are involved in, and dependent on, many biological systems that in turn are affected by climate that we need to take note. Such climate change will certainly affect virtually all our lives both individually and on a societal level. To recognise the serious nature of this possibility one needs only see how the more modest climatic changes have affected human activities from the medieval climatic anomaly to the present.

5.2 Human change arising from the Holocene climate

5.2.1 Climatic impacts on early human civilisations

Human civilisations (settlements and cities) arose early on in the Holocene interglacial. These settlements were small, minute by modern standards. Importantly, sea-level rise following the last glacial continued for a few thousand years after the Holocene's commencement and continued through to about 7000 years ago. Consequently some of the early coastal civilisations became inundated, as is evident from their settlements' now submarine remains in the Mediterranean and off the coasts of India. This is probably the earliest major impact of global climate change on human civilisation. That is to say, an impact as a direct result of climate on the most advanced type of human activity at the time (unless one were to argue that, prior to settlements of buildings, the development of the use of fire during some of the preceding Quaternary glacials–interglacials amounted to civilisation).

Early civilisations were vulnerable to climate change. For example, it is likely that the civilisation in the Indus valley 4500–3500 years ago, and their principal cities of Harappa and Mohrnjo-daro, died due to drought. There is no evidence of war but we do know that there was a major decline in rainfall across the Middle East and northern India 3800–2500 years ago. The same drought may also have affected civilisations, such as the Hittite empire, Mycenae and Egyptians, around the eastern Mediterranean. I say 'may have' because their decline is sometimes attributed to a raiding, aggressive, sea-going people. However, it is interesting to note that the Roman Empire thrived over a period that was coincident with a benign climate. Indeed it was a Roman, the writer Saserna, who was one of the first to suggest that a biologically related human activity, viticulture, was a proxy indicator of climate change.

Later the decline of many civilisations after AD 536 could also have been climate-related, due to a major volcanic eruption (perhaps Rabaul, in New Guinea). Chroniclers of the day from Rome to China report that the Sun dimmed for up to 18 months, enough to have a marked effect on some temperate species, and there were widespread reports of poor harvests.

Meanwhile in Central America, ^{18}O sediment analysis from Lake Chichancanab reveals that over the past 8000 years it was driest in AD 750–900 (Hodell *et al.*, 1995). This may explain the collapse of the Mayan civilisation that had previously thrived for over 2000 years. It culminated in a civilisation covering much of Guatemala, Belize, Honduras and Mexico, sustaining a population at a level higher than today's. Although the evidence is not absolutely conclusive, the pattern of Mayan decline reveals that cities next to rivers survived the longest, which is what you would expect in a drought-driven decline. It has also been suggested that in South Africa the medieval anomaly around AD 900–1300 may have been responsible for changes in settlement patterns (Tyson *et al.*, 2002).

5.2.2 *The Little Ice Age's human impact*

The Little Ice Age (see the Holocene section in the previous chapter), and the period that led up to it, is probably one of the best-recorded historic impacts of climate on human activity. Prior to the Little Ice Age, the warm medieval climatic anomaly (MCA; or medieval warm period) had enabled viticulture in England. Since then and up to the late twentieth century viticulture's main northern boundary (other than a few pockets) existed in Normandy some 480 km to the south. Because in Europe the human impact of this transition from medieval warmth to Little Ice Age is so well recorded we know what took place. It is less well documented outside of Europe, although it has been thought that possibly the decline of the Mapungubwe state (in the current semi-arid Shashi-Limpopo basin, the area where the Botswana, Zimbabwe and South African boundaries meet) may have been a Little Ice Age impact. As noted in the previous chapter the palaeo-evidence for the severity of the Little Ice Age outside of the northern hemisphere is not so clear, although there is evidence of other climatic change, other than temperature, such as changes in aridity/precipitation. Similarly, in the Kuiseb River Delta of Nambia settlements were abandoned between 1460 and 1640 (Tyson *et al.*, 2002). Was this due to the Little Ice Age? There is oxygen isotopic evidence from stalagmites in Makapansgat that the termination of what are assumed to be Little Ice Age effects in South Africa was abrupt. Also, as discussed in the last chapter, irrespective of its extent, even in the northern hemisphere the reasons for the Little Ice Age are not entirely clear. This should alert policy-makers regarding future climate change that distinctive climatic-forcing events are not necessary for climate change associated with considerable human impact: combinations of lesser forcing factors may result in marked human impact.

The Little Ice Age saw Western Europe experience a general cooling of the climate between the years 1150 and 1460 and a very cold climate between 1560 and 1850 that brought dire consequences to its people. The colder weather impacted agriculture, health, economics, social strife, emigration and even art and literature. Increased glaciation and storms also had a devastating effect on those who lived near glaciers and the sea. In the northern hemisphere the more severe climatic changes were not restricted to Europe, however. In China the population fell from around 100 million in the mid-thirteenth century by about 40% over the next century or so.

What happened in terms of human impact has been pieced together from a variety of sources. Records of English, French and German harvests and/or prices survive to this day and these correlate well with the weather in good and bad years. So, it is easy to see that each of the peaks in prices corresponds to a particularly poor harvest, mostly due to unfavourable climates, with the most notable price peak in the year 1816 – 'the year without a summer' – that followed the Tambora eruption of 1815 (see Figure 2.1).

The Little Ice Age harvest losses not only caused price increases but death too. Famines became more frequent and one in 1315 killed an estimated 1.5 million in Europe. This was probably the first big famine of the Little Ice Age and may have been aggravated because agricultural systems were still partly embedded in medieval climatic anomaly (MCA) mode that lasted through much of the thirteenth century. This happened elsewhere in the northern hemisphere. In the east, as mentioned, there was famine in China. In the west of the northern hemisphere, at the century's end the Anasazi abandoned their long-established communities in the now desert of the south-western USA. At the time, North American climatic belts shifted, including those suitable for economically important crops: in what is now Wisconsin the northern limit of maize cultivation moved southwards by up to 320 km.

At the beginning of the fourteenth century, Northern Europe saw two exceptionally severe winters, in 1303 and 1306. However, it was the run of cool and wet summers between 1314 and 1317 that particularly decimated harvests, hence precipitating the aforementioned great famine of 1315. Records show that this reduced the food supply from Scotland to the Pyrenees and as far east as Russia. In some places, even when plant growth was good, the wet, as is clearly recorded in England, made the harvest difficult. In London in the spring of 1316 the price of wheat rose eight-fold over that of late 1313: these price levels were not to be reached again until later in the Little Ice Age in the late fifteenth century. Meanwhile, across the English Channel in France in just 6 months grain prices rose by more than three-fold up to May 1316.

The climate also had military impacts. In France King Louis X's attempt to combat Count Robert of Flanders came to a halt when Louis's army got bogged down in mud. Things were so bad that horses could not pull wagons. Meanwhile, for the average person matters were dire. People ate dogs and reports of cannibalism were not uncommon; graves were robbed and criminals cut down from gibbets for food. Similar reports emerge from other Little Ice Age famines but those from 1314–17 were particularly bad and this famine undoubtedly exceeded the others (Burroughs, 1997).

It is worth emphasising that the aforementioned human impacts experienced early in the fourteenth century were not simplistically proportional to the poor climate. Their severity was augmented by communities having grown during the preceding warmer climatic anomaly and expanding on to lands that were ideal for the warm anomaly but which became marginal during the Little Ice Age. Such demographic pressures can exacerbate the impacts of climate change. This is a theme we will return to again when looking at future climate change (Chapters 6 and 7) as currently the early-twenty-first-century global population is at a record-breaking high, at over 6 billion and counting.

It is not known how many died in total due to the 1314–17 famine but in Ypres in Belgium (an advanced city of the time) records suggest a mortality of over 10%. Because the Black Death (or the Great Mortality as it was known then) ravaged through England in 1348 and Scotland in 1349, fourteenth-century mortality from both plague and various climate impacts reduced England's population by around a third. Looking at the fourteenth century as a whole, much of Europe and Eurasia was similarly affected.

Climate (or in this case perhaps the weather) also affected the Black Death. While the Little Ice Age saw many cool years, there were also occasional warm ones. One of the warmest was that of 1348 and this may account for the rapid spread of plague that year throughout England. But more typical of the Little Ice Age was cold and wet, and this too, under certain circumstances, helped spread plague. Great floods in China in 1332 not only killed several million people but disrupted large areas of the country, causing a substantial move-ment of both people and wildlife. Regarding emergent diseases, this was a recipe for disaster and so it is not unreasonable to conclude that climate played its part in the way that the plague spread (Burroughs, 1997).

Although the 1314–17 famine was probably the worst in terms of human impact, it was not the only one during the Little Ice Age. There were vagaries of the weather but harvests were not usually so bad as to provoke sharp and major price rises in grain, although there were some, such as in 1439, 1482, 1608, 1673 and 1678. Another severe famine, particularly according to French records, took place in the seventeenth century due to the failed harvest of 1693. Millions of people in mid-western Europe – France and surrounding countries – died. This time was again one of poor climate (see Figure 5.2) and the 1690s saw a number of very cold winters and cool, wet summers affecting much of western Europe. Indeed, 1693 was one of the worst famines since 1314–17, although the impact was worse in continental Europe than in England. 1697 was also a bad year and it is estimated that in Finland roughly a third of the population died.

As noted, the effect of the harsher climate on those already on marginal land was far worse than those on formerly good lands of the MCA that the Little Ice Age in turn made marginal. Those already living at high altitudes in mountain pastures were very badly affected. Consequently Scandinavian, Scottish and

Swiss farmers suffered particularly. Problems were not solely confined to direct climatic impacts. The new environmental conditions favoured some crop pathogens. Since snow covered the ground deep into spring in the mountainous parts of Europe a fungus known as pink snow mold (*Fusarium nivale*) devastated crops. It thrives under melting snow cover or prolonged, cold, drizzly weather. Additionally, due to the increased number of days of snow cover, the stocks of hay for the animals ran out so livestock were fed on less-nutritious straw and pine branches. Many cattle had to be slaughtered. Others still died in the heavy snow. Marginal lands at high latitudes were similarly pressured. In northern Norway many farms were abandoned for better land in the bottoms of valleys. By 1387, Norwegian agricultural production and tax yields were far less compared with what they had been around 1300. By the 1460s it was being recognised that this change was long-term. As late as 1665 the annual Norwegian grain harvest is reported to have been only 67–70% of what it had been in the year 1300 (Lamb, 1995.)

The Little Ice Age did not just affect marginal lands but also sensitive crops; crops that required specific climatic conditions at specific times in the thermal growing season. Illustrative, as previously noted, is Saserna's suggestion of viticulture as a climate proxy. Grapes require a narrow range of conditions for ripening and so are particularly sensitive to climate change. The Little Ice Age saw many countries with a strong viticulture industry, such as France, have bad years for wine coincident with climate swings. Naturally those areas, such as southern England, that had benefited from the MCA and which had become the new northern margin for viticulture, were not able to grow vines much after 1400. German wine production between 1400 and 1700 was never above 53%, and at times was as low as 20%, of the production before 1300.

The sociological impacts of the Little Ice Age have also been suggested to have resulted in one particularly gruesome aspect of European life of the time and particularly around the 1590s: persecution. It is human nature to want to point the finger of blame when detrimental events take place. Food, being a key biological driver, and food security, were more important locally than today, when developed nations have access to harvests from different parts of the planet. A poor harvest for a farmer in the sixteenth century caused personal hardship and poor harvests for a village or town would undermine the economy. It was therefore easy to blame someone for casting a spell that ruined crops, and so the 1590s were particularly notorious for witch hunts and executions. Indeed, as would be expected from a climate perspective, southern Europe saw fewer witch executions than more northern countries such as Germany, where there were many. Indeed, warmer Spain, which is famed for the Inquisition, had comparatively few executions. Here crop damage was far less of a factor than the desire to convert to the 'true faith'. Crop damage was worse in the north and again catastrophic on the northern marginal lands.

So much was this so that the descendents of those that previously took advantage of the northern migration of marginal lands at the beginning of the MCA found their fortunes reverse with the Little Ice Age. This is particularly highlighted by the pattern over the centuries of international migrations. During the climatic anomaly years, AD 800–1200, Iceland and Greenland were settled by the Vikings from Scandinavia. By around AD 1100 the population of Iceland peaked at some 70 000 and this is clearly climate-related. Human remains in Norse burial grounds located in Greenland have been found which are now in permanently frozen soil that is virtually impossible to dig by hand. This suggests, for the bodies to have been buried, that the average local temperature at the time of Norse occupation was some 2–4 °C higher than at present. In reality, the coastlands of the pleasantly named Greenland were not that much different from Iceland and, if anything, worse, but it was felt that a country with a nice-sounding name would be more likely to attract new settlers.

Archaeological evidence has revealed that Icelandic Vikings had large farmsteads with cattle, pigs, sheep and goats, and there were ample pastures to sustain these as well as crops such as barley. However, the growing season in Greenland, even in the MCA, was very short. Frost typically occurred in August and the fjords froze in October. Nonetheless, by the 1300s more than 3000 colonists lived on 300 farms scattered along the west coast of Greenland.

This was all to change with the Little Ice Age. Indeed, well into this cool period, in 1492, the Pope complained that no bishop had been able to visit Greenland for 80 years on account of the sea ice. In fact, by that time his Greenland congregation was in all likelihood either already dead or had migrated. Hermann (1954) notes that during the mid 1300s many Greenlanders had moved on to Markland (today's Newfoundland) in search of a more benign home. Formal communication between Greenland and Europe had indeed ceased back in 1410 and was not re-established until the 1720s. Europeans did not recolonise there until the 1800s. Meanwhile, Icelandic grain production was given up in the fourteenth century and in 1695 sea ice completely surrounded Iceland except for one port. This ties in with the aforementioned reports that in western Europe in the 1690s crop failures were common. It was the coldest decade of the seventeenth century.

As noted in the previous section, systemised meteorological recording began in the eighteenth century and has continued to the present day. Figure 5.3 shows the 3-year-smoothed global temperature change since 1861 relative to the 1961–90 average as determined by systemised recording.

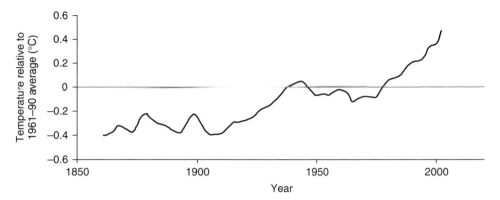

Figure 5.3 Global temperature (3-year-smoothed) relative to the 1961–90 average and covering the period 1861–2002.

Not surprisingly the coldest year on record was early on and, as one might expect, occurred during the Little Ice Age. It was 1816 and followed the eruption in 1815 of Mount Tambora in Indonesia. The eruption distributed aerosols into the stratosphere which cooled the Little Ice Age climate further. 1816 saw the latest grape harvests ever on record: November. Meanwhile, in eastern Canada and New England three cold waves in June, July and the end of August (with frosts) devastated harvests and damaged crops as far south as Boston, MA, USA. Harvests were down across the northern hemisphere and there is even a record of famine in Bengal that triggered an outbreak of cholera. The 1816 cholera epidemic even became global (the world's first pandemic of the disease), reaching western Europe and North America by 1832.

In terms of human mortality, correlation of the CET record between 1665 and 1834 with mortality reveals that mortality increased both in cold winters and hot summers. Quantitatively, warming of 1 °C in winter and cooling of 1 °C in summer reduced annual mortality by 2 and 4% respectively. The combination of such mild winters and cool summers was to raise life expectancy by about 2 years. We will revisit this when discussing health impacts of future climate change in Chapter 7, and it will be important to remember that the age profile of early twenty-first-century developed nations is far different to their eighteenth-century counterparts, with a higher proportion of elderly people. Nonetheless, climate-related health effects, albeit to an increasingly different degree, were still relevant through the nineteenth and twentieth centuries. As recently as 1940 UK mortality from bronchitis and pneumonia was, at over 850 000, double the remainder of the decade's average, and correlated with that year's cold winter. 1941 also saw a harsh winter, as did 1942, but there were successively fewer pneumonia and bronchitis cases as the most vulnerable had already died (Burroughs, 1997). This last point is also relevant to longer-term climatic change over a number of years compared to the impact of an isolated atypical year.

Returning to the question of food and the security of supply, one famine is particularly notorious in the nineteenth century: the Irish potato famine. This famine of 1845 was caused by a fungus, *Phytophthora infestans*. It is true that the climate of the autumns of 1845 and 1846 was particularly conducive to its spread. However, that a million Irish peasants died (and the other million and a half emigrated as a result) cannot solely be laid at climate's door. Little then was known about such fungi, let alone their life-cycle and habitat preferences: these were to be later elucidated by the German biologist Anton de Bary. The use of monocultures, the few potato cultivars (strains) available, the high population density as supported by near subsistence farming (living on the edge) and a virulent new strain of the pathogen, together with the weather, all combined synergistically to aggravate the famine. As with climate and seasonal mortality, there are parallels with the position in which we find ourselves today. True, today the developed nations are not dependent on subsistence farming, but this just means that the ante has been upped: we have today a far higher population (over 6 billion at the end of the twentieth century as opposed to under 2 billion at its beginning) that is supported by globally connected, intensive agricultural systems. Today we are reliant on a broader base of crop cultivars, but the conservation of genetic strains and wildlife biodiversity (the source of new natural genetic material) are still problematic. We are still close to the edge; possibly too close? As we shall see, climate change and weather can still serve to aggravate agricultural problems in the twenty-first century.

5.2.3 Increasing twentieth-century human climatic insulation

The climate's impact is not confined to more organised societies, although these societies may better document such episodes. Less-developed nations have always been subject to the vagaries of the weather and vulnerable to climate change. This is as true in the high-tech present as it was a century ago and earlier. However, those living in developed nations, while not completely escaping the ravages of weather extremes and climate, have become increasingly insulated from the weather and seasons. At the beginning of the twentieth century in Europe, North America and Japan (the centres of industrialisation at the time), even those living in urban areas would need to note the weather forecast and dress accordingly: wrap up well in winter and wear waterproof clothing when appropriate. Travel for most in these nations was still by foot or animal power. Steam was beginning to provide occasional long-distance travel for some, but most travel miles were by those exposed to the elements.

Venturing outside meant preparedness dictated by the weather, seasons and climate. Similarly, the diet in these nations was also largely climate-determined. The fresh produce available was largely local and very seasonal. Indoors most dwellings had just one or two primary sources of heat, usually a fire in the main living room, sometimes with a back-boiler to provide hot water. Gas cookers in urban areas were common, as denoted by the term town gas, which actually was largely a mix of methane and carbon monoxide derived from heating coal.

Jump forward to the end of the twentieth century and things were markedly different. The developed world consisted of more nations, not to mention that the Earth's population had roughly quadrupled. Venturing out commonly involved using the car, the global passenger fleet having grown from virtually nothing in 1900 to over 500 million vehicles by 2000 and proportionally with a nation's wealth per capita (Worldwatch Institute, 2003). By this time cars came with heaters as standard and many included air-conditioning, which is fast becoming a standard fitting, especially in wealthier countries. Nor was diet in these nations any longer restricted by season or indeed locality. Visit any supermarket in a wealthy country today and the range of out-of-season and foreign foods is readily apparent. Meanwhile, homes themselves were increasingly cocooned by central heating and, where summers are regularly warm, air conditioning. Many families in north-western Europe celebrated the turn of the millennium at home in a shirt-sleeve environment. In short, the mundane lives of those in developed nations have become apparently increasingly insulated from the environment. 'Apparently' because those in developed nations are still vulnerable to extreme weather events.

Yet the very means that enabled this independence – energy use – was itself responsible for the generation of greenhouse gases that were, and are, altering the climate. Notwithstanding, this insulation from the environment has its limits. If there is ice on railway lines or on the road system, even those railway carriages or cars with the best heating systems are impeded. As for the future, as we shall see, developed nations will not be immune to impacts arising from likely climate change. Meanwhile, those in the less-developed nations have been, and are, less fortunate and vulnerable to direct impacts from climate change.

5.3 Climate and business as usual in the twenty-first century

5.3.1 IPCC Business as Usual

No one can see into the future, but we can attempt to forecast what might happen given certain circumstances. One set of circumstances is known as

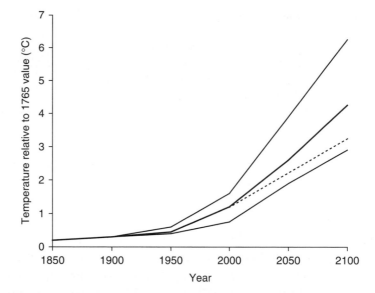

Figure 5.4 The 1990 IPCC Business-as-Usual (B-a-U) scenario simulating the anticipated increase in global mean temperature with best estimate (thick line) and upper and lower estimates (thin lines) showing the temperature rise above the 1765 value. The IPCC subsequently revised the way it presented its scenarios and its forecasts in 1992 and adhered to these for its 1995 review. The nearest equivalent to the IPCC 1990 B-a-U scenario are the IPCC IS92 scenarios a and b and they are portrayed approximately by the dotted line. As these scenarios are within the upper and lower estimates of the 1990 B-a-U forecast, and as all the IPCC forecasts are decided by committee (hence are based on scientists' opinions rather than scientific fact) there was, at the time of the second report, little advantage in considering other forecasts than the 1990 B-a-U: the take-home message over the intervening half decade was essentially the same, be it regarding energy-policy questions raised or the possible environmental impacts to consider. In part after IPCC (1990).

the Business-as-Usual scenario or B-a-U scenario (Figure 5.4). The next chapter looks at our greenhouse gas contributions, especially from energy, at the moment and for the rest of the twenty-first century. For now, while looking specifically at climate, it is worth appraising the scientific consensus for global-warming forecast under the B-a-U scenario. Back in the 1980s there was such concern as to possibilities of global warming, as well as confusion, with doubt that there was even a likelihood of anthropogenic climate change, that politicians, through the United Nations (UN), called for international scientific guidance. And so in 1988 two UN agencies – the World Meteorological Organization (WMO) and the UN Environment Programme (UNEP) – established the IPCC under the chairmanship of Professor Bert Bolin.

The panel was charged with:

1. assessing the scientific information related to the various components of the climate-change issue, such as the emissions of greenhouse gases and the modification of the Earth's radiation balance, as well as the science needed to assess environmental and socio economic consequences;
2. formulating realistic response strategies for the management of climate-change issues.

It needs to be emphasised that there were some (especially in what came to be known unofficially as the fossil-fuel lobby) who did not believe – or perhaps did not wish to believe – that human-induced climate change was real. This belief still exists among some energy-related and political quarters today, whereas many in the public are still uncertain, or unaware, of the science underpinning current climate issues. Some countries too (again especially those consuming prodigious quantities of fossil fuels) felt that the issue was overblown. Nonetheless, politicians needed clear unambiguous guidance. The IPCC aimed to provide this by ascertaining the scientific consensus. For its first report (1990) the IPCC brought together 170 scientists from 25 countries through 12 international workshops. A further 200 scientists were involved in the peer review of the report. Although there was a minority of opinions that could not be reconciled with the panel's conclusions, the extensive peer review helped ensure a high degree of consensus among the authors and reviewers. Even so, reaching this consensus was difficult. One of the ways the panel did this was to present both a best estimate as well as high and low estimates of the likely changes in global temperature. Further, they looked at a number of scenarios assuming that politicians introduced three levels of controls, as well as the B-a-U scenario which supposed that nothing was done and the late-twentieth-century trend in emissions continued to grow. It has to be acknowledged that this methodology was as much to bring the disparate views of the scientists underpinned by various degrees of climatological understanding on board as it was to provide margins for error. Consequently, as we shall see, the difference between the high and low estimates under the B-as-U scenario for twenty-first-century warming was greater than the difference between the low estimate and nothing happening at all (zero climate change). Furthermore, the best estimates for the various scenarios all fell within the range between the B-a-U scenario's high and low estimates.

For these reasons the IPCC's 1990 B-a-U estimate (Figure 5.4) was of central importance to international policy-makers, whose initial problem was whether or not to acknowledge the problem and then, if so, to establish the environment in which subsequent detailed climate-change policy might be

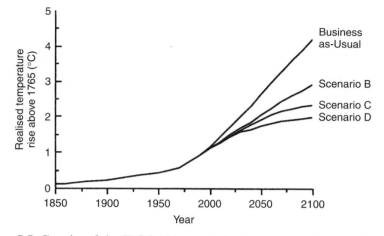

Figure 5.5 Graphs of the IPCC 1990 scenarios. Redrawn with permission.

formulated and enacted. A degree more or less warmer on top of the IPCC's 1990 anticipated twenty-first-century warming simply did not figure in policy discussions within the political arena: the high–low range was just stated and the policy debate in the early 1990s focused on whether doing anything about such warming actually mattered. Even today policy-makers are not concerned with the degree of climate change, but what should they do about managing or controlling change per se. As we shall see in Chapter 8, even when leading politicians met in December 1997 in Kyoto, Japan, the policy questions largely boiled down to whether countries would sign up to such a modest international agreement that (according to the IPCC scientific conclusions) would not halt the increase in global warming but might slightly slow down the rate of increase in warming.

Meanwhile, the IPCC's 1990 B-a-U envelope has not yet (in 2006) been *effectively* superseded. Even in each of the IPCC's subsequent reports, in 1992, 1995 and 2001, the range of scenarios had the best estimates for warming within the 1990 high–low envelope of the B-a-U scenario. Only the high–low envelopes of just a few of the special 2001 policy scenarios went outside the 1990 high–low B-a-U envelope and that was just at the low end (1.15 °C less warming for 2100 since 1750 than the low B-a-U estimate in 1990). So, while Figures 5.5 and 5.6 provide best estimates for various 1990 scenarios and 2001 all-scenario and all-model envelopes respectively, it is important to note that Figure 5.4, the original 1990 B-a-U scenario, is so similar to the 2001 all-model best-estimates envelope (Figure 5.6) that we will use the 2001 best-estimates envelope against which to set the policy discussion for much of the rest of this book. Specifically, the discussion of this all-model best-estimate envelope will be compared with what we know actually happened from palaeoclimatic

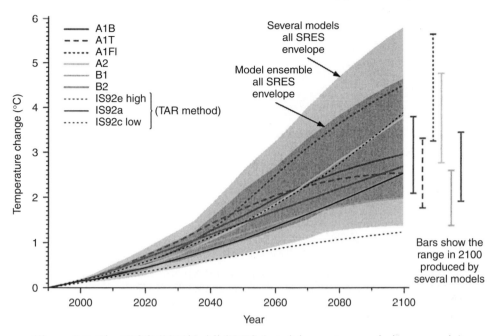

Figure 5.6 The IPCC (2001b) All SRES (special report on emission scenario) temperature-change forecasts, which provide an envelope of predicted twenty-first-century global climate change. Reproduced with permission of the IPCC.

indicators of the real Earth (and not an expensive computer model) at times of past warming and from when it was actually warmer than today. The former is analogous as to how biomes react to warming (climate change), while the latter is analogous to where we are heading with global warming.

These various IPCC forecasts portray a similar picture. Figure 5.7 is a 3-year-smoothed graph (as is Figure 5.3) of the global temperature since 1861 to 2002 with the IPCC 1990 B-a-U forecast added to the end of the twenty-first century. From this it can be seen that should we continue to increasingly add greenhouse gases to the atmosphere as we have been, then the planet will end up warmer than it has been since before the beginning of the Industrial Revolution. Even if we were to have used the lowest of the best estimates in the various IPCC 2001 scenarios, the forecast still would have been far warmer at the end of the twenty-first century than at any other industrial time. Indeed, the greatest likelihood is that the twenty-first century will see the Earth warmer than at any other time in our Holocene interglacial. In fact, using IPCC 2001 high estimates (as opposed to best estimates) the end of the twenty-first century could see temperatures not seen on the Earth since the last interglacial some 140 000 years ago if not longer. However, the Earth is a very different place today compared to 140 000 years ago. To begin with,

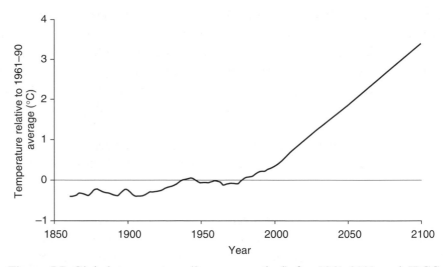

Figure 5.7 Global temperature (3-year-smoothed) for 1861–2002 and IPCC 1990 B-a-U scenario for the twenty-first century, relative to the 1961–90 average.

back then there were not over 6 billion humans, and rising. Nor was the planetary landscape so managed or the global commons, both atmosphere and oceans, perturbed by human action. Therefore the biological and human ecological consequences of such a climate change will be marked.

If we carry on this century much as we have throughout the last, and so end up with marked climatic change, the question arises as to what would happen if greenhouse gas emissions were curbed? The IPCC looked at a range of such scenarios in each of their assessments (1990, 1995 and 2001), which assumed various controls of gas emissions, improvements in energy use and so forth. These scenarios were realistic in that they took into account that the starting point was (then 1990) the present, so humanity was stuck with our then actual current pattern of energy production and use, and could only modify matters as old power-station plants came to the end of their economical (and even carbon-economical) life. (By 'carbon economical' I mean if we scrapped a young coal-fired power station and built a wind farm then the energy investment in building the coal-fired station would not have been recouped.) The IPCC knew that the present time also includes a growing global population and they used demographers' quite reliable ranges of forecasts through to the middle of the twenty-first century and beyond. They took into account that it would take time and investment to introduce alternatives. Finally, their scenarios varied in that some assumed that there would be only modest emission reductions compared to the B-a-U scenario, while others assumed that more aggressive policies would mean greater savings. Among the strongest climate-combating IPCC 1990 scenarios, carbon dioxide emissions would be reduced to 50% of their 1985 levels by 2050 and even this unlikely scenario still resulted in a climbing global temperature through to 2100 and beyond. So, all the IPCC scenarios showed that global warming would continue. Obviously, if the cuts in greenhouse-gas emissions were sufficiently great there must be a point at which the global climate would stabilise without further warming. The IPCC did look at such climate stabilisation but did not include this as one of their detailed scenarios. The reason was that the magnitude by which we would need to change the global economy to realise these stabilising cuts was so great. For to stabilise the climate we would need to reduce greenhouse-gas emissions planet-wide by 60–80% of 1990 levels! To do this, against a backdrop of global economic growth and growing population (both multiplicative and not just additive factors; see discussion on population and Figure 5.9, below) such cuts would have an even greater socio economic impact. This cut to 60–80% of 1990 levels is therefore not even remotely likely. It would appear virtually certain that we are committed to seeing the Earth warm up.

For its 2001 assessment, the IPCC not only reviewed the literature as to the latest science and looked at improved computer models, it developed a range of more complex greenhouse-gas-emissions scenarios arising from different policy assumptions. These it developed in a special report on emission scenarios (or SRESs), of which six were highlighted but a full set of 35 were used in the IPCC's 2001 report to provide a high–low envelope inside which all the SRES best-estimate forecasts lay. Further, an outer envelope was ascertained, representing the scenarios as used by several different climatic models. This outer envelope provided the IPCC with new high and low estimates for anticipated climate change in the twenty-first century (see Figure 5.6).

The IPCC applied a similar methodology to sea-level rise. The seas are expected to rise due to the melting of glaciers, the Greenland and Antarctic ice caps and the thermal expansion of the oceans. (Antarctica *may*, however, offset part of this rise by trapping more ice from the increased snow precipitation expected in a warmer world.) However, while its 2001 best-estimate range for 1990–2100 warming essentially captured its 1990 best estimate, the 2001 estimates for sea-level rise were lower (see Table 5.1). This was because of improved computer models. The IPCC refers in its texts to average SRES forecasts and a high–low range for the same. Yet they also wisely included in their graphs a higher–lower estimate due to uncertainties in land-ice changes, permafrost and sediment changes. Importantly, even with this extension of range, they did not allow for uncertainty relating to the West Antarctic Ice Sheet's (WAIS) stability. This point is often overlooked. This is because the

Table 5.1. *Summary conclusions of the IPCC's 1990 and 2001 reports for forecast temperature and sea-level changes 1990–2100. Subsequent research suggests that these may be low, although this is an area of considerable debate.*

Change 1990–2100	1990 B-a-U	2001 Several models, all SRES envelope
Warming		
High estimate	5.25 °C	5.8 °C
Best estimate	3.25 °C	2.0–4.5 °C (model ensemble)
Low estimate	1.8 °C	1.4 °C
Sea level		
High estimate	110 cm	88 cm (including land ice uncertainty)
Best estimate	65 cm	22–70 cm (all models, all SRES averages)
Low estimate	22 cm	9 cm (including land-ice uncertainty)

IPCC's 2001 summary for policy-makers and the technical summary, published together in *Climate Change 2001: The Scientific Basis*, (IPCC, 2001b) only touched briefly upon this uncertainty, even if this brief mention is perfectly clear. The WAIS 'dynamics are still inadequately understood', and a part of a lengthy IPCC graph caption states that its 'range does not allow for uncertainty relating to ice dynamic changes in the WAIS' (IPCC, 2001b). These uncertainties and caveats are worth noting for just half a decade later we are seeing unexpected degrees of melting in both Antarctica and Greenland. These suggest that the IPCC's sea-level rise estimates in 2001 may well be low.

The IPCC's 1990 and 2001 conclusions for the anticipated change forecast between 1990 and 2100 are summarised in Table 5.1. We will return to sea-level rise when discussing possible 'surprises' in the next chapter.

The IPCC 1990 report summarised that it was 'certain' that the forecast emissions would result 'on average in both an additional warming of the Earth's surface' and sea-level rise. The 1992 IPCC supplement concluded that the subsequent scientific understanding 'either confirm[ed] or [did] not justify alteration of the major conclusions of the first IPCC Scientific Assessment'. The IPCC's 1995 second assessment was if anything more cautious of its conclusion, in part due to the fierce political debate surrounding the first assessment's conclusions. In 1995 the IPCC concluded that, 'the balance of evidence suggests a discernable human influence on global climate'. By 2001, the IPCC had regained some confidence and concluded that 'an increasing body of observations gives a collective picture of a warming World and other changes in the climate system'. Further, that 'there is now new and stronger evidence that most of the warming observed over the past 50 years is attributable to human activities'.

The reasons for the IPCC's caution (note the prudent use of the word 'most') were four fold. First, the IPCC was presenting a scientific consensus of a range of views and this range's breadth had to be signalled. Second, the IPCC only assessed work that had been published in peer-reviewed journals. Preliminary work, work underway and even work that had been accepted by eminent peer-reviewed journals but not yet published was not considered. Third, as a consequence of this personal assessment of the emerging picture, discussions over lunch at symposia and scientists' other unpublished insights were simply not admissible for consideration by the IPCC. Finally, fourth, IPCC researchers, and leading climate-change programmes in a number of countries (such as in the UK and USA), spent much of their research investment on constructing incomplete – but ever more sophisticated and hence useful – climate computer programmes.

In economic terms, 1990s investment in whole-organism biological palaeo-responders to change, and even monitoring of current change, came very much second to investment in computer models. In the UK at least, in the early twenty-first century this is beginning to change, and even regarding computer use the biological components are increasingly being incorporated into models (in contrast, the climate models of the 1980s, on which the first IPCC assessment was based, essentially modelled the planet's physicochemical dimensions as if it were lifeless). The 2001 report did include model outputs that incorporated some biologically determined factors. However, there is still much biology to include in climate models. Indeed, although policy-makers continue to place great faith in computer models, that models still have a long way to go is emphasised occasionally in the scientific literature. Considerable development is required before policy-makers should consider them a reliable tool for the detailed regional analysis that is required for a nation's climate-change policy.

One recent (at the time of writing) example of such a warning in a high-impact journal was a letter in *Nature* from three US researchers (MacCracken *et al.*, 2004) from the Climate Institute and the H. John Heinz Center for Science, Economics and the Environment. They concluded by saying, 'We strongly agree that much more reliable regional climate simulations and analyses are needed. However, at present, as [the *Nature* news report] makes clear, such simulations are more aspiration than reality.'

One of the biggest problems with global computer models is that to date (2006) they do not successfully reflect climate change at high latitudes, both when modelling the present Earth and in the past with a different continental distribution. Their failure to capture the mid-Cretaceous warm Arctic climate (see Chapter 3) is a case in point, although we know from palaeoclimate proxies (fossil remains, biomolecular climate proxies and oxygen isotope analysis) that the Arctic was warm at that time. It was even warmer later, during the initial Eocene thermal maximum (Chapter 3), and global models also have difficulty in capturing this. The implication is that high latitudes will warm far more than models currently suggest. Indeed, Antarctica is currently the fastest-warming place on Earth, having warmed some 3 °C in the 50 years up to 1990 (which compares to a global average warming of about 0.6 °C for the entire twentieth century). Whereas model failures may not so markedly affect the IPCC's climate-scenario forecasts for tropical and temperate zones, implications for the IPCC's projections of sea-level rise are considerable, as the major ice caps (Antarctica and Greenland) capable of contributing to sea-level rise are at high latitudes.

Part (though not all) of this high-latitude problem, especially with regard to the northern hemisphere, which has a greater land area just outside the Arctic Circle compared to the Antarctic Circle, is the effect of biology on albedo. Computer climate models have been, and are, getting ever more sophisticated. Up to the end of the twentieth century models did include surface cover in a simplistic sense and surface-cover albedos so that land without snow (and sea without ice) absorbed more, and reflected less, solar energy than land with snow (or sea with ice). Forests too, in areas prone to snow, have a lower albedo than snowfields without trees. Conifers have evolved to allow snow to fall off of them comparatively easily. All well and good, and most of the more sophisticated models at the beginning of the twenty-first century accounted for such factors and included a (coarse) model map of land cover. These models naturally took into account one of the positive-feedback cycles of warming reducing the area of snow cover, so reducing the solar energy reflected, so in turn causing warming, which then further reduces snow cover (Figure 1.8b). However, only a handful of models in the first half of the decade included the effects of a changing regional ecology and hence a changing and dynamic landscape. For, in addition to the above feedbacks (that included a static biological landscape), there is the biological dynamic. As the region warms then so it is open to colonisation from species that could not have survived the previously cooler climate. So, even if the ground is snow-covered, if it has shrubs and trees that have migrated into the area due to warming, then the albedo will change. No doubt, as we come to the end of the decade in 2010, such biological dynamic factors will (at least on a coarse level) be included in models. But in 2005 it was becoming clear that such factors were important in the field, in real life (not so much models). At that time a wide range of innovative Alaska data-sets were assembled covering the decades from the 1960s (and in some cases from the 1930s) to 2004. They included surface-temperature records, satellite-based estimates of cloud cover, ground-based estimates of ground cover and albedo, field observations in changes of snow cover and, importantly, changes in vegetation cover. Alaska was a worthy study area for it has seen a number of thresholds crossed with regional warming that relate to abrupt physical and ecological change near the freezing point of water. And the region has warmed. Palaeoclimatic evidence shows that Arctic Alaska was in 2005 warmer than at any time in at least the past 400 years and so there should be a large impact on water-dependent processes. The warming, and especially summer warming, cannot be attributed to climatic cycles such as the North Atlantic Oscillation or the Arctic Oscillation, while cycles such as the El Niño mainly affect Alaskan winter temperatures and not summer ones. Changes in sea ice around Alaska would have an effect in the spring and autumn, when the ice melts and freezes, and so, whereas this time may have changed by days or weeks with regional warming, it would not affect matters in the depth of winter or in the middle of summer. The study team (Chapin *et al.*, 2005) state that the degree of change in Arctic Alaskan summer warming, above that of the global average, is best explained not by landscape drivers but by the lengthening of the snow-free season. Indeed, snow melt has advanced by 1.3 days/decade at Barrow (a coastal site) and an average of 2.3 days/decade over several other coastal sites. Inland, in the northern foothills of the Brooks Range, warming has taken place at a rate of 3.6 days/decade, and 9.1 days/decade for the entire Alaskan North Slope. Since 1950 the cover of tall shrubs within the North Slope tundra has increased by 1.2% each decade from 14 to 20%. They pooled, in a meta-analysis, field-warming experiments showing that increasing the summer temperature by 1–2 °C generally triggers increased shrub growth within a decade (which is consistent with both recent field observations and the palaeo record of Holocene shrub expansion). However, it is thought that the increase in shrubs only accounts for 2% of recent warming. A further 3% may be attributable to increased tree cover, especially that of the white spruce (*Picea glauca*) whose cover has increased, decreasing the treeless area by 2.3%. Although this makes only 5.3% of Arctic Alaskan summer warming accountable to changes in vegetation cover (and the increased snow-free season accountable for much of the rest) the researchers warn that the region is on the cusp with a number of thresholds about to be crossed. The conversion of Arctic Alaskan tundra to spruce forest (a possibility by the early twenty-second century) has never before occurred during previous Holocene warm intervals: that is, the past 11 500 years since the end of the last glacial. Given that Alaska would have been covered in ice during the glacial, it means that it could not have been covered by forest for over 100 000 years since the brief height of the last interglacial. In addition to ice-water-related thresholds, as shrubs become established so the nitrogen cycle is enhanced, which in turn encourages further vegetation. This is another positive feedback serving to accelerate matters. In short, the research suggests that the already anomalous increased warming of Arctic Alaska is likely to become even more pronounced in the future (Chapin *et al.*, 2005). Environmental change thresholds are currently a priority area of climate research and, as the above-mentioned research suggests, are central to high-latitude warming over and above those that computer models have predicted to date.

Including biology and landscape change factors in computer models is now beginning to show that there are differences between the various of the IPCC's 2000 SRESs for the twenty-first century (IPCC, 2000). Even between scenarios with similar carbon dioxide at a given time there can be differences in temperature due to the biological landscape. The IPCC's A2 storyline of these SRESs is one of more regional resource use and self-sufficiency with a slightly slower economic growth compared to the B1 storyline with the same population up to the mid twenty-first century (but which declines thereafter), and again increased sustainability, but through more global solutions and without extra climate initiatives. In 2005 Johannes Feddema and colleagues showed that incorporating differences in the biological landscape from these two scenarios resulted in somewhat different temperature profiles compared to the IPCC 2001 assessment. The agricultural expansion in the A2 scenario appeared to result in significant additional warming over the Amazon and cooling of the upper air column as well as nearby oceans. These effects in turn influenced the Hadley and monsoon circulations and so modified some climate regimes outside of the tropics. Meanwhile the model of Feddema and colleagues showed that agricultural expansion in mid latitudes produced cooling and decreases in the mean daily temperature over many areas (Feddema *et al.*, 2005). Although it is early days for such biological landscape inclusions in global models, they will undoubtedly soon become an increasingly routine, if not fundamentally intrinsic, addition to global climate computer simulations.

Not only do models need to accommodate more biological components and in considerable detail, modellers need to keep abreast of biological research outputs. For example, a recent (2004) paper, from a team led by Colin O'Dowd from Ireland, showed that the contribution of organic matter to marine aerosols over the North Atlantic during plankton blooms made up 63% of the sub-micrometre aerosol mass. However, in winter, when biological activity is at its lowest, the organic fraction decreases to just 15%. Further, their simulations suggest that organic matter can enhance cloud-droplet concentration by between 15% and more than 100% and so is an important component of the aerosol-cloud climate-feedback system involving marine biota.

Computer modellers also need to continually reappraise the physical processes they are modelling, especially when new evidence and ideas come along. For example, at the end of 2004 a team of US geologists (Unfar *et al.*) drew attention to field evidence that mid- and high-latitude rainfall in the mid-Cretaceous was far higher than today. They suggested that this might have warmed high latitudes, as ocean evaporation causes cooling and precipitation results in latent heat release, hence regional warming. In the warmer Cretaceous there would have been far more evaporation from the tropical oceans and this could provide another mechanism for transporting heat to high latitudes (in addition to atmospheric and ocean circulation mechanisms).

Having said all this, while computer models still have a long way to go before detailed regional forecasts can be made with sufficient confidence for local policy-makers and planners, a rough-and-ready near-future or even past (historic) global approximation is another matter. The latter is particularly revealing (and important in the climate-change debate) because running models of the historic climate, during which meteorological measurements of the real world have been made, means that the model can be tested against reality. If this reality test works then it is possible to run the model again but without the factors of anthropogenic climate change. This has been done. By 2005 the Hadley team in the UK had a model that successfully captured the vagaries of the climate from the middle of the nineteenth century to the early twenty-first century. Its output seemingly tracks with reasonable accuracy actual meteorological measurements of global temperature. Both the Hadley model and real measurements show that the planet from the middle of the nineteenth century through to the early twentieth century had a varying climate but the overall trend was more or less steady (an output not too dissimilar from the relevant years portrayed in Figure 5.2). But after the early twentieth century *both* the Hadley computer model and real meteorological measurements show a rising temperature. Clearly, the model works broadly. However, take out the anthropogenic greenhouse component from the model and its output continues the nineteenth-century stable trend (albeit fluctuating a bit year to year) through to the present. This is good corroborating evidence that human factors are behind the twentieth- and twenty-first-century warming of climate.

5.3.2 Uncertainties and the IPCC's conclusions

The above discussion demonstrates that we need to be aware of uncertainties and view the IPCC's conclusions with these in mind. Related to the mid-Cretaceous warm-Arctic problem above, one of the major concerns is why the greatest warming actually measured during the twentieth century took place at high latitudes? A concern doubly worrying because the global climate models do not yet (2006) properly reflect such warming. This is likely to be reconciled once we have a better idea as to how heat is distributed about the planet: the previous subsection's suggestion is but one idea.

Another concern is that of a phenomenon nicknamed global dimming. This at first may seem paradoxical in a world that is supposed to be warming up. Yet those concerned with agricultural biology have known for some years that the amount of sunlight reaching parts of the Earth's surface has been in marked decline. Normally weather instruments for temperature and humidity are kept in the shade, but water evaporation is not just a matter of temperature and humidity: it is also a matter of sunshine. The relevance of this has been demonstrated by a series of long-term (multi-decade) measurements of what is called pan evaporation taking place in several countries. Pan evaporation refers to the daily evaporation of water from a pan exposed to the elements including sunlight and is of importance to biologists studying agricultural systems, as pan evaporation is relevant to irrigation methods. It has therefore been monitored for a number of decades in various countries. What has been found is that pan evaporation has been decreasing markedly in many countries for over 50 years. Sunlight, separate to temperature, excites water molecules and so enhances evaporation. In 2002 two Australian researchers, Michael Roderick and Graham Farquar, discovered that reduced pan-evaporation rates appeared to be linked to a decrease in sunshine: hence the expression global dimming.

What appears to be happening is that fine aerosol pollution from urban areas and industry is providing seeds for the formation of clouds that consist of finer-than-usual droplets (the indirect aerosol effect as opposed to the direct aerosol effect of the aerosols without the atmospheric water). These clouds are more reflective than usual. Also, the finer droplets are less likely to result in rain.

The consequences of dimming are considerable. First, there is the human implication: it is now considered by some that global dimming caused less ocean evaporation in parts of the northern hemisphere, such that in the 1980s and 1990s it resulted in the failing of the monsoon and mass famine in east Africa. But it also means that if there is a greater cooling force than we thought over much of the temperate northern hemisphere, then there must be an even greater warming force than we thought for there to have been the net result of warming in the twentieth century. In 2004 a small team of US researchers led by Joyce Penner reported that they had a measure of this indirect aerosol effect by comparing conditions in two separate areas of that country, one of which had clean air and the other of which was polluted with aerosols. They showed that this effect was markedly greater than estimates used up to that time by IPCC climatologists. Aeroplane con (condensation) trails also cause significant global dimming and this was demonstrated in a rare experiment following the attack on the World Trade Center in New York on 11 September 2001, when for 3 days virtually all of the US civilian jet air fleet was grounded (Travis *et al.*, 2002). Furthermore, the IPPC estimates for aerosol forcing for its first three assessments (1990, 1995 and 2001) were based on satellite measurements only over the ocean. Subsequently, with improved orbital remote sensing, it was possible to make estimates of aerosol forcing over land. This indicated that aerosol forcing was in fact at the high end of the earlier range the IPCC used. These results suggest that present-day direct radiative cooling forcing due to aerosols is stronger than that represented in the IPCC assessments up to and including its third report. The implication is therefore that future atmospheric warming due to an increase in anthropogenic greenhouse gases will be greater than the IPCC had predicted (Bellouin, 2005).

As for the longer-term trends in global dimming, 2005 saw the publication of the summaries of two sets of data. One was from a set of hundreds of terrestrial sunlight measurements gathered by an international team led by Martin Wild, an atmospheric scientist at the Swiss Federal Institute of Technology in Zurich. The other study looked at satellite data and was led by Rachel Pinker from the University of Maryland. Both found a similar trend. It would appear that since the 1950s up to the early 1990s global dimming had been taking place. However, since the early 1990s there had been some recovery, with global brightening. This is thought to be due to the decline in the Soviet Block's dirty industries as well as the increased use of improved cleaner technology by western developed nations. Currently work is underway to estimate how much global dimming has offset global warming.

If global dimming is something the IPCC have (prior to 2007) not fully taken into account then this could, at least in part, explain why high-latitude warming has actually been greater than their models predicted. It could be that the cleaner air of the poles (protected by circumpolar air currents) is not as globally dimmed. This is but one possibility. In this instance much of the academic research relating to global dimming and brightening was published after 2000. The IPCC only considered science research published prior to its latest 2001 report, so the IPCC's first three assessments could not take this dimension into account. The IPCC's anticipated report for 2007 should begin to include such considerations (see Appendix 4).

So how robust are the IPCC's conclusions? As noted, the IPCC arrives at its decisions by committee, which in turn bases its conclusions on its members' own interpretations of a variety of evidence, both palaeoclimatic and

palaeoclimatic combined with computer models. Even so, these interpretations are subject to human psychological factors that make it additionally
difficult to assess how reliable the IPCC's conclusions may be. Nonetheless, it
is possible to begin to estimate how robust the IPCC's conclusions are against
the types of computer model the IPCC uses. Following the publication of the
IPCC's 2001 assessment, researchers from the University of Bern, Switzerland,
attempted to do just this (Knutti *et al.*, 2002). They took a simple climate
model (therefore, one whose output was subject to greater error margins) but
calibrated to the same three-dimensional ocean–atmosphere models as used by
the IPCC in their third assessment. Other than ensuring that the model's
responses reflected known past climatic change (i.e. that the model's responses
to established historic circumstances were 'realistic'), they varied the model's
parameters by the estimated uncertainties of its various climate-forcing factors. In this way they obtained a number of climate forecasts for the twenty-
first century. The conclusions were clear. Just 5% suggested that the IPCC's
2001 high–low estimates were too high. In other words, there was a 5% chance
that the IPCC was being too alarmist in predicting greater warming than is
likely. Conversely, the researchers concluded that there is a 40% probability
that the increase in global mean surface temperatures will exceed the range
predicted by the IPCC. In short, they estimated that the balance of probability
would be that global warming in the twenty-first century will be greater than
the IPCC forecasted in 2001. If the researchers' conclusions were valid, it
means that the consequences of the warming discussed in subsequent chapters
will be greater. One literature review in 2005 on aerosol cooling, by Meinrat
Andreae, Chris Jones and Peter Cox, suggests that warming forcing since the
Industrial Revolution up to the end of the twenty-first century might be between
2 and 8 °C, although a mid-estimate of 6 °C is unlikely as there will probably be
some natural background aerosol cooling in addition to that from human activity.
Even so, it is likely that future (post-2001) IPCC estimates for twenty-
first-century warming are likely to increase unless they find other, hitherto unrecognised, significant cooling agents affecting the climate-forcing balance. Past
strong aerosol cooling will not increase as much as future greenhouse warming.

This does not mean that models have no value, rather that they need
(considerable) improvement. The best models (including the ones the IPCC
used) may not be accurate on a sub-regional basis but they do *broadly* paint
similar pictures to each other on a global scale. Getting similar results by
different computer methods lends a certain credibility to the results, hence the
models. This is worth noting, as it is model development and the increase in
their sophistication that will one day surely prove most useful for determining
change at increasingly local levels in numerous ways.

A second instance of concern that the IPCC's forecast may be under-estimating possible future climatic change was presented by a European research team led by Hans von Storch in 2004. They were examining the validity of previous interpretations of the palaeoclimatic proxy record for the northern hemisphere that were in turn used by the IPCC. Their conclusions were that such past analyses of noisy (highly variable) proxy data may well have under-estimated past climate change. If the IPCC had been relying on under-estimates of past change to make their future forecasts, then their future forecasts would also be under-estimates. Past variations may have been at least a factor of two larger (see section 2.5).

The above gives just two recent examples of an examination of the confidence that one might place in the 2001 IPCC scenarios. These examples are scientific attempts to quantify confidence in scenario conclusions that, in addition to the previous subsection, provide further reason to treat the climate models on which the IPCC relies with caution.

Conversely, again as noted previously, policy-makers are concerned with a bigger socio economic and political picture than the exactness of warming magnitudes. It actually is of negligible importance to a government that warming 100 years from now may be half a degree or so more or less than a current estimate. The key conclusion of importance (to policy-makers) is that warming is almost certain to take place within such-and-such a range. This seems sacrilegious to some scientists who (rightly) take pride in their work and who in conversation with politicians in Westminster, or on Capitol Hill, correct policy-makers in conversation by half a degree, or x centimetres of sea-level rise, or whatever. This point may seem trivial but it is not. It is increasingly important (and not just regarding climate change) for scientists to have good communications with politicians. Pedantry to politicians of the current vogue of predictions, however scientifically worthy, can give the impression of fact and importance when in actuality there is greater uncertainty as to detail. In truth there is considerable uncertainty as to whether a specific x degrees warming has taken place since the Industrial Revolution let alone whether there will be a specific y degrees on into the future. Conversely, there is great certainty as to the bigger picture, of which we can be confident: the Earth has been getting warmer since the Industrial Revolution and will continue to do so in the twenty-first century. Indeed, the IPCC reports themselves are very good as to the various degrees of certainty and confidence they can attribute to their conclusions, but one has to read the full text and not just the executive summaries.

For the full range of its SRES scenarios the IPCC 2001 assessment concludes that the Earth will warm by between 1.4 and 5.8 °C from 1990 to 2100, and it has considerable confidence in this. Improved environmental understanding (much of which will come from the biosciences) and computer modelling will further refine our view of the future, as indeed will our travelling into the future with the passage of years. However, the IPCC do give one additional warning and that is to be prepared for surprises. We need to be aware that a volcanic eruption, or abrupt changes in ocean circulation and other events, can happen, and that these will have considerable climatic consequences that take us outside the IPCC B-a-U scenario and other scenario forecasts. This may be an unsettling conclusion but in terms of policy it suggests the need for us to plan for the worse-case scenarios rather than easier options (see section 6.5).

It would be prudent – despite it being early days – to consider options outside the IPCC forecast envelope. So are we moving into an even warmer world than the IPCC forecast or not? A major volcanic eruption, such as of Yellowstone, would cool the world for several years. However, currently changes in most of the other probable climate-forcing variables are likely to force the climate the other way, suggesting that the IPCC forecast is conservative and that global warming will be greater. (Indeed, even Yellowstone would generate carbon dioxide, whose warming effect may well outlast its cooling impacts.) Since 1990, global temperatures, atmospheric concentrations of carbon dioxide and carbon emissions have all been marginally higher than the IPCC's SRESs. (Only the global population is marginally lower, and that only if you take that statistic to a greater degree of significance than the IPCC.) Indeed, impacts on human well-being from extreme weather events have already begun to become manifest, as the IPCC predicted (see Chapter 7).

Finally, notwithstanding the real possibilities of an even warmer world than the IPCC predict, there is additional concern over sea-level rise. The IPCC was always cautious as to its sea-level estimates in each of its reports and its main texts repeatedly refer to the considerable uncertainties in the science. This is for a variety of reasons, including that the component factors (such as ice-melt and thermal expansion) of existing estimates of the past two centuries of sea-level rise (measured at various sites in Europe) each have large estimated errors associated with them. In short, we do not know for certain what has caused the annual slight 1–2-mm changes over this time. With regards to future rise, there is considerable uncertainty as to how Antarctica will behave with warming.

We are now beginning to see, through monitoring the Greenland and Antarctic ice caps in the early twenty-first century (a body of work not available for consideration by those compiling the 2001 IPCC report), far greater discharge from both these than anticipated. Indeed, recent research (Thomas *et al.*, 2004) suggests that Antarctica, far from having a negative to +0.02-m change in sea level between 1990 and 2100 (as suggested by the

IPCC in 2001), had a discharge in 2002–3 from West Antarctica that was much greater than that observed during the 1990s. This discharge rate is of some 250 km^3/year, so by itself (without other countering factors) it provides a rising factor to sea level of more than 0.2 mm/year. Even if the rate of discharge ceases to continue to increase and remains the same for the rest of the twenty-first century, then West Antarctica's contribution to sea-level rise could – and note that this is a speculation – be a factor of 10 higher than the IPCC forecast for all of Antarctica up to 2100.

However, even these high rates, should they come to pass, may be unlikely to remain the same in the future, as research has now revealed that most of the West Antarctica's off-shore-but-grounded ice shelves (continental ice shelves) are hundreds of metres deeper than thought. Without the near-surface under-pinning rock that was until recently assumed to be there, and with the shelves collapsing, ice will flow more freely from further inside Antarctica (Thomas *et al.*, 2004). For this reason alone Antarctica's possible contribution to sea-level rise may even be found to become greater than the 10-fold excess above the 2001 IPCC rate now being observed, unless (and this caveat is important), as mentioned above, increased precipitation (snowfall) in a warmer world on East Antarctica freezes, so having the net effect of removing water from the Earth's oceans.

There is a second reason for the concern that Antarctica may contribute more to sea levels, and that is, if global warming is itself greater, then Antarctica's margins (though not so much its interior, which retains consider-able thermal inertia) could become even more unstable and experience greater ablation. Research on Antarctic ice using satellite data gathered between 1992 and 2003 reveals a mixed picture (Davis *et al.*, 2005). It shows that West Antarctica is losing ice but that East Antarctica is gaining it, with its ice sheets mainly becoming thicker (Figure 5.8). This is because of increased snowfall, which itself is an expected feature of global warming. The gains in East Antarctic ice between 1992 and 2003 north of 81.6 °S (below which the satellite never orbited) was about 45 ± 7 billion t. This is enough to offset Antarctica's net contribution to sea-level rise by 0.12 ± 0.02 mm/year. However, whereas a far more detailed assessment of Antarctica's contribution to sea-level rise is required, the picture that emerges is that the continent, being the fastest-warming continent on the planet, may well be on the cusp. If this is so then detailed, on-going monitoring will be required. We will return to this and the related research published just prior to the IPCC's forthcoming 2007 report in the next chapter's subsection on sea-level rise (section 6.5.3).

To sum up, the uncertainties over the IPCC's 2001 conclusions do not centre on whether the Earth is getting warmer: the IPCC is confident that global

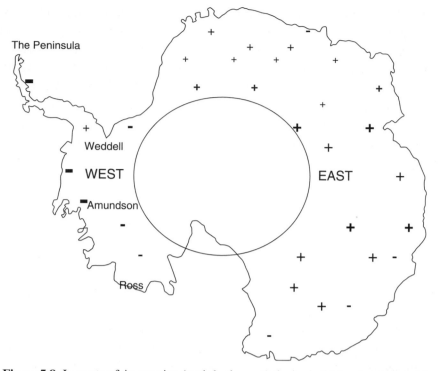

Figure 5.8 In parts of Antarctica (mainly the east) the ice is thickening (+) and in other parts the ice is thinning (−). Sketch map loosely based on Vaughan (2005), which is in turn based on satellite data from Davis *et al.* (2005). The circle in the centre represents the high-latitude zone not covered by the satellite's orbit.

temperatures will rise throughout the twenty-first century and are 'very likely [90–99% chance] to be without precedent during at least the past 10,000 years' (IPCC, 2001b). Instead, uncertainties centre on how much warming and sea-level rise will take place. But, because warming and sea-level rise are projected to continue well beyond the end of the twenty-first century the question really is not so much by how much will the Earth warm and its seas rise, but by when?

One thing we do know is that during the last interglacial, temperatures briefly peaked to within the window that the IPCC forecasts for the twenty-first century. We also know that maximum sea levels during the last interglacial were about 5–6 m higher than they are today (we will briefly return to this in section 6.5.3). However, how long after the peak in the last interglacial's temperature the peak in sea level took place is not known with the certainty we need for the purposes of planning for the twenty-first century. The computer models (which are continually being developed) currently suggest that we will not see a 5–6-m rise in sea level during this century. Conversely, although the geological evidence does not have the temporal resolution one might like it does suggest a rapid rise (geologically speaking). Resolving this question (be it through improved computer models and/or geological evidence) is a research priority. Meanwhile, policy-makers should not be complacent and should take due heed of the IPCC's caveats and warnings of climate-related surprises. We will return to surprises at the end of the next chapter.

This brings us on to what will happen beyond the year 2100. In theory, in three or so centuries time, should we, as may reasonably be expected, have exploited virtually all of our present-day economically recoverable fossil fuels, but assuming without adopting climate abatement measures, temperatures could rise comparable to the initial Eocene thermal maximum (IETM) described in Chapter 3. (And this does not take into account any warming from marine methane hydrate (clathrate) carbon release, as was likely in the Eocene, or re-organisation of biosphere carbon pools (such as in boreal soils).) Whereas sea-level rise over a timescale of two or three centuries could perhaps be 20 m (or even more) it would be far greater still over a timescale of many centuries, as in the Eocene event. The IPCC in 2001 only briefly considered events beyond 2100 which then was based on science from the 1990s (which is how the IPCC – rightly – works). This is not to decry the IPCC, science has its limits and research itself functions within constraints. Indeed, the first IPCC report (1990) did much to spur investment in climate-related research.

Finally, looking centuries ahead it is thought that once the sea level has risen due to Greenland and Western Antarctica, there would be little scope for levels to lower completely to present values without the onset of a proper glacial. For example, once the Greenland ice sheet has melted then sea levels from this alone (ignoring thermal expansion or any other contribution such as from Antarctica) would raise sea levels by 7 m. Furthermore, both geological evidence and computer models (run in 2002) suggest that even if the Earth returned to twentieth-century temperatures, the Greenland ice sheet would not reform (Toniazzo *et al.*, 2004).

Given the latest science, the palaeoanalogues, and separately that much of the global economy is based on settlements that have been located for many centuries – and so which are ill-suited to address some climate-change consequences – the impact of climate change in purely human terms cannot be ignored. The IPCC attempts to provide a comparatively near-future picture; the uncertainty relates as to how near a future while planning arguably needs to be undertaken with a longer-term perspective. We have considerable confidence that both warming and sea-level rise will continue beyond the twenty-first century. So while today we may build sea defences to cope with the twenty-first-century anticipated rise together with a safety margin, we should perhaps construct them in anticipation of their needing to be enhanced to cope with further rise in the future. After all, many of today's rail networks are grounded in centuries-old constructions. Such planning will be far more cost-effective in the long term as well as allowing us to better cope with a surprise rise as warned of by the IPCC.

The above multi-century scenario may sound sensationalist but it is not one of doom and disaster. Many species on Earth survived the IETM even if the planet changed and there was a (primarily) marine extinction event. Indeed, modern humans (*H. sapiens sapiens*) survived the considerable environmental changes that have taken place since the LGM even if some civilisations have not survived subsequent regional climate change. To max-imise the longevity of human and some type of biological system (but not necessarily their location) *all* that is required is understanding (research), anticipation (planning) and application (implementation). Application is beyond the scope of this text, but planning and policy concerns will be addressed in Chapter 8, while climate-related biological and human impacts will be summarised in Chapters 6 and 7 respectively. Meanwhile, there are further matters arising out of the 2001 IPCC report, and subsequent science, to note.

5.4 Current human influences on the carbon cycle

Human influences on the carbon cycle will, the IPCC concluded in 2001, continue to change atmospheric composition throughout the twenty-first century. As the IPCC provides the official scientific consensus, it is worth summarising their conclusions with regards to the principal greenhouse gases in the current century.

5.4.1 Carbon dioxide

As noted in Chapter 1, the level of human contribution to atmospheric carbon dioxide is increasing. The breakdown of the principal sources of this carbon dioxide is given in Table 1.3.

The IPCC's 2001 assessment concludes that carbon dioxide emissions from fossil-fuel burning are 'virtually certain' to be the dominant influence on changes in atmospheric concentrations for the twenty-first century. Further to the greenhouse-gas information given in Chapter 1, the reason for this dominance is two-fold. First, the global population is becoming more energy-intense: people are consuming more energy on a per capita basis. Second, the global population is growing (see Figure 5.9). Both these factors are not additive but more

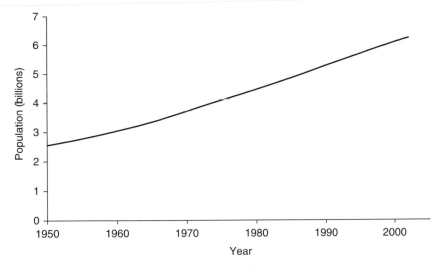

Figure 5.9 The world population 1950–2002.

multiplicative. We will return to global population when looking at the likely future population as part of human ecology in the next chapter.

With reference to the carbon cycle (Figure 1.3), the IPCC (2001b) concludes that as atmospheric carbon dioxide increases the ocean and land will absorb a smaller fraction of subsequent releases. This means that burning a litre of petrol in 10 years time will result in more carbon dioxide residing in the atmosphere for a century or so than a litre burnt 10 years ago. This in turn means that the warming effect from burning a single litre will be greater. (In other words the release of a unit of carbon dioxide has a greater warming ability, even if its global warming potential (GWP) remains by definition as 1.) With regards to Table 1.1 and the current concentration of carbon dioxide, the IPCC (2001) concludes that while pre-industrial levels of carbon dioxide were around 280 ppm, and the present (2006) concentration is around 380 ppm (a 34% rise), the projected 2100 concentration is likely to be 540–970 ppm (or 90–240% over pre-industrial levels). The anthropogenic carbon dioxide forcing of climate is therefore going to be some three times greater (possibly more) for the twenty-first century alone than it was two and a half centuries ago!

In terms of palaeoclimatology, today's carbon dioxide concentration has not been exceeded during the past 420 000 years and likely not during the past several million years. The rate of increase since 1900 is unprecedented in terms of human history and, indeed, evolution. This means that current and near-future (a human generation or two) climate change due to anthropogenic greenhouse forcing will be of a nature and degree that humanity has yet to experience.

The IPCC (2001b) also notes that 'hypothetically' if all the carbon released due to historic land-use change (forest clearance and such) could be reversed then projected 2100 atmospheric concentrations could be reduced by 40–70 ppm. However, this would not only be a proportionally minor reduction compared to the aforementioned anticipated rises (less than four decades' worth of 1990s-rate atmospheric rise), it would also be a temporary one. For once forests are grown, or soils managed so as to absorb the maximum amount of carbon they can, then the forests and soils (under climatically stable and managed conditions) will at best remain greenhouse-neutral. 'At best' because the necessary continued proper management of such forests and soils cannot be assured (notwithstanding that we still have much to learn about soil carbon). Indeed, soils managed at the beginning of this century to absorb carbon might well see the carbon released again as the climate warms. Given that atmospheric carbon dioxide stabilisation (as discussed above) is unlikely, so that the global climate will continue to warm, and that management cannot be assured (in some circumstances a single ploughing can release many years of a soil's carbon sequestration), it would be most unwise to rely on land management as a principal tool of carbon sequestration. This is just one reason why the IPCC use the word hypothetically. Having said this, whereas the prospects for land management to counter atmospheric increases are very limited, greenhouse-gas-sensitive land use can at least slow potentially significant further additions of carbon to the atmosphere. (Deliberate, human-managed sequestration will be dealt with in Chapter 7.)

Aside from climate-change concerns, as noted in Chapter 1 the oceans absorb some of the increased atmospheric carbon dioxide from human activities. There are impacts arising from changes in the oceans' chemical environment. The carbon dioxide absorbed by the sea reacts with water to form carbonic acid, so making the oceans more acidic. In 2004 a research team for an international 5-year programme, led by US researcher Chris Sabine, reported on a compiled data-set of 9618 hydrographic stations and their dissolved inorganic carbon (DIC) measurements. The global data-set gathered was based on 10 times more observations than the last time such a survey was conducted, in the 1970s. The researchers estimated that the cumulative oceanic sink from anthropogenic carbon dioxide from about 1800 to 1994 was 118 ± 19 Pg (or Gt) of carbon and that this is possibly about a third of the oceans' long-term sink potential. However, this does not mean that we are a third of the way in time to this ocean-saturation point; we are closer, as anthropogenic emissions per year are greater now than during the nineteenth and early twentieth centuries. Further, as we shall see when discussing the human ecology of climate change, human emissions of carbon have been steadily increasing since 1800, so that instead of four centuries (had the 1800–1994 atmospheric concentration of carbon dioxide been a constant average) we have far less time before saturation is reached. Emissions for the last 50 years of the twentieth century were very roughly equal to total emissions for the 200 years up to 1950. Furthermore, as we continue to approach saturation so the oceans become *ever more* acidic: the reduction in pH of a more acidic ocean has a greater biological impact than the same lowering of pH in a less acidic ocean. Alongside the paper by Chris Sabine *et al.* was another from a team led by Sabine's co-worker, Richard Feely (2004). They concluded that, under IPCC's 2001 forecasts for the end of the twenty-first century, atmospheric carbon dioxide levels could be around 800 ppm. At that level ocean pH would further decline, by about 0.4, to a level not seen for more than 20 million years. At lower pH values calcium carbonate shells of forams will begin to dissolve faster and growth will be compromised. Calcification rates can drop by as much as 25–45% at carbon dioxide levels equivalent to atmospheric concentrations of 700–800 ppm. These levels will be reached by the end of this century if fossil-fuel consumption continues at projected levels. Effects will be worse in those parts of the ocean where carbonic acid concentrations are highest. They would have a profound effect on ocean acidity and the oceans' biological community that relies structurally on carbonate. Already one of the researchers on the team (Victoria Fabry) has observed partial shell dissolution in one of the pteropods (a group of marine gastropod molluscs), a live pyramid clio (*Clio pyramidata*) in the sub-arctic North Pacific. Groups and phyla at possible risk include calcareous green algae, echinoderms (the phylum of starfish and sea urchins), bryozoans (those of a phylum of colonial animals also known as ectoproctans) and the deep-sea benthic forams.

About half of the anthropogenic carbon dioxide taken up over the last 200 years – the 118 ± 19 Pg of carbon – can be found in the upper 10% of the ocean. In turn, this 118 Pg very approximately represents under 40% of the carbon released as carbon dioxide by human action (both fossil-fuel burning and land-use change). If this carbon had not been absorbed by the oceans, atmospheric concentrations would be higher. Recalling that pre-industrial atmospheric carbon dioxide was 280 ppm (Table 1.1), with the 2003 concentration being 376 ppm, this last would be some 55 ppm higher still were it not for oceanic uptake. So, in one sense what is bad for the ocean is good for the atmosphere (although do note that this use of good and bad is unscientific).

Of the ocean DIC from anthropogenic carbon dioxide, more than 23% is found in the North Atlantic, which might be surprising considering its small size (15% of the global ocean surface) compared to other oceans such as the Indian Ocean and North and South Pacific. However, the North Atlantic is bounded by the north-western European and the North American countries, which together have historically (up to 2000) been responsible for very roughly half the planet's anthropogenic carbon dioxide emissions. This in turn suggests that a good proportion of carbon dioxide enters the ocean through being washed out by rain, as opposed to ocean-surface diffusion and wave mixing. However, because so much of the southern hemisphere is ocean, some 60% of ocean DIC from anthropogenic carbon dioxide is stored in that hemisphere's seas. Taking all this together, we might expect signs of a major impact on carbonate species to first begin to manifest themselves in the cool North Atlantic and in waters arriving from it, such as in the Arctic Ocean or the sub-arctic North Pacific.

Since the Industrial Revolution sea-surface pH has dropped by 0.1 and this is over a 200-year period. A further fall in pH of 0.4 over the next 100 years represents a far greater rate of change. The impact will not just be on calcareous species but, because of ecological relationships, reverberate throughout much of the food web. In August 2004 Britain's Royal Society (the British Government-sponsored learned society for scientific excellence) declared ocean acidification to be a research priority. Ocean acidification could well be one of the surprises of which the IPCC warn as a caveat to their twenty-first-century forecasts (see the end of Chapter 6). In 2005 the Royal Society published a report of their working party, assembled to look at the matter. It noted that there appears to be no practical way of removing additional carbon dioxide once it has been absorbed by the oceans and that such absorption, should it continue as it has, would impact on the marine ecosystems with possible substantial socio economic impacts. It concluded that action needed 'to be taken now to reduce global emissions of carbon dioxide to avoid the risk of large and irreversible damage' (Royal Society, 2005).

5.4.2 Methane

Atmospheric methane concentrations have increased by 150% since 1750 (see Table 1.1). It was thought that more than half the current methane flux to the atmosphere is anthropogenic (agriculture, natural gas and landfills). The remainder is natural, such as from wetlands. The atmospheric concentration of methane continues to increase but in the 1990s the rate of this increase declined. The reasons for variability of the rate of increase in atmospheric methane in the 1990s are not entirely clear as quantitative measurement of global sources and sinks is far from complete. Indeed, one recent discovery is that plants seem to generate methane in aerobic conditions (see Chapter 1). Although the levels of methane production from individual aerobic plants are low, globally they could be significant (Keppler *et al.*, 2006). Consequently our lack of knowledge of methane sources and sinks is a source of error (or uncertainty) for climatologists' computer models.

5.4.3 Halocarbons

Halocarbons are carbon compounds, though not strictly part of the carbon cycle due to their resilience and their being human-made. Following international agreements, principally the Montreal Protocol (a rare success story), the atmospheric concentrations of many stratospheric ozone-depleting and greenhouse halocarbons are either decreasing (CFC-11, CFC-113, CH_3Cl_3 and CCl_4) or the rate of their increase has markedly slowed. The combined abundance of ozone-depleting gases peaked in 1994 and is slowly declining, (although ozone depletion will be with us for decades to come). The atmospheric concentrations of these gases tie in with recorded emissions and so in this regard we can be more confident in the latest computer models of the global climate.

If carbon dioxide and methane concentrations were declining, as is the combined abundance of ozone-depleting gases, then greenhouse concerns would not necessitate nearly so much attention. However, as stated above, this is not the case and carbon dioxide and methane reductions seem unlikely. In this context halocarbons are of importance as they can further enhance warming. This in turn can affect the carbon cycle through the release of carbon from soils and other carbon pools (see Chapters 1 and 7).

One problem with halocarbons is that some of the replacements for the ozone-depleting ones are themselves greenhouse gases. The abundance of the hydrochlorofluorocarbons and hydrofluorocarbons is rising as they are increasingly used as chlorofluorocarbon substitutes. Although their atmospheric concentrations are very small (hence their mass impact on the carbon cycle is small), compared to methane and carbon dioxide, they are very powerful greenhouse gases.

Fully fluorinated chemicals have no biological function (hence are not a major focus of this book) and cannot be removed by biogenic action. Furthermore, they are chemically stable, so they have a long atmospheric residence time, typically of a number of thousands of years. However, they are at very low atmospheric concentrations but have typically high GWPs. As such, they will have a significant role in greenhouse warming, albeit in a lesser way compared to methane or carbon dioxide, for a number of centuries.

5.4.4 Nitrous oxide

We briefly touched on nitrous oxide (N_2O) in Chapter 1. As noted, there are many uncertainties as to quantifying the sources and sinks of this long-lived and significant, albeit comparatively minor, greenhouse gas. It is related to the carbon cycle in that nitrous oxide is part of the nitrogen cycle, which includes all living things, as the amino-acid building blocks of proteins contain nitrogen as well as carbon. Hence plants rely on nitrogen and plant activity affects carbon sequestration and soil storage. Consequently, understanding the components of carbon-cycle dynamics helps us relate to nitrous oxide concentrations (and vice versa), as well as greenhouse warming through carbon dioxide, methane and of course nitrous oxide itself.

Again, as with carbon dioxide, methane and halocarbons, the elevated post-industrial atmospheric concentrations of nitrous oxide are to do with human activity. Human activity, and climatic impacts thereon, will form the subject of Chapter 7.

5.5 References

Andreae, M. O., Jones, C. D. and Cox, P. M. (2005) Strong present-day aerosol cooling implies a hot future. *Nature*, **435**, 1187–90.

Augustin, L., Barbante, C., Barnes, P. R. F. *et al*. (2004) Eight glacial cycles from an Antarctic ice core. *Nature*, **429**, 623–8.

Bellouin, N., Boucher, O., Haywood, J. and Reddy, M. S. (2005) Global estimate of aerosol direct radiative forcing from satellite measurements. *Nature*, **438**, 1138–41.

Berger, A. and Loutre, M. F. (2002) An exceptionally long interglacial ahead. *Science*, **297**, 1287–9.

Burroughs, W. J. (1997) *Does The Weather Really Matter? The Social Implications of Climate Change*. Cambridge: Cambridge University Press.

Chapin, III, F. S., Sturm, M., Serreze, M. C. *et al*. (2005) Role of land-surface changes in Arctic summer warming. *Science*, **310**, 657–60.

Cowie, J. (1998) *Climate and Human Change: Disaster or Opportunity?* London: Parthenon Publishing.

Davis, C. H., Li, Y., McConnell, J. R., Frey, M. M. and Hanna, H. (2005) Snowfall-driven growth in East Antarctic ice sheet mitigates recent sea-level rise. *Science*, **308**, 1898–1901.

Feddema, J. J., Oleson, K. W., Bonan, G. B. *et al*. (2005) The importance of land-cover change in simulating future climates. *Science*, **310**, 1674–6.

Feely, R. A., Sabine, C. L., Lee, K. *et al*. (2004) Impact of anthropogenic CO_2 on the $CaCO_3$ system in the oceans. *Science*, **305**, 362–6.

Hermann, P. (1954) *Conquest by Man*. New York: Harper.

Hodell, D. A., Curtis, J. H. and Brenner, M. (1995) The role of climate and the collapse of the Mayan civilization. *Nature*, **375**, 391–4.

Intergovernmental Panel on Climate Change (1990) *Climate Change: the IPCC Scientific Assessment*. Cambridge: Cambridge University Press.

Intergovernmental Panel on Climate Change (1992) *Climate Change 1992: the Supplementary Report to the IPPC Scientific Assessment.* Cambridge: Cambridge University Press.

Intergovernmental Panel on Climate Change (1995) *Climate Change 1995: the Science of Climate Change.* Cambridge: Cambridge University Press.

Intergovernmental Panel on Climate Change (2000) *Emission Scenarios – IPCC Special Report.* Geneva: WMO.

Intergovernmental Panel on Climate Change (2001a) *Climate Change 2001: Impacts Adaptation and Vulnerability – a Report of Working Group II.* Cambridge: Cambridge University Press.

Intergovernmental Panel on Climate Change (2001b) *Climate Change 2001: the Scientific Basis – Summary for Policymakers and Technical Summary of the Working Group I Report.* Cambridge: Cambridge University Press.

Keppler, F., Hamilton, J. T. G., Brab, M. and Röckmann, T. (2006) Methane emissions from terrestrial plants under aerobic conditions. *Nature*, **439**, 187–91.

Knutti, R., Stocker, T. F., Joos, F. and Plattner, G.-K. (2002) Constraints on radiative forcing and future climate change from observations and climate model ensembles. *Nature*, **416**, 719–26.

Lamb, H. H. (1995) *Climate History and the Modern World*, 2nd edn. London: Routledge.

Luterbacher, J., Dietrich, D., Xoplaki, E., Grosjean, T. and Wanner, H. (2004) European seasonal and annual temperature variability, trends and extremes since 1500. *Science*, **303**, 1499–1503.

MacCracken, M., Smith, J. and Janetos, A. C. (2004) Reliable regional climate model not yet on the horizon. *Nature*, **429**, 699.

Manley, G. (1974) Central England temperatures: monthly means 1659 to 1973. *Quarterly Journal of the Royal Meteorological Society*, **100**, 389–405.

Noren, A. J., Blerman, P. R., Steig, E. J., Lini A. and Southon, J. (2002) Millennial-scale storminess variability in the northeastern United States during the Holocene epoch. *Nature*, **419**, 821–4.

O'Dowd, C., Facchini, M. C., Cavalli, F. *et al.* (2004) Biogenically driven organic contribution to marine aerosol. *Nature*, **431**, 676–80.

Penner, J. E., Dong, X. and Chen, Y. (2004) Observational evidence of a change in radiative forcing due to the indirect aerosol effect. *Nature*, **427**, 231–4.

Pinker, R. T., Zhang, B. and Dutton, E. G. (2005) Do satellites detect trends in surface solar radiation? *Science*, **308**, 850–4.

Roderick, M., L. and Farquar, G. D. (2002) The cause of decreased pan evaporation over the past 50 years. *Science*, **298**, 1410–12.

Royal Society (2005) *Ocean Acidification due to Increases in Atmospheric Carbon Dioxide.* London: Royal Society.

Sabine, C. L., Feely R. A., Gruber, N. *et al.* (2004) The oceanic sink for anthropogenic CO_2. *Science*, **305**, 367–71.

Thomas, R., Rignot, E., Casassa, G. *et al.* (2004) Accelerated sea-level rise from West Antarctica. *Science*, **306**, 255–8.

Toniazzo, T., Gregory, J. M. and Huybrechts, P. (2004) Climatic impact of a Greenland deglaciation and its possible irreversibility. *Journal of Climate*, **17**, 21–33.

Travis, D. J., Carleton, A. M. and Lauriston, R. G. (2002) Climatology: Contrails reduce daily temperature range. *Nature*, **418**, 601.

Tyson, P., Odada, E., Schulze, R. and Vogel, C. (2002) Regional-global change linkages: Southern Africa. In Tyson, P., Fuchs, R., Fu, C. *et al.*, eds. *Global-Regional Linkages in the Earth System*, chapter 2. Berlin: Springer.

Ufnar, D. F., González, L. A., Ludvigson, G. A., Brenner, R. L. and Witzke, B. J. (2004) Evidence for increased latent heat transport during the Cretaceous (Albian) greenhouse warming. *Geology*, **32(12)**, 1049–52.

Vaughan, D. G. (2005) How does the Antarctic ice sheet affect sea level rise? *Science*, **308**, 1877–8.

Von Storch, H., Zorita, E., Jones, J. M. *et al.* (2004) Reconstructing past climate from noisy data. *Science*, **306**, 679–82.

Wild, M., Gilgen, H., Roesch, A. *et al.* (2005) From dimming to brightening: decadal changes in solar radiation at the Earth's surface. *Science*, **308**, 847–50.

Worldwatch Institute (2003) *Vital Signs: 2003–2004*. London: Earthscan.

6

Current warming and likely future impacts

6.1 Current biological symptoms of warming

As noted in Chapter 2, climate change impacts on living species and so in turn these impacts can be used as climatic proxies. As discussed in Chapter 3, climate change throughout the Earth's history has been complex and features a number of distinct and characteristic episodes as well as periods with defining trends. However, as was observed in Chapters 3 and 4, the biological response of species and natural systems to climate change is also complex, as is the biological response of (and impact on) humans. A good proportion of this last is reflected in recorded history. Further, these impact elements – of species response and climatic-event and natural-systems response – are all discernable when applied to climate at the end of the twentieth and early twenty-first century and all exhibit varying degrees of complexity. Some responses are sufficiently complex that they appear as the opposite to what might initially be expected. Even in such cases, when examined carefully, they all clearly reflect the fact that the planet is warming up.

6.1.1 Current boreal dendrochronological response

One illustrative example was briefly mentioned in Chapter 2 when looking at dendrochronology. Figure 2.1 portrays the pooling together of several dendrochronological series from the northern hemisphere's high latitudes. This clearly shows that in the twentieth century the north of the northern hemisphere has warmed above the 1601–1974 average. However, some dendrochronological series (Figure 2.1, the dashed line on the far right) seem to exhibit a late-twentieth-century return to average temperatures even though others reflect continued warming (in line with instrumental measurements). As indicated, this deviation from what would be expected may be due to increased snowfall with the volume of the extra snow delaying the melting of snow cover and so delaying the onset of spring greening. This is not unexpected, because, as previously noted, a warmer planet would have increased ocean evaporation and hence increased precipitation (be it rain or snow). Another explanation mentioned is that some other environmental factor impedes growth. As noted in Chapter 2, species respond to a range of environmental factors, some of which act synergistically, sometimes in a positive way and others in a negative way.

The response of the Alaskan white spruce (*Picea glauca*) to late-twentieth-century warming exemplifies these complexities. Our current warming period has at high latitudes extended the thermal growing season (TGS) and this is clear from observations from instruments both on the ground and in satellites. Of course, there being more days in the year in which plants can grow, primary biological productivity (plant biomass increase per unit area) increases. This means that more carbon dioxide is sequestered by photosynthesis and this in turn increases water consumption (above and beyond that from climate warming itself). Of course, increased water demand is dependent on there being the necessary extra water present. True, in a warmer world we would expect more rainfall, but this is only an expectation (although there is certainly more water vapour; see Chapter 1) and an expectation we would expect to be reflected in, on average, more precipitation across the whole planet. Yet, again as previously noted, there are frequently regional and even comparatively local marked deviations from the average expectation. So consider a species that is on the border of its environmental range. Whereas one environmental factor may increase (in this case temperature) it does not necessarily mean that other environmental factors will increase (even if we might expect them to) so as to accommodate a species' healthy growth. Then again, whereas in a warmer world we would expect more precipitation, equally we should expect more evaporation. In short, in some places and for some species whose environment has changed, water availability could be a problem. In 2000, researchers from the University of Alaska Fairbanks found that, whereas warming had extended *P. glauca*'s TGS, growth had not increased as expected (Barber *et al.*, 2000). The problem was one of temperature-induced drought stress. The interior of Alaska is semi-arid and in many places the potential evapotranspiration is equal to annual precipitation. Increase the potential evapotranspiration (through increased temperature and hence extended TGS) without an appropriate increase in precipitation and species experience drought stress.

This phenomenon is not just of intellectual interest as far as climate change is concerned, but of practical relevance too. *P. glauca* is one of the most productive and widespread forest tree species in the boreal forest of western North America. Consequently it plays an important role in the north northern-hemisphere terrestrial sink, which soaks up (sequesters) some of the excess (over natural fluxes of) carbon dioxide released through fossil-fuel burning and other human activities. If this changes with warming then the sequestration rate will in theory change which in turn will affect the rate of future warming from carbon emissions, although in practice the scale of such increased carbon sequestration is highly unlikely to make a meaningful differ-ence at the policy level. Nonetheless, such biological work is timely for com-puter modellers of climate, who, having begun to incorporate some terrestrial and marine biological factors into their simulations, are now doing so with increasing complexity, which is beginning to include the dynamics of bio-logical response to the climate change being modelled. Such developments are vitally important, for, as discussed previously, whereas computer models have improved they are still a long way from reflecting climate change in a detailed way in both space and time, or in recreating past major change in anything other than an approximation.

6.1.2 Current tropical-rainforest response

Tropical rainforests are among the most species-rich tree communities on Earth and the most productive of terrestrial ecosystems. Consequently it is not too surprising that tropical rainforests are affected by global warming and concomitant increases in atmospheric carbon dioxide. A recent medium-term study, over two decades long in 2004, of 18 tropical rainforest plots that were otherwise undisturbed, has shown a number of changes (Laurance *et al.*, 2004).

All trees within each plot that were over 10 cm in diameter at breast height were tagged and mapped. In total nearly 13 700 trees were recorded. Because rainforests are so biodiverse some 88% of tree species were too rare for there to be a robust statistical analysis. Furthermore, there were so many species and subspecies that their analysis was done at the genus level so as to statistically capture the changes. There were 244 genera, of which 115 were sufficiently abundant for statistical analysis.

They found that within the plots the rates of tree mortality, recruitment and growth have all increased to some degree or other over the 20 years of the study. Of the 115 relatively abundant tree genera, 27 changed significantly in population density or basal area to a degree that was nearly 14 times greater than might be expected from chance alone. This result is similar to other studies. However, contrary to some predictions there was not an increase in pioneer species. Pioneer species are those that are commonly found colonising new or disturbed ground, and higher mortality associated with ecological change would be expected to disturb ground with tree falls.

The researchers also found that there were an increasing number of faster-growing trees, including many canopy and canopy-emergent species. These increased in dominance or population density. Conversely, a number of genera of slower-growing trees, including many sub-canopy species, were declining.

So, what is causing these changes? The researchers noted that it could be that the plots were recovering from some previous disturbance. However, their plots' location being away from areas of human land use as well as in the depth of forests (so avoiding edge effects) and over an area of $300\,km^2$ (so avoiding purely local factors) meant that previous disturbance was an unlikely cause of the changes seen.

Second, the researchers wondered whether the changes might be due to regional climate cycles, such as El Niño-related droughts. These have increased over the last century and this itself is thought to possibly be related to global warming. (Remember from the Pliocene discussion (Chapter 4) that there is evidence for permanent El Niño-like conditions around 5–3.5 mya and, irrespective of this, this time may be considered a palaeoclimatic analogue to a late-twenty-first-century global climatic mode.) However, countering this two separate analyses suggest that El Niño-related droughts were not the cause of these tropical-rainforest changes, although the researchers advise further checks, so this matter has yet to be resolved. Third, rainfall changes (as distinct from drought, which sees a decline in rainfall, river flow and ground water) might be the cause, but this too was discounted on checking meteorological records.

The final possible reason was that the changes might be driven by accelerated forest productivity. The researchers believed that this was the most likely reason, due to rising atmospheric carbon dioxide, although increased cloudiness and higher airborne nutrient transfer from increased forest fires (both climate-related factors) could be the cause. Either way, greenhouse changes seem to be the spur.

These changes could have both local and global implications. Currently, undisturbed tropical rainforests appear to function as carbon sinks and so they help slow global warming. Yet such ecological changes as seen in the rainforest study by Laurance *et al.* (2004), if pervasive, could modify this effect. Increases in carbon storage by such tropical trees may be slowed by the tendency of canopy and emergent trees to produce wood of lower density as their size and growth rate increases. Further, such species shade the more densely wooded sub-canopy species that were seen to have lower growth and were declining. Indeed, a sufficiently warmer climate may turn a tropical forest into a carbon source as opposed to a sink. Currently there is a research priority to ascertain whether such carbon changes are taking place in the tropics and to confirm their cause as well as to try to forecast the effects of warmer climates on existing tropical forests.

6.1.3 Some biological dimensions of the climatic-change fingerprint

The above current dendrochronological response (section 6.1.1) is but one example of the complexities of biological response to climate change. We have previously noted others, including with regard to climate-driven species migration (both horizontal and vertical), extinctions and evolution. We have also seen (Chapter 2) how individual species can serve as indicators of past climate. So the question arises as to whether there is any general effect of current warming across natural systems. In other words, is it possible to discern a global-warming fingerprint?

In 2003 biologist Camille Parmesan (who had contributed to the IPCC's 2001 Working Group II report), and economist Gary Yohe, had the results of their attempt to discern such a fingerprint published. They noted that the discussion surrounding the IPCC's Working Group II's analysis (IPCC, 2001a) of climate impacts and vulnerability had featured a 'divergence of opinion'. In no small part this was because of the differences in the way economists and biologists work. Economists tend to look at aggregate data over a long time. Biologists, conversely, recognise that biological change is multi-factorial and much is to do with non-climatic human intervention (such as land-use change) and so they tend to examine unperturbed systems. This the economists view as data selection, whereas biologists consider it a quality-control filter. However, Parmesan and Yohe combined biological and economist approaches, adopting IPCC criteria, to look at natural (as opposed to human-managed) systems. They looked at data-sets as well as individual cases found in the scientific literature. They then assessed these using three variables: the proportion of observations matching climate-change predictions, the numbers of competing explanations for these observations and the confidence of relating each observation to climate change. Finally, to help overcome any literature bias (scientists might prefer to write up only positive correlations), they used only multi-species studies that reported neutral and negative climate correlations as well as positive ones. The brunt of their focus was on phenological effects – those of season changes on the lives of plants and animals.

They found that for range boundaries of 99 species of northern-hemisphere temperate bird, butterfly and alpine herb, at the end of the twentieth century these have moved on average 6.1 km ($\pm$2.4 km with 95% confidence) northwards or the same number of metres upwards per decade. Phenologically a total of 172 species of herb, butterfly, shrub, tree and amphibian revealed an earlier spring timing of 2.3 days per decade with a 95% confidence range between 1.7 and 3.2 days' advancement per decade. With regards to phenology (see the following subsection), purely on the basis of either advancement or delay of spring events (i.e. no quantification, just direction of change), such as frog breeding, bird nesting, first flowering, tree burst and arrival of migrant birds and butterflies, of 677 species assessed from the literature, 27% showed no change, 9% showed delayed spring events (the opposite of what you would expect with global warming) and 62% showed spring advancement. In summary, about 87% of these species that showed any change, showed a change one might expect with climatic warming.

Interestingly, published along with this study was another that also looked at the fingerprint of global warming on wild plants and animals (Root *et al.*, 2003). Again there was an association with the IPCC, in that four out of the six researchers were previously members of the IPCC's 2001 Working Group II and the study by Root and colleagues built on some of the literature examined by that group. There is little doubt that their motivation to express a separate analysis arose out of the Working Group II's own internal debate, of which Parmesan and Yohe were aware (see above), that there was a 'divergence of opinion'. It is important to note that Root *et al.* only included those studies that (1) spanned 10 years or more, (2) showed that at least one species changed over time and (3) found either a phenological change associated with temperature or the reverse. However, in addition they also attempted to see whether there were any correlations between climatic cycles (see section 5.15), such as the North Atlantic Oscillation (NAO) and the El Niño/Southern Oscillation. Of over 2600 species covered in 143 studies that met the researchers' criteria they found that more than 80% of animal (molluscs to mammals) and plant (grasses to trees) species exhibited change in the direction that one might expect with global warming. They conclude that the balance of evidence is that there is a discernable fingerprint of recent decadal warming on animal and plant populations. Whereas the authors admit that their study selection is open to possible bias, it is interesting to note that this figure of more than 80% appears to broadly corroborate Parmesan and Yohe's figure of 87%.

To take just one example of how a regular climate oscillation affects species, the NAO index has been shown to have a correlation with UK populations of a number of species of butterfly. Specifically, the correlations are between a so-called collated index of the butterfly species and the NAO index.

The collated index of butterflies is calculated by the UK Centre for Ecology and Hydrology (CEH), which has been maintaining a database since 1976. CEH researchers monitor the populations of butterfly species at 130 sites and data from these is collated into an index. Meanwhile, the NAO index is calculated from the difference in air pressure between Iceland and the Azores, Lisbon or Gibraltar (the choice of the southern grouping of weather-station sites seems to make little difference). The NAO index can be positive or negative. If positive then depression systems tend to take a more northerly route across the Atlantic, so giving the UK a warmer and wetter winter and autumn. A negative index is associated with drier and cooler weather in the UK as moisture goes more towards the Mediterranean basin. A number of UK butterfly populations have been shown to fluctuate with the NAO index (Westgarth-Smith *et al.*, 2005), including the common blue (*Polyommatus icarus*) and small cabbage white (*Pieris rapae*).

It is important to note that these periodic fluctuations in species populations that correlate with climatic oscillations are distinct from species population change due to long-term climate change. This means that those studying the biological impacts of climate change have to disentangle these two related phenomena. However, the biological impact of climatic oscillations is relevant to the impact of longer-term climatic change. First, regionally part of a climatic cycle may go with the direction of longer-term climatic change, so that part of the cycle can provide insights into the regional manifestations of longer-term climatic change. Second, understanding the biological impact of climatic cycles is intrinsically important to regional biomes. Finally, longer-term climatic change may result in changes in ocean and atmosphere circulation (see later on in this chapter) and these circulation changes can themselves change the characteristics (or even existence) of regional climatic oscillations. If we are unaware of the current effects of regional oscillations then it will be difficult to discern what biological changes are due to contemporary regional climate systems and what changes are due to longer-term climatic change.

Finally, in a warmer world with elevated levels of carbon dioxide we would expect increased levels of photosynthesis. Although regionally there may be differences, with some areas seeing decreases in photosynthesis (for example, due to water restriction), and regionally there would be differences in fixing and retaining carbon (both from plants and soils), in purely photosynthetic terms the reasonable expectation would be to see an increase in global primary production (see the definition of production in the Glossary, Appendix 1). A satellite observation of global vegetation between 1982 and 1999 suggests that globally terrestrial (excluding oceans and freshwater systems) net primary production has increased by 6% (3.4 GtC) over 18 years. The largest increase was in tropical ecosystems and the study suggested that the Amazon rainforest accounted for 42% of the global increase in net primary production (Nemani *et al.*, 2003). Here we should remember that whereas global climate-induced changes in primary production will not meaningfully affect atmospheric concentrations of carbon dioxide for policy-makers (who seek large changes over a short time), changes in primary production are a regionally important mechanism for carbon transportation, via the atmosphere, between ecosystems.

6.1.4 Phenology

Aside from changes in species ranges (spatial distributions), phenological studies are, as mentioned, the other main indicator of a climatic fingerprint, and phenological records have been built up over the past few centuries. (Phenology being the study of a plant or animal's progression through its life cycle in relation to the seasons.) In Britain the oldest phenological record known is a diary entry for the first cuckoo of spring in Worcestershire in 1703. But the association of the arrival of spring with certain species is much older. For example, the Romans in Britain knew all about the swallow and spring. However, it was many hundreds of years before phenological data began to be assembled systematically. Arguably the first, which began with observations from 1736, was the English Norfolk country gentleman Robert Marsham's collection of '27 indications of spring'. Others, such as Thomas Barker of Rutland and Gilbert White of Selborne, Hampshire, were also keen observers of nature and kept phenological diaries. Gilbert White even compared his own records with those of a friend in Sussex. Phenological observation has been a popular activity for professional biologists and amateur natural historians ever since, so much so that between 1875 and 1947 the Royal Meteorological Society managed a voluntary national network of recorders (Sparks & Collinson, 2003).

As for the changes in phenology that these records reveal in Britain, as a very loose rule of thumb it appears that flowering and leafing occur 6–8 days earlier for every degree Celsius rise in temperature. However, species do react differently. For example, oak (*Quercus* spp.) responds twice as fast as ash (*Fraxinus* spp.) to an increase in temperature.

European and North American phenological sets have predominantly reflected changes that are probably associated with climate change, especially with regards to earlier spring phenology (Table 6.1). But there is also some indication of later autumnal events, which again would be expected in a warmer world; however, these are less marked. Indeed, birds that show an earlier migration arrival do not show as late an autumnal departure. Autumnal leaf colouring is even more ambiguous within regions. On a long timescale, European autumnal leaf colouring has shown to be delayed at a rate of about 0.3–0.6 days/decade, whereas the length of the thermal growing season (TGS)

Table 6.1. *Some phenological evidence of advances in spring events (from Walther et al., 2002).*

Taxon	Location	Changes	Period observed
Many plant species	Europe	Flowering and leaf unfolding 1.4–3.1 days/decade earlier	Past 38–40 years
	North America	1.2–2.0 days/decade earlier	Past 35–63 years
18 Butterfly species	UK	Earlier appearance by 2.8–3.2 days/decade	Past 23 years
Amphibians	UK	Earlier breeding	Past 25 years
Numerous bird species	Europe, North America	Earlier spring migration 1.3–4.4 days/decade and breeding by 1.9–4.8 days/decade	Past 30–60 years

has increased by 3.6 days/decade over the past 50 years. This correlates well with independent satellite data.

Here it is worth emphasising that although a warming world may be a dominating driver of current phenological change, as mentioned in the previous subsection, a number of the major climate-related oscillations have a phenological effect. In 2002 Gian-Reto Walther and a small group of mainly European scientists published a review of ecological responses to climate. They report that an analysis of 50 years of data on 13 plant species in 137 northern-hemisphere locations revealed responses to the NAO in 71% of the data. Early-blooming herbaceous species showed greater response to winter warming than late-blooming and woody plants. Knowledge of regional climatic cycles is fundamentally important when studying biological response to climate change.

Finally, knowledge of a species' primary requirements is also fundamentally important given that these primary requirements may in turn be driven by a climatic dimension or dimensions. These can be complex. Consider the migratory black-tailed godwit (*Limosa limosa islandica*), a shore bird that winters between Britain and Iberia while breeding in the summer almost exclusively in Iceland. Breeding pairs exhibit a high degree of partner fidelity. Whereas most migratory pairs winter together, and so migrate together, the black-tailed godwit is one of a number of species where male and females winter in different locations many hundreds of miles apart. Yet breeding pairs arrive in Iceland typically within about 3 days of each other. The question then arises as to how this degree of synchrony is maintained when the environmental conditions at the different sexes' wintering sites are dissimilar (Gunnarsson et al., 2004). The answer will be key in beginning to understand how such species will respond to future climatic change. Indeed, respond to climatic change they will, for clearly as little as 15 000 years ago, towards the end of the last glacial, their migratory patterns would have had to have been most different and so a successful response to glacial–interglacial climate change must have taken place on a number of occasions.

Today climate change seems to be affecting another migratory bird species, the pied flycatcher (*Ficedula hypoleuca*). Its migration is timed to the availability of food for its nestlings. However, phenological changes have meant that in some parts of the Netherlands the caterpillar's that are its food peak early in the season. Here the flycatcher populations are in decline (Both *et al.*, 2006). Whether such species will adapt with time in the modern world remains to be seen.

6.1.5 Biological communities and species shift

Of course, some species do not migrate as part of their life cycle but will shift their geographical position, or range, in response to climate change. In the previous chapter key characteristics and examples of species' response to glacial–interglacial climatic change were summarised. However, because climate is but one factor (indeed, climate itself is multi-factorial) of many that determine a species' spatial distribution, species rarely move uniformly with each other in response to climate change. For example, a calcareous soil species prefers soils with a high pH and this preference might dominate that of minor climate change. Figure 6.1 illustrates uniform and non-uniform vertical and horizontal migration. (See also Chapter 4 and the discussion on biological and environmental impacts of the last glacial, as well as Figure 4.10.)

Added to the above, and again as noted in the previous chapter, different species migrate at different rates. Consequently, not only is a period of climate change ecologically dynamic but it takes time for ecological communities to stabilise after a period of climate change. Furthermore, again as touched upon in Chapter 4, species at the leading edge of shifts/migrations tend to migrate faster than those already established. As ecologists Walther and colleagues (2002) note, changes in distributions are often asymmetrical, with species invading faster from lower elevations or latitudes than resident species receding upslope or poleward. The result is often an increase in species richness of communities at the leading edge of migrations. However, this biodiversity 'benefit' is transitory. Furthermore, whereas such biodiversity may be seen in natural systems (and so found in the palaeontological record), it is less likely in many of today's systems, which are either managed or bounded by land

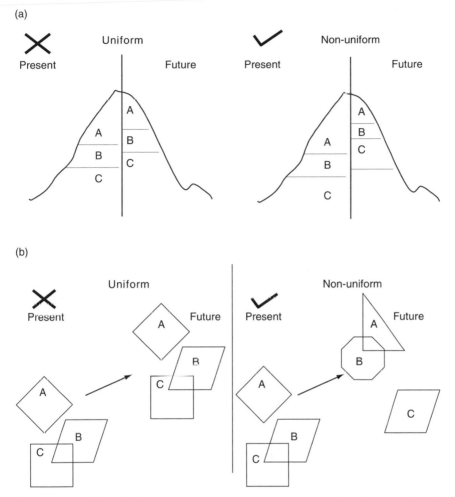

Figure 6.1 (a) Vertical migration of species and communities due to climatic warming is frequently not uniform. Individual species respond to different environmental factors differently. Consequently there is little reason for uniform migration. An example is given relating to Figure 4.10. (b) Horizontal migration of species is rarely uniform (see Chapter 4 with reference to glacial–interglacial change). As with vertical migration, this is due to different species responding with different sensitivity to different environmental dimensions of which climate (itself multi-dimensional) is among many other factors, which together will determine a species' spatial distribution. This makes prediction of ecological response to climate change difficult.

which is managed by humans. Such managed lands often make for an effective barrier to species migration. This in turn poses problems for the future of species, let alone ecological communities; for even, as noted, if old ecological communities are disrupted, impeding species migration impedes new communities being formed.

This problem of synchrony of biological community species migration as a response to climatic change is loosely analogous to the synchrony problem for migratory bird pairs over-wintering in different locations (see the end of the previous subsection). The big difference, though, is that whereas successfully breeding pairs are fundamentally necessary to a species' continued existence – and even a synchronous response across species is necessary for the continued existence of a biological community – the continued existence of a biological community is not always necessary for the continuity of a species. Species may continue, but in new biological communities, as represented in Figure 6.1b. Here the community of species B and C is disrupted by climate change and a new community of species 'A' and 'B' is created.

Such changes can be particularly profound. Changes in response to climate change at the lower trophic levels of a community are likely to have profound effects at higher trophic levels (towards the top of the food chain). This can happen in time (as part of the annual cycle) as well as in space.

This is aptly illustrated, for example, with the results of a long-term (1958–2002) monitoring of marine pelagic (middle levels or surface) species in the North Sea (Edwards & Richardson, 2004). The study looked at the following five functional groups (with their trophic level in parentheses): diatoms and dinoflagellates (the primary producers in this case), copepods (the secondary producers) and non-copepod holozooplankyton (secondary and tertiary producers) and mere plankton, including fish larvae (some of the North Sea secondary and tertiary producers). Because the tertiary producers feed off of the secondary producers which in turn feed off of the primary producers, which photosynthesise, species in each trophic level are dependent on the one below. As is fairly common in temperate species, abundance varies throughout the year. Consequently, if the abundance of one species changes at a particular time then the others would have to similarly change if the ecosystem's structure is to remain unaltered, maintaining proportionality between populations. Edwards and Richardson found that species responded differently over the four-decade study period, with some species advancing their abundance peaks and others barely changing. This mismatch in time and between trophic levels as well as functional groups could disrupt energy/nutrient flow to the highest trophic levels. This may explain why (in addition to fishing pressures) North Sea cod (*Gadus morhua*) stocks have declined so markedly in recent years.

Of course, not all marine change can be attributed to climate change. Other things, including anthropogenic factors such as fishing, affect marine ecosystems, which is why the aforementioned North Sea study did not look at the adult commercial species. Also note that the fishing of a few species can affect others in the ecosystem that they feed off of or otherwise interact with. Again, as with much climate-change biology, the picture is complex and so there is often a reluctance to draw too firm a conclusion as to what is driving change. One instance of debate has been the marine populations in the eastern Pacific associated with the Californian current. US research early in the twenty-first-century of planktonic Foraminifera, whose remains accumulate over the years in sediments (the research covered 250 years' worth of sediment), has shown that in the late twentieth century there were decreasing abundances of tempe-rate sub-polar species such as *Neogloboquadrina pachyderma*. At the same time there was an increase in tropical and sub-tropical species of forams such as *Globigerina bulloides* and *Neogloboquadrina dutertrei*. Such change seems to reflect twentieth-century warming of the current, so lending credence to it being due to climate change and not some other factor (Field *et al.*, 2006). Yet, rather than cite this as hard fact, it is better to view it as one of a number of pieces of corroborative evidence of climate-driven ecosystem change.

As noted previously, phenological change is a common biological response to climate change but also there is altitudinal and latitudinal migration. Here there are many recorded examples of such shifts in species' ranges, and a number are summarised in Table 6.2.

Table 6.2. *Examples of recent latitudinal and altitudinal shifts in species' ranges (from Walther et al., 2002).*

Species/ecotone	Location	Observed changes
Treeline/mountain zonation	Europe, New Zealand, Alaska	Rising in altitude
Artic shrubs	Alaska	Expansion of range
Alpine plants	European Alps	Altitude rise of 1–4 m/decade
Antarctic plants and invertebrates	Antartica	Distribution changes
Zooplankton, intertidal invertebrates and fishes	Californian coast, North Atlantic	Increase in abundance of warm-water species
Thirty-nine butterfly species	North America, Europe	Northward shifts of up to 200 km over 27 years
Range of 12 bird species	Britain	18.9 km average movement northwards over a 20-year period
Red fox (*Vulpes vulpes*), Arctic fox (*Alopex lagopus*)	Canada	Northward expansion of red fox range and retreat of Arctic fox range

The summary review of ecological response by Walther and colleagues (2002) exemplifies climate-change impact on ecosystems by noting the change at the extreme ends of an equatorial polar bio-environmental gradient. They note that on one hand tropical oceans have increased in temperature by 1–2 °C over the past 100 years, but this average warmth has had superimposed on it short-term warming episodes due to oscillations such as the El Niño/Southern Oscillation. Here the trend has been for these episodes to increase in intensity and duration over the past century. (Again, see Chapter 4's section on the Pliocene, its permanent El Niño-like conditions and the Pliocene being a possible palaeoclimatic analogue to the end of the twenty-first century.)

Of the biological impacts due to warming found in tropical seas, the bleaching of corals has been among the most noticeable. Already near their upper thermal limits, mass bleaching events have taken place whenever sea temperatures have exceeded the long-term summer average by more than a degree. Six periods of mass bleaching took place between 1979 and 2002, with their incidence increasing progressively in both frequency and intensity. The most severe took place in 1998, in which it is estimated that some 16% of our world's reef-building corals died. When the thermal stress is not so extreme only the thin-tissued, branching acroporid and pocilloporid corals have been bleached, leaving larger coral species (such as *Porites* spp.).

At the other end of the gradient, Antarctic terrestrial ecosystems have seen visually marked changes in response to warming. These include the colonisation of previously bare or newly exposed ground by mosses (in the main), as well as by other plants. Associated alongside this colonisation, soil invertebrate animals have also spread. At the moment there are few examples of exotic species colonising Antarctica, although those that have are mainly found close to geothermal sources of warmth and the continent's margins. In addition to this climate impact there are a number of human-imported species. Here climate change favours increasing the chances of imported species becoming established. As for the native species, because many Antarctic species are already at their lower thermal threshold, future warming is likely to see their numbers and range increase, spreading away from the coast. Having said this, imported species will offset part of this range expansion as coastal colonisation takes place and the imported and native species compete.

Meanwhile, in the northern-hemisphere peri-Arctic environment of Siberia, satellite analysis of some $515\,000\,km^2$, containing Arctic lakes (about a seventh of this Siberian biome), since the early 1970s compared with recent

(1997–2004) images saw a decline of 1170 large lakes (those >40 ha) from 10 882: that is, approximately 11% (Smith *et al.*, 2005). However, most did not disappear altogether but shrank to sizes below 40 ha. Total regional lake surface area decreased by 93 000 ha (approximately 6%). One hundred and twenty-five lakes vanished completely. None of these have refilled since 1997–8. However, where the permafrost has remained continuous (more to the north) lake area has increased by 13 300 ha (+12%). Nonetheless, the more southerly declines in lake area have outpaced the more northerly gains. What appears to be happening is that the more southerly permafrost soils have ceased being permanently frozen, so allowing the lakes to drain. Then again, further to the north the increased precipitation expected with global warming (see section 6.5.6 on ocean and atmosphere circulation) is enabling lakes to become larger. Both the decline in the more southern *and* the increase in the more northern Siberian Arctic lake areas appear to be caused by climate change and warming. Such changes have significant implications for soil carbon storage.

It needs to be remembered that it is not just near-polar high-latitude eco-systems such as Arctic lakes that depend on ice. Mountain snowpacks affect the quantity and timing in the year of water in streams supplying ecosystems in the surrounding lowlands. Even in a warm interglacial mode, nearly all moun-tains of sufficient height on Earth, save those near the equator, have snow caps. In a warmer world these will be reduced in volume, especially at lower latitudes, although, note that smaller mountain snow caps may be seasonally thicker due to extra precipitation. This reduction in snowpack volume is something we are seeing already, not just at lower latitudes but mid latitudes, including those in part occupied by China, North America and Europe. Furthermore, in a warmer world snow-cap melt run off will shift away from summer and autumn, when biological, and indeed human, demand for water is greatest compared to winter and early spring. In short, the annual cycle of water supply for many terrestrial and human systems will see reduced tem-poral buffering. The human ecology dimension is indicative of the likely biological impact of such a shift, and here this is exemplified by the broad statistic that one-sixth of the human population relies on glaciers and seasonal snowpacks for their water supply. In terms of terrestrial ecosystems in which over 50% of river flow is dominated by snow melt this area includes nearly all of Canada's catchment, the north-western Rocky Mountain states of the USA, nearly all of Scandinavia with the exception of Denmark, Alpine Balkan and Carpathian Europe, nearly all the Russian nations and north-eastern China, much of Chile and south-western Argentina and the south of New Zealand's South Island. In these regions ecosystems downstream rely on snow to act as a reservoir to regulate river flow and water supply throughout the year (Barnett *et al.*, 2005a). If in a warmer world there is a smaller volume of snow melt then either other water reservoirs will need to be constructed (or have increased aquifer abstraction) to meet human and ecological need or we accept the impact. This may be a theoretical and an expensive choice (especially when considering the currently free service that snow provides). It may be that action is taken to help with meeting human needs, but wildlife outside of semi-natural systems is onlikely to see similar investment and will probably be the hardest hit.

In terms of absolute human numbers, perhaps the most critical region in which vanishing snow, ice and glaciers will negatively affect water supply will be China and parts of India, or the Himalaya–Hindu Kush region. This area not only supports some 2 billion people but also currently represents the largest volume of ice outside of polar and peri-polar regions. Nearly 70% of the Ganges' summer flow and 50–60% of other of the regions' major rivers comes from melt water (Barnett *et al.*, 2005a).

Once again, as with virtually all areas of the climate–biology relationship, there are complications in the detail. Here, it is undeniable that with global warming we are moving away from a world with many local areas of season- ably variable snow cover (similarly affecting many water catchments), to a world with far fewer areas of snow cover. However, this reduction in cover does act to reduce carbon loss from soils around the margins of snowfields. Such mountain areas of thin snow cover tend also to be ecotones, which are special biological communities in border areas between two habitat types. Because, with climate change, biological communities migrate and change, eco- tones are particularly sensitive. As has been mentioned previously (Chapters 3, 4 and 5) carbon in soils can either accumulate or be released, in no small part depending on climate and climate change. Areas of thin snow cover are less insulated from the cold and this in turn affects the rate of carbon release from the soil (Monson *et al.*, 2006). Such detail informs climate computer models.

Of course, all the changes noted in this section relate to change observed to date. Further change is anticipated with further warming. The scientific con- sensus (from the IPCC) is clear: warming will continue through the rest of this century and beyond. The only questions that remain are how much warming and at what rate. This in turn depends on the future emissions of greenhouse gases. Nonetheless, future species migration and other climate-related bio- logical impacts will continue beyond those already seen.

For example, in the UK biological surveys between 1940 and 1989 have shown that a species of butterfly, the speckled wood (*Pararge aegeria*), occupies much of the UK south of the central midlands. However, given business-as-usual climate forecasts to the end of the century, the species is expected to be found virtually throughout the UK up to the Shetland Isles, with only the Scottish highlands and the highest North Pennine peaks being excepted. So, in the UK *P. aegeria* is finding that warming is extending its northern limits. Meanwhile, at the other end of a species' thermal range the opposite can happen. For example, in North America, where the Edith's checkerspot (*Euphydryas editha*) has a range that extends in places from Mexico to Canada, there is a real risk that with warming its southern limits will move north, and in the south its lower altitudinal limits will rise (Hill and Fox, 2003).

It is important to note once again that such shifts in species' ranges due to climate change are likely to be confounded by human land use and landscape fragmentation, and that this is likely to affect some species more than others. Continuing with the example of butterflies, a study of all 46 non-migratory British butterfly species that have their northern range margins in the UK over the last 30 years of the twentieth century illustrates this climate–landscape fragmentation problem (Warren *et al.*, 2001). Being at their northern limit, it being too cold for them further north; it might be assumed that climate warming would favour an extension in their range. The researchers noted that the butterflies could be described in one of two ways: sedentary specialists or mobile generalists (there is an advantage for specialists to be sedentary so that they are always near the specialist ecological community with which they are associated, whereas generalists are able, being mobile, to realise the advantages of new ecosystems; indeed, 26/28 of the specialist species in the study were sedentary, whereas 18/18 generalist species were mobile). The study also noted that, from the 1940s to the end of the century, agricultural intensification in Britain led to a reduction of about 70% of semi-natural habitats that butterflies use. So, on one hand, on the basis of habitat alone it would be expected that most species would decline. On the other, climatic warming in Britain would be expected to benefit these northern margin species. They found that three-quarters (34/46) of the species declined in their range. Further, habitat specialists fared worse than mobile generalists. The researchers took weekly counts of adults from 120 fixed sites between 1976 and 2000 and demonstrated that mobile generalists tended to increase compared to sedentary specialists. Looking at the broader picture, most species of non-migratory butterflies that reach their northern limits of geographical range in Britain have declined, as they have elsewhere in northern Europe. Conversely most sedentary species have not expanded because their prospective habitat is too isolated. Consequently the scale of expansions of butterflies that took place following the LGM through to the twentieth century is unlikely to be repeated with further warming in the twenty-first century. Indeed, the researchers conclude, if sedentary specialists continue to decline then British (and northern European) butterflies will increasingly become dominated by mobile generalists (Warren *et al.*, 2001)

Species migration problems are great, and complex, in a modern landscape that is highly fragmented by human management and land use. However, they become critical to the level of extinction if the species' thermally determined spatial range becomes restricted and reduced to nothing. Typically this takes place at the high-latitude margins of continents. For example, a terrestrial species in Eurasia continuing to migrate northwards as the climate warms will ultimately reach the continent's edge and the Arctic Ocean. Faced with nowhere else to go, extinction becomes virtually inevitable. ('Virtually' because translocation to another part of the globe may be an option, but species translocation and introductions to new environments are not taken lightly.) Many of the species at risk are those that evolved during the Pleistocene or earlier, when the Earth's climate was cooling. Of the peri-Arctic species at risk one of the ones most often cited is the polar bear (*Ursus maritimus*). It evolved in the mid-Pleistocene some 100 000–250 000 years ago, it is thought, from a group of brown bears that became stranded by glaciers. They speciated rapidly, evolving sharper canine teeth (brown bears have far more vegetation in their diet), a longer neck (more suited for keeping their heads above the water line), larger paws (for spreading weight on the ice and for swimming) and of course thicker fur that is lighter in colour (from light brown to white). Polar bears require ice platforms from which to hunt and here is the problem with global warming: Arctic ice is currently reducing at a considerable rate. One early-twenty-first-century estimate has been a 7% reduction in ice cover in 25 years and a 40% loss of thickness. Because polar bears are at the top of the food chain, changes in their numbers are likely to affect the populations of many other species. For example, some seal populations have increased in response to fewer polar bears and this in turn is likely to affect fish predation.

The walrus (*Odobenus* spp.) is also declining, possibly because they too have to work harder to find food due to there being less sea ice. Walrus mothers nurse their young on sea-ice floes. As ice recedes the walrus do too. Further from the coast the mothers must dive longer and deeper from the ice to the sea floor to find clams. Of the three walrus species the population living in the Russian Arctic (*Odobenus laptevi*) has the smallest population, thought at the turn of the century to be between just 5000 and 10 000.

In the southern hemisphere there are analogous problems. In 2001 French researchers, Christophe Barbraud and Henri Weimerskirch, reported on a long-term survey (1952–2000) of the emperor penguin (*Aptenodytes forsteri*) colony near the Antarctic station of Dumont d'Urville and linked this to meteorological data gathered just 500 m from the animals. They noted that the breeding population was stable until the mid 1970s but then declined abruptly in the late 1970s to around 50% of its former level, coinciding with a time of high winter temperatures. They contemplate that declines in Antarctic krill (*Euphausia superba*) – which form much of the emperor penguin diet, along with fish and squid – may be linked to the decline in the emperor penguin population. Krill also sustain a number of commercial fisheries and concomitantly are fundamental to much of the Antarctic marine food web and populations. Yet lower krill abundance is associated with less winter ice. That krill populations around the Antarctic are changing with warming waters, and changes in sea ice, has been corroborated by a body of research. For example, in 2004 a research team, led by Angus Atkinson of the British Antarctic Survey, reported that between 1976 and 2003 the Antarctic south-west Atlantic sector, which contains over half of Southern Ocean krill stocks, has seen stock densities decline by over 50%, while elsewhere those of salps (mainly *Salpa thompsoni*) have increased. This represents a profound change in the Southern Ocean food web.

The other place where species' spatial options become constrained with warming is the tops of mountains. As noted, with global warming thermally determined zonation on mountains changes and rises. If the warming becomes too great then there is nowhere for alpine species to go as species cannot migrate above mountain summits. This is of some ecological and wildlife concern. The alpine biome is just 3% of the vegetated terrestrial surface and such ecological islands are shrinking. For instance, in the Ural Mountains, where temperatures have risen by as much as 4 °C in the twentieth century, tree lines have risen between 20 and 80 m upslope, so reducing the regional alpine zones by between 10–20%. Meanwhile, an example of an animal under severe threat comes from the southern hemisphere and is the mountain pygmy possum (*Burramys parvus*). It is a marsupial found in montane heathland from 1400 to 2230 m in south-eastern Australia. It is not a possum in the American sense and is unrelated to such true opossums. (The word opossum was taken from the Algonquian Indian word for the American animal.) At the beginning of the twenty-first century, with just some 2600 adults remaining, the mountain pygmy possum is considered to be under serious threat (and indeed prior to 1970 was thought to be extinct). The species is further pressured as its habitat is favoured by skiers. As previously noted human land use fragments the landscape and further aggravates the conservation problems posed by current climate change. Land-use concerns recur in many discussions about current climate change and wildlife conservation.

Another mountain example, but this time from the changing environmental conditions that accompany climate change, comes from local endemic animal extinctions in the highland forests of Monteverde, Costa Rica. The situation became apparent following a step-like warming of the tropical ocean in 1976. Twenty species of 50 anurans (frogs and toads) in a 30 km^2 study area went extinct, which suggests pressure on the broader population. In conservation terms, one of the most serious of these local extinctions was that of the locally endemic golden toad (*Bufo periglenes*) in 1987. The population crashes were all associated with a decline in dry-season mist frequency (which had been measured) and it is presumed that this in turn was due to a raising of the cloud-bank base (which unfortunately was not measured). Furthermore, as noted at the beginning of this section, climate change causes modifications in the way that different species populations associate. In this instance, along one stream as the habitat dried harlequin frogs (*Atelopus varius*) gathered near waterfalls, so increasing the probability of their being attacked by parasitic flies, and this in turn resulted in increased mortality (Pounds *et al.*, 1999). The researchers concluded that the aforementioned population crashes belong to a constellation of demographic changes that have altered communities of birds, reptiles and amphibians in the area and which are clearly linked to regional climatic warming. Later work more clearly linked *Atelopus* population crashes with climate change. There were some 110 species of *Atelopus* endemic to the American tropics; a number that made it possible to make a statistically valid local extinction study.

The idea was put forward that temperatures at many highland habitats of *Atelopus* spp. were shifting towards the growth optimum of a chytrid fungus, *Batrachochytrium dendrobatidis*, which is known to infect many species of frog. In short, in this instance climate change was promoting infectious disease in highland areas. Looking at the records of when species were last recorded in highland areas in the wild and comparing these to climate records the researchers were able to show that most extinctions (78–83%) occurred in years of higher tropical temperature and that this was statistically significant to 99.9%. Further, the extinctions varied with altitude (which is also related to temperature) and it is thought that this reflected the optimal conditions for *Batrachochytrium* to flourish (Pounds *et al.*, 2006). This means that not only does global warming reduce individual mountain habitat islands as habitat zones migrate upwards, but that the environmental conditions of the complex mosaic of habitats in highland areas can change so that some areas see an increase in disease organisms.

There is a further twist to this story. Around a third (1856) of amphibian species in the Global Amphibian Assessment are classified globally as threatened and 427 species as critically endangered. One question therefore is whether the *Atelopus* studies meaningfully relate to the global amphibian picture. The earliest known hosts of *Batrachochytrium* were a continent away, the African clawed frog (*Xenopus*) of South Africa. These frogs became economically important in the 1950s as their tissue was used in some pregnancy-testing kits. Museum records also suggest that *Batrachochytrium* achieved a global distribution after the 1960s. The possibility exists therefore that the expansion in one frog species due to commercial activity might perhaps have led to the extinction of other amphibian species (Blaustein and Dobson, 2006).

This is not the only interaction known to take place between climate and disease. Similarly the warmer climate conditions in upland regions of the western USA favour the mountain pine beetle (*Dendroctonus ponderosae*), so that it can complete its life cycle in 1 year rather than 2 years. These beetles act as a vector, enabling the transmission of pine blister rust (*Cronartium ribicola*), and this is having a serious impact on pine trees on some of the Rockies' highest mountains (Blaustein and Dobson, 2006). The issue of economic activity, globalisation, human population growth, disease species and climate interactions is a theme which will be returned to when we look at human ecology in Chapter 7.

Returning to climate change and altitude, an example of a plant species restricted by altitude and which may see a population decline with warming is Britain's Snowdon lily (*Llodia sevotina*).

Of course, species' shifts across landscapes are not just associated with future climatic warming; we have already discussed (Chapter 4) how species relocated with past climate change. Notably, this took place to and from refugia where a species (or species community) survived either a glacial or an interglacial.

One significant concern in all of this, with regard to our present-day situation and future climate change, is the plight of our nature reserves. As has been noted (and will be again) these mainly sit among landscapes that have become highly fragmented by human use and transformation. This fragmentation impedes species migration and so means that many nature reserves can be considered to be ecological islands: an idea that sprang from two North American ecologists, MacArthur and Wilson, in their *Theory of Island Biogeography* in 1967. The problem is that because the climate is changing these island nature reserves are in the wrong place. Alternatively, they could be considered as being in the right place but not featuring the communities of species likely to flourish under future climates. Either way, these nature reserves will not, in a climatically changed future, sustain the species they have helped conserve in the past. Species will be stranded. So the current places of refuge will not be sufficient for our present conservation goals in the future. What is to be done?

One solution could be to ensure that there are ecological corridors to allow species to migrate. Of course, a highly fragmented landscape tends not to lend itself to such corridors. In such circumstances it may just be possible to create a sufficient number of scattered small reserves or mini-communities of species in the landscape. These might be carefully managed hedgerows, strips of green along motorways, parks (and especially managed areas within parks) and even clusters of urban gardens. Such areas may provide stepping stones for some species. Altogether the landscape, though still fragmented, could feature escape routes. The landscape, with its scatter of mini and temporary refuges, could be viewed as a matrix. (Although many may associate the term matrix with mathematics, it actually comes from the Latin meaning a breeding animal or mother but with particular respect to propagation, and womb. Nonetheless, this conservation use does chime with its other use to denote a complex assemblage.)

Of course, as part of creating an assemblage of mini-communities of species, many of these communities will need to be managed to encourage diversity. This is not just for the obvious goal of maximising biodiversity within the managed ecosystem (which may in some cases not be ecologically desired), but within the assemblage of ecosystems within the reserve, as there are often knock-on effects between ecosystems. Effects can cascade across ecosystems. For example, the presence of fish in a pond can have the effect of increasing plant diversity on the surrounding land. As has been shown in a Florida reserve, fish can reduce, through predation, the abundance of dragonfly larvae (such as *Anax junius* and *Erythemis simplicicollis*) in ponds, which leads to fewer adult dragonflies in the air that would in turn prey upon pollinator insects. Fewer predating adult dragonflies means more pollinator insects and so more plants pollinated, hence, in theory, more plants (Knight *et al.*, 2005). So not only does conservation management need to be conducted with regard to the landscape level so as to facilitate migration between sites, it also needs to be done at the ecosystem level, not just for the ecosystem but to facilitate trophic cascades between ecosystems so as to realise biodiversity synergies.

Having said this, because human pressures are so great and becoming greater, the management of reserves and the landscape to create such a matrix, as we shall see in Chapters 7 and 8, will be difficult. Even so, this is a possible conservation tool currently being advocated by a number of ecologists (Lovejoy and Hannah, 2005). Certainly, the trend towards the end of the twentieth century among ecologists was to have regard for managing ecology on the scale of the landscape, rather than purely site by site.

The biological impacts of climate change discussed in this section have been thematic – generalities that apply to many places. To re-cap, these include phenological changes, species migration (and limitations), changes in ecological communities and the problems of an increasingly anthropogenically fragmented landscape. But because the determinants of biological effects of climate change are so multi-dimensional, they can only really be considered in the context of specific regional climate change. The next two sections provide brief summaries of likely change and impacts for two places on Earth, the USA and the UK.

6.2 Case study: climate and natural systems in the USA

In 2000 a major policy study into the *Climate Change Impacts on the United States* was published by the National Assessment Synthesis Team from the US Global Change Research Program (USGCRP). It was, in effect, a stock-take of US scientific understanding at the turn of the millennium has to how climate change might affect the USA. It concluded that the USA would very likely get substantially warmer, with temperatures rising over the twenty-first century more rapidly than in the previous 10 000 years. There would also be more rain, some of it in heavy downpours, even though some parts of the USA would become more arid. Drought in some parts may become more common as the effects of temperature rise on water evaporation outstrip increases in precipitation. It noted that reduced summer runoff coupled with increased summer demand for water would probably stress some parts of the US water-supply system, while increased heavy precipitation would increase the need for flood management. These changes would disturb freshwater ecosystems. Some species (both aquatic and terrestrial) would in all probability adapt to climate changes by shifting their ranges, but invasive species and land use would together undermine this, resulting in a risk to biodiversity. There were particular worries expressed for 'rare ecosystems', but not so for managed ones. Forest growth would possibly increase due to the increase in carbon dioxide and temperatures, spurring growth; however some forests would probably become more susceptible to fire and pests. With regards to agriculture, there was no concern for any decline in food security and the assessment considered that agricultural profit margins would continue to decline for non-biological reasons. Human health impacts were also noted, especially in the summer with heat stress. Finally the assessment warned of the likelihood of extreme events.

The importance of the USGCRP's National Assessment Synthesis Team (2000) report was that not only was it the US Government's first (and at the time of going to press the only) major attempt to assess the likely impacts of climate change on the nation, but that it also underpinned the US Department of State's 2002 *US Climate Action Report* (covered in Chapter 8.)

The assessment painted a very much broad-brush picture. This was due to the coarse nature of (the then) cutting-edge computer models, which by 2000 had not the sufficient agreement in terms of forecasting the future to portray a spatial picture with much certainty. However, the two models (the Canadian model from the Canadian Climate Centre and the Hadley Centre model from the UK) that the report frequently referred to did, on a state-by-state scale, give a broadly similar pattern of temperature change throughout the twenty-first century, even if in some places they disagreed by a few degrees Celsius.

The assessment noted that over the past century temperatures over much of the USA have risen, but the states of, and around, Mississippi and Alabama had cooled. With regards to the twenty-first century, both the Canadian and the Hadley models agreed that the USA would warm up but disagreed as to the exact intensity of this warming on a state-by-state basis. The Canadian model predicted warming across the USA but with a large centre of stronger warming in the central states of South Dakota, Wyoming and Nebraska. Conversely, the Hadley model predicted an almost similar, but marginally milder, back-ground of warming compared to the Canadian model, without a central island of stronger warming and with slightly milder warming of the south-eastern states around Florida, Alabama, Georgia and North and South Carolina.

With regards to changes in precipitation, the USGCRP report (National Assessment Synthesis Team, 2000) again presents outputs from both the Canadian and Hadley models. These suggest that much of the USA will see an increase in precipitation of 20–40% and that California (especially southern California) might experience more than this. However, they differed in that the Canadian model attributed far more rainfall to the states to the east of California and also up to 20% less rain in the central and eastern-seaboard states. Of course, as previously noted, increased rainfall does not necessarily mean more water for plants. If evaporation due to a temperature rise is more than the extra rainfall then soils become drier. Here both models forecast that the area around the Colorado/Oklahoma border will experience a decline in soil moisture and California an increase. Other than this, the models give contrasting forecasts. With reference to the medieval climatic anomaly (MCA; see Chapter 4) being an analogue to a degree or so of medium-term future climatic warming that we would expect, only the Hadley model suggests that part of the Sierra Nevada will experience drought. Both models do indicate that Montana, Wyoming and North and South Dakota may retain less soil moisture, as would be expected from looking at past climates when it was a little warmer than today.

These temperature and rainfall factors suggest increases in vegetation with climate change across much of the USA, with the exception of the Mississippi floodplain and Florida. However, the Hadley model suggested a marked decline in vegetation in Arizona. This is not as big a difference between the two models as it seems. If an area is already vegetation-poor, in this case due to aridity, then only a small proportional change in an already small biomass per unit area equates to a big difference in the ecosystem. So much depends on which parameter is represented more in the output.

At the moment one of the larger features in an ecosystem map (technically, a biome map) of the USA is the eastern states, with their temperate deciduous and mixed forest, and the grasslands of Oklahoma, Kansas, Nebraska and South Dakota, with a border between these two ecosystems running through eastern Texas, and up to the Iowa/Illinois boundary. This boundary line lies north by north-east. However, at the end of the twenty-first century both models suggest that this will become north–south, still beginning in mid-Texas but instead ending close to the Minnesota and North Dakota border (see Figure 6.2). These last two states, together with Iowa and Missouri, will see considerable ecosystem change.

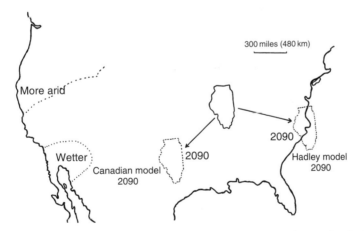

Figure 6.2 Difficulties in predicting twenty-first-century change in the USA, in this case by the Hadley and Canadian climate models (National Assessment Synthesis Team, 2000). Both models suggested that southern California would become wetter and that parts around Colorado and Kansas would become drier. However, differences between the models include whether Illinois's climate will become like North and South Carolina's or like Oklahoma's.

All this is not to say that there will not be change elsewhere. These are the broad changes across large areas that both models predict. However, they also predict much change (but differently) on smaller scales. The Rocky Mountains, already a highly fragmented natural landscape in terms of ecosystem type, are predicted to change. However, it is possible that the small lateral distances of such change might be accommodated through vertical migration, but the picture is not that clear for definitive predictions, let alone the national determination of ecosystem-management policy at the local level.

The agreement, or lack of, between the two models used in the USGCRP report is aptly illustrated by how they each predict the summer climate of Illinois will change by 2090 (see Figure 6.2). Both models predict that the climate at the end of the twentieth century approximately along the latitude 35°N would shift 5° further north. However, they disagree as to whether Illinois's climate will become like North and South Carolina's or will be more like Oklahoma's. The main difference between the two is the amount of summer rainfall: Oklahoma gets less.

The USGCRP report did not just look at possible US climate and ecosystem change but also how its population and economy would develop throughout the twenty-first century. It predicted that the population would increase by between just 20% and almost a doubling, depending on various emigration scenarios. Employment forecasts were also variable, from barely rising at all through to rising up to 2070 but then levelling off or slowing down in the century's final quarter. Either way, the US gross domestic product (GDP), it was thought, would increase, although there was considerable uncertainty as to by how much.

Putting the above population and wealth forecasts into the context of extreme climatic events such as hurricanes (which in a warmer world we might reasonably expect to see more of), the report notes that although the number of hurricanes per year has been constant in the last century, the number of deaths by hurricanes has declined (through presumably better facilities affordable in a richer society). Yet conversely, property losses from hurricanes increased (presumably because a wealthier society has more to lose).

That there is a link between climate and the destructiveness of cyclones has become increasingly apparent. In 2005 Kerry Emanuel from the Massachusetts Institute of Technology published results based on surface wind measurements taken during storms over the past 30 years. He found a high correlation between the destructive potential of storms and sea-surface temperature. He suggested that future warming might lead to an upward trend in destructive potential and predicted a substantial increase in hurricane-related losses in the twenty-first century. Meanwhile, in the same year Peter Webster, of the Georgia Institute of Technology, and colleagues, including one from the National Center for Atmospheric Research, Colorado, published a study covering a 35-year period of cyclone activity. They found no statistical increase in the global number (frequency) of tropical storms and hurricanes but, in a way affirming Emanuel's results, a large increase was seen in the number of the most powerful storms (categories 4 and 5 on the Saffir–Simpson scale). Over the 35-year study period the number of the most powerful hurricanes had increased by 80%. Interestingly, the survey showed no statistical increase in the actual intensity of the most intense hurricanes: the maximum intensity appears static but the number and proportion of these most powerful hurricanes have risen. In this precise sense the two reports disagree and at the moment there is much debate as to the exact link between climate change and the nature and/or frequency of hurricanes.

Meanwhile, the 2000 USGCRP report did look briefly at each region. Alaska, with its ecologically important arctic tundra and sub-arctic environments, is likely to see much change in its wildlife conservation and biodiversity concerns. This applies to much of Canada too. Elsewhere in the USA the question largely becomes one of having to manage change. However, in some places the change is not one of movement along a broad ecotone (a gradual environmental gradient, such as a hill) but catastrophic change on a comparatively isolated ecosystem type. For example, the encroachment of salt water due to sea-level rise on the coastal forests of the USA around the Gulf of Mexico has led to trees dying. This problem is particularly acute in parts of south Florida and Louisiana, with the creation of dead or dying areas called ghost forests. In other areas there is the problem of increased summer heat which, despite increased rainfall, is likely to lead to greater aridity that will raise the likelihood of wildfires. This again is likely to be of particular concern to those in the south-east USA. The report's regional conclusions are summarised in the following paragraphs.

The midwest, with Chicago being the nation's rail hub, is likely to see transportation problems due to extensive flooding, which might in turn result in nutrient enrichment (eutrophication) of aquatic systems. Ironically, despite increased rainfall (much of which will be in seasonal bursts), it is anticipated that there will be a reduction in river and lake levels. These in turn will result in transport impacts as docks, locks and associated navigation structures will require re-engineering. The report cites the anticipated reduction of water levels in the Great Lakes.

The west, the USGCRP report concludes, will continue to see an increase in precipitation, as it has already done during the twentieth century. In some parts river flow may double by the end of the twenty-first century. However, some parts of the west are arid and it is not clear whether increased temperature would offset any extra precipitation. The USGCRP report is ambiguous, although palaeoclimatological evidence suggests caution and the need to be prepared for drought.

The Pacific north-west is likely to see far greater temperature changes, at 4–4.5 °C by the end of the twenty-first century, than the 1.5 °C rise experienced in the twentieth century. Rainfall across the region is likely to increase, although decreases in some areas may be experienced. Much of the north-west's extra rainfall is likely to take place in winter months. There is considerable uncertainty associated with this extra rain, which could be as little as a few per cent or as much as 50%. The overall concern is about prospective deficiencies in water supply during the summer. In Columbia, where water-allocation conflicts are already acute, matters are likely to become increasingly difficult. Erosion from sea-level rise in areas already threatened (such as in low-lying communities of southern Puget Sound) will need to be addressed.

Alaska, one of the states with the lowest population density, is an area of important wildlife resources. It is also a large state, with an area nearly a fifth of the size of the entire southern 48 states. It contains 75% (by area) of US National Parks, 90% of wildlife refuges and 63% of wetlands. It also has more volcanoes and glaciers than the other states combined. Alaska has already warmed considerably: by 2 °C between 1950 and 2000. This considerable warming, far more than the global average, is likely to continue and is part of a trend seen over much of the Arctic. Permafrost thawing and sea-ice melting are likely to be at the heart of many of the climate-change problems and Arctic summer sea ice is unlikely to exist by the end of this century. Meanwhile, the precipitation/warming duality occurring elsewhere in the USA is also likely to apply in Alaska, with soils (without permafrost) becoming easier to drain. Together this adds up to considerable environmental change.

The change to the Alaskan soils, particularly those of tundra, have potentially extreme carbon-cycle implications (see Chapter 1). They are also likely to have profound ecological consequences regarding their flora; hence there are probable nutrition consequences for caribou (*Rangifer arcticus*) and its more domesticated counterpart, the reindeer (*Rangifer tarandus*). The loss of sea ice will have a dramatic effect on polar bears (*U. maritimus*), which rely on ice rafts from which they hunt (see the previous section). The loss of protective sea ice combined with sea-level rise is likely to result in increased erosion, with consequences for some Alaskan settlements.

The south-east USA is already vulnerable to weather, let alone climate change. During the last century's final two decades the south-east saw half of the nation's costliest weather-related disasters, mostly floods and hurricanes. These had cost some US$85 billion. Over the past century precipitation has increased and this is forecast to continue. Then there is sea-level rise. The USGCRP anticipates increased flooding in low-lying counties from the Carolinas to Texas and this is likely to increase the area of the aforementioned ghost forests, and have impacts on human health and property. Because coastal flooding concerns in the region were already an issue (for example, with Hurricane Betsy in 1965), some protection proposals have already been made. In 1998, prior to the 2000 USGCRP report, a 30-year restoration plan was proposed called *Coast 2050*. However, to date (2007) there has been political stagnation on its implementation due to Congressional disagreement as to who should pay the estimated bill of US$14 billion. One proposal has been to restore the wetlands around New Orleans, one of the biggest property flood risks in the south-east USA, due to both sea-level rise and increased precipitation, as a result of being at the head of one of North America's largest rivers. New Orleans lies in a basin, much of which is below sea level, and which at its lowest is 6 m below sea level. Past water management is in no small part to blame. Since 1930 much of the lower Mississippi has had extensive embankments to control flooding. These prevent water and sediment from reaching the delta, which is steadily eroding.

Events subsequent to 2000 make it worthwhile to look at this last USGCRP conclusion more closely. It was thought that New Orleans could survive surges caused by a category-3 hurricane, but not by any higher categories. Yet on 16 September 2004 hurricane Ivan (a category-5 hurricane with winds in excess of 240 km/h) hit the coast some 160 km to the east of New Orleans. Despite this narrow escape a serious flood incident appeared to be just a question of time. Then, on the 29 August 2005, the category-5 hurricane Katrina struck. (Bearing in mind the possible relationship between hurricanes and sea temperatures discussed above interesting – but not scientific – to note that Gulf of Mexico sea temperatures at the time were, for August, the highest ever recorded.) Aside from Katrina's immediate wind damage, sea surges flooded much nearby coastal property. The levy defences around New Orleans itself were breached so that most of the city was flooded: the flood area exceeded 400 km^2. The city effectively shut down and nearly all its inhabitants either left or subsequently were evacuated. President Bush declared a state of emergency for Louisiana and surrounding states. The economic losses associated with Katrina were in excess of US$125 billion. Many would have wanted the authorities to have heeded the warnings given years in advance in reports such as that from the USGCRP.

The USGCRP's National Assessment Synthesis Team (2000) report also looked at four sectors, and, although there is not the space here to review all of them, the section on agriculture deserves special attention, as the USA is a major supplier of food globally. In 2000 US products accounted for more than

25% of the global trade in wheat, maize, soya beans and cotton. Furthermore, cropland occupied some 17% of the total US land area and in addition pastures and grazing lands accounted for a further 26% of land area. Economically, agricultural commodities annually exceed US$165 billion (1999) at the farm gate and over US$500 billion (10% of US GDP) after processing and marketing.

Historically, US agricultural productivity has improved by roughly 1% per annum since 1950. This success could cause an economic problem for the sector, as profit margins are tight with little scope for further efficiency gains. The USGCRP report focused primarily on plant crops and not animal husbandry, even though (as the report itself acknowledged) climate change will also effect this dimension of food production.

The USGCRP report notes that region-by-region climate change will have both positive and negative effects on different types of agricultural activity. For instance, yields in the northern midwest are currently limited by temperature, so agricultural production, it is thought, would increase with warming. Indeed, the anticipated longer TGS would allow double cropping. The carbon dioxide fertilisation effect would also be an additional positive influence on yields. However, these would be balanced by yield decline in the southern part of the region by as much as 10–20% due to a lack of sufficient water that would be needed in the warmer climate.

Much of the nation's meat, wheat and fibre is produced on the Great Plains, including 60% of the country's wheat, 87% of sorghum and 36% of its cotton. In the twentieth century temperatures across the northern and central Great Plains increased by over 1 °C and by up to 2 °C in parts of Montana and North and South Dakota. The eastern portion of the Great Plains has seen precipitation increased by over 10% and Nebraska and Texas both had higher levels of rainfall at the century's end compared to its beginning. Climate models predict that in the twenty-first century temperatures would rise throughout the region, with the largest rise in the western parts of the Great Plains. Precipitation generally is likely to continue to increase, except in the lee of the Rocky Mountains. Furthermore, the effects of temperature increases on potential evapotranspiration (the theoretical maximum level of evaporation and plant transpiration with unlimited water and optimal nutrients) are likely to outstrip the extra supply of water received through increased precipitation. The result is likely to be a net decline in soil moisture from levels in 2000, especially in Louisiana and Mississippi. Maintaining water supplies through supply management throughout the year and ensuring water quality are likely to become key issues for the region, especially in the south.

The USGCRP report (relying, as it does, on the sometimes contradictory Hadley and Canadian models of the late 1990s) does not raise concerns as to lack of water in some parts of the west. Instead, it concludes that more atmospheric carbon dioxide, higher temperatures and increased precipitation are likely to increase crop productivity and extend the TGS. However, it does warn that the patterns of potential crop productivity are likely to shift. It notes that fruit and nuts make up 32% of the west's crop value (with a third of that from grapes alone) and that changing the principal locations of such crop production may well be problematic, bearing in mind that such crop plants can take decades to become established. Here, if anywhere, there is an urgent need for accurate local forecasting of likely climate change.

The USGCRP's report in 2000 concludes that without adaptation many crops will grow better by the end of the twenty-first century, especially cotton and grapefruit (but not oranges), and pastures will improve. Barley and oats are unlikely to improve or fare badly under future conditions: which perspective depends on which model (Hadley or Canadian) is determining the forecast. Potato and orange production experience minor declines under forecasts from both models. Both models agree that if adaptation is made then most crops could see a 10–100% increase in productivity. Such adaptations importantly will necessitate changing which crop grows where and changes in water management. This last will often involve inter-state cooperation and considerable cost, whereas a minority of farmers will have to accept abandoning their land.

Adaptations needed are likely to include the following.

- Changing sowing dates and other seasonal activities with phenological changes.
- Introducing new cultivars (crop varieties), including those created through biotechnological methods as opposed to conventional breeding.
- Improved water-supply management and water use.
- Changes in tillage practice to retain organic matter in soils.
- Increase use of short-term (6-month) forecasts relating to key climatic cycles, such as El Niño events.
- Tweaking of management systems from plough to gate, and even plough to market.

The USGCRP's report also has a sector report on the related topic of forestry. Many of the concerns focus on the Pacific north-west, which is likely to suffer from changes in water availability throughout the year. The Pacific north-west west of the Cascades coniferous forest covers about 80% of the land and, more importantly, this accounts for about half the planet's temperate rainforest. Because of the length of forest crop turn-around (decades rather than within a year for most agriculture) the need is to plant species adapted to the forecasted climate rather than the climate as it is now. Careful management is also required to reduce water stress and the risk of forest fires.

The net economic effect of the forecast climate change is thought to be marginal but minor economic improvement is anticipated, with most of the gains relating to hardwood as opposed to softwood. However, this does necessitate the adaptations being undertaken in a timely manner and so climate-driven agricultural change is unavoidable.

The other important sector the USGCRP's report covered was water. As seen from the above summary of the report's key points, water in the USA was considered a key dimension to the biological impacts of climate change on both natural and managed systems. The report notes that population in the USA increased by nearly two-thirds (well over 60%) in the latter half of the twentieth century. However, whereas water consumption did increase between 1950 and 1979 by around 60%, it stabilised thereafter to the century's end, at around 1.5 billion m^3/day. Farming today, at 85%, dominates water use with 82% being used for crops and 3% for livestock (see Figure 6.3). This highlights the importance of water management in food production. However, water management in those areas relying on seasonal mountain snow melt will find that warming makes things harder for management throughout the year. With the amount of snow due to decline as snow lines rise, and there being warmer

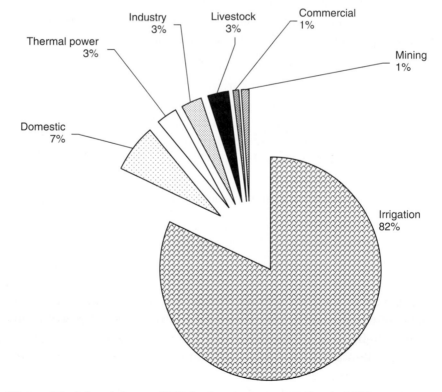

Figure 6.3 A breakdown of US freshwater consumption in 1995.

years, snow melts will occur earlier in the year and summer melt-fed streams will be drier. Models suggest that Southern Rock, Sierra Nevada and central Rocky Mountain snowpack volumes are likely to decline by between over 50 and 90% by the end of the twenty-first century. Given this, and that a number of groundwater aquifers (such as the Great Plains Ogallala) are already diminishing due to extraction rates exceeding replenishment rates, water storage combined with increased efficiency of use will be the important management measures required by the main users of water in the USA, the farmers.

The USGCRP report concluded that climate-change impacts 'will be significant for Americans', although 'for the nation as a whole the direct economic impacts are likely to be modest'. There were many uncertainties regarding the precise nature and locality of specific impacts, even though themes such as the need for water management were clear.

We will turn to the human ecology needed to underpin climate policy matters briefly in Chapter 7 and policy itself in more depth in Chapter 8. For now it is worth noting that the USGCRP report underpinned the 2002 *US Climate Action Report* from the US Department of State, which outlines what the USA must do to address greenhouse (energy-policy and climate-change) concerns. But this report was light on biological specifics and stated that 'one of weakest links in our knowledge is the connection between global and regional projections of climate change'. And with regard to potential impacts that we have a 'lack of understanding of the sensitivity of many environmental systems and resources – both managed and unmanaged – to climate change'.

Whereas the USGCRP report effectively represented a stocktake in 2000 of US climate-change science on the likely future climate of the USA, research naturally still continued. Although there is no space, nor is it appropriate, to update this report here, of note among more recent contributions was an assessment as to whether the current multi-year drought being experienced in the western USA has any precedent. The research team was led by Edward Cook of the National Oceanic and Atmospheric Administration (NOAA) and their work was published in *Science* in 2004. The drought experienced by the western USA at the end of the twentieth century and now is both severe and unprecedented over the period of detailed hydroclimatic measurements, which mainly covers the twentieth century. If the early twenty-first century has its palaeoclimatic analogue back in the medieval warm period (MWP) or medieval climatic anomaly (MCA), to see whether the current drought is a likely response to global warming what is needed is to ascertain conditions during and prior to the MWP, so as to capture all of the MWP episode. Edward Cook and colleagues looked at palaeoclimatic dendrochronological indicators in a geographical grid across the USA and southern Canada. They found that the current drought does appear to be analogous to one that took place in the western half of the USA and western half of southern Canada during the MWP. However, the recent drought up to July 2004 (when the paper was submitted) was not as severe as the one during the MWP between AD 900 and 1300. They concluded the western US drought is likely to get worse with continued global warming.

6.3 Case study: climate and natural systems in the UK

It was the UK Prime Minister, Margaret Thatcher (herself inspired by the UK's former UN ambassador Crispin Tickell), who did much to promulgate climate-change concerns among international politicians. Whereas researchers in the USA had done important climate work (for example at the National Center for Atmospheric Research in Boulder, Colorado), it is probably fair to say that US research into likely impacts was not pursued as vigorously as in the UK. Already by the late 1980s (before the first IPCC assessment) UK state-funded science had climate change as a research priority and was communicating research outputs to a broader audience, such as with the Natural Environment Research Council's (NERC's) former Institute of Terrestrial Ecology (ITE) booklets (the ITE subsequently became the NERC's Centre for Ecology and Hydrology; CEH). The first of these booklets was *Climate Change, Rising Sea Level and the British Coast* (Boorman *et al.*, 1989). This considered the likely impact of a sea-level rise of up to 1.65 m over the next 100 years. It noted that estimates for maintaining sea defences to withstand (as opposed to retreat from) this sea-level rise would cost some £5 billion in the money of the day. It cited stretches of the coast and land deemed at risk from sea-level rise (see Figure 6.4). It also considered what would be the general ecological impacts of various response options, from raising present sea walls and building storm-surge barriers to new sea walls constructed landward, as well as abandonment and retreat. Each has an effect on biomes; specifically, changing their size. Of greatest wildlife concern are the likely effects on salt marshes and mudflats. Here the effects depend on which coast-management option is chosen. However, in addition to whether the UK adopts protection or retreat, there are likely to be transition problems which could greatly restrict a biome before expansion as coastal biotic and abiotic systems adjust to new sea levels. In addition to biome-related flora and invertebrates, bird populations were considered to be at greatest risk. Unfortunately the likely impacts are too numerous (due to many possible management options) for them to be summarised here.

Figure 6.4 Coast and land in the UK deemed to be at risk from sea-level rise.

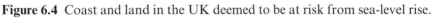

Turning to recent research outputs, early in twenty-first century there were two major reports on climate-change impacts on British natural systems. The first was the 2000 Government departmental report *Climate Change and UK Nature Conservation: a Review of the Impact of Climate Change on UK Species and*

Habitat Conservation, published by the then Department of the Environment, Transport and the Regions (DETR) in 2000 as part of the UK Climate Impacts Programme. The second was *Climate Change and Nature Conservation in Britain and Ireland* for the UK Climate Impacts Programme (UKCIP; 2001) compiled by leading Government departmental agencies and leading UK wildlife charities led by the agency English Nature, again as part of the UKCIP, and also known as MONARCH I (Modelling Natural Resource Responses to Climate Change).

The DETR report based its assessment on the UK's Biodiversity Action Plan (BAP), which the UK Government developed as part of the nation's commitment to the 1992 UN Conference on Environment and Development (in Rio de Janeiro) and its Convention on Biological Diversity (CBD), but not its Framework Convention on Climate Change (which in the USA prompted that country's *Climate Action Report* (Department of State, 2002) – see the previous section).

The UK departmental report (DETR, 2000) addresses many of the points regarding species impact covered in this and the preceding chapter. However, it also notes that the UK, being comprised of islands, has restricted population pools and that the sea inhibits species migration (unlike the USA, which has Canada to the north and Mexico to the south). The report also comments on human-imported species, currently residing in parks and gardens, noting that some of these may find future conditions favourable for expansion.

Landscape fragmentation is also an issue highlighted by the report. It noted that some species, which otherwise might benefit from a warmer climate, such as the dormouse (*Muscardinus* spp.), may find that landscape fragmentation hinders its expansion.

The report's consideration as to the vulnerability of UK biomes to climate change is summarised below.

Low vulnerability

Arable and horticulture: the change of species will need to be managed and a management regime for existing species in existing locales established.

Low-to-medium vulnerability

Mixed broadleaved and yew woodland: damage will occur due to extreme weather events. Climatic effects may become manifest in 100–150 years. There is a particular risk to lowland woodland in southern England due to dry summers.

Fen, marsh and swamp: there will be a change in species composition and a risk of soligenous (communities relying on water that has drained through geological strata) fens drying out in summer. Drought may exacerbate air-pollution impacts. Coastal fens will be affected by sea-level rise and a sea-defence response to sea-level rise will be necessary.

Standing water and canals: isolated high-altitude lakes are vulnerable to temperature increase. High-summer temperatures could restrict cold-water species. There could be changes in the frequency of algal blooms in eutrophic (nutrient-rich) waters. Warmer winters could increase populations of over-wintering plankton.

Medium vulnerability

Improved grassland: there may be changes in the thermal growing season (TGS). Risks include an increase in moisture-loving species (such as rushes; *Juncus* spp.) and surface damage to soil and vegetation due to hoof action on wetter ground (called poaching). Coastal grazing marshes will be affected by sea-level rise and changes in sea defences.

Acid grassland: there may be possible changes in competitive ability with other grassland types and the potential for increased bracken invasion in the north (with a possible reduction in the south). Soil-nutrient limitations may restrict adaptations to changes in the TGS.

Dwarf shrub heath: a decrease in summer rainfall may change community types in the south. There may also be an increase in insect damage.

Medium-to high vulnerability

Coniferous woodland: there may be extreme weather events, encroachment by southerly species and limitation in altitudinal migration due to land-use change and specifically grazing.

Calcareous grassland: well-drained chalk soils increase the likelihood of drought stress in dry summers. Species community change and/or structure alteration are likely.

Bogs: there may be possible peat erosion and vulnerabilities due to changes in the water table. Lowland bogs will be at risk from decreased summer rainfall, whereas upland bogs will face erosion pressures from increased winter rainfall.

High vulnerability

Rivers and streams: low summer and high winter flows, together with high summer temperatures, are likely to have major impacts on species and the river environments.

Montane habitats: species migration is likely but some altitudinal limitations may result in population declines/extinction. Less snow may affect snow bunt-ings (*Plectrophenax nivalis*), which rely on insects in snow patches, and some bryophyte species (a group of small, rootless, non-vascular plants), which rely on snow cover for insulation and/or spring moisture. An increase in average climate of just 0.5 °C would eliminate many montane biomes in the UK.

Unknown vulnerability

The effects of climate change on the following biomes are unknown: inland rock, built-up areas and gardens, maritime cliffs and coastal sand dunes.

Finally, the DETR report highlighted the need for climate impacts to be considered proactively within UK conservation policy.

The second UK assessment was the UKCIP (2001) report on *Climate Change and Nature Conservation in Britain and Ireland*, or MONARCH I. It was a suitable follow-on from the 2000 departmental report. It linked established environmental impact models to a national bioclimatic classification. As such, the report took on board climatic changes but not impacts arising from elevated levels of carbon dioxide. (You will recall that increased carbon dioxide levels have an effect separate to that of climate change, even if there may be interactions and synergies between the two.) MONARCH I was a detailed report and its overall conclusions were that a more proactive and flexible management of ecological sites is required. It recognised that many existing protected sites were not in the right place to fulfil current protection goals in the future but that such sites will continue to be important for maintaining certain species populations.

Virtually echoing the US GCRP report discussed in the previous section, the MONARCH I report's authors also pointed to the difficulty in being able to develop response strategies when the regional and local nature of future climate change was so uncertain. They noted that, up to the report's publication, the only detailed monitoring of biological climate indicators have been those commissioned by the UK Government's environment ministry, the then DETR. There the focus had been change across a wide range of sectors and not just natural ecosystems. However, MONARCH I also made research recommendations, including the following:

- refining models predicting change and the use of field studies for model validation;
- applying the MONARCH methodology to different scales;
- improving understanding of ecosystem dynamics;
- identifying possible natural adaptations to climate change.

The MONARCH report noted that for wildlife conservationists to address climate change their response must lie somewhere between two extremes. At one end there is a 'King Canute' (*sic*)-type approach, resisting change and trying to maintain existing natural communities. (An historical analogy that does not bear too close a scrutiny as Canute 'attempted' to hold back the tide to show that even he, a King, was not impervious to natural order and not, contrary to popular belief, in a bid to demonstrate his regal power.) At the other end of the spectrum is 'wildlife gardening', with species translocation and habitat-type creation in suitable areas, or areas due to become suitable through change climate.

The MONARCH authors also drew attention to the plight of the natterjack toad (*Bufo calamita*) as an example of special conservation interest, as its habitat area will become reduced before climate change creates new areas for it. The authors therefore recommended consideration to purchasing likely areas to facilitate such species translocation, and even provide 'holding bays'.

A second report, MONARCH II, was published in 2005. It took the MONARCH I climate space approach but downscaled it for use at the local level. A dispersal and ecosystem function model was also developed to run alongside the climate space models. MONARCH II focused on methodological development and testing in four case-study areas. A third study, MONARCH III began in April 2004 and was due to report in 2006. It is refining and applying the output from MONARCH II in the context of the UK BAP. (BAPs themselves were a result of the aforementioned 1992 international Biodiversity Convention (see Chapter 8) and so relate MONARCH III to Britain's UN obligations.)

As in the USA, the British series of climate reports tended to leapfrog each other. This was particularly clear with the MONARCH reports as effectively this was an evolution of a specific policy concerns; those of wildlife conservation and climate change. Understanding in both areas evolved in parallel so that any reference from one to the other tended to become a little dated in any given report. (This is an almost inevitable problem common to a number of areas of cutting-edge science.) Just as the USGCRP report recognised the limitations of computer models, so MONARCH used the latest models in each of its successive reports. The need to do this is aptly demonstrated when looking at the anticipated change in precipitation: as with the USA, water issues are of concern when considering likely climate change. The climate model used in MONARCH I suggested that by the middle of the century the south of England (particularly the central south) would become drier in the summer compared to the 1961–90 average, by up to 110 mm, and the northwest of Scotland slightly wetter, by 60 mm. Conversely, in winter Scotland and west England down to and including Wales would get wetter by up to 60 mm whereas the south of England and East Anglia would only get wetter by 40 mm. This was based on the UKCIP model UKCIP98. Compare this to the UKCIP02 model forecasts used in 2002, which predicted the opposite for winter 2050, that the south east of the UK would have more rain and the north-west of Scotland would remain within the natural variability of 1961–90. Because of this successive MONARCH reports, and other UK policy reports, did not build on the results of previous models but built on policy methodologies and understanding, while starting afresh each time using the latest computer climate models.

In 2002 the UKCIP updated its first (1998) climate-change scenarios for the UK (Hulme *et al.*, 2002). That in Britain (more so than in the USA) climate-change issues were of considerable political concern, both by the Government and the opposition parties, is indicated by the Secretary of State Margaret Beckett personally endorsing the report with an introduction. UKCIP02 took four carbon-emission scenarios (low, medium-low, medium-high and high), which were based on the IPCC SRES scenarios (2000). These in turn were related to change from a 1961–90 baseline to 2011–40 (labelled the 2020s), 2041–70 (the 2050s) and 2071–2100 (the 2080s).

It concluded that by the 2080s

- temperatures across the UK may rise by between 2 (low-emission scenario) and 3.5 °C (high-emission scenario), with the greatest warming being in the south and east and greater warming in the summer than winter (parts of the south east may be up to 5 °C warmer in the summer);
- high summer temperatures will become increasingly frequent, and cold winters rare;
- winters will become wetter and summers drier. Overall the UK will become drier as the winters are expected to become only 30% wetter while the summers may have only 50% of the precipitation;
- days of heavy rain are likely to increase in frequency;
- the sea level will rise between 26 and 86 cm in the south east of England but, due to geological (isostatic) rise, the sea level may fall by 2 cm (low-emissions scenario) or rise by 58 cm (high-emissions scenario) in western Scotland;
- the Gulf Stream is unlikely to weaken its warming of the UK, but there is uncertainty due to lack of detailed knowledge of ocean-circulation–climate interactions.

UKCIP02 and MONARCH are being used to underpin national policy. In the UK, as in the USA, Government-funded research is supported both by scientific agencies (such as NERC) and Governmental departments such as the DETR and now its successor the Department for Environment, Food and Rural Affairs (DEFRA). The DETR also had agencies to undertake environmental monitoring and protection work and, of these, English Nature (which was soon to be integrated in 2006 with the Countryside Agency) was responsible for wildlife and the Environment Agency was responsible for waterways and coastal protection. UKCIP scenarios have alerted the Environment Agency to the problem of winter flooding due to excessive rain bursts becoming a chronic condition (see also Chapter 7), although flooding instances in the late 1990s and at the beginning of the twenty-first century effectively ensured that the agency would develop a long-term strategy. Some of the areas at specific risk are shown on Figure 6.5. However, there are other impacts on society. As already seen in Chapter 5, there have been past climate-change impacts on human activities and Chapter 7, on human ecology, will look at likely future impacts.

One of the indicative ways in which climate change will affect UK natural systems was discussed in 2005 at a symposium at Sussex University jointly run by the Royal Horticultural Society, Forest Research and the UKCIP entitled 'Trees in a Changing Climate'. There are some 4 billion trees in the UK. In summary, it was noted in the north and west of the UK, where mainly there is little water limitation, trees would thrive. Conversely, as the south and east of the UK is likely to become drier, it is likely that some species will suffer. Sycamore (*Acer pseudoplatinus* spp.) is one species at risk and on well-drained soils birch (*Betula* spp.) in particular will suffer. Finally, beech trees (*Fagus sylvatica*) generally do not fare well in hot summers.

In a hypothetical warming Britain without humans, the mix of species in woodlands would change, with tree species moving to the north and west. Furthermore, woodland understorey species would also change and so with them the mix of invertebrate systems. However, as previously noted, our real modern world sees many parts of developed and even less-developed nations have an increasingly fragmented landscape that does not facilitate species migration. Britain is no exception. Biologists have for many years argued that biological conservation should not solely be carried out on a site-by-site basis, laudable as such initiatives are. What is needed is active management at the landscape level. Present-day concerns about climate change underscore such needs. Even so, one confounding factor may well be the speed and magnitude with which anthropogenic climate change is taking place. This, as has been noted, is greater than anything seen for the past 10 000 years of the Holocene and is likely, over the coming two centuries, to cause a return to a warmer world that has not existed for tens of millions of years.

Shifts in species ranges are not just confined to well-known birds, animals and butterflies, although naturally these form the basis of many studies. Many other invertebrates are also affected. In 2006 a telling study of many UK species was published (Hickling *et al.*, 2006), demonstrating that current climate change is impacting on many classes and families of species. It looked at dragonflies and damselflies (Odonata), grasshoppers and related species (Orthoptera), lacewings (Neuroptera), butterflies (Rhopalocera), spiders (Arachnida), herptiles (Amphibia and Squamata), freshwater fish (Teleostei), woodlice (Isopoda), ground beetles (Carabidae), harvestmen (Opiliones),

Figure 6.5 Principal areas of England and Wales at risk from river flood as a result of an extreme weather event with a 1% chance of the event happening in any given year and from sea with a 0.5% chance in any year. This data is for 2005 and does not include the effects of future climate change. © The Environment Agency, reproduced with permission.

millipedes (Diplopoda), longhorn beetles (Cerambycidae), soldier beetles and related species (Cantharoidea and Buprestoidea), aquatic bugs (Heteroptera) as well as birds (Aves) and mammals (Mammalia). For each group, species were only included in the analyses if they were southern/low-elevation species, as these species would be expected to increase their range sizes, move northwards and/or shift to higher elevations if they were responding solely to temperature. Northern species were excluded from the analyses because of a lack of data, with the exception of birds. The results revealed that out of a total of 329 species analysed across 16 taxa, 275 species shifted northwards, 52 species shifted southwards and two species' range margins did not move, with an average northwards shift across all species of 31–60 km. Similarly, with regard to altitudinal shifts, 227 species moved to higher altitudes and 102 species shifted to a lower altitude, resulting in a mean increase of 25 m overall.

In addition to biological change, change in the soil carbon in England and Wales has also been detected. Indeed, in 2005 the results of a detailed Department for Environment, Food and Rural Affairs (DEFRA; formerly DETR and the Ministry for Agriculture, Fisheries and Food (MAFF))-funded survey was published. The survey was conducted by Pat Bellamy and colleagues and covered the 25-year period from 1978 to 2003. Its importance was that up to then evidence from temperate climates had virtually entirely come from small-scale field experiments, laboratory studies and computer modelling. However, the Bellamy team looked at samples of 15 cm depth from 5662 sites across England and Wales that had originally been taken as part of the National Soil Inventory. They then re-visited sites 12 or 25 years on. As such, up to then this was the only soil inventory on such a scale to have been re-sampled anywhere on Earth. In addition to the soil analyses they also looked at each site's land use. They found that carbon loss from soils across England and Wales over the survey period took place at a mean rate of 0.6% a year. They also found that the relative rate of carbon loss increased with a soil's carbon content and was more than 2% a year for soils with greater than 100 g of carbon/kg of soil. They also found that the relationship between the rate of carbon loss and carbon content was irrespective of land use. Their findings suggested that the losses of soil carbon in England and Wales are irrespective of land use, which suggests a link to climate change. However, they could not say where the carbon had gone but likely destinations are a combination of transfer to the atmosphere as carbon dioxide and leeching deeper into the ground. Nonetheless, considering the area of temperate terrestrial ecosystems internationally, the implications for changes in carbon cycling are significant.

Putting this top-soil carbon loss into the context of Britain's anthropogenic emissions of carbon dioxide, the carbon loss from England and Wales is around 13 million t/year. This is equivalent to about 8% of UK carbon dioxide emissions (which includes those from Northern Ireland and Scotland, whose soils were not included in the survey). It is also more than equivalent to the past reductions in UK carbon dioxide emissions between 1990 and 2002 (12.7 million t of C/year).

This chapter's British and US case studies illustrate the types of terrestrial biological change that are taking place and the likely threats due to current climate change. Other than temperature, central to the changes in climate factors affecting biology is water availability. This also applies to many other terrestrial areas subject to climate change outside the USA and UK. For instance, one example is that the arid desert biome of northern Africa may extend and jump the Mediterranean into Spain. In 2005 the Spanish Government reported, when announcing a £50 million programme to combat desertification, that more than 90% of the land between Almeria in the south to Tarragona in the north east is considered at risk. This includes the provinces of Murcia, Valencia and southern Catalonia. Over-grazing and development pressures have exacerbated the problems.

Spain, though, is not alone in facing both climate and development pressures given both a warming world and growing global population. Access to water will become a far bigger issue as the century progresses. Indeed, not only will African hot dry biomes probably spread across the Mediterranean, deserts are likely to spread within parts of Africa. This is even though the balance between deserts spreading and areas of high rainfall spreading, or rain getting more intense over a similar land area, remains uncertain. Nonetheless, there is some evidence to suggest that areas that were previously drier and more sandy during (geologically) previous climatic regimens may once more become desert. In 2005 researchers from Oxford and Salford in Britain looked at southern Africa's dune systems, much of which today are covered with grass and indeed trees (Thomas *et al.*, 2005). This is an important biome because desert dunes, whether dry and active, such as in much of the Sahara Desert, or inactive, covered by vegetation north of the Kalahari Desert, account for 5% of the global land surface and 30% in Africa. Knowing how moisture levels determined by precipitation and plant evapotranspiration, together with wind speeds, combine to affect dune mobility, they were able to apply climate models to see the effect on sand dunes. If, due to global warming and regional climate change, temperatures rise without a counteracting increase in rainfall, and factoring in wind speeds, then vegetation that currently stabilises dunes could die, and the dunes could mobilise. This would also kill off other vegetation previously sheltered by the dune, and so the sandy desert would return. Such an analysis suggests that the current situation is precarious. It would appear that regardless of which IPCC future scenario is considered, the southern dunes bordering the dry Kalahari Desert that are already mobile will become prone to greater movement by 2040. Meanwhile, the eastern and northern dune fields that are currently covered by trees and/or grass could become mobile by 2070. By 2100 all dune fields from northern South Africa to Angola and Zambia may become mobile. The authors conclude, 'There are uncertainties within the modelled Kalahari scenarios but the general trend and the magnitude of possible change in the erodibility and erosivity of dune systems suggests that the environmental and social conse-quences of these changes will be drastic' (Thomas *et al.*, 2005).

6.4 Biological response to greenhouse trends beyond the twenty-first century

Biome shifts and ecosystem change with their consequential impacts on human activities have been a theme of this chapter. As we shall see when looking at the human ecology of anthropogenic global warming, it is most likely (some other global catastrophe aside) that fossil-fuel burning will continue throughout this century, so raising atmospheric carbon dioxide levels from around 400 ppm (to be soon reached early in the twenty-first century) to 700–1000 ppm during the early twenty-second century.

There is therefore little doubt that biome shifts and ecosystem change of the sort described so far will continue to be seen both while atmospheric carbon dioxide continues to increase and also after carbon dioxide levels have stopped increasing and while ecosystem communities stabilise, adapting to the new conditions. However, there will be other changes. As noted in Chapter 2 with, for example, leaf shape and climate, species themselves will change. Some of these transformations will manifest themselves genetically and others be phy-siological, with regards to the way carbon is metabolised (see Chapter 1).

A recent instance of genetic change under circumstances analogous to twenty-second-century carbon dioxide conditions (namely carbon dioxide at 1050 ppm) was found in 2004 when Canadian biologists, Sinéad Collins and Graham Bell, grew a protoctist species of Chlorophyta (green algae) called *Chlamydomonas reinhardtii* over about 1000 generations. *C. reinhardtii* was used because it has been studied extensively to elucidate photosynthetic mechanisms. Like most other eukaryotic algae, it is known to have a carbon-concentrating mechanism (CCM) that increases the concentration of carbon dioxide near Rubisco (see Chapter 1). After 1000 generations at elevated carbon dioxide levels there was less need for a CCM. So *Chlamydomonas'* CCMs became less efficient and its cells became smaller but with higher rates of both photosynthesis and respiration. The researchers tentatively concluded that as carbon dioxide levels increase so many Chlorophyta will have less-effective CCMs and changes in photosynthetic and respiratory rates. This, they note, would affect global processes by changing the rate of carbon turn-over in aquatic and perhaps terrestrial ecosystems. It needs to be remembered that very roughly a quarter of the Earth's photosynthetic primary production is undertaken by algae, so even if other photosynthetic species do not see such a radical response to elevated carbon dioxide (though some change is likely), there will be a major change in carbon transferences around sources, sinks and reservoirs and not just changes part-mediated by changes in physiology. These unknowns further confound those developing computer models that fore-cast climate change from the effects of greenhouse gases. They will, though, be important should we enter a warming period analogous to the Initial Eocene thermal maximum.

6.5 Possible surprise responses to greenhouse trends in the twenty-first century and beyond

The IPCC recommendations providing the scientific consensus on climate change were summarised in the last chapter (see section 5.3). Despite the broad picture (greenhouse gases are increasing, the Earth is warming and sea levels are rising) the IPCC acknowledge that there are uncertainties and that in addition to heeding their main conclusions we also need to be aware of surprises. As we have seen from the above two case studies, at the present time the IPCC broad picture globally at best translates to very basic scena-rios, as do other perspectives that are able to draw on additional regional research. These regional scenarios will undoubtedly improve with further research both in the field and with development of more sophisticated models. But what of the IPCC warning as to surprises? This section looks at some key areas.

6.5.1 Extreme weather events

One swallow a spring does not make, and so a single extreme weather event (be it an abnormally heavy downpour or a particularly dry month) does not necessarily signify climate change. But how many swallows do? An increase in the frequency of weather abnormalities will, at some point, signify that the climate has altered. Recent years have seen an increasing frequency of strong hurricanes (it is thought that the total number of all hurricanes has broadly remained the same) and record-breaking global warmth. This last is only to be expected with (indeed this is inherent in) global warming. There have also been local and regional times of unusually intense rainfall.

Such events are surprising, especially for those who experience them. When the 'hurricane' hit southern England one night in 1987 it caught the population by surprise and disrupted lives of virtually everyone for at least a day with power cuts, telephones not working and roads closed due to fallen trees. Not least surprised was one meteorologist working for BBC Television as a weather presenter, Michael Fish, whose evening broadcast prior to the storm was erroneously reassuring. Now, whereas this particular single swallow cannot itself be attributable to climate change, it was the sort of event that climatologists were contemplating. Indeed, the incident was referred to by one UK governmental department as climate-change-related, even though that claim could not be substantiated. Nonetheless, the Department of Energy used the 1987 storms in an energy-conservation promotion, with the slogan 'Global warming, we have been warned'. The department was subsequently reprimanded in March 1992 by the Advertising Standards Authority (Cowie, 1998). Today there is arguably less chance that such a reprimand would be issued.

Having discussed increases in the number of strong hurricanes earlier in the chapter, let us now focus on temperature. A warmer world will, by definition, have more thermal energy within the atmosphere- and ocean-circulation system. Some of this energy will become manifest in the form of greater evaporation (hence rainfall) and some in wind speeds or storm size. Therefore, as the world warms so the frequency of extreme weather events is likely to increase. Although worldwide ocean evaporation and hence precipitation are two of the key indicators of global warming, arguably the factor of greatest psychological impact is the intensity and frequency of heatwaves. These too are likely to become more intense and frequent as the world warms. For example, in 2004 Gerald Meehl and Claudia Tebaldi of the National Center for Atmospheric Research in Boulder, Colorado, published the results of a coupled model together with field observations to see whether the extreme heatwaves of Chicago and the south-western area of the Great Lakes (1995) and London–Paris region (2003) were one-offs or alternatively more likely in a greenhouse future. Their conclusions were that the heatwaves over those parts of the USA and Europe coincided with a specific atmospheric circulation pattern that is intensified by ongoing increases in greenhouse gases; this suggests more intense and severe heatwaves in those regions in the future.

Do not be surprised by extreme weather events in our greenhouse future. But equally do not be quick to attribute them to climate change as there is a difference between an anomaly within background noise and a statistical trend. To overcome this, statistical analysis is required of what might be expected (1) with natural variability assuming no carbon dioxide forcing and (2) with climate forcing from additional anthropogenic greenhouse gases, both compared with reality.

In 2004 a small UK team led by Peter Stott performed such an analysis to examine whether the 2003 western European heatwave was due to either (1) natural variability or (2) global warming, as per the previous paragraph. These two scenarios were derived using one of the more sophisticated climate models available at the time, namely the HadCM3 climatic model developed by the UK's Meteorological Office at the Hadley Centre for Climate Prediction and Research. The model took into account the impact of actual volcanic eruptions and also solar variability specifically for the south and western part of the European peninsula. It was run three separate times simulating June–August temperatures as if there were human additions of greenhouse gases and over the period from 1900 to 2003 and then on to 2100: these are scenarios of type 2 as per the previous paragraph. It produced three temperature lines with a typical annual variation of about ± 1 K and this followed both actual meteorological measurements and also a smoothed, few-year-filtered, line to 2000. These, the actual meteorological line and, consequently, the smoothed line, all lay within the model's annual variability band through to the 1990s. Consequently, the smoothed line also rose, albeit erratically due to actual annual, volcanic and solar variation, by about a degree along with the simulated lines, so suggesting that the model effectively captured anthropogenic greenhouse warming.

The model was also run again but under a scenario assuming no additional greenhouse forcing over this time – that is, as per scenario 1 in the preceding paragraphs – and so only included anthropogenic greenhouse forcing up to 1900 (and not beyond that year). This non-global warming line largely followed (intertwined with) both the natural and smoothed natural lines for the twentieth century's first three-quarters. However, for the century's final quarter this simulation continued horizontally, albeit erratically as if with natural variation, so that by the century's end there was nearly a 1 K difference between it and both the actual instrumental record as well as the anthropogenic greenhouse simulation. The inference clearly was that this 1 K difference at the end of the twentieth century was due to anthropogenic emissions of greenhouse gas.

However, looking at the actual meteorological line, not only did it broadly follow the model's global-warming scenarios, the 2003 summer stood out as being about 0.75 K warmer than any other previous June–August period on record and is thought to have given parts of western Europe the warmest summer for over 500 years. Indeed, the next-warmest summers all took place since the mid 1980s and this too was reflected by the three global-warming model scenarios. It suggests that the model successfully captures the annual *likelihood* of an exceptional summer heatwave in western and southern Europe even if it cannot predict the *actual* year in which such a heatwave might take place.

Having established that the HadCM3 climatic model captures both global warming and the annual likelihood of extreme events, what did the model predict for the twenty-first century? The indication is that over 2000–2100 the summers in this part of Europe will on average gradually warm by 6.5 K. In addition, extremely hot summers will continue, at least once or twice a decade, above this average warming. However, because the average summer temperature is set to increase during the twenty-first century, an exceptionally warm summer up to 2020 will become a normal summer by the 2040s. From the 2040s exceptionally warm summers will all be far warmer than summer 2003, so that by the end of the century exceptional heatwaves can be expected to be around 6.5 K warmer that those at the century's beginning. The researchers' conclusion was that it seems likely that past human influence (greenhouse emissions) has more than doubled the risk of European mean summer temperatures as hot as those in 2003 for the early twenty-first century, and further that the likelihood of such warmth is projected to increase 100-fold over the next four decades. They said, 'It is difficult to avoid the conclusion that potentially dangerous anthropogenic interference in the climatic system is underway' (Stott *et al.*, 2004).

The agricultural and health impacts of such extreme weather events will be considered in the next chapter.

6.5.2 Greenhouse gases

The IPCC scenarios are based on a number of global economic possibilities ranging from business as usual to prudent policies curbing emissions and introducing new low-fossil and/or alternative technologies. These, from the later 2001 reports, result in a range of predictions (see Figure 5.6). Fortunately, for simplicity the clear majority of these fall within the IPCC's original 1990 Business-as-Usual scenario's high and low estimate boundaries (Figure 5.4 and Table 5.1). Yet even so there may be surprises. For example, these models assume that the biosphere's response to the additional atmospheric carbon dioxide will tomorrow be proportionally similar to today. Yet, what if some threshold were reached? It may be that circumstances are such that the oceans cease to absorb as much of this additional carbon as they have done in the past.

(Perhaps, as will be discussed shortly, due to increasing ocean acidity being unable to buffer so much carbon dioxide.) In such a case the build-up of atmospheric carbon dioxide would be greater than expected and, as a result, so would climate change. Alternatively, if natural (as opposed to human) systems do operate as expected then there may be unforeseen developments in human systems. Then again, what would happen if China's population of 1.3 billion people were to start increasing their fossil-fuel consumption at rates previously not contemplated? If either natural or human systems do not behave in the future as anticipated then there will be unforeseen consequences.

Such unforeseen circumstances could take us outside of the IPCC forecasts in either a positive or a negative way. At the most fundamental level, with regard to natural systems, we know that in the past atmospheric carbon dioxide has sometimes been both markedly higher and, at other times, lower than the IPCC window. However, since the Industrial Revolution the trend has been upwards, so it would be prudent to consider surprises that increase climate change. Indeed, in 2004 it was realised that, from direct measurements, atmospheric carbon dioxide has recently been increasing faster than contemplated. The current question revolves around whether this is an atypical blip or part of an unforeseen trend.

6.5.3 Sea-level rise

The IPCC predictions for sea-level rise have if anything to date become more conservative with successive reports (see the summary in Table 5.1). To some, within the UK scientific community at least, this has been itself surprising as a body of research since the late 1980s (the time from which the IPCC first drew on existing peer-reviewed academic literature) has pointed to higher sea levels of up to 6 m from the last (Sangamon or Eemian) interglacial, even though much of that interglacial was climatically comparable to the present one and only part of it marginally warmer than today. Further, there is recent evidence that Antarctic ice flows in some places are faster than thought and that Greenland's ice cap is melting more rapidly than predicted. Having said this, there are also possible counteracting factors but there is uncertainty, and hence the possibility of surprise.

It has to be again said that the IPCC came to its conclusions through a committee process and that its committees were not always as unified as might be preferred, even if a good proportion of its members were in general accord. It is not appropriate, or possible, for those not on the committees to explore exactly how they came to their conclusions, although we can be certain that their conclusions do arise from a variation in opinion and/or confidence in forecasts (hence one reason for the upper and lower estimates in addition to the best estimates the IPCC provides). We also have to acknowledge, with a significant proportion of the global population being located in areas likely to be affected by sea-level rise, that flooding attributable to climate change could have major political implications. In other words, irrespective of the risk (probability), the potential hazard is high. So, given the comparative reassurance of a low forecast on one hand, and potentially high costs in the event of high sea levels on the other, it may be tempting for policy-makers to play down sea-level concerns when in fact there should be a greater apprehension and attention to the precautionary principle. Psychological perceptions and political drivers are one thing; but scientific understanding is another, even if the latter can contribute to the former, (which is what was hoped with the formation of the IPCC).

We know that during the last interglacial sea levels were up to 6 m higher than today. This therefore might be considered at least a realistic case scenario of concern that would be *currently* foreseen to take place *beyond* the 2100 horizon frequently used by policy-makers. Conversely, the total sea-level rise due to mass contributions if all terrestrial ice melted would be somewhere around 80 m. (This does not include the sea-level rise from thermal expansion and it would probably take a few thousand years.) In 1999 the US Geological Survey published the following estimates as to the breakdown of this 'all-melt' potential (see Table 6.3 summary).

On one hand a worse-case 6 m rise (as in the last interglacial) over a few (two, three or four) centuries seems modest against a potential rise of around 80 m in

Table 6.3. *Sea-level rise from terrestrial ice melt from estimates by the US Geological Survey (1999).*

Source	Volume (km^3)	Percentage	Level rise (m)
Glaciers, etc.	180 000	0.55	0.5
Greenland	2 600 000	7.90	6.5
Antarctic*	30 109 800	91.49	73.5
Total	32 909 800	100	80.5

* Of which the West Antarctic Ice Sheet would contribute 5–6 m.
Note: this excludes any rise from thermal expansion of the oceans. Also note that Antarctic land ice is largely stable and unlikely (unless there are surprises) to make a major contribution to sea-level rise, if at all, in the twenty-first century. However, since 1999 new evidence suggests that terrestrial ice melt may be faster, especially from Greenland (see text). The US Geological Survey data have been rounded to the nearest half metre.

the event over very many centuries of *all* terrestrial ice melting. The long-term all-melt scenario is far greater that the IPCC's 2001 forecast for the end of the twenty-first century but would be one that we might well encounter should an event like the initial Eocene thermal maximum (IETM) be triggered (see Chapter 3). Meanwhile, the 6 m rise seen in the last interglacial took place at a time when temperatures were not far off what they are anticipated to become later in the twenty-first century; although of course, due to thermal inertia it will take a long time for Greenland, let alone Antarctica, overall to melt. On the other hand, at over 375 ppm, current atmospheric carbon dioxide levels are already exceeding the peak level of just over 300 ppm seen during the last interglacial. The reason for this disparity of the last interglacial sea-level peak (approximately 5–6 m) and the potential rise if all our planet's ice melted (approximately 80 m) is due to the aforementioned huge thermal inertia inherent in very large ice caps: it takes a while for such ice caps of a very large mass to adjust to extra warming. Conversely, if we stopped further global warming geo-biosphere inertia would ensure that some melting would continue through to the end of the century due to the increased warmth already in the system.

In addition to the above terrestrial ice-melt factors, roughly a quarter to a third of recent sea-level rise has not been due to melt but thermal expansion of the oceans. As the oceans warm, they expand and the sea-level rises. This means that the sea level rise in a warmer world thousands of years from now, with all terrestrial ice melted, would exceed the 80 m or so anticipated rise from melt alone.

Before dealing with possible surprises, it is worth reminding ourselves that our current understanding of historic, let alone possible near-future, sea-level rise is fraught with uncertainties (see the previous chapter). Uncertainties are not the same as surprises, even if one can lead to, or spring from, the other. As previously mentioned our estimates for the various components of historic sea-level change include a considerable margin of error so that in one sense we could say that we can only explain about half the change seen over the past couple of centuries. Another major reason is that we not only do not understand how Antarctica is likely to behave in a warmer future but we do not know entirely what is going on today.

What we do know is that the marine ice shelves of the Antarctic Peninsula are vulnerable to collapse, but because much of the bulk of these is already under water their future melting will not contribute as much to sea-level rise as the continental ice shelf (grounded on rock). The Antarctic Peninsula is particularly sensitive to climate change as it juts away from the main body of the continent and the South Pole. We also know that much of West Antarctica consists of floating marine shelves and so will certainly contribute to sea-level rise in a warmer future, long before East Antarctica. It appears that parts of the West Antarctic Ice Sheet (WAIS) contributed to the (5–6 m) higher sea levels seen in the previous interglacial. Conversely, East Antarctica is largely composed of grounded continental ice shelves and is higher (hence cooler) than West Antarctica. On one hand these ice shelves are more stable and so will not melt except with warming over a considerable period of time (hundreds of thousands of, if not 1 or 2 million, years) or unless global warming is really exceptional such as perhaps with an IETM analogue event.

Yet uncertainty abounds as to what is currently happening in much of Antarctica (let alone what might happen in the future). There were in 2006 18 long-term monitoring stations in Antarctica. Eleven of these show warming over the past few decades and seven show cooling. This may suggest that Antarctica is in fact warming up. Yet, it must be remembered that Antarctica is a continent the size of Europe and the USA combined, and so 18 monitoring stations are not sufficient for detailed monitoring. This insufficiency is even more evident when one considers that 16 are arranged around the coast and just two (at the South Pole and at Vostok) are in the continent's interior. Indeed, even the South Pole station is not particularly useful for getting an idea of what is going on in East Antarctica, as it is not that far from the WAIS and in the transition border zone between the part of the continent that is thought to be losing water and the part that is accumulating it. Consequently, that station does not inform us as to how the continent's interior is faring as we would like. Remote sensing via satellite has also been limited as no satellite has passed directly over the South Pole (see Figure 5.8). However, this will change shortly with the launch of the second cryosat satellite, following the first cryosat's crash due to a launcher failure in 2005.

It is worth briefly summarising what we currently know of what is going on in Antarctica, as it is so important to sea-level rise, which will affect low-lying regions worldwide, although it is decidedly geology, not biology. There are broadly three Antarctic zones. The East Antarctic (see Figure 5.8) contains some 90% of the continent's ice and this is virtually (apart from a few small coastal areas) all overlying surface ground and so are continental shelves that are very stable but not immune to climate change. It is thought that this ice is getting thicker because snowfall (precipitation) – although low – exceeds melt. The second area is the Antarctic Peninsula which, as mentioned, sticks out away from the continent and is sensitive to warming, with its marine shelves (resting on, or floating over, the sea floor) on its eastern side particularly so. These include the Larsen ice shelves that are being undermined by submarine currents from the east that have warmed a little in recent decades (at a rate of around 0.01–0.02 °C a year) and recently subject to warmer surface air. Some of the peninsula's ice-shelf retreat is unique to the current Holocene inter-glacial, although melt appears to have previously been slower but continual over much of the Holocene. What is now known is that the collapse of Antarctica's Larson B ice shelf from the mid 1990s appears to have been a unique event. One possibility that had previously been considered was that the ice shelf may have collapsed previously during other warm periods in the current interglacial (the Holocene) and then returned in cooler times. After all, we know that the Holocene has had warm periods (such as the MCA or MWP) and cooler times (such as the Little Ice Age); see Chapters 4 and 5. However, in 2005 the results were published of research by geoscientist Eugene Domack and colleagues. They had analysed core samples taken in 2002 from various sites that had been beneath the former ice shelf that had existed up to 2001. They also looked at a core taken where the ice shelf had existed up to 1995. A few large boulders were also recovered and only some showed evidence of recent colonization by organisms. Carbon analysis of these organisms gave modern ^{14}C ages (it was possible that older carbon could have been recycled). Their conclusions, based on the biological and geological evidence, was that the Larsen ice shelf had been in continual existence throughout our Holocene interglacial so that its disintegration from the mid 1990s was a unique event since either the last interglacial or earlier. Domack and colleagues said that the ice shelf's demise was prob-ably the consequence of long-term thinning (by a few tens of metres) over thousands of years and of recent short-term (multi-decadal) cumulative increases in surface air temperature that have exceeded the Holocene's natural variation in regional climate.

Then there is the WAIS, other than the Peninsula, which consists of the Ross, Amundson and Weddell Zones. Of these, some of the Amundson ice shelf has melted before in previous interglacials and so will play an important part in near-future sea-level rise. Indeed, we know from foram remains under some of the Amundson shelves that these have retreated in the past and reformed during glacials. (The foram remains detected were of species that evolved in the past 600 000 years and so could not have been left over from before Antarctica began to freeze a few million years ago.)

Where does this leave us with regards to surprises? The IPCC's 2001 (Special Report on Emission Scenarios (SRES)-based) best estimate for sea level rise from 1990 to the end of the twenty-first century is 22–70 cm. This is within a high–low window – due to land-ice uncertainties – of 9–87 cm. For the IPCC, and hence policy-makers, the surprise would be if we were taken outside this window, which in effect means if the rise exceeds 87 cm. Is this likely? The reasons to suggest that it might be are three-fold.

1. Some of Greenland's margins and Antarctica's ice shelves appear to be disintegrating faster than anticipated.
2. Antarctic and Greenland interior ice is now moving faster to the coast.
3. Mass additions to past twentieth-century sea-level rise appear to be greater than thought.

First, with regard to points 1 and 2, even though from Table 6.3 the potential sea-level rise from Antarctica greatly exceeds that from northern-hemisphere ice caps, recent research suggests that during the last glacial the sea-level rise of 30 m (when the sea was much lower than now) over the period 65 000–35 000 years ago was due to equal contributions (±10%) of melt from northern-hemisphere ice caps and Antarctica, and that these were pulsed with 5000–8000-year spacing between events. The suggestion is that while the then glacial Earth warmed 2–3 °C the sea-level rise of 30 m took place at a rate of 2 m per century (Rohling *et al.*, 2004). This is over three times the IPCC's best estimate (IPCC 2001a, 2001b) of the rate of rise forecasted for the twenty-first century.

Furthermore, many had assumed that just 25 m of the total 120-m inter-glacial–glacial difference in sea level came from Antarctica. Consequently this more recent research suggests much of the past 25-m Antarctic (100 000-year glacial interglacial) melt was released in episodes within a 30 000 year period in the depth of the ice age. Whereas *past* melt is not necessarily a guide to *future* melt (the twenty-first century does not have large Laurentide or Fennoscandian ice sheets, which accounted for much of the rest of the 120-m glacial–interglacial rise), it does suggest that past melt was not smooth, and so future melt might not be as linear or smooth as the IPCC predict. We should not allow current slow sea-level rise to lull us, as a period of faster rise may yet take place. Of course, while comparisons of a present interglacial Earth warming of 2–3 °C are not directly analogous to a past glacial Earth warming 2–3 °C, we might at least bear in mind the past when considering possible future surprises that the IPCC warn of.

Second, with regards to point 2 above, Antarctic (specifically West Antarctica) and Greenland ice may be less stable than previously thought. There has already been much concern over the stability of Antarctic ice sheets, especially those grounded beneath the sea along the coast (continental ice shelves). (Those ice sheets – marine ice shelves – already floating, will not when melted contribute to sea-level rise.) In 2004 concern was raised that the Greenland ice cap was melting faster than thought. More evidence came to light in 2006 (to which we will return shortly). Indeed, around the turn of the millennium large parts of Antarctica's peninsula marine ice shelves broke free. Among the most spectacular collapses were those of the Larsen A and B ice shelves in 1995 and 2002. In 2004 it was reported that a survey in West Antarctica, using ice-penetrating radar, had revealed a change in ice flow compared to that about 1500 years ago, as deduced from geological evidence (Siegert *et al.*, 2004). It is now thought that coastal ice was/is physically supporting ice further inland and that with this coastal ice gone, ice inland was/is more free to flow to the coast. It is separately known that some Antarctic glaciers more than 100 km inland are now accelerating (Kerr, 2004). If ice caps are moving faster than thought then IPCC estimates of sea-level rise (which are based solely on previously published work more than a year prior to each report) will need to be revised. At the moment the IPCC consider that, as previously mentioned, the Antarctic contribution to sea-level rise in the twenty-first century will be negligible due to ocean evaporation falling as Antarctic snow, which will then become trapped as ice. This provides a countering effect to Antarctic ice melt. The question is whether Antarctica is taking more water from the oceans, via the evaporation/snowfall route, than it is contributing to the oceans through melt. This is a current research priority.

With regards to estimates of the mass of meltwater contributing to sea-level rise, the IPCC (2001a, 2001b) refer to twentieth-century sea-level measurements to assist them in forecasting the future. The IPCC do warn that their twenty-first-century sea-level forecasts have considerable uncertainty to them (a warning not often highlighted, or even mentioned in much other documentation). Indeed, there has been a discrepancy between direct estimates of twentieth-century rise compared to estimates based on contributions of mass (meltwater) and volume (thermal expansion). Research published subsequent to the IPCC's 2001 assessment looked at temperature and salinity close to sea gauges. Sea temperature affects the amount of thermal expansion and salinity is an indication of how the sea is being diluted by freshwater. They found that the mass (melt) contribution to twentieth-century rise was greater than previously estimated (Miller and Douglas, 2004). This picture has been reinforced by other research published just prior to the IPCC's forthcoming 2007 report (Overpeck *et al.*, 2006). This compared sea-level rise during the last interglacial with computer models and current trends. This again suggests that the IPCC's 2001 estimates for sea-level rise may need to be revised upwards and further that during the twenty-first century we may pass thresholds that enhance Antarctic warming as well as – as others have thought (such as Toniazzo *et al.*, 2004; referred to in the previous chapter) – Greenland. For example, if the collapse of West Antarctic ice sheets and other melt changes albedo sufficiently then atmospheric-circulation patterns may also change, so altering East Antarctic warming, not to mention that of the continent as a whole. This would increase the likelihood of sea-level rise reaching more quickly that of the last interglacial 127 000–130 000 years ago, 6 m above that at the beginning of the twenty-first century. The big question is how quickly would this happen and so how great a revision of the IPCC's 2001 sea-level rise estimate would be required?

Again if there is more melt, and if we have yet to experience much of the thermal expansion anticipated from global warming, then both these factors have the possibility of increasing potential twenty-first-century sea-level rise. This has conceivably serious consequences for the resulting impacts on human activity. Policy-makers and planners need to take this, if you will, certain uncertainty into account.

Finally, what of beyond the twenty-first century? While the 6-m sea-level rise during the last interglacial may arguably be a very (possibly unconceivable) worse-case scenario for the twenty-first century, it is not so in the longer term. Conversely, it is most unlikely (indeed virtually impossible), for *all* Antarctic, Greenland and glacier ice to melt by the end of the twenty-first century or even the next few centuries: so unlikely, in fact, that this particular possibility can really be discounted. However, unless greenhouse emissions are curbed considerably in the near future, and indefinitely, on the scale of many centuries if not millennia hence, a significant proportion of such melt becomes far more probable. Indeed, if current global warming triggered an event analogous to the IETM then such melt would become more likely. If the Earth warmed sufficiently over the millennia sea levels could well rise by over 80 m (Table 6.3). This would have a profound effect on our modern global civilisation. All current ports, and many capital cities, would be under water. It would also re-shape coastlines in a manner close to (because of plate tectonic movement) that not seen since the Pliocene. Such extremely long-term thinking is not currently on the political agenda. It is also beyond the IPCC's current agenda. Yet at the moment, with greenhouse-gas emissions currently growing (and not being curbed), this is the long-term future into which we are probably heading.

Turning away from Antarctica, the climatic monitoring of the Greenland ice cap is far better. Furthermore computer modellers seem to have greater confidence in their forecasts of possible future Greenland melt under a variety of IPCC SRES scenarios. Taking the high IPCC climate-change scenario, with carbon dioxide rising to some 1000 ppm up to the year 2200 and then stabilising (assuming it is not pushed higher still by other surprises from other (non-fossil) carbon pools), the Greenland ice cap will disappear over the next 2000 years (although there might be a small amount of ice remaining on top of its eastern mountains). This melt alone will raise sea levels by 6.5 m; though of course should this happen then sea-level rise will be higher than this due to thermal expansion and other contributions from Antarctica (Alley *et al.*, 2005).

Although our understanding of the Greenland ice cap is better than that of Antarctica, our knowledge is still far from complete. The above 2000-year scenario is largely based on melt factors. However, there are other mechanisms that serve to release water from Greenland ice. One of these is increased glacier movement. In 2006 it became apparent from satellite images that glacier movement had increased more than had previously been thought. It was already known that glaciers near the coast had been moving faster, but the new satellite data showed that there had been a marked increase some way inland and so the flow rates of Greenland's largest glaciers had doubled in the 5 years since the beginning of the twenty-first century. This in turn means that Greenland's estimated loss of ice mass has increased from a little over $50 \, \text{km}^3 \, \text{year}^{-1}$ to in excess of $150 \, \text{km}^3 \, \text{year}^{-1}$. The mechanisms behind this change are thought to include removal of the formally stable glacier ends (so reducing the glacier ends' buttressing effect), and the percolation of meltwater from the top of glaciers through to their bases, so lubricating them (Dowdeswell, 2006; Rignot and Kanagaratnam, 2006). Consequently the IPCC 2001 and the US Geological Survey estimates (Table 6.3) will need revising and so the IPCC's next assessment (due in 2007) may well contain increased estimates for sea-level rise.

Such a remote future may seem to have little bearing on present-day life. Yet even a small proportion of this long-term rise would in the shorter term have a significant effect on humans and human-managed ecosystems in the present and on land we currently consider utilising (from our human perspective) indefinitely. What we have inherited from the past we largely expect to either use, or have its use evolve, in the future. Today much of our infrastructure had its practical day-to-day foundations laid centuries ago. For example, much of Europe's rail network was established well over a century ago; in Britain Brunel's bridges are still being used today. Much of Europe's principal road networks were established centuries earlier still. Yet again, the geographical location of many of humanity's major settlements were established thousands of years ago. But consider a future a thousand years hence. London, for instance, would either be submerged or protected, either as an island or as part of the south east region of England, needing hundreds of kilometres of towering sea defences scores of metres high. London would not be alone. New York, Washington and indeed many of the Earth's major urban centres would need similar protection. Similarly, low-lying countries, such as Bangladesh, would be submerged, so they would effectively vanish. Even low-lying developed nations, such as the Netherlands, with a reputation for excellent sea protection, would have to bolster these several-fold or be submerged. The big question is what will happen between now and this far future? When exactly will a 1-m rise become manifest, let alone a 2- or 3-m rise?

Sea-level rise is a recognised dimension to global warming. Yet, other than to very low-lying states (such as the Maldives), it is either not considered at all by planners above and beyond immediate risks (for instance, storm surges based on historic levels), or proper investment fails to follow policy initiatives (such as in the southern USA). Yet such is the way of likely climate impacts that future sea-level rise will almost certainly receive more serious attention as this century progresses.

6.5.4 Methane hydrates (methane clathrates)

Methane is a far more powerful greenhouse gas than carbon dioxide (see Chapter 1) and releases from the dissociation of semi-stable methane hydrates in marine sediments have been thought to contribute to past warming and some mass-extinction events (see Chapter 3). So what are the prospects of such a methane release today? There is still considerable uncertainty; this is of course the reason for methane releases being included as a possible surprise confounding the IPCC forecasts. There is still much to learn and research continues to point to this being a risk warranting attention.

In 2004 Hornbach *et al.* reported the discovery of methane gas being trapped in a layer below some methane hydrate provinces. The researchers state that if such a layer is common, and thick along passive continental slopes, then a 5 °C increase in sea temperature at the sea floor could release as much as around 2000 GtC. This is the same order of magnitude of carbon as was involved in the Eocene thermal maximum. Whereas a 5 °C rise is unlikely during the twenty-first century (but is likely for the early twenty-second under the IPCC 2001 Business-as-Usual scenario), such a release would have a major (high-hazard) warming effect. 2000 Gt of methane is approximately equivalent to the climate effect of 120 000 Gt of carbon dioxide on a 20-year time horizon (see global warming potentials in Chapter 1). Given that in 2001 fossil-fuel burning additions of carbon dioxide to the atmosphere were just over 8 GtC/year with a further 2 GtC from land use and industry, the climatic change associated with a methane release equivalent to 120 000 Gt of carbon dioxide would be dramatic.

In addition to trapped methane gas, there are the methane hydrates themselves. There could be as much as between 2000 and 4000 times as much methane locked up as hydrates as is currently in the atmosphere, although it is unlikely that all of this would be released without a far greater warming of the ocean than is anticipated even with the worst IPCC forecasts, even beyond the twenty-first century. However, the release of a proportion of this is likely if not inevitable. The question is how much, especially as a small proportion of even a few per cent, would present a major change in atmospheric methane concentration and a major climate hazard?

Low-probability, high-hazard surprises pose great difficulty for policy-makers. Investment in monitoring a low-probability event is – virtually by definition – unlikely to see a return, yet should the event actually happen the costs would be high. A recent illustration of failing to address such a low-probability, high-hazard event was that of the 2004 Pacific tsunami. Prior to 2004 it was not considered worthwhile investing in an Indian Ocean seismic-event monitoring and alarm system. After the tsunami a warning system was developed. The difficulty of investing in low-probability events is also high-lighted by some failures to invest in countering likely-probability, high-hazard events, a recent example of which would be the flooding of New Orleans in 2005. This risk was previously identified and the economic value of New Orleans was also (obviously) known to be high.

So what are the prospects of a mega-large release of methane hydrates? As previously mentioned, the big concern is whether global warming beyond anything known in the Quaternary will trigger something analogous to the Initial Eocene Thermal Maximum (IETM) or, worse, the Toarcian event (see Chapter 3). Methane releases from the ocean are considered the source of the Eocene carbon isotope excursion (CIE) and the warming that took place then (55 mya). It is thought that the release was at least triggered by volcanic action (from what is now a large igneous province) heating organic-rich (fossil-fuel type) strata, so releasing greenhouse gases.

From the CIE we have a very rough estimate of the total ^{12}C involved. What we do not know is how much came as a result of the volcanic action on organic strata and how much came from marine methane hydrates. Yet estimates are that during the twenty-first century we are likely to release a similar amount of carbon from the atmosphere that took place during the IETM. Two possibilities therefore present themselves.

First, if *all* the Eocene ^{12}C came as a result of volcanic action warming organic-rich strata (and none initially from methane hydrates, but only far later in the event as the oceans warmed), then, because the amount of carbon involved is similar to what we are releasing now, in a worse-case scenario we would be looking at an early-Eocene-like thermal event. Here the methane hydrates would only destabilise later in the event and prolong the warming as carbon cascaded (via the atmosphere) between various bio-geosphere carbon pools.

The second possibility is if just *part* of the Eocene ^{12}C came as a result of volcanic action on organic strata early on in the event (during the first few thousand years), and the remainder early on from methane hydrates, then the current prospect for future warming would be worse. We already know that the *total* ^{12}C during the Eocene event is broadly comparable to the amount we expect to release over the next century under IPCC Business-as-Usual (BaU) scenarios. This then can be considered to be analogous to the carbon released due to volcanic action from the organic-rich strata. Subsequent carbon from marine methane hydrates would be extra. If this was so then as today we are already in the process of releasing carbon from an Eocene-comparable fossil-fuel burden, it means that any future release from hydrates *must be above and beyond this*. In short, we could be heading for a climatic event that is greater than that during the Eocene.

These two possibilities of mega-releases of methane from organic sediments, or organic sediments and hydrates, are two ends of a spectrum. One would be a faster-to-initiate but longer IETM-analogue event. The other would be a more extreme IETM event. Should an IETM analogue take place then its severity would in all likelihood be somewhere between these two extremes. This is something of little comfort. Whatever may transpire, it is worth remembering that at the moment the prospect of such a mega-release is considered low even if of high hazard. Such mega-releases are only likely *if* we continue (as we currently are) down the BaU route. There are other options, but as we shall see (Chapter 8) realising these will be difficult.

Finally, it needs to be emphasised that Eocene warming and the Eocene atmospheric carbon built up over several thousands of years. In the present we are adding the same carbon in just a century or so. Whereas the ocean warming will be far slower, the terrestrial warming will be far faster than during the Eocene. The Eocene thermal maximum cannot therefore be considered as analogous as might be thought to a BaU scenario for the late twenty-first and early twenty-second centuries. Our current predicament is worse, in that terrestrial biological systems would not have the time to adapt as they had in the Eocene. Further, even though the oceans will take longer to warm because *complete* circulation-driven ocean turnover takes the order of 500–1000 years, there are also bound to be surface-ocean acidity concerns due to the same, slow, complete ocean turnover. This may well be why we are already seeing the early effects of surface-ocean acidification.

Leaving aside the above risk of a mega-release of methane, what are the chances of smaller releases from semi-stable marine methane hydrates that might drive up atmospheric methane by, say, 50–150 parts per billion by volume (ppbv), as happened during the last glacial? (For comparison the concentration in 2001 was 1750 ppbv.) This could happen as a result of either decreased water pressure or increased temperature. The chances of it being due to sea-level reduction are quite small (if not virtually zero as sea levels are rising, not falling). However, the chance of a release due to temperature increases are greater. The Earth is, after all, getting warmer and warmth increases methane hydrate instability. Methane releases during the last glacial have been deduced for interstadials (short, comparatively warm periods within glacials), which have atmospheric concentrations that are increased by 100 ppb or more (Kennett *et al.*, 2000). Here it is not so much that the Earth is warmer during interstadials per se, but that the warmer interstadial Earth has different atmospheric- and ocean-circulation patterns and these can suddenly bring warmer water to an area that was previously cooler and so is conducive to hydrate build-up. This relates to another possible surprise in the change in circulation patterns (see below).

Yet, whereas a small methane release (boosting atmospheric methane by only 50–150 ppbv) might possibly change global circulation patterns that themselves might greatly change the climate in a few regions, the overall effect of such a methane release itself in terms of raising the average global temperature would be not that great, and far smaller than the several degrees difference between a glacial and an interglacial. It might, though, be as much as the temperature difference between the depth of the Little Ice Age and today, or possibly more. Even so, the concern would be that such a sudden release would have a rapid effect on the global climate, possibly equivalent to a few decades of IPCC-anticipated, twenty-first-century BaU warming impacting within a year or so.

Whether a big release or a small one, from our understanding of past events it does appear that methane releases from marine hydrates can either trigger climate change or exacerbate an episode of climate change that has already begun. Consequently, it is not surprising that some concerned with gathering science to underpin policy are looking seriously at such risks. Indeed, in 2004, when the UK Department of Environment, Food and Rural Affairs (DEFRA) was considering its science requirements over the next decade, it specifically included research into reasons behind sudden climate change, such as release of methane hydrates (DEFRA, 2004).

So what do we need to look for? Methane hydrates form at depths roughly exceeding 400 m in some places and exceeding 1 km and more in others. Aradhna Tripati and Henry Elderfield (2005) of the University of Cambridge, using fossil forams and employing a magnesium/calcium ratio temperature proxy, ascertained the (sea) bottom water-temperature rise at the Paleocene–Eocene boundary. Their conclusion was that the water temperatures then rose by 4–5 °C and that this was caused by a change in ocean circulation (see section 6.5.6). To put this in perspective in terms of temperature, since 1960 much of the ocean surface has broadly risen by 0.2 °C. One of the places where surface warming has been greatest is the North Pacific (0.25 °C). As such, this is broadly in line with the IPCC estimate of global surface warming and so greater warming should be expected if we go down the IPCC BaU route. However, this warming fingerprint has only penetrated about the first 75 m of the sea column and this penetration seems to reasonably reflect current computer models (Barnett *et al.*, 2005b). Even with further warming, as the IPCC BaU scenarios suggest, it is likely that a change in ocean circulation to bring warm surface waters to a depth that had previously remained cool will be necessary to trigger a major methane pulse in this century. Nonetheless, this possibility cannot be dismissed.

Taking all this together it may initially look as if marine methane releases are unlikely. However, we do know that the Earth is warming up (even if we cannot forecast future warming exactly). Further, as shall be discussed below, with warming there is an increased chance of changes in circulation patterns. The question not so much becomes one of whether there will be such methane releases, but when and by how much? To address this, research on Earth-system thresholds is a current priority.

6.5.5 Volcanoes

Major volcanic activity has been associated with mass-extinction events in the past (see Chapter 3). Specific events are the end-Permian or Permo–Triassic extinction 251 mya and the Cretaceous–Tertiary extinction 65.5 mya. In the former some 90% of ocean species vanished and 70% of vertebrate families on land became extinct, and with the latter 75–80% of terrestrial species. We have to remember that asteroid impacts have also been associated with both these extinction events (especially the Cretaceous) but it is thought that the impacts provided the *coup de grâce* and that many species were in decline beforehand with vulcanism implicated in the cause.

So what are the chances of a major or super-volcanic event today? There are a number of places where such significant volcanic events might take place (albeit smaller than those that created igneous provinces in either the Permian or Cretaceous periods). These include Long Valley in eastern California, Toba in Indonesia (which last erupted 74 000 years ago, and which is thought to have lowered temperatures 3–5 °C for up to 5 years) and Taupo in New Zealand. However, one of the sites of greatest concern is at Yellowstone National Park in the USA. Starting 2.1 mya, Yellowstone has been on a regular eruption cycle of 600 000 years. The last big eruption was 640 000 years ago. Yellowstone has continued in a quieter fashion since, with a much smaller eruption occurring 70 000 years ago, so a large one is overdue. Indeed, today it is clear that volcanic activity beneath the park is taking place and this is signalled by marked surface deformation.

The next question is whether a Yellowstone eruption would affect the global climate and have a global biological impact. There are two ways that we might begin to roughly estimate impacts. First, the Yellowstone eruption 640 000 years ago takes us to long before the Vostok ice-core record and to seven or eight glacials ago. However, the marine ^{18}O isotopic evidence does go back this far and it suggests that the eruption was roughly contemporary with the height of a particularly cold glacial (there has been none colder since). Milankovitch factors (see Chapter 1) would have made it a cold glacial anyway so the question remains as to whether that eruption made that glacial even colder than it would have otherwise been, and if so by how much and for how long?

Second, the last great Yellowstone eruption is thought to have released about 1000 km^3 of lava. This compares with a release of 1–4 million km^3 of lava in Siberia at the time of the Permo–Triassic extinction and some 512 000 km^3 from the Deccan Traps in north-west India with the Cretaceous–Tertiary extinction. As stated, whereas these eruptions were not the sole cause of the mass extinction, it is now well established (especially with regards to the Cretaceous–Tertiary event) that species were declining at the time of the eruptions. So, by this standard alone a large Yellowstone eruption may have roughly a five-hundredth the impact of Cretaceous–Tertiary vulcanism. Here current global warming might actually help a little in offsetting the cooling effect of the eruptions. Even so, an eruption at Yellowstone would most likely have a climatic impact far greater than that of large eruptions of normal volcanoes that do not have large volcanic traps.

So, what of a large eruption of a standard-type volcano? The Huaynaputina in February 1600 (see Chapter 4 and Figure 2.1) eruption was not the biggest eruption in the Little Ice Age. Nonetheless, the following few years saw very low temperatures. In Europe the summer of 1601 following the eruption was cold, with freezing weather in northern Italy extending into July and overcast skies for much of the year, while in parts of England there were frosts every morning throughout June. Yet despite this the economic impact of this climate change was marked, although not as clear cut as one might expect. The eruption happened to follow some years of the late 1590s that also saw climate-related food shortages. By 1601 people were adapting (as they will in the future once they have experience of global warming in their area). In England in 1600–10 documented wheat prices were not high. However, the purchasing power of non-agricultural workers, as revealed by the purchasing power of English builders' wages, not only dipped due to the earlier shortages of the 1590s but failed to rise significantly throughout much of the first half of the seventeenth century, after which there was a steady recovery and upward trend lasting a century (Burroughs, 1997). Builders' wealth is a useful indicator of societal economic well-being, as building is inevitably associated with a wealthy society, but builders themselves do not generate food. Consequently builders' wealth is in no small part an indicator of a society's agricultural surplus. So the Huaynaputina eruption during the Little Ice Age, while not having a devastating effect on the economy, does coincide with a period of little food surplus (if anything, a marginal food deficit).

While the economically discernable effect on agricultural systems from standard (as opposed to super-) volcanoes lasts only a few years, their thermal impact on the bio-geosphere lasts longer. Evidence comes from today's computerised climate models that are sufficiently good to simulate at least most continental-scale climate events of which we have knowledge (be it through direct measurement or from several sets of palaeoclimatological data). Of course, there is still work to be done regarding modelling of high latitudes and increasing the level of spatial resolution, not to mention in reducing uncertainty relating to, and increasing the number of, the various climate forcing factors. Nonetheless, that many of today's global models do successfully capture past climatic change suggests that, even if much refinement is still possible, they do have a very meaningful role to play in informing us as to likely climate factors (and indeed policy). There is some evidence that volcanic eruptions, such as Krakatoa in 1883, have a small, but clearly discernable, impact on deep-ocean temperature that lasts for between 50 and 100 years.

Twelve state-of-the-art global climate models were analysed, both including and omitting the impact of the Krakatoa eruption and its injection of particulates, gases and aerosols into the atmosphere. Each of these models differed in their physics, resolution, initialisation and ocean–atmosphere coupling as well as different combinations and strengths of forcing factors. However, each successfully captured the principal climate features of the past couple of centuries, albeit coarsely and within certain error. However, the analysis did show that simulations that included a Krakatoa-type eruption noticeably differed from those that did not. The eruption's thermal fingerprint is clearly seen in the simulated oceans, with a small but clear temperature difference of around $0.01\,^{\circ}\mathrm{C}$ in the top kilometre of the ocean column that lasted for about 50–100 years. This analysis also revealed that present-day (non-super-) volcanoes have a smaller cooling event. The simulated heat-content recovery after the 1991 Pinatubo eruption (which was comparable to Krakatoa in radiative climate-forcing terms) was far quicker. This was actually measured by satellite monitoring data. The reason for this more speedy oceanic thermal recovery was that at the time of Pinatubo (and more so now) there were rising greenhouse-gas emissions and global warming taking place. Anthropogenic greenhouse-gas emissions at the time of Krakatoa were roughly a tenth of those at the time of Pinatubo. The thermal impact on the oceans of each volcano may be small, and shorter lived in today's faster-warming planet than the past, but taken together over time they do have a more pronounced effect. The importance of this work is that it helps validate components of climate models; in this case the inclusion of volcanic activity (Gleckler *et al.*, 2006). Including volcanic eruptions in models also helps current forecasts, as there will be a different likelihood of global warming if there is a run of volcanic activity as opposed to if there is not. Remember that much of the twentieth century up to the 1960s were largely free of major (standard) volcanic eruptions, whereas the period from the 1960s to the mid 1990s saw considerable climate-forcing volcanic activity.

Although standard-sized volcanic eruptions do seem to impart a discernable thermal fingerprint on the oceans and large volcanoes may impact on food prices for a couple of years, a super-volcanic eruption has a different order of impact that noticeably affects biological systems. By comparison today, should the Yellowstone trap erupt, it would impart a much greater level of impact than a standard Huaynaputina-type volcano, so that there would almost certainly be significant food-security implications. These would be global, so there would be no major bread-basket country from which to import food. Furthermore, eruption size aside, the lack of food security would be far worse than the European food shortages experienced periodically during the Little Ice Age, due to the twenty-first century's high population density (and the concomitant smaller areas of woodland in which to forage for alternatives) and the lack of recent experience of such conditions.

In 2005 the Geological Society of London established a working group to examine the probability and likely global effects of a super-eruption. Its report's conclusions called for the completion of a database of super-eruptions over the past 1 million years. It also recommended that the UK Government establish a task force to consider the environmental, social and political consequences of large-magnitude volcanic eruptions and that this involve international collaboration (Geological Society, 2005).

6.5.6 Oceanic and atmospheric circulation

As the Earth warms it is possible that new ocean and atmospheric circulation patterns may become manifest. For example, the Broecker thermohaline ocean circulation (see Chapter 4) is one of the key mechanisms by which heat is transported away from the tropics to higher latitudes. It is driven by warm water from the tropics moving northwards and becoming saltier due to evaporation. At higher latitudes this denser water becomes even more dense through cooling and so sinks. A current in the ocean depths then carries it between oceans, where it surfaces elsewhere. One of the key places where a warm current from the tropics sees considerable evaporation and then sinking at high latitudes is the Gulf Stream in the North Atlantic, with sinking taking place around Iceland and Greenland. However, freshening of this water, such as by iceberg discharges, Arctic ice melt or changes in regular precipitation patterns, would make the current less dense and so reduce sinking. If the reduction was sufficiently great it could shut the conveyor down. This might block the Gulf Stream's northern route. Of course, the Coriolis force from the Earth's rotation would still continue, as would trade winds, so a warm current would still leave the Gulf of Mexico but it would not travel so far north, but rather circle around at a lower latitude. Without the Gulf Stream off of north-western Europe, the British Isles, the North Sea continental nations and the western edge of Scandinavia would see far harsher winters, just as are today experienced in Labrador, which is on the same latitude as Britain.

Although we do not have a particularly thorough understanding of the conveyor's precise operation, the current understanding is that a complete shut-down of the North Atlantic driver of the conveyor is unlikely (but not impossible). However, a weakening of the conveyor by around 30% is (again, currently) considered a reasonable possibility. If this happened then seasonality in north-west Europe is likely to increase as winters would not warm in pace with summers due to long-term climate change.

Whereas such a change in ocean circulation would result in agricultural dislocation and the need for infrastructure investment, the effect would be regional. Overall the Earth would still be experiencing global warming, it is just that the climatic patterns would change more markedly on a regional basis in many places. This would have profound consequences for north-west Europe.

It is known from salinity studies that more fresh water has been added to the North Atlantic in the twentieth century. In particular during the late 1960s a large pulse of fresh water entered the Nordic Seas (between Greenland and Scandinavia) through the Fram Strait. This freshwater pulse was of the order of around $10\,000\,km^3$ at a rate of about $2\,000\,km^3$/year, implying a net flux anomaly of approximately 0.07 sverdrups (Sv) during a 5-year period (where $1\,Sv = 10^6\,m^3\,s^{-1}$). It is known as the Great Salinity Anomaly (GSA). Naturally there is always a freshwater addition (during interglacials) into the Arctic from rivers and ice melt, but the GSA is thought to represent roughly around a 40% increase of normal flow. In 2005 Ruth Curry, from the Woods Hole Oceanographic Institution, and Cecilie Mauritzen, from the Norwegian Meteorological Institute, used these long-term salinity data-sets to ascertain the likelihood of an effect on the Broecker thermohaline circulation. In a carefully worded conclusion they said that assuming the rates of fresh water observed between 1970 and 1995 were to continue then it would take a century for enough fresh water (9000 km^3) to accumulate to affect ocean that is exchanged across the Greenland–Scotland ridge and two centuries to stop it. Consequently disruption of the ocean thermohaline circulation did not appear imminent. However, they noted that this assumption might not hold, indeed would not hold, if enhanced high-latitude northern-hemisphere precipitation and/or glacial melt increased. That these are predicted in many climate-model forecasts for the twenty-first century means that disruption of the thermohaline circulation cannot be ruled out.

If the currently prevailing view is that the North Atlantic driver of the conveyor will only modestly weaken, what is the evidence for this and for the notion of perhaps an even greater conveyor shutdown with further warming? As stated above, it is known that in the twentieth, but also early in the twenty-first, century key parts of the North Atlantic have been freshening. Part of this is thought to be due to Greenland ice-cap melt. However in 2005 research, funded by the UK Government's Department for Environment, Food and Rural Affairs and carried out by the Met Office's Hadley Centre for Climate Prediction and Research in Exeter by Peli Wu and colleagues, showed that the Arctic Ocean is freshening due to an increasing hydrological cycle (Wu *et al.*, 2005). Remember, in a warmer planet there is more ocean evaporation and hence rainfall. What Wu and colleagues did was to look at data from the six largest Eurasian rivers flowing into the Arctic Ocean since the mid 1930s, namely the Yenisey, Lena, Ob, Pechora, Kolyma and Severnaya Dvina. This showed that river flows had increased by about 2 ± 0.7 mBe year^{-1} (1 Be or Bering $= 10^3$ km^3) over a 140-year historic average of 2.3 Be year^{-1}. This was a very good agreement with the predictions from runs of the HadCM3 climate model of 1.8 ± 0.6 mBe year^{-1}. With this qualified validation of the model, they then ran the model to forecast the future to the end of the twenty-first century assuming standard IPCC predictions. It showed that the discharge from Eurasian rivers may increase by between 20 and 70% during the course of this century. This is quite a range and so the modelling still needs to be improved. Nonetheless, such results have implications for the Broecker circulation and this line of research is among that being considered by compilers of the fourth IPCC assessment due in 2007.

Changes in atmospheric circulation are also possible and again these would mainly have regional consequences and be less likely to have as pronounced an effect on the Earth's overall temperature. Nonetheless, regional consequences could be severe, especially if climate change disrupted water-distribution events such as the monsoon. Even if major individual water-distribution events remained unaltered, globally water distribution about the planet via the atmosphere will almost certainly change. Indeed, this is connected to the predictions of regional precipitation change and water availability noted earlier in the case studies of climate and natural systems of the USA and UK. Water availability is not only important for ecosystem function and geophysical processing but (as shall be explored in the next chapter) it also fundamentally affects human ecology and therefore economic activity. Further to the above work on the hydrological cycle as it affects Arctic rivers, a global assessment of stream flow and water run off suggests that there will be considerable change on many parts of the Earth by the mid twenty-first century compared with the twentieth century.

In 2005 an ensemble of 12 climate models was used to see how coherently they forecast run off change; run off being defined as precipitation less evapotranspiration. The picture emerging is that as the climate warms there is more ocean evaporation hence precipitation, but also that this precipitation is subject to more evaporation. The picture is therefore somewhat complex, with some areas having more precipitation but paradoxically less run off due to this increased evapotranspiration. Overall, most areas are likely to see change in run off and this change is likely to continue as climate change (in the thermal sense) progresses through to the middle of the twenty-first century (which is as far as this analysis forecast). Initial increases of run off seen in the twentieth century are projected to reverse in some regions in this century, namely eastern equatorial South America, southern Africa and the western plains of North America. The modelled drying of the Mediterranean extends north from Spain and Greece (which we were beginning to see by the end of the twentieth century) to south-eastern Britain, France, Austria, Hungary, Romania and Bulgaria as well as the sub-Russian states. In many regions *all* the models in the ensemble agreed as to the likely mid twenty-first-century change. These agreements include increases of typically 10–40% in the high latitudes of North America and north Eurasia (as in Wu *et al.*, 2005), in the La Plata basin of South America and in eastern equatorial Africa as well as some of the major islands of the eastern equatorial Pacific. Agreements on decreasing run-off include southern Europe, the Middle East, mid latitude western North America and southern Africa (Milly *et al.*, 2005).

The possibility of changes in ocean and atmospheric circulation is one current focus of climate research. It appears that circulation systems interact with the global climate system through a number of feedback cycles (see Chapter 1). These result in semi-stable global climate modes with each mode having its own characteristic regional distribution of local climate types. Consequently the stability of a global climate mode is related to the stability of circulation patterns such as ocean thermohaline circulation (Clark *et al.*, 2002). Perturb the climate a little too much and this may perturb thermohaline circulation (as discussed above). Perturb the thermohaline circulation and this can affect ocean circulation, creating new regional patterns of warm and cool localities. Create new regional temperature patterns and this will affect atmospheric circulation, and so the Earth system will settle into a new semi-stable state until further climatic perturbation takes place. Note that in addition to immediate regional climatic impacts, a change in the thermohaline circulation could possibly disrupt marine methane hydrates (see section 6.5.4) and these would have an additional and global climatic effect.

So, how likely is a circulation change? The IPCC's projections for the twenty-first century are fairly linear. Most of its scenarios result in near-linear projections; projections that do not appear to contain a quantum jump. Yet the palaeoclimatic record is full of sudden jumps and this alone suggests that we need to take changes in ocean and atmospheric circulation seriously.

If the question of rapid change deviating from the linear has any credence, and if such fluctuations are discernable at the global level, then they should be seen within major biosphere subsystems (remembering that the biosphere includes the geosphere). In 2005 a team led by Chih-hao Hsieh of the Scripps Institution of Oceanography in San Diego, California, published work suggesting that large-scale marine ecosystems (such as oceans) are dynamically non-linear. As such they have the capacity for dramatic change in response to stochastic fluctuations in basin-scale physical states. This, they say, is exemplified by fish-catch data and larval fish abundance that are prone to non-linear responses and do not simply track a dimension of physical change such as sea-surface temperature. In short, ecological catastrophes (such as a species' regional population crash) may be the result of a modest change in, say temperature, at a critical point. This has implications for adopting the precautionary principle when estimating ecosystem sustainable yields in a time of change.

Of course, as noted a number of times above, biological systems feed back into climate. The area of wetlands affects atmospheric methane, which in turn affects climate. Vegetation cover, be it a change in or decline of cover, affects regional albedo and so local climate. So, if the biological response to modest linear environmental change is not necessarily itself linear, and biology affects environment, then environmental change such as climate change will also be non-linear. On one hand this does not lend confidence to the IPCC's near-linear forecasts. On the other, the IPCC do provide high and low near-linear forecasts and so can be said to provide a window in which non-linear responses can take place. This theoretical line of argument for non-linear change is all very well, but what is happening in the real world at the macro scale?

There was much concern with regard to changes in atmospheric and ocean circulation in the late 1990s and early twenty-first century. Computer modellers have undertaken much of the work in possible atmospheric circulation changes. Conversely, work on ocean thermohaline circulation has been undertaken more equally by those making field observations whose results then feed into modellers' work. Changes in thermohaline circulation are of particular concern as field observations have shown that some factors potentially affecting the circulation are beginning to take place. For example (just one of many), there has been a decrease since 1950 of cold, dense water from the Nordic Seas across the Greenland–Scotland ridge. The decrease is of the order of 20% and if this reduction is not compensated for by increased cold, dense water from elsewhere then the thermohaline circulation could be weakening (Hansen *et al.*, 2001). What is known is that the North Atlantic around Iceland is freshening. The thinning of Arctic ice (itself a result of climate change) could account for part of this freshening. Sea-level rise has not been as great (considering contributing factors such as ocean thermal expansion and mountain glacier melt) yet glaciers almost everywhere around the world are in retreat so some of this freshening could come from North Atlantic continental glacier melt (with the missing sea-level rise perhaps being accounted for by increased evaporation and precipitation as snow over Antarctica taking water out of the ocean system). If Arctic melt accounts for much of this freshening then the problem will be comparatively short-lived as it is likely that the Arctic could become largely ice-free during the summer in the course of the twenty-first century. Ice-free summers mean that the Arctic annual carry over store of freshwater ice has been depleted and so major freshening of the North Atlantic from this source would cease.

Meanwhile, in the southern hemisphere the Ross Sea has indeed become fresher during the late twentieth century, which itself has thermohaline implications for one of the drivers of the Broecker conveyor in that hemisphere (Jacobs *et al.*, 2002). Of course, any change in any single driver of the conveyor has implications for the conveyor as a whole; hence, remote impacts even in another hemisphere are important. The possibilities of changes in the Broecker circulation in the North Atlantic are the subject of research initiatives in Britain and Norway and these nations have engaged in joint research programmes, whereas European, North American and Australasian researchers have undertaken work on the thermohaline circulation in the southern hemisphere.

6.5.7 *Ocean acidity*

Human additions of carbon dioxide to the atmosphere affect flows to other carbon-cycle reservoirs and notably flows into the ocean (see section 5.4). This has resulted in increased ocean acidification, much like that thought to happen early in the Eocene. However, there are limits as to how acid the ocean is likely to become; after all, there have been times both tens and hundreds of millions of years in the past when atmospheric carbon dioxide was higher than today and calciferous sediments such as chalk did not dissolve. In part this is because should the ocean (or rather the ocean/atmosphere interface) become saturated then sea uptake of carbon dioxide would be reduced. This would mean that the atmospheric increase in carbon dioxide associated with the burning of a unit of fossil fuel would rise and with it the climatic impact. Also, acidity would be stabilised by buffering.

With little other than clues from the Eocene thermal maximum and Toarcian events (see Chapter 3), we do not know exactly what will happen to ocean pH if an IPCC BaU scenario were followed for just three centuries to the year 2300, but a handful of researchers are beginning to look at the problem. In 2003 Ken Caldeira and Michael Wickett of the Lawrence Livermore National Laboratory ran the IPCC 1992a scenario (IS92a) through Livermore's ocean general-circulation model. This assumed that fossil-fuel consumption (and greenhouse-gas emissions) would continue to grow much as they have from 1950 to date through to the year 2100 and then begin to stabilise logistically through to 2300 as the economically recoverable reserves are used up. The IS92a scenario is, in terms of the IPCC's Special Report on Emission Scenarios (SRESs) used for their 2001 assessment (and with respect to their twenty-first-century forecasts), a middle-of-the-road BaU scenario.

Their results suggest that the atmospheric carbon dioxide level would exceed 1900 ppm by 2300. At first this might seem acceptable; after all, some 500 mya the biosphere had atmospheric concentrations up to some 7500 ppm (see Figure 3.1) and calcareous marine species survived that. However, the current anthropogenic carbon dioxide pulse is different from that early Palaeozoic rise. That peak was reached after many millions of years of carbon dioxide build-up; conversely, the current human BaU rise would be literally millions of times faster. When an atmospheric carbon dioxide change of such magnitude takes place in less than 1000 years then geological weathering and the resulting buffering cannot take place. For greater buffering it is thought that such a fossil-fuelled carbon dioxide increase would need to take place over tens of thousands of years or longer. Consequently, Caldeira and Wickett's simulation predicts that the pH reduction would be 0.7. Currently there is no evidence that ocean pH has been more than 0.6 pH units lower than today. Their conclusion is that unabated carbon dioxide emissions over the coming centuries may produce changes in ocean pH that are greater than any experienced since 300 mya, with the possible exception of those resulting from some mass-extinction events. This suggests that a marine mass-extinction event, certainly of calcareous forams, would be a possibility.

Then in 2005 an international team led by James Orr used 13 ocean carbon-cycle models to assess ocean chemistry, especially with respect to calcium carbonate and its associated ions. As with the Caldeira and Wickett team, the Orr team were again relating ocean chemistry to the IPCC's IS92a scenario. Using pteropods, a group of species of gastropod mollusc that have calcareous shells, in tanks with water with chemistry matching the IS92a scenario they attempted to see whether there were any adverse effects. They found that water simulating the IS92a scenario up to the year 2100 did affect species. This was particularly a problem at high latitudes when the sea water becomes undersaturated to aragonite (a metastable form of calcium carbonate). According to the models, aragonite undersaturation is likely to take place around the middle of the century at some high latitudes and by 2100 throughout the entire Southern ocean and in the sub-Arctic Pacific. The researchers exposed a small population of a live pteropod (*Clio pyramidata*) to IS92a forecast, high-latitude ocean waters. They observed 'notable dissolution' of their calcareous shells (see section 5.4).

The ecological consequences of ocean acidification are likely to be significant. The researchers suggested that 'conditions detrimental to high latitude ecosystems could develop within decades [of 2005], not centuries as [previous work] suggested' (Orr *et al.*, 2005).

Pteropods, especially at high latitudes, form integral components of food webs. Species high in the food web also rely on pteropods and include North Pacific salmon (*Oncorhynchus*), mackerel (family Scombridae), herring (*Clupea*), cod (*Gadidae* spp.) and baleen whales (Cetacea of the sub-order Mysticeti). In the Ross Sea, for example, the prominent sub-polar pteropod, *Limacina helicina*, sometimes replaces krill as the dominant zooplankton. Indeed, in this sea pteropods account for the majority of the carbonate and organic carbon-export flux. If pteropods cannot grow their shells then they are unlikely to survive. If carbonate and organic carbon fluxes are altered then this could also conceivably have climatic repercussions.

Calcite is another form of calcium carbonate that is a little more stable than aragonite. According to the ocean-chemistry models that the researchers developed from the IS92a scenario, calcite undersaturation will lag that for aragonite by some 50–100 years. Species that use the calcite form of calcium carbonate include Foraminifera and coccolithophorids. Moreover, some Antarctic and Arctic species use magnesian calcite, which can be more soluble than aragonite. These include sea urchins (echinoderms). Although all these species may survive in warmer latitudes, nearer the poles ecosystems are likely to be severely disrupted. Further, should acidification continue well beyond this century then oceanic refugia for vulnerable species will become more scarce and the likelihood of an extinction event more probable.

6.5.8 *The probability of surprises*

It may seem oxymoronic to discuss the probability of surprises, for if it were possible to calculate any such events with accuracy then strictly the surprise would not be a surprise. However, we do know a few things about the surprises discussed so far.

First, regarding vulcanism, we know that over a period of centuries there is a certain level of volcanic activity that includes large eruptions of standard-type volcanoes and that over far longer timescales the eruption of volcanic traps. That, for example, a Yellowstone eruption is overdue on a geological timescale of 100 000s of years translates to such an eruption in the next 100 years as being not particularly surprising. We also know that in the past other surprises, such as methane hydrate releases, have happened. With further elucidation of their mechanisms we may better be able to estimate the probability of future happenings.

Second, we know we are not in a steady-state situation. We are not just talking about the early twenty-first-century world being warmer than the nineteenth century, but a world that is continuing to get warmer.

If the second point was not significant enough there is a third, that anthropogenic releases of carbon have been rising almost exponentially since the Industrial Revolution (a graph of historic carbon releases on a logarithmic scale against time is nearly linear). We are therefore steadily accelerating the greenhouse forcing of climate year on year. This means that the increase in warming noted in the previous paragraph is not linear but has an exponential component.

Fourth, some greenhouse surprises (for example ocean-circulation changes and methane from hydrate releases) are linked. Surprises can compound and sometimes in surprising ways.

Finally, we are reaching the limitations of Quaternary experience. The twentieth century saw the global climate warm to levels not seen for over 500 years. Early in the first quarter of the twenty-first century we are likely to see temperatures not previously reached for 750 years since the MCA (MWP). Sometime before the middle of the twenty-first century could well see temperatures rise to those of the Holocene maximum or above, which took place some 8000–5000 years ago (Chapter 4). Shortly after that, sometime around the middle of the century, we are likely to reach a temperature comparable to that experienced at the height of the last interglacial. Some decades before the end of the twenty-first century (should the IPCC forecasts be largely correct) the Earth is likely to experience a greenhouse regime broadly analogous to a time at least 1.0–3.5 mya, and by the century's end there could be conditions not seen since the Pliocene. ('Broadly analogous' because these palaeo-analogues did not have quite the same greenhouse forcing as today due to circulation reasons discussed in Chapter 4.) The passing of each of these successive stages is likely to increase the prospects of a climatic surprise.

One purpose of current climate research is to try to ensure that IPCC surprises do not surprise. Climate-related policy needs to reflect this. Furthermore, just because the IPCC focused on a 100-year forecast it does not mean that greenhouse problems will cease or become static in the year 2100. Global fossil-fuel reserves (especially dirty oils such as orimulsion and lower-grade coal) could continue to provide a significant proportion of civilisation's energy budget well into the twenty-second century. This may seem like a long time hence and not worth worrying about but it is not an especially long time in geological and evolutionary terms. Most nations' capital cities have many buildings and infrastructure that were built two or more centuries ago. In London commuter trains use bridges built by Brunel in the nineteenth century, so it would not be ridiculous – indeed it would be most prudent – to plan with this time frame in mind at the very least.

Of course, encouraging policies that are likely to offset the worst effects of future climate surprises is easier said than done. After all, by definition, surprises are surprising. However, a clear priority to inform such policies is research into natural processes that may become the drivers of abrupt climate change. Here special attention needs to be given when these drivers function close to sensitive thresholds that may become critical. So, identifying such thresholds is equally important.

To engage in a longer-term perspective the likely climatic impacts on human society, and civilisation's probable energy requirements for the remainder of this century, need to be assessed. This is not a straightforward task as the human ecology of the twenty-first century will inevitably be as different from that of the twentieth century as that century was from the nineteenth. Human ecology as it relates to climate change and greenhouse-gas emissions in the twenty-first century is outlined in the next chapter.

6.6 References

Alley, R. B., Clark, P. U., Huybrechts, P. and Joughin, I. (2005) Ice-sheet and sea-level changes. *Science*, **310**, 456–60.
Atkinson, A., Siegel, V., Pakhomov, E. and Rothery, P. (2004) Long-term decline in krill stock and increase in salps within the Southern Ocean. *Nature*, **432**, 100–3.

Barber, V. A., Juday, G. P. and Finney, B. P. (2000) Reduced growth of Alaskan white spruce in the twentieth century from temperature-induced drought stress. *Nature*, **405**, 668–72.

Barbraud, C. and Weimerskirch, H. (2001) Emperor penguins and climate change. *Nature*, **411**, 183–5.

Barnett, T. P., Adam, J. C. and Lettenmaier, D. P. (2005a) Potential impacts of a warming climate on water availability in snow-dominated regions. *Nature*, **438**, 303–9.

Barnett, T. P., Pierce, D. W., AchutaRao, K. M. *et al.* (2005b) Penetration of human-induced warming into the World's oceans. *Science*, **309**, 284–7.

Bellamy, P. H., Loveland, P. J., Bradley, R. I., Lark, R. M. and Kirk, G. J. D. (2005) Carbon losses from all soils across England and Wales 1978–2003. *Nature*, **437**, 245–8.

Berry, P. M., Harrison, P. A., Dawson, T. P. and Walmsley, C. A. (eds) (2005) *Modelling Natural Resource Responses to Climate Change II (MONARCH II): a Local Approach*. Technical Report. Oxford: UKCIP.

Blaustein, A. R. and Dobson, A. (2006) A message from the frogs. *Nature*, **439**, 143–4.

Boorman, L. A., Goss-Custard, J. D. and McGrorty, S. (1989) *Climate Change, Rising Sea Level and the British Coast*. ITE Research Publication No. 1. London: HMSO.

Both, C., Bouwhuis, S., Lessells, C. M. and Visser, M. E. (2006) Climate change and population declines in a long-distance migratory bird. *Nature*, **441**, 81–3.

Burroughs, W. J. (1997) *Does the Weather Really Matter? The Social Implications of Climate Change*. Cambridge: Cambridge University Press.

Caldeira, K. and Wickett, M. E. (2003) Anthropogenic carbon and ocean pH. *Nature*, **425**, 365.

Clark, P. U., Pisias, N. G., Stocker, T. F. and Weaver, A. J. (2002) The role of the thermohaline circulation in abrupt climate change. *Nature*, **415**, 863–9.

Collins, S. and Bell, G. (2004) Phenotypic consequences of 1,000 generations of selection at elevated CO_2 in a green alga. *Nature*, **431**, 566–9.

Cook, E. R., Woodhouse, C. A., Eakin, C. M., Meko, D. M. and Stahle, D. W. (2004) Long-term aridity changes in the western United States. *Science*, **306**, 1015–18.

Cowie, J. (1998) *Climate and Human Change: Disaster or Opportunity?* London: Parthenon Publishing.

Curry, R. and Mauritzen, C. (2005) Dilution of the northern North Atlantic Ocean in recent decades. *Science*, **308**, 1772–4.

Department of Environment, Food and Rural Affairs (2004) *Evidence and Innovation: DEFRA's Needs from the Sciences over the Next 10 Years*. London: DEFRA.

Department of Environment and Transport for the Regions for the UK Climate Impacts Programme (2000) *Climate Change and UK Nature Conservation: a Review of the Impact of Climate Change on UK Species and Habitat Conservation Policy*. London: DETR.

Department of State, US (2002) *US Climate Action Report: the United States of America's Third National Communication Under the United Nations Framework Convention on Climate Change*. Washington DC: US Department of State.

Domack, E., Duran, D., Leventer, A. *et al.* (2005) Stability of the Larsen B ice shelf on the Antarctic Peninsula during the Holocene epoch. *Nature*, **436**, 681–5.

Dowdeswell, J. A. (2006) The Greenland ice sheet and global sea-level rise. *Science*, **311**, 963–4.

Edwards, M. and Richardson, A. J. (2004) Impact of climate change on marine pelagic phenology and trophic mismatch. *Nature*, **430**, 881–3.

Emanuel, K. (2005) Increasing destructiveness of tropical cyclones over the past 30 years. *Nature*, **436**, 686–8.

Field, D. B., Baumgartner, T. R., Charles, C. D., Ferriera-Bartrina, V. and Ohman, M. D. (2006) Planktonic foraminifera of the California current reflect 20th-century warming. *Science*, **311**, 63–6.

Geological Society (2005) *Super-Eruptions: Global Effects and Future Threats*. London: Geological Society of London.

Gleckler, P. J., Wigley, T. M. L., Santer, B. D., Gregory, J. M., AchutaRao, K. and Taylor, K. E. (2006) Krakatoa's signature persists in the ocean. *Nature*, **439**, 675.

Gunnarsson, T. G., Gill, J. A., Sigurbjörnsson, T. and Sutherland, W. J. (2004) Arrival synchrony in migratory birds. *Nature*, **431**, 646.

Hansen, B., Turrell, W. R. and Østerhus, S. (2001) Decreasing overflow from the Nordic seas into the Atlantic Ocean through the Faroe Bank channel since 1950. *Nature*, **411**, 927–30.

Hickling, R. W., Roy, D. B., Hill, J. K., Richard Fox, R. and Thomas, C. D. (2006) The distributions of a wide range of taxonomic groups are expanding polewards. *Global Change Biology*, **12**, 450–5.

Hill, J. K. and Fox, R. (2003) Climate change and British butterfly distributions. *Biologist*, **50(3)**, 106–10.

Hornbach, M. J., Saffer, D. M. and Holbrook, W. S. (2004) Critically pressured free-gas reservoirs below gas-hydrate provinces. *Nature*, **427**, 142–4.

Hsieh C., Glaser, S. M., Lucas, A. J. and Sugihara, G. (2005) Distinguishing random environmental fluctuations from ecological catastrophes for the North Pacific ocean. *Nature*, **435**, 336–40.

Hulme, M., Turnpenny, J. and Jenkins, G. (2002) *Climate Change Scenarios for the United Kingdom: The UKCIP02 Briefing Report*. Norwich: Tyndall Centre.

Intergovernmental Panel on Climate Change (2000) *Emission Scenarios – IPCC Special Report*. Geneva: WMO.

Intergovernmental Panel on Climate Change (2001a) *Climate Change 2001: Impacts, Adaptation and Vulnerability – a Report of Working Group II*. Cambridge: Cambridge University Press.

Intergovernmental Panel on Climate Change (2001b) *Climate Change 2001: the Scientific Basis – Summary for Policymakers and Technical Summary of the Working Group I Report*. Cambridge: Cambridge University Press.

Jacobs, S. S., Giulivi, C. F. and Mele, P. A. (2002) Freshening of the Ross Sea during the late 20th century. *Science*, **297**, 386–9.

Kennett, J. P., Cannariato, K. G., Hendy, I. L. and Behl, R. J. (2000) Carbon isotope evidence for methane hydrate instability during Quaternary interstadials. *Science*, **288**, 128–33.

Kerr, R. A. (2004) A bit of icy Antarctica is sliding toward the sea. *Science*, **305**, 1897.

Knight, T. M., McCoy, M. W., Chase, J. M., McCoy, K. A. and Holt, R. D. (2005) Trophic cascades across ecosystems. *Nature*, **437**, 880 3.

Laurance, W. F., Oliveira, A. A., Laurance, S. G. *et al.* (2004) Pervasive alteration of tree communities in undisturbed Amazonian forests. *Nature*, **428**, 171–4.

Lovejoy, T. E. and Hannah, L. (eds) (2005) *Climate Change and Biodiversity*. Yale: Yale University Press.

MacArthur, R. H. and Wilson, E. O. (1967) *The Theory of Island Biogeography*. Princeton, NJ: Princeton University Press.

Meehl, G. A. and Tebaldi, C. (2004) More intense, more frequent, and longer lasting heat waves in the 21st century. *Science*, **305**, 994–7.

Miller, L. and Douglas, B. C. (2004) Mass and volume contributions to twentieth century global sea level rise. *Nature*, **428**, 406–9.

Milly, P. C. D., Dunne, K. A. and Vecchia, A. V. (2005) Global pattern of trends in streamflow and water availability in a changing climate. *Nature*, **438**, 347–50.

Ministry of Agriculture, Fisheries and Food, Department of Environment Transport and the Regions, and ADAS (2000) *Climate Change and UK Nature Conservation: a Review of Climate Change on UK Species and Habitat Conservation Policy*. London: DETR.

Monson, R. K., Lipson, D. L., Burns, S. P. *et al.* (2006) Winter forest soil respiration controlled by climate and microbial community composition. *Nature*, **439**, 711–14.

National Assessment Synthesis Team (2000) *Climate Change Impacts on the United States: the Potential Consequences of Climate Variability and Change*. US Global Change Research Program. Cambridge: Cambridge University Press.

Nemani, R. R., Keeling, C. D., Hashimoto, H. *et al.* (2003) Climate-driven increases in global terrestrial net primary production from 1982 to 1999. *Science*, **300**, 1560–3.

Orr, J. C., Fabry, V. J., Aumont, O. *et al.* (2005) Anthropogenic ocean acidification over the twenty-first century and its impact on calcifying organisms. *Nature*, **437**, 681–6.

Overpeck, J. T., Otto-Bliesner, B. L., Miller, G. H., Muhs, D. R., Alley, R. B. and Kiehl J. T. (2006) Palaeoclimatic evidence for future ice-sheet instability and rapid sea-level rise. *Science*, **311**, 1747–50.

Parmesan, C. and Yohe, G. (2003) A globally coherent fingerprint of climate change impacts across natural systems. *Nature*, **421**, 37–42.

Pounds, A. J., Fogden, M. P. L. and Campbell, J. H. (1999) Biological response to climate change on a tropical mountain. *Nature*, **398**, 611–15.

Pounds, A. J., Bustamante, M. R., Coloma, L. A. *et al.* (2006) Widespread amphibian extinctions from epidemic disease driven by global warming. *Nature*, **439**, 161–7.

Rignot, E. and Pannir Kanagaratnam, P. (2006) Changes in the velocity structure of the Greenland ice sheet. *Science*, **311**, 986–90.

Rohling, E. J., Marsh, R., Wells, N. C., Siddall, M. and Edwards, N. R. (2004) Similar meltwater contributions to glacial sea level changes from Antarctic and northern ice sheets. *Nature*, **430**, 1016–21.

Root, T. L., Price, J. T., Hall, K. R., Schneider, S. H., Rosenzweig, C. and Pounds, A. (2003) Fingerprints of global warming on wild animals and plants. *Nature*, **421**, 57–60.

Siegert, M. J., Welch, B., Morse, D. *et al.* (2004) Ice flow direction change in interior west Antarctica. *Science*, **305**, 1948–51.

Smith, L. C., Sheng, Y., MacDonald, G. M. and Hinzman, L. D. (2005) Disappearing Arctic lakes. *Science*, **308**, 1429.

Sparks, T. and Collinson, N. (2003) Wildlife starts to adapt to a warming climate. *Biologist*, **50(6)**, 273–6.

Stott, P. A., Stone, D. A. and Allen, M. R. (2004) Human contribution to the European heatwave of 2003. *Nature*, **432**, 610–11.

Thomas, D. S. G., Knight, M. and Wiggs, G. F. S. (2005) Remobilization of southern African desert dune systems by twenty-first century global warming. *Nature*, **435**, 1218–21.

Toniazzo, T., Gregory, J. M. and Huybrechts, P. (2004) Climatic impact of a Greenland deglaciation and its possible irreversibility. *Journal of Climate*, **17**, 21–33.

Tripati, A. and Elderfield, H. (2005) Deep-sea temperature and circulation changes at the Paleocene-Eocene Thermal Maximum. *Science*, **308**, 1894–8.

UK Climate Impacts Programme (2001) *Climate Change and Nature Conservation in Britain and Ireland*. Oxford: UK Climate Impacts Programme.

US Geological Survey (1999) *U.S. Geological Survey Professional Paper – Estimated Present-Day Area and Volume of Glaciers and Maximum Sea Level Rise Potential*. Reston, VA: US Geological Survey.

Walther, G.-R., Post, E., Convey, P. *et al.* (2002) Ecological response to recent climate change. *Nature*, **416**, 389–95.

Warren, M. S., Hill, J. K., Thomas, J. A. *et al.* (2001) Rapid response of British butterflies to opposing forces of climate and habitat change. *Nature*, **414**, 65–8.

Webster, P. J., Holland, G. J., Curry, J. A. and Chang, H.-R. (2005) Changes in tropical cyclone number, duration and intensity in a warming environment. *Science*, **309**, 1844–6.

Westgarth-Smith, A. R., Leroy, S. A. G. and Collins, P. E. F. (2005) The North Atlantic Oscillation and UK butterfly populations. *Biologist*, **52**, 273–6.

Wu, P., Wood, R. and Stott, P. (2005) Human influence on increasing Arctic river discharges. *Geophysical Research Letters*, **32**, L02703.

7

The human ecology of climate change

Given that climate change has been part of, and has helped shape, the biosphere's development, it should not have been surprising that climate change affected human evolution and humanity's historical affairs (see Chapter 5), or that it will do so in the future. That humanity itself has affected the climate – modern technological civilisation significantly so – all the more demonstrates the interconnections between climate and our species. If we are to begin to assess this relationship's future it is necessary to understand the relevant fundamentals of human ecology (the way our species as a population develops and its relations with other species). This includes population demographics (broadly our numbers reflect the amount of biomass of, and hence impact on, the species to which we relate), energy supply (which relates to carbon-cycle short-circuiting), health (mainly with regard to those species that prey on us) and food supply (those species which we harvest). Finally, given all of this, what are the prospects for limiting, or ameliorating, our short-circuiting of the biosphere's carbon cycle?

7.1 Population (past, present and future) and its environmental impact

7.1.1 Population and environmental impact

Generally, if one person has an environmental impact of some arbitrary quantification, x, then it is not unreasonable to suppose that two people will have an impact of $2x$, three people an impact of $3x$, etc; in other words, that environmental impact is proportional to population and so may be represented in equation form.

One of the first environmental-impact equations widely considered, where impact is proportional to population, was formulated by Paul Ehrlich and John Holdren (1971). It was expressed as:

$$I = PAT$$

where *I* is the impact on the environment resulting from consumption; *P* is the population; *A* is the consumption per capita (affluence) and *T* is a technology or, as it is more recently sometimes known, an inefficiency (inverse efficiency of use) factor.

Today environmental impact is considered in a more complex and multi-factorial way. Most often, impact is not considered in terms of arising from a human population, but occasionally it is considered in this way: one example would be the concept of a community's, or a nation's, ecological footprint. More often than not, impact is considered as relating to a development project, be it an airport or the construction of a road, when an environmental impact assessment is carried out.

Policy-makers rarely consider environmental impact as arising out of human-population terms. Instead, much of the focus is on either the affluence or technological dimensions; that the policy in question will have to generate this much wealth for that much cost and generate this much pollution. Population and attempts to curb population through lowering fecundity[1] are rarely considered as a means to curb impact or frame economic and sustainability policies. There are exceptions, the most notorious being China in the late twentieth century; 'notorious' because China's draconian measures to curb population growth greatly infringed individual freedom and distorted its population's balance of the sexes. However, in the main, up to the early twenty-first century addressing population issues has been low on policy agendas.

More recently there has been a move away from the above lack of interest in population; population concerns have begun to come to the fore. Politicians in the developed nations have begun to realise that current population demographic changes (the changes in a population's composition, such as by age) mean that there will not be enough younger people to care for the increasing number of elderly (and especially partially infirm elderly, for we are living longer more in an unhealthy than a healthy state). The early twenty-first century has seen a number of developed nations, such as the USA, UK, Italy and Australia, actively introduce policies to address this problem, be it through immigration or fiscal encouragements to increase fecundity for couples wanting children. Consequently, following the Ehrlich–Holdren equation, such policy trends are likely to increase environmental impact rather than

---

[1] Up to recently many demographers used fertility to mean both the actual level of human reproduction and the potential for human reproduction, even though these concepts are quite different. More exact terminology is required (especially when communicating across disciplines) and increasingly the ecological term fecundity is used to relate to the amount of offspring a human couple actually have, while fertility is used to reflect the potential to have offspring. It is this way that these terms are used in this book.

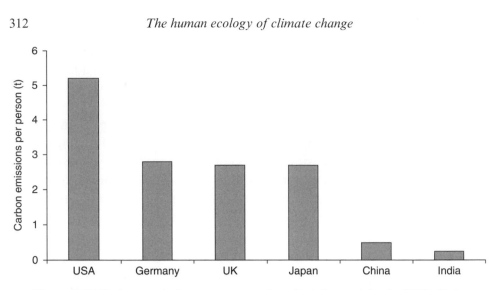

Figure 7.1 Carbon emissions per person in selected countries in 2002. Data from the Worldwatch Institute (2003).

reduce them. Indeed, because developed nations' citizens are more affluent the environmental impact arising from each one is greater than that from citizens from less-developed nations, so actively encouraging a growing developed nation's population will often (but not always) tend to increase that population's environmental impact. However, some academics do present detailed arguments as to why population growth might be seen in a positive light (Simon, 1996) and, as we shall see below, the addition of an extra person to a population can result in either an improvement or a decline in the wealth-to-impact ratio.

As per the Ehrlich and Holdren impact equation carbon dioxide emissions do broadly (but again, not always) relate to affluence in terms of gross domestic product (GDP; the sum total of that population's earnings) and GDP per capita. In the late twentieth century, affluent North America (Canada and USA), with approximately 2% of global population, was responsible for about 25% of global anthropogenic greenhouse-gas emissions and had a GDP of a similar order of magnitude. Conversely, less wealthy China, with some 20% of the Earth's population, generates just 12% of global emissions. The carbon emissions per capita for selected countries in 2002 are depicted in Figure 7.1.

Affluence does not just vary between national populations but within them, and it varies for populations over time. Overall, citizens of the Earth are getting wealthier (see Figure 7.2); although some remain more wealthy than others and for a minority wealth is decreasing. The differences in wealth across the planet do have a bearing on environmental impact but as this relates to international development it is a subject too great to discuss here, other than as

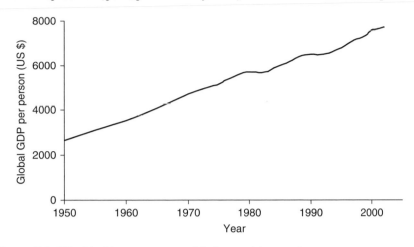

Figure 7.2 Worldwide average wealth (annual income) per person.

a brief reference. Nonetheless, broadly speaking there is some apparent correlation between financial wealth and carbon emissions. For example, as noted, the USA is the biggest economy on the planet and also, with less than 5% of the global population, generates nearly a quarter of all greenhouse-gas emissions, so making it (currently) one of the largest fossil-fuel emitters on both a per capita and a national basis. Whether this apparent correlation will continue remains to be seen. This is because the per-capita carbon emissions of a number of less-developed nations are also increasing, even if they are currently low. Here the concern is that the populations of countries like India and the aforementioned China are large, so that if their per-capita emissions continue to rise then on a national basis this will have a major impact in the way global trends in atmospheric carbon dioxide concentrations change.

Notwithstanding all this, according to our current global society's accounting system (there are others) we are getting wealthier and, for the moment at least, our global greenhouse-gas emissions are seemingly similarly increasing. This overall increase with time in affluence is reflected in parameters such as household size (number of occupants), which is one of the key indicators of a populations' environmental impact, although the relationship is not a linear or a proportional one. The broad reason for this is that a building or apartment consists of a fixed mass, volume and area so that even if the food and water and other consumables used by a household of three are approximately half those of a household of six, the same volume of air and bricks needs to be warmed with both households, irrespective of the number of people living there. In short, there is a decrease in energy intensity per person as the number of a household's occupants increases.

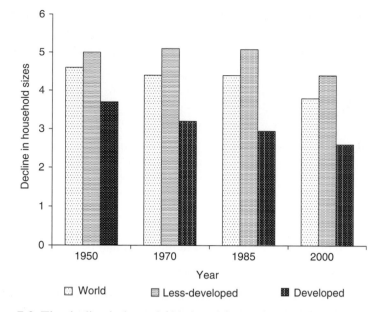

Figure 7.3 The decline in household size. After Keilman (2003).

Furthermore, a halving of household size for a fixed population means that the number of houses required for that population doubles. Take both factors (energy intensity and number of households) together and it can be seen that a halving of household size more than doubles the environmental impact. As globally there is a clear overall trend of increasing affluence, so this is reflected in decreasing household size. This decrease in household size is manifest in both less-developed and developed nations, albeit more markedly in the latter (see Figure 7.3).

Decline in household size is just one aspect of the way affluence generates environmental impact irrespective of population size. Affluence, among other things, also affects annual per capita distance travelled and consumption of material goods, water and food. In terms of greenhouse impact, goods and services each have an inherent energy content (energy intensity) associated with their production and delivery. If this energy comes from fossil fuel then there will be an associated greenhouse cost. This cost might be met (say by chemically capturing the carbon dioxide emitted from the power station (see Chapter 8), through planting new woodland, etc.) and the cost covered by the price of the energy sold. Or, alternatively, this greenhouse cost may not be met, in which case the greenhouse gas escapes to warm the planet. In this instance the environmental cost is not covered by energy sales and so is external to the price of energy. In economic terms such environmental impacts are known as environmental externalities.

Accurately quantifying environmental externalities globally is extremely difficult. There are two main reasons for this. First, assessing the cost of the impact and, second, assessing affluence. Assessing the cost of an impact is difficult because, for example, how does one price the loss of a species let alone an ecosystem? The financial value of a species that has no known economic value is obviously going to be hard. Valuing an ecosystem in one sense is a little easier. Ecosystems have function. For example, a wooded valley will retain rainwater falling on it and release it slowly so providing a more steady water supply through the catchment. The same valley devoid of life with just bare rock will see the water run off immediately, so scouring the lower catchment and exposing it to periods of drought. This particular function of the ecosystem can be part quantified in the value of the water that would have been purchased commercially to replace that lost in the drought as well as the erosion-protection mechanisms to prevent scouring.

This idea of ecosystem services is not purely theoretical. Furthering the above ecosystem-function example, in 1996 New York City invested between $1 and 1.5 billion in ecosystem management in the Catskill Mountains, which provide its main water supply. Up to the 1990s plant root systems, soil structure and its ecology largely served to purify the water to US standards. However sewage, fertiliser use and pesticide use caused water standards to fall. New York either had to restore the Catskill ecology and ecosystem function to what it had been or to build a water-purification plant at an estimated capital cost of $6–8 billion and then annual running costs of $300 million. By comparison the ecosystem-management option was more attractive (Chichilnisky and Heal, 1998).

However, valuing ecosystems can be harder than for species because it is difficult to quantify all the benefits arising out of an ecosystem's function and not every ecosystem has a human population with an easily identifiable ecosystem demand and associated opportunity cost (see Glossary, Appendix 1). This does not mean that some far-flung tract of tropical forest has no ecosystem service to offer, because vast tracts of tropical forests for instance affect the regional climate and the (fast) global carbon cycle.

In short, even though many developed nations theoretically embrace the polluter pays principle, the reality is that much of such costs remain environmental externalities if only because we are not, when undertaking economic transactions, valuing all existing ecological services.

Nonetheless, rough valuation attempts have been made to value ecosystem services globally. These suggest that the global value of services provided by ecosystems, but external to our human economy, is of the same order of magnitude as the 'real' human economy, if not greater. One 1997 estimate was in the range of US$16 $\times$ 10^{12}–54 $\times$ 10^{12} and this range was considered by the researchers to be a minimum one. It does, though, compare with the then global annual economy of US\$18 $\times 10^{12}$ (Costanza *et al.*, 1997) even if some of the assumptions and methods caused some debate in the ecological economics community. Subsequently, in the run-up to the 2002 UN Conference on Environment and Development (UNCED) +10 summit (see section 8.1.9), another assessment concluded that the benefit/cost ratio of conserving the Earth's then remaining wildlife and natural systems was at least 100/1 (Balmford *et al.*, 2002).

Although attempts to quantify the (largely unrecognised) value of ecological services have proven difficult – other than to say that they probably contribute roughly as much value as the conventional economy – it is possible to identify how such services might be affected in a region due to climate change. One such assessment of change in ecosystem service supply due to climate change in Western Europe identified both positive and negative service changes (Schroter *et al.*, 2005). Positive ones included increases in forest area and productivity. Negative ones included declining soil fertility (due to temperature and extreme rain-burst erosion) and increased risk of forest fires. These respectively provide opportunities (biofuel production or increased agriculture) or remove options (such as traditional water use). This assessment particularly noted that among the Western European countries, those around the Mediterranean appeared to be most vulnerable, with multiple projected impacts primarily related to increased temperature and reduced precipitation with resulting loss of agricultural potential.

Assessing affluence (*A* in eqn 1) can be hard because, in the sense of Ehrlich and Holdren impacts, it does not equate with money, although frequently it can approximate to it. Yet the currency of money is at the heart of our economic system. A person's affluence, for example, might be considered in terms of the size of their home (harking back to a previous example), together with diet, material possessions and so forth. Yet property and even food prices across a nation vary, and between nations they vary considerably. The price of a four-bedroomed house in, say, Washington DC, USA, is different to the price of a similar-sized and -constructed house in Timisoara, Romania. The same thing applies to a loaf of bread in the two places. Yet an American living in an analogous dwelling and consuming analogous goods and services to a Romanian will be paying far more in dollar terms than a Romanian, even though the two are consuming a similar level of environmental resources. The Romanian may even pay the same proportion of their financial income for a loaf of bread as the American, and so have similar so-called purchasing parity. The problem comes when the Romanian wishes to purchase goods only sold in dollars, as the dollar is stronger than the lei (the unit of Romanian currency).

To get around this problem economists sometimes use parity purchasing prices.

Even parity purchasing prices are not a perfect way of measuring affluence and all monetary, indeed all other, means have their own limitations and this subject has been the focus of considerable debate among economists for many decades. Any monetary measure of a society's GDP, even compared with making parity-pricing allowances with another country's GDP, does not accurately reflect that society's affluence let alone enable true comparisons with other societies both economically and in terms of environmental impact. This is in part because it is possible to employ someone to create something, or provide a service, that enhances citizen's well-being as well as someone who does not. For example, those involved in oil-spill accidents as well as those who clean them up are both paid, but all these citizens' salaries contribute to their nation's GDP yet clearly their actions, taken together, have a zero effect on well-being and a negative effect on the environment, which has been disrupted. This is somewhat analogous to the problem (see below) as to what size of population the Earth can support. Using genuine progress indicators (such as literacy levels, longevity, health, etc.), as opposed to purely GDP-type indicators (of economic wealth and growth), does lead to a different perspective. For example, one such indicator suggests that genuine progress in the USA remained broadly static between 1970 and 2000, even though US GDP in total has more than doubled, while GDP per capita increased by over 60% (Worldwatch Institute, 2003).

Purely in greenhouse terms (which is not satisfactory for a broad range of other economic purposes) using the emissions of carbon per person, as shown in Figure 7.1, is one way to begin to address the difference in the monetary value of goods globally (note, it is important to check whether it is carbon by weight or alternatively carbon dioxide being used, as the latter is 3.666 times heavier than the former and yet refers to the same amount of fossil-fuel energy consumed). However, because there is no one perfect measure of affluence, its inclusion in the Ehrlich and Holdren equation must be notional, and therefore so must the equation itself. Even so, notions and theories do have merit and can help illuminate issues – in this case that of environmental impact – and so help frame solutions.

The final component of the Ehrlich and Holdren equation is T, the technological factor. This too is difficult to quantify. Ehrlich and Holdren meant it to reflect the way different technologies impart different environmental impacts. For example, generating a kilowatt hour of electricity from coal will impart a greenhouse impact as well as involve part of the environmental impact of the coal mine. Whereas generating the same electricity from a hydroelectric scheme will have a smaller greenhouse impact (there will be a greenhouse impact from creating the cement and the energy used in the dam's construction), there will also be the environmental impact of the volume of dammed water on land use and so forth. So the different technologies have different impacts and the T remains hard to quantify. More recently T has been replaced by some users with a term reflecting the inverse of environmental efficiency of that population. Energy that is produced with high environmental efficiency will have a low inverse value and so contribute to a low environmental impact.

All of the above demonstrates that quantifying environmental impact and hence the cost of global warming is difficult, if not impossible, to assess with any accuracy. Indeed, there is an entire sub-discipline of environmental economics that attempts to come to grips with such problems. As we shall see, this problem affects one type of greenhouse policy whereby permits to release carbon dioxide can be sold by those who have invested in low-fossil-fuel technology to those whose investments are already tied up in high-fossil-fuel technology. Because the estimation of impacts, the identifying of externalities and the allocation of costs and benefits to energy technologies is so fraught with difficulties, any system of greenhouse permit trading is likely to be very complex and fail to reflect properly the impacts. Such systems also can easily lead to encouraging gamesmanship whereby energy firms, and countries, play the system, so undermining their contribution to greenhouse solutions (we shall return to this).

If such difficulties were not enough, to return to the heart of this sub-section's theme, population and environmental impact, the consequences of all of this is that the impact of one population of a similar size and affluence to that of another can be quite different. This is again related to the T, or inverse of the population's environmental efficiency factor. What it means is that similar ecosystems can support different numbers of people depending on what they do. This notion did not begin with, but was developed by, the American demographer Joel Cohen (1995). Specifically, Cohen suggests that it is the choices a society (or a population) makes that affect the carrying capacity of that nation's environment. It can be explained thus. Consider two self-sufficient populations of the same size. The birth of an extra person (in the absence of any other deaths, etc.,) in each will increase those populations by one. Yet that extra person in one population might go on to invent greenhouse-efficient systems or something else that lowers environmental impact without reducing the goods and services generated by that population. In the other population that person might become, to take an extreme example, a tyrant bent on domination through aggression. This last population will divert resources to those ends as opposed to improving environmental efficiency. Such a population lowers the carrying capacity of its environment. The inescapable conclusion in greenhouse terms is that it is the choices we, as a global society, make that determine our ability to handle climate-change concerns, and that there is not some fixed quantity of global level of fossil-fuel consumption that can be considered as ideal. This may seem counterintuitive but can be illustrated by considering a power station that burns fossil fuels whereby all the carbon dioxide generated is isolated through carbon capture (see Chapter 8). Such a station will not through its emissions (though admittedly may in other ways) affect the climate as a normal power station would.

To sum up, although it is extremely difficult to quantify the costs of the global population's greenhouse impact, the level of population and affluence do very broadly relate to the level of impact. However, above and beyond this, it is the environmental efficiency and, specifically, the choices made to realise this efficiency, that the global population makes which affects climate change. This last lies at the heart of much of the next chapter.

7.1.2 Past and present population

It is commonly known that our global population of *H. sapiens sapiens* is currently growing and for many (there are exceptions) this is an accepted and often unremarked-upon fact of modern life. This is in no small part because all those alive today have lived in a world that has seen continuous growth, so that such growth seems unremarkable: after all, the continuation of the line above and below one point on an exponential curve looks just like elsewhere on the same line above and below any other point on that curve. A person today in their late middle age will have seen the global population grow from a little over 2 billion to over 6 billion, while a typical undergraduate student (aged 19 or 20) will have seen the population rise some 25% from over 5 billion (see Figure 5.9). Yet our post-Industrial Revolution period of population growth is not typical.

Regarding early human population size, Joel Cohen (1995) cites the independent estimates for over a million years ago of demographers Edward Deevey Jr and Michael Kremer. They both approximate the then global population as being around an eighth of a million. However, it is known that Kremer based his estimates on those of Deevey Jr for the global population up to 25 000 years ago. After that time their estimates for population differ substantially up to the birth of Christ. After AD 1650 their estimates are more in agreement, being no more than 3% apart each other. Furthermore, from that date their figures are joined by similar estimates from five other demographers and together these present a coherent picture.

If these estimates are to be believed then it would appear that from 1 million to 300 000 years BC the Earth's population of *Homo* spp. grew slowly from an eighth of a million to a million. The rate of increase required for this is of the order of a little over one person a year. However, 1 million to 300 000 years ago was a period of Pleistocene Quaternary glacial–interglacial cycles so that the idea of a fairly linear growth in population is unlikely. More probable is that the global population was reduced at times of extreme climate change (such as at the beginning and end of glacials and interstadials) and grew when the climate was stable. Furthermore, this meta-population of *Homo* spp. saw its various component populations (of *Homo erectus*, *Homo sapiens neandertha-lensis* and *Homo floresiensis*) fluctuate and, of course, there was an overall trend of succession from one to the other, albeit with considerable overlap, before *H. sapiens sapiens* rose (see Chapter 4).

By the time of the domestication of barley and wheat, early in our current Holocene interglacial some 10 000–9000 years ago, *H. sapiens sapiens* was the sole hominid extant with a global population probably somewhere between 3 and 5 million. This increased over the next 7000–8000 years to the birth of Christ. It is not known for certain whether or not this growth was slow and steady, at a fraction of a percent a year, or whether it increased dramatically when and where agriculture replaced hunting and gathering as the primary food source, but the latter is more likely. Either way, this length of time was sufficient for the global population to grow to somewhere between 100 and 300 million.

The next one and half millennia to AD 1500 saw the population roughly treble to below half a billion. But it was after AD 1700, and the Renaissance and then the Industrial Revolution, that the population began to grow at a pre-viously unprecedented rate.

In terms of human population the development of science and its resulting technology have proven hugely successful. Indeed, many of the problems we (as a species and the planet) now face are the problems of success. Namely, how do we maintain a large population that is used to harnessing considerable environmen-tal resources? This success in population is demonstrated by the leap in popula-tion that largely began with the Renaissance and the Industrial Revolution, and can be best seen by looking at a graph of the estimated global population over a few millennia (see Figure 7.4a). Much of the past couple of millennia has been characterised by low population and low population growth by today's stan-dards. But it is clear that since the Industrial Revolution the global population has experienced a radical demographic change. This has been fuelled by

- lower mortality, especially child mortality,
- increased longevity (by around 50% since the onset of the Industrial Revolution) and
- a lag in the decline of fecundity (which counters the lower mortality).

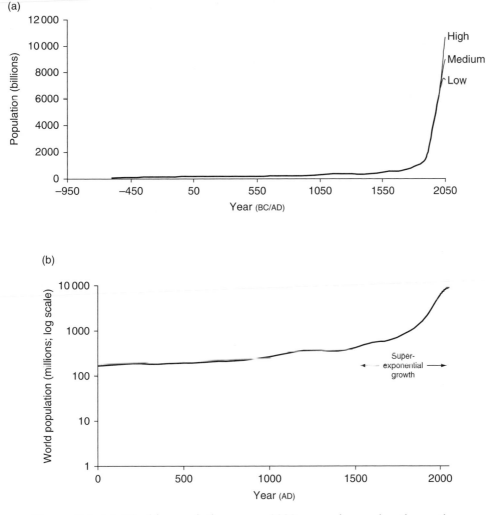

Figure 7.4 (a) World population over 3000 years drawn by the author compiled from various demographic, historic estimates presented in a table by Cohen (1995) and future forecasts from the UN Population Division of the Department of Economic and Social Affairs (2002). The bold line depicts the actual population up to the early twenty-first century. Thin lines depict the future UN low, medium and high projections to 2050. (b) World population drawn using a logarithmic scale. This indicates the rate of exponential growth varying over time. A straight line would represent constant growth and its steepness the rate of growth. Note that the graph is not a straight line but an upward curve through to the late twentieth century, reflecting super-exponential growth for part of that century, after which it becomes less steep, indicating that population growth is slowing down.

The population curve looks almost exponential in nature. (That is to say, that the curve is similar to an equation of the type of $y = x^n$, where n is a constant.) So, another way to look at how the global population has evolved is to plot it on a logarithmic scale. If population growth truly were exponential then such a graph (as in Figure 7.4b) would be a straight line, the slope of which would reflect the rate of growth. Indeed, up to the Industrial Revolution the population on a logarithmic scale does approximate a straight line, which suggests a slow but steady rate of compound growth. However, the time after the Industrial Revolution through to shortly after the mid twentieth century sees this line become an upward curve, indicating that the rate of population growth was itself growing. That this upward curve itself looks as if it too has an exponential shape, signifies that for a time the population was growing at a super-exponential rate (see Figure 7.4b). Not surprisingly, population demographers of the 1960s realised this and it is from this time that modern concerns as to the population explosion (or bomb) came to the fore. Finally, the latter part of the twentieth century saw a return from super-exponential growth to a more logistic rate of growth, where the population stabilised.

Figure 7.4a helps explain a good part of the reason why human impact on the global commons (the atmosphere and oceans) has been most marked in recent decades, rather than in pre-industrial times. Nonetheless, there is an argument that human impact thousands of years ago on the biosphere might be discerned (Ruddiman, 2003) and that much of the current Holocene inter-glacial might be better termed the Anthropocene (see Chapter 4).

From the perspective of the Ehrlich and Holdren equation, the post-Industrial Revolution escalation in human population has had a profound effect on likely environmental impact. Indeed, future population growth will be one of the drivers of future greenhouse-gas emissions.

7.1.3 Future population

The UN forecast for the future of the Earth's population is based on three scenarios: high, medium and low. These are shown at one end of Figure 7.4a, which covers some 3000 years, but are seen more clearly when the Earth's population for the last decades of the twentieth century are plotted together with the UN forecast for the first half of this century (see Figure 7.5).

Between 1980 and 2002, the UN's various global population estimates for the twenty-first century began to decline to their present level. This was because initially the super-exponential growth of the mid twentieth century gave many demographers cause to suggest that higher rates of population growth would continue. Furthermore, it was not fully appreciated how

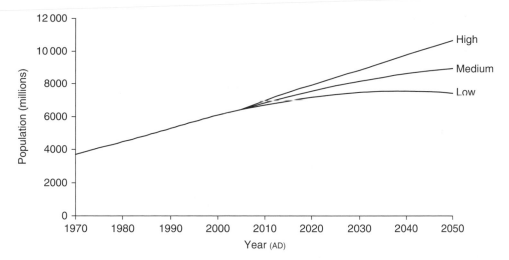

Figure 7.5 The last 80 years of the 3000 years represented in Figure 7.4a. This details the Earth's population of late twentieth century to 2002 and also forecasts (high, medium and low scenarios) the future of population up to 2050 (UN Population Division of the Department of Economic and Social Affairs, 2002).

increased individual wealth would actually reduce fecundity. With increased individual wealth a range of factors come into play. These include increased control over personal reproduction (through late twentieth-century biomedicine), lower childhood mortality (so increasing future parental security), increased aspirations for offspring and, from this last, increased cost of raising offspring to realise such aspirations. Each of these factors serves to lower the need and/or desirability for offspring and so together provide a powerful motivation to lower the rate of population growth. However, past population growth still impacts on the present and the future because typically a birth in one decade will see a contribution to the population for a number of decades (through the individual's lifespan and own reproductive contribution).

Such individual wealth factors show up as drivers of past actual and near-future estimates of wealthy and poor countries' populations. Figure 7.6a depicts the contributions to the Earth's population. The three decades up to the end of the twentieth century show that nearly all of the increase in global population has come from the poorer, less-developed or developing nations and virtually none from the wealthy industrial nations (as might be classified in 1970). However, whereas, according to this line of reasoning, one might expect the wealthier urban population within the developing nations to contribute less than their rural counterparts to population growth, this has not been the case, and it is not considered that it will be in the first half of the twenty-first century. Instead there has been considerable migration in less-developed

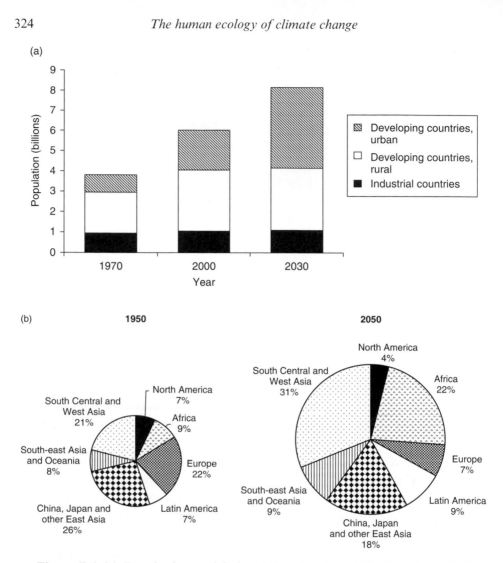

Figure 7.6 (a) Developing and industrial nations' contributions to global population growth (past actual values for 1970 and 2000 and UN medium-scenario future forecast for 2030). (b) Geographical distribution of the global population in 1950 and as predicted for 2050. After Raleigh (1999).

nations, from the countryside to towns, so swelling these developing nations' urban populations.

Figure 7.6b compares the geographic location of the actual global population in 1950 with its anticipated location in 2050. It is expected that by that year North America and Europe, although in absolute number their populations will only fall marginally in the first half of the twenty-first century, will have a markedly smaller proportion of the total world population. The largest increase in world population will take place in regions with high poverty and unemployment (Raleigh, 1999). Irrespective of climate-change concerns, this larger future population will mean that resources will have to be used more effectively and sustainably if human well-being is to be maintained without increasing environmental impacts.

Meanwhile, in a number of developed nations the decline in fecundity of their ageing populations has given some cause for concern as to the size of the future economically active cohort necessary to sustain a larger elderly population. In other words the ratio of active workers to retired people is decreasing. Just one instance of the policy response to such concerns came in 2003 in France where the Government announced an €800 grant to each baby born to a mother from 2004.

With regards to climate change, the growth in developing nations' population has driven, and is expected to continue to drive, an increase in their contribution to global greenhouse-gas emissions. This will be explored in section 7.2, on energy.

7.1.4 Food

Agriculture enabled early civilisations to secure a regular surplus of food. Even so, the negative pressures of disease on the population remained, albeit in a changed manner: the gathering of humans into larger local population centres increased the opportunities for disease transmission even if an improved diet reduced the diseases associated with poorer nutrition. Consequently, despite an agricultural surplus there was not entirely a commensurate increase in early civilisation's population. Indeed, the agricultural surplus of such early civilisations was not great, and although more of this surplus could have supported more children, hence greater population growth, instead it was largely used for other purposes. For instance, Egypt, during the time of the Pharoahs (2780–1625 BC), had a population of around 3 million, of whom 95% were engaged directly in food production, so being able to support a smaller part of the population (5%). This consisted largely of slaves, some military as well as a small aristocracy. Calculations suggest that at that time the 20-year construction of the Cheops pyramid absorbed nearly all of this agricultural surplus. In later periods the emphasis of support from this proportionally small agricultural surplus shifted from slaves to the military. At other times there was little agricultural surplus and these are known to be times of social strife (Cottrell, 1955).

The historic period (the time since written records began) has seen a gradual accumulation of knowledge and technical ability. This enabled there to be a similar development of the aforementioned small agricultural surplus and so gradually this freed more of the population for other endeavours. Virtually all these early societies relied on (non-fossil) biological energy and as such their impact on the biosphere's global commons was negligible and, in terms of climate, was swamped by other variables such as solar and volcanic climate forcing.

For thousands of years plant and animal breeding contributed hugely to the success of early agrarian societies. Prior to the Industrial Revolution, under-standing the need to lay fields fallow and the use of animal and vegetable wastes as fertiliser improved agricultural productivity wherever such practices were conducted. The Industrial Revolution itself enabled agriculture to slowly become more mechanised, so reducing the need for such a large rural work-force. It also enabled agricultural goods to be shipped in bulk over long distances. Finally, the mid twentieth-century development of fertilisers and pesticides in the Green Revolution further increased agricultural productivity to levels comparable with the present day.

By the end of the twentieth century in industrialised countries the propor-tion of the population directly employed in agriculture had become a small minority. In the UK, which was then nearly three-quarters self-sufficient in food, just 2% of the population was directly employed in agriculture, con-tributing to about 1% of GDP (Cabinet Office, 1994), with food processing, packaging and transport contributing an additional 2% of GDP. Allowing for the overseas contribution to current UK food supply, a reasonable estimate for the total equivalent proportion of the UK population involved in food production, processing and transport is 7–8%. Although this excludes retail and catering commercial activities, it is still a dramatic contrast to ancient civilisations populations' involvement in food production. Today 92% or more of many industrialised-nation populations are free to engage in activities other than basic food supply, which is almost a complete reverse of the position in ancient Egypt.

What enabled this radical change from ancient to modern societies to take place was the increased contribution made by energy. Indeed, with regards to nitrogen fertiliser (which commonly comes in the form of urea), fossil fuel is consumed in two ways. First, in the energy-intensive fixing of nitrogen from the atmosphere. This has in the past mainly been done using energy derived from fossil fuels, which are finite or so-called fund resources (the fund being finite). However, in the future nitrogen fixation might be done by energy from renewable (flow resources) and/or nuclear alternatives. Second, fossil fuel is used to provide urea's carbon content (mainly from natural gas, methane). With regards to this last, the prospects of not using fossil carbon for fertilisers are limited, although biogas is a part possibility.

The provision of tremendous quantities of energy has today resulted, in the industrialised world, in our being used to supermarket goods being largely independent of the seasons and consisting of produce from around the planet (i.e. with high food miles). Furthermore, as the global economy becomes wealthier, so the meat content of the average human diet increases: the point being that meat is an inefficient food source in both energy and land terms compared to humans consuming grain directly, rather than through feeding the animals that we in turn consume. World meat production per person has increased more or less linearly throughout the twentieth century through to the present day, from 17.2 kg per capita in 1950 to 39.0 kg in 2002 (FAO, 2002).

Consequently, the food dimension is not just a fundamental part of human ecology and the way we as a species relate to other species, it is also an important driver of human activity – which depends on a societal food surplus above subsistence – as well as being a sector that consumes energy.

This energy relationship is two-way. Not only does the agricultural production of food, its processing, storage and delivery consume energy which, if fossil fuel, in turn has a climate impact, but these climate impacts also impact back on agriculture. For example, the vagaries of our Holocene climate have affected European agricultural activity (see Chapter 5), and so future climate change is also likely to affect agriculture (see Chapter 6 and likely US impacts).

If human food supply is fundamental to the nature of human activities, if it is dependent on fossil energy and if food supply is shaped by climatic patterns then (against a backdrop of an increasing and unsurpassed global population) what of the question of food security? Will future climate change improve or weaken the global security of food supply? This question will be addressed in a section later in this chapter.

7.1.5 Impact on other species

Because the Earth's population is now so large, and the commensurate need for food is so great, food production dominates much of the most biologically productive temperate land. Together with the land required to house, transport and provide a location for economic activity, this has seen our species take much land away from purely natural systems. Indeed the (perceived market) value of economic activity can be such that it surpasses that of purely agricultural value, which in turn leads to housing and urban settlements being constructed on fertile land. As mentioned in Chapter 4, the extent of this domination over natural systems is such that it is estimated that just under a third of global terrestrial net primary production (the biomass arising from the energy trapped by photosynthesis) is appropriated for human use (Imhoff *et al.*, 2004). Another perspective on the degree of human management of global ecosystems is provided by the amount of nitrogen fixed by humans (primarily for fertilisers) which already exceeds that fixed globally through natural means, or the fact that over half of all accessible fresh water (i.e. excluding sources such as the Antarctic ice cap) is put to use by humanity (Vitousek *et al.*, 1997) or that 10–55% of terrestrial photosynthesis production is managed by humans (which includes not just food for humans or farm animal feed, timber, biofuels, etc. but also amenity value and active wildlife conservation management; Rojstaczer *et al.*, 2001). Such natural resource management has necessitated the modification of many ecosystems and this in turn has resulted in a high degree of fragmentation of the natural landscape (see the previous chapter). Again it is important to remember that this fragmentation is unique to modern times and so will greatly exacerbate the impact of climate change on biodiversity compared to nearly all other times in the Earth's history.

At this point it is worth considering the benefits of intensive agriculture. Intensive agriculture has had some bad press. Whereas much of this may well be justified there are also clear benefits. Of course, land that is used to produce food intensively for agriculture is almost certainly to have a negative effect on biodiversity, if only because primary production will be geared to that of nurturing food species at the expense of others. What is less considered is the effect of a plot of intensive farming on the biodiversity of the landscape as a whole. Which would be better: 1 ha of natural environment next to 1 ha of highly intensively farmed land or 2 ha of far less intensively farmed land? The exact answers tend to be site-specific within the broader landscape and in no small part dependent on the natural environment and the type of intensive farming being considered.

Bringing the above into context with the earlier sections on human population, the past 100 years have seen a tremendous increase in the demand and consumption of food, commensurate with our increasing global population. If farming had not become more intensive over this time then inevitably more land would have been switched from being habitat for wildlife to agriculture and the (potential) impact of climate change on natural systems more acute (see Figure 7.7).

The impacts of this competition on other species, for and with natural systems as well as on land and related landscape fragmentation, are many and in the main largely negative. This topic itself is so great that it is the focus of a considerable body of work that is far too great to meaningfully summarise

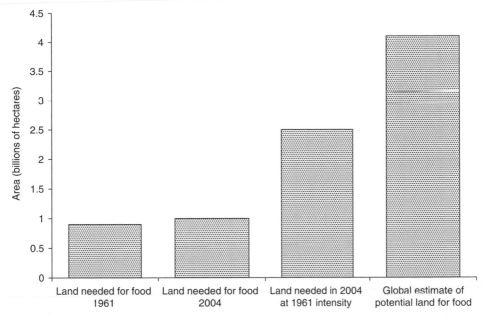

Figure 7.7 Land for global food production; cropland versus intensification. After Mooney *et al.* (2005).

here. However, it is worth citing some recent work suggestive of the scale of the problem.

In 2005 a team of researchers led by Gerado Ceballos of the National Autonomous University of Mexico conducted the first global analysis of the conservation status of all known land mammals with regard to their spatial distribution. The International Union for the Conservation of Nature (and Natural Resources; IUCN; now also known as the World Conservation Union) and the United Nations Environment Programme (UNEP) have for many years identified and/or sponsored the identification of endangered species as well as listing protected areas, but this is different to a global all-mammal analysis of spatial distribution. The Ceballos team compiled the geographic ranges of all 4795 known species of land mammal. They then linked this to cells, rectangles of $10\,000\,\text{km}^2$ each. They found that 1702 cells of land would conserve at the very least 10% of the range of each species; this 10% is a critical threshold that many conservation biologists use as the minimum of a species' natural range needed to conserve a viable population. The spatial detail of the 1702 cells is, unfortunately, too coarse for it to be used in practical conservation management. However, 1702 cells is roughly equal to 11% of the Earth's ice-free land surface and provides an indication of the scale of the problem. That around 80% of the critical 1702 cells are already used for agriculture demonstrates the conservation impression imposed by human food demands. Conversely, looking at some 250 conservation scenarios, 95% of these cells can be replaced by other cells without loss of conservation value, and only 135 cells (<1%) were irreplaceable – that is, were required in all the 250 conservation scenarios. Of these, those located within wealthy countries (that have the resources for implementing developed, sustained, conservation programmes) are most likely to succeed, whereas extinction risks are higher for cells in less-developed nations.

Another key question is how, as opposed to where, should finite conservation effort be focused? A long-running debate in conservation biology has been whether efforts should be area-based (say, conserving a region as a national park or World Heritage Site) or based on a species: such are often called flagship species (a famous example is the giant panda, *Ailuropoda melanolueca*). Marcel Cardillo, Georgina Mace and colleagues (2005) looked at 4000 species of terrestrial mammal. They looked at three principal groupings of predictors of extinction risk: (1) environmental factors such as size and range of species, (2) species' ecological traits such as population density and (3) species' life-history traits such as gestation period and weaning age. They found that extinction risk exhibited a positive relationship to a species' body mass. Importantly, they noted that extinction risk increases sharply when a species' adult body mass exceeds 3 kg. Second, whereas extinction risk in small species is driven by environmental factors, in larger species it is driven by both environmental factors and the species' intrinsic traits. Consequently, they conclude, 'the disadvantages of large size are greater than generally recognized, and future loss of large mammal biodiversity could be far more rapid than expected' (Cardillo *et al.*, 2005). Interestingly, this is in line with separate work done on mass-extinction events demonstrating that it is the large species that are most vulnerable in such events (Hallam, 2004).

Linking such research to the spatial manifestation of climate change is likely to receive greater attention in the coming years and no doubt will continue to do so as geographic climate models become more detailed with regards to both climate and geography. Meanwhile, competition for land for humans as opposed to habitat for wildlife continues irrespective of climate change, which is a compounding factor.

7.2 Energy supply

Over half of global warming in recent decades is attributable to anthropogenic carbon dioxide (see Figure 1.2) and the clear majority of this comes from the burning of fossil fuels (see Chapter 1). Consequently, the fuels we use to supply commercial energy are central to climate-change policies.

7.2.1 Energy supply – the historical context

Ultimately nearly all energy on Earth – except for hydrogen (and perhaps theoretically potential exotic energy such as zero-point) – comes from the Sun and other stars. Most mechanical energy on the Earth arises in one, or a combination, of two ways: first, from the solar energy driving movement in the atmosphere, wind and in turn waves, waving tree branches and so forth; second, gravitationally through Earth–Moon interactions and, of course, motion within the Earth's gravitational field.

Solar energy is also trapped by photosynthesis and stored chemically. Chemically stored energy may be harnessed by life itself through the various trophic levels of ecosystems or geologically trapped as fossil fuels. Geothermal energy (in no small part) comes from the decay of radionuclides deep in the Earth: the Earth would have cooled to a solid long ago if its internal heat arose solely from its formation, notwithstanding the energy arising from the Moon's tidal drag. These radionuclides were formed from supernovae and so again have stellar origin.

Humans as a species are unique in that it appears that we learnt quickly to use fire to supplement the biological energy of our own bodies. However, for many millions of years for this supplement to our energy we were reliant on basic biofuels such as wood and animal oils and dung. Even so, the use to which this energy was put was almost invariably confined to cooking, space heating and illumination, and not motion, or information processing and transference. As discussed in the previous section, this meant that the energy surplus which early societies had above the level of subsistence was small and largely dependent on bio-energy in the form of the excess food the early agrarians could provide to support others engaged in non-food tasks. Mechanical energy was harnessed from animals, as well as wind and water, and this too was a substantial contribution to the energy surplus enjoyed by early *H. sapiens* over that in a hunter–gatherer existence. The benefits of harnessing energy are clear. This leads to two ultimate societal consequences.

First, to realise these benefits there was and still is a drive to increase energy consumption. If such consumption increases globally in an uninterrupted fashion over time then sooner or later the magnitude of this energy surplus will affect the biosphere's thermodynamic balance. This is inescapable. It is exactly what is happening with anthropogenic global warming from fossil-fuel burning and its resultant climate change, although – almost ironically – it is not the energy surplus itself that is causing the problems but the thermodynamics of its chemical carriers. Prior to this, energy-related environmental impacts existed but were not climatically significant. For instance, the early human use of wood as a fuel and construction material impacted on ecosystems and species as woodland was cleared and replaced by fields, but did not impact as it does today on global energy flows.

Second, given clear benefits, it follows that there must be equally clear disadvantages to a society that has enjoyed a certain level of energy surplus if that surplus is reduced. We tend to think of energy crisis as a modern problem but this is not so. Early societies were constrained by the limitations they had to harnessing and deploying energy resources. Consider what the Egyptians would have done had their society's surplus been 10 or 15% rather than just 5%. More recently, in seventeenth-century England there was an acute shortage of fuel wood and charcoal. This led to more than a century of iron shortage. By the late nineteenth century the use of wood as a major fuel had declined to negligible levels and coal dominated. However, coal had its limitations and, while suitable for steam locomotives, was not nearly as suitable as liquid fossil fuel, oil, for personal transport. That fuel made a major contribution to society's global energy budget in the twentieth century.

Our species' use of energy has therefore been characterised by fuel-switching at an ever faster rate. Aside from the various early biofuels and the early use of wind and water, this fuel-switching is clearly seen from the advent of the Industrial Revolution to the present day in the country where the revolution began. Figure 7.8a depicts the proportional contribution of various fuels made to Britain's energy budget. It can also be seen in other developed nations outside of Europe but with a delayed start and hence in a contracted form. Figure 7.8b depicts the proportional contribution of fuels to the US energy budget. These show that in the twentieth century it took the order of three or four decades to develop a new fuel stream that made a major proportional contribution to a nation's energy budget. Such lead times need to be considered when contemplating future energy resources.

The global energy picture that emerges has been one of diversification. There was a proportional decline in the importance of coal over the last half of the twentieth century, from being the dominant fuel accounting for a little over 60% to less than 25% by the end of the century (see Figure 7.8c). The proportion of global energy demand met by oil over this time increased from under 30% to around 46% by 1980 when price and energy-security concerns began to cause a shift to other energy sources such as nuclear power and gas. Most of the global energy diversification has been at the expense of coal, much in the way that fossil fuels replaced wood in the USA and much of the New

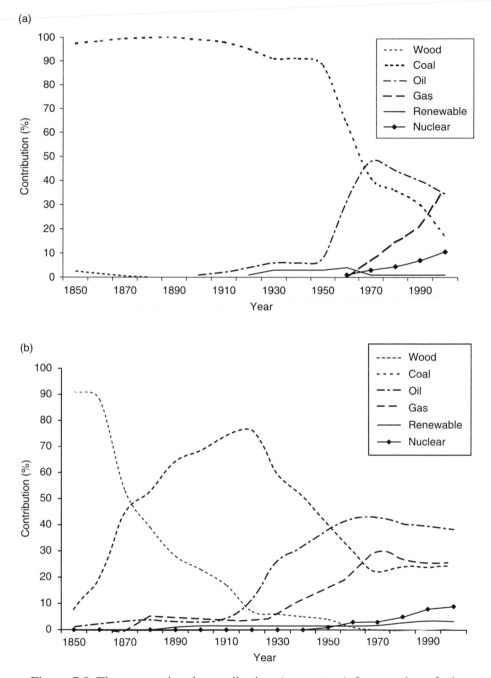

Figure 7.8 The proportional contribution (percentage) from various fuels made to the energy budget of the (a) UK and (b) USA. (c) Summary of the proportional contribution (percentage) from various fuels to global energy supply. Before 1950 theoretical estimates are used and after 1950 data are based on estimates derived from commercial figures with a 5-year filter; see Appendix 3 for sources.

(c)

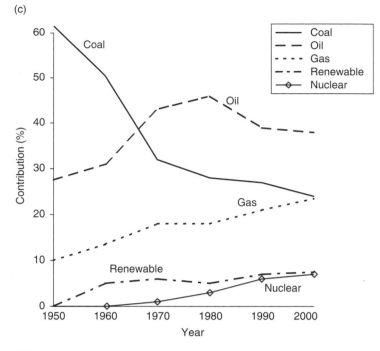

Figure 7.8 (cont.)

World and a little later less-developed nations in the late nineteenth and early twentieth centuries. However, individual energy-resource longevity (the economically recoverable resources) will, as we shall see in the next chapter, determine the economic sustainability of respective fuels.

At the end of the twentieth century nearly all countries were dependent on fossil fuels for the majority of their energy needs, although nuclear and renewables energies, as a proportion of the total, were beginning to grow.

Although a look at the breakdown of a nation's (or the global) commercial energy budget informs as to fuel-switching and is suggestive of lead times, it does not indicate how much carbon dioxide will be generated in the future: it all depends whether the total fossil energy consumed remains constant or whether it increases or decreases. In fact, commercial energy consumption overall has been increasing globally with time and markedly so (see Figure 7.9). This exerts a pressure to increase fossil-fuel consumption. Indeed, by the end of the twentieth century most of our global commercial energy consumption (around 87%) came from fossil fuels, this having increased by over 350% since the middle of the century. Even though the contributions from nuclear and renewable energy have grown slowly, their growth – as a proportion of the global commercial energy budget – has

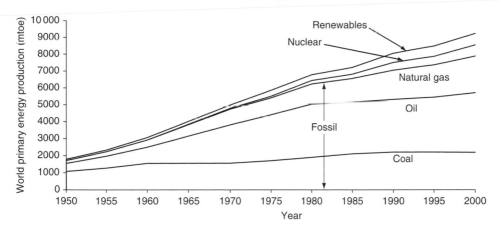

Figure 7.9 Broad summary (5-year filter) of global primary energy production/ consumption in million tonnes of oil equivalent (mtoe). Because of the 5 year filter short marginal dips in production/consumption (such as 1979–82) do not show.

increased the annual non-fossil contribution only marginally compared to fossil exploitation in absolute terms. Consequently there has been a commensurate rapid increase in carbon dioxide emissions in the latter half of the twentieth century (see Figure 7.10).

Reflecting the nature of human population growth (and the Ehrlich and Holdren equation, linking environmental impact to population), most of the nineteenth- and early twentieth-century carbon emissions were made by the industrialised countries and little from the less-developed nations. In the twentieth century the populations of the developed industrialised countries had largely ceased to grow (Figure 7.6), whereas those of the less-developed nations grew markedly. Consequently, by the late twentieth century the position reversed and more than half the emissions were coming from the less-developed (or developing) nations.

As we shall see in the next chapter, this has coloured international policy discussions. On one hand some have argued (as did the USA around the turn of the millennium) that, as the less-developed nations were contributing more carbon than the developed nations, these countries needed to either demonstrate a commitment to cutting emissions or to make some other contribution to the international effort to curbing greenhouse-gas emissions. On the other hand (argued at the time mainly by the less-developed, or developing, nations) the industrialised countries need to carry the brunt of costs for international greenhouse controls as, on a per-capita basis, each person in an industrialised

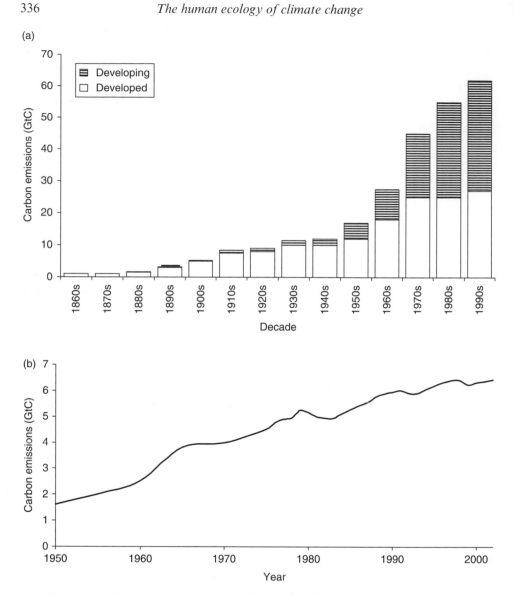

Figure 7.10 (a) Outline of developed and developing nations' emissions of carbon (GtC) by decade from fossil-fuel burning and cement production. (b) Annual global emissions of carbon (GtC) from fossil-fuel burning since 1950 – the period of most emissions. Source: see Appendix 3. Annual emissions prior to 1990 were less than 6 GtC and emissions prior to 1850 less than 0.1 GtC (IPCC, 1990).

developed nation contributes far more than a person in a less-developed nation (see Figure 7.1 and relate that to Figure 7.6). Remember that North America (USA and Canada), with roughly 5% of the global population, was contributing about a quarter of global emissions. The argument here is that greenhouse-gas emissions need to be tackled on a per-capita basis. Such policy concerns will be discussed further in the next chapter.

Finally, the Intergovernmental Panel on Climate Change (IPCC, 1990) estimates that the total emissions of carbon from fossil fuels between 1850 and 1987 were around 200 GtC ($\pm$10%). This is nearly double the IPCC's estimate of 115 GtC ($\pm$10%) for carbon release due to historical changes in land use (deforestation, etc.). This, and that two-thirds of fossil-fuel emissions over this 120-year period took place in its last three decades, explains the importance of fossil-fuel carbon releases to climate change and why this climate forcing has dominated matters only towards the end of the twentieth century. With the prospect of ever larger fossil-fuel releases in the twenty-first century, this means that the thrust of twenty-first-century policies to control atmospheric concentrations of carbon dioxide needs to focus on the energy dimension (see Chapter 8). It also explains why the political manifestation of climate concerns became more marked at that century's end, with more severe impacts anticipated in the twenty-first century, even though Svante August Arrhenius identified the problem as far back as 1896.

7.2.2 *Future energy supply*

Nobody can predict the future. However, we do know with certainty the historic pattern of energy production and the present situation. We also know, as described earlier in this chapter, the likely future population and economic trends. Overall this suggests that, barring major disruption of our global society, demand for energy will increase. The question is, if energy demand is to increase, how will this be met and what will this mean in terms of carbon dioxide emissions?

The IPCC's 2001 reports and their forecasts for twenty-first-century global warming are based on a series of future scenarios it developed in 2000: the *Emission Scenarios – IPCC Special Report*. These scenarios vary from their lowest (which forecasts marginal growth in emissions to the middle of the century followed by a decline to just below 1990 levels) to their highest (which predicts an increase in emissions up to five times 1990 levels by the end of the century). Again remember, carbon dioxide accounts for over half of anthropogenic warming (see Figure 1.2).

The pattern the IPCC emission scenarios for methane, the next major greenhouse gas contributing to anthropogenic warming (about 15%), proportionally follows that of their carbon dioxide scenarios. This is because it too is either population-driven (needing an increased area of rice paddies and number of enteric fermenting animals) and/or energy-related (needing more gas drilling and venting). However, in the future this may not be so, as the technical and energy options can vary.

Whatever the future, the IPCC predicts a range of possible futures and it is this that drives their core warming forecasts (see Table 5.1). So, what are the energy possibilities and options? Excluding environmental costs, the cheaper options (excluding long-term environmental and social costs) include the Business-as-Usual scenario; that is to say, humans broadly continue the present pattern of energy-resource consumption. The reason for this is that no extra cost is required to develop and implement new energy technologies. In this scenario we would continue to rely heavily on fossil fuels as long as reserves will allow throughout the twenty-first century. The caveat about reserves allowance is an important one. For whereas there is plenty of economically recoverable coal worldwide, over two centuries' worth at current rates of production, there is less oil and gas. From the beginning of the twenty-first century British Petroleum (BP, 2005) estimated that there was enough commercially available proven reserves of gas to last a little over six and a half decades at 2004 levels of production, and only enough oil for around four decades; of these, nearly two-thirds are found in the Middle East. It needs to be remembered that the pressures for global consumption of these resources to increase are plainly evident, as discussed earlier in this chapter.

Of course, should prices change (as they will), today's economically recoverable reserves may not be so tomorrow. Indeed, with increasing scarcity prices will rise, so enabling reserves that are currently either harder to get at, or more expensive to refine (such as sulphur-rich deposits), to be economically recovered. Then again, future innovation and new technology may enable reserves that were previously uneconomical to recover to become economical. Furthermore, as prices rise so non-fossil technologies, formerly considered too expensive, will become economically viable. Finally, proven reserves are a subset of total reserves, which includes deposits yet to be discovered as well as exotic sources (possibly such as marine methane hydrates) that are not usually considered. In short, estimates as to ultimate economically recoverable reserves vary.

To get a feel as to how these reserve caveats work, consider the analogous question as to how long US food stocks may last. One might say that adding up all the food in people's homes, shops and warehouses, and dividing it by the rate of consumption, might give a figure of, say, x months. However, in addition to this there will be crops growing in the fields not yet harvested. There will also be the food wasted due to, say, combine harvesters missing some of the crop, mishandling or poor storage. Then there are species of food plant outside of the economic system, from apple trees in people's gardens to raspberries growing in the wild. There are also exotic food species, that is to say perfectly edible plants and animals, that do not normally feature in most Americans' cuisine. Finally, there will be species not normally edible that will require some sort of processing before they can be used. In short, what can be considered the total US food stock depends on what exactly is being considered. So, such an estimate, when expressed in terms of a reserve/production (r/p) ratio, is therefore not as exact a figure as one might think. Food in the USA will not run out on the last day of the xth month. Quantifying oil reserves is fraught with analogous difficulties and so estimates have to be viewed with extreme caution.

An example of how reserve assessments can be misleading comes from the Kern River oil field. It was originally discovered in California in 1899. Calculations in 1942 suggested that 54 million barrels of economically extractable oil remained. However, in subsequent years some 736 million barrels were recovered and a new estimate later in the century suggested that 970 million barrels remained. What had happened was not that the oil field had changed but that knowledge and technical ability had increased (Maugeri, 2004). Having said this, it does not mean that oil is an inexhaustible resource. Nor does it mean that current estimates as to ultimate economically extractable reserves do not so badly reflect the true position: petroleum geology and assessment are today far more advanced than in the mid twentieth century. However, this example does demonstrate that estimates have their limitations.

From this it can be seen that r/p ratios are at best only a broad indication of future potential. However, with regard to fossil fuels, they do suggest a magnitude of resources for continued, if not increasing, exploitation throughout the twenty-first century irrespective of whether or not they fully meet energy demand. They can tell us whether the resource is likely to be a major economic one for years, decades or longer, after which it will become more scarce/expensive, or be less economically competitive, and perhaps indicate a time when other (possibly more expensive) energy resources may become more competitive. In greenhouse terms r/p ratios are suggestive of the *minimum* time in which under Business-as-Usual fossil resources are likely to make major contributions to atmospheric carbon dioxide. Of course, 'business' is unlikely to remain 'as usual' the further into the future one projects a forecast and so no meaningful estimate can be made as to fossil-fuel consumption in the twenty-second century or beyond. Among the various wild cards are the roles the various alternative resources will play.

Alternative (non-fossil) energy resources are in the main currently more expensive than fossil fuels and/or have other problems associated with them. Renewable energy from so-called flow resources such as tidal, wind and wave power, have a lower energy density and so require a larger area for their exploitation, hence a large land-use impact ('flow' because unlike 'fund' resources, there is a continual energy flow to be harvested). Conversely, finite fund resources such as fossil fuels and uranium have high energy intensities. Biofuels which, although carbon-based, do not short-circuit the deep carbon cycle, do take up productive land that could be used for food crops. This needs to be taken into account. We need to remember, against a backdrop of a growing global population, that land for which there will be an increasing food-crop demand has competing use options. This is to say that biofuels have a significant biological opportunity cost, this being the cost of a venture in terms of lost opportunity of another venture. In short, agricultural land can be used for food or energy crops, or even a mix, but one will always tend to offset the yield from another. There are also non-biological opportunity costs, for the agricultural land might be sold and used in a non-biological way such as to provide a site for a factory.

Other than sources of renewable energy, the other principal energy alternative to fossil fuels is nuclear power. The cost of nuclear generation is generally considered at best comparable with the cleanest and most efficient of coal stations but is usually more expensive. Its advantages are that it furthers a nation's energy security and, of course, has a small greenhouse impact. Its disadvantages (other than cost) are that it has other environmental impacts, chief among which is the problem of long-term disposal of radioactive waste. This is a financial unknown and has not generally been completely solved to public and political satisfaction. Further, much of the unknown component of the cash cost of long-term disposal will not be incurred until many years into the future so that, unless a provision is made today to put aside some of the income from the sale of nuclear-generated electricity, there will be an intergenerational transfer of externality ('externality' because this cost is external to the price of the electricity sold). In this sense nuclear power is similar to power arising from fossil fuels in that the consequential global warming generates an environmental impact with economic consequences for future generations to bear. Economists therefore say that both fossil-fuel use and nuclear power currently have environmental externalities associated with them that in turn have intergenerational transfer consequences. (The issue of nuclear waste will be briefly touched upon in the next chapter.)

Then there is the question of how much fossil carbon is used by nuclear power. It has been said that nuclear power is not a zero-carbon energy source as fossil-fuel energy is required to mine, transport, process and refine (enrich) uranium ore. This is true. While it is not this book's purpose to look at energy resources in detail, because of the biological and human ecology consequences of climate change it is necessary to at least examine them in a broad-brush way to ascertain the approximate greenhouse-gas-emitting potentials of differing energy resources. Fortunately, comparing UK and France's energy consumption and their respective fossil-fuel/nuclear balance can make a very simplistic but informative assessment. This comparison is all the more valuable because France and the UK are geographically close, share the same European economic grouping, have a similar population size (60.6 million and 60.4 million respectively; CIA estimate for 2005; www.cia.gov/cia/publications/factbook/index.html) and have a similar annual total (fossil and non-fossil) energy consumption, which for 2004 was 262.9 and 226.9 mtoe respectively (British Petroleum, 2005). Yet France consumes some 29% (60.4 mtoe per annum) less fossil-fuel energy than the UK. This fossil saving comes from France's nuclear-power programme which is far bigger than the UK's. That there is a fossil saving means that France itself is not expending more fossil energy (offsetting the fossil savings) in processing or enriching uranium ore or processing nuclear waste. It can be argued that there is a considerable amount of energy expended in mining and transporting uranium ore. Looking at nations from which uranium ore mining takes place it is difficult to discern any major difference in fossil-fuel consumption on a per-capita basis above that of their economically comparable neighbours. This is not to say that there is no energy expended in uranium extraction and refining but that it does not put anywhere near as heavy a fossil-fuel burden on that nation as nuclear power saves. Indeed, one estimate, admittedly assuming efficient fast-breeder reactors, of a low-grade uranium ore (Chattanooga black shale, with just 60 g of uranium/t of shale) has an energy equivalent to 1000 t of coal (McMullan, 1977). Higher-grade ore is better still. So, even without using breeders, we can see that a tonne of higher-grade uranium ore is worth thousands (or tens of thousands for even higher grades of ore) of tonnes of coal, hence the energy used for mining and refining it is nearly proportionally less per tonne of fuel (hundreds or thousands of times less, as uranium is more energy-intensive to refine than oil). This was reflected in a 2006 estimate provided by a UK 'green' policy forum, the Sustainable Development Commission. It cited an estimate for carbon emissions from (non-fast-breeder) nuclear power as being 4.4 tC GWh^{-1} compared to 243 tC GWh^{-1} for coal. In short, while nuclear power does not have a zero-fossil carbon cost it does have a very significant net fossil carbon-saving impact. The exact magnitude of such savings will become more certain in the coming decade as assessments are made in the run up to the second phase of the Kyoto Protocol (see Chapter 8) that is due to begin in 2012.

The other difficulty with the nuclear option is that civil nuclear power can be used to further military nuclear technologies that in turn undermine international stability. This is likely to become an increasingly pertinent issue as developing nations on one hand legitimately seek to become more energy sustainable, and so explore the nuclear option, while on the other they may be tempted by the international leverage a nuclear military option confers. (Iran in the years subsequent to 2003 is but one example.)

There is also another similarity between nuclear fission and fossil-fuel use, which is that strictly both are finite resources or fund resources. This is not always appreciated. Currently the level of the nuclear contribution is such that there is enough known uranium-containing geological strata to fuel our current nuclear power industries for over a century, and more considering that uranium prospecting has not been as intensive as that for oil. This broad estimate does not take into account fast-breeder options that would greatly enhance uranium's longevity as a resource. But equally, as mentioned, this comes with greater technological as well as military security challenges (see the next chapter's subsection on the prospects for nuclear power). Furthermore, since the mid 1980s (specifically the Chernobyl power-plant disaster in 1986) the growth of the nuclear industry has slackened markedly. If it were internationally resumed in a major way for the long-term then global uranium resources would need to be reassessed.

With regards to the global commercial energy budget, the factors of resource availability (or scarcity) and cost compete. Together these largely determine the global energy resource mix, the fossil contribution of which will determine anthropogenic carbon releases from energy (sequestration technologies aside). Although reserves of overall fossil fuels (coal, oil and gas) are enough to theoretically enable carbon emissions to increase throughout the twenty-first century, the low r/p ratios for gas and oil suggest that the consumption of at least oil and gas might peak early in the twenty-first century. Estimates vary but there is a rough consensus from a number of analysts suggesting that the peak consumption of oil and gas might be around the middle of the twenty-first century (Bentley, 2002). The options for coal are less restricted and this could make up for any shortfall in gas or oil should additional alternatives fail to be developed.

In short, such is the likely magnitude of future energy demand that unless non-fossil energy dominates the global energy mix, fossil fuels are likely to continue to potentially add significantly to the atmospheric reservoir of carbon throughout the twenty-first century and well into the next century. Given that to stabilise temperatures we would need to lower greenhouse-gas emissions by around 60% of their 1990 level, and that the atmospheric residence time of a carbon dioxide molecule is between one and two centuries, it seems unlikely that global warming will cease over the next few centuries. Further, in all likelihood human society will have to live with the impacts of further global warming beyond the twenty-first century. (Indeed, there will be many centuries' worth of impact if global warming causes carbon to cascade from biosphere pool to biosphere pool via the atmosphere in an early Eocene-like (IETM) event.)

In terms of human biology and ecology these climate impacts can largely be grouped into the related categories of health and food supply.

7.3 Human health and climate change

Extreme weather events, such as the heatwaves of north-east North America and Western Europe in August 2003, according to some climate models, are thought to be associated with global warming (see section 6.5.1; Meehl and Tebaldi, 2004). The weather, it is said, is a traditional topic of conversation associated with the British. It should therefore not be surprising that when extreme weather events take place the British press come up with some colourful copy. And so *The Times* informed that in heatwaves the clock of the Palace at Westminster in London (colloquially known as Big Ben) slows down but also that recorded sexual offences were 5–15% higher during the unusually warm summers of 1976, 1989 and 1990 (Coates and Cohen, 2003). While the psychology of climate change is itself a fascinating subject, and has been outlined elsewhere (Cowie, 1998), global warming has more immediate effects on human biology. Indeed, during the 2003 heatwave it was reported that the UN Early Warning and Assessment Centre was considering concerns that global warming would impact on human health through changes in the geographical distribution of disease, flooding and sea-level rise. This prompted the then British Environment Minister, Elliot Morely, to comment that climate change 'is a major threat to our global community and one that has to be taken seriously'. The relationship between climate change, development and health impacts is summarised in Figure 7.11.

Climate change affects human health through processes mediated by disturbances of ecological systems (Chapters 4 and 5). The degree and nature of recent and future health effects from climate change is related to the nature of the said climate change (Chapters 5 and 6). However, in terms of energy, climate and human health, these dimensions have only recently come to the fore in academic discussions even though there were earlier greenhouse warnings arising from fossil-fuel use (Chapter 1).

In the early twentieth century the concern of industrial countries, such as it was of energy and environment matters, focused on ensuring that people were warm enough in the winter and, to a lesser degree, that they were cool enough

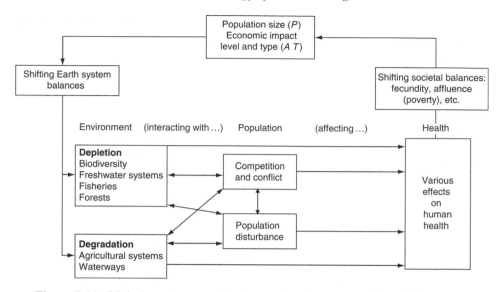

Figure 7.11 Global environmental change, development and health impacts. Developed from McMichael and Haines (1997).

in the summer. Here the energy conclusions of Sir (Sydney) Frank Markham's sociological analysis of 1942 are typical and are reflected in its title, *Climate and the Energy of Nations*, which he further developed in a 1944 edition in the middle of World War II. Sir Markham had an interest in history, having a humanities degree from the University of Oxford and being President of the Museums Association, and also in politics and science, being a Parliamentarian for a 35-year period, during which his duties included terms being Parliamentary Secretary to the Prime Minister and Chairman of the Parliamentary Science Committee. His original goal was to discern what it was that made nations great. He was prompted, one day in 1933, when despondently Prime Minister Ramsay MacDonald had said to him that 'Government has become no more than an ambulance, and its success is measured by the speed with which it deals with disaster.' (This was during the global economic depression of the 1930s.) After years of on and off contemplation Markham concluded that climate and climatic independence was central to national success. His assessment focused on how nations through the ages had striven to become more independent from the environment and he argued (albeit obliquely) for the need to address what is known today as fuel poverty. The one area of environmental degradation from energy and health consequences he touches on concerns the urban smogs that arose out of the then prodigious domestic coal consumption. Nonetheless, his overall conclusion was that, to be successful, nations need to ensure (through coal, oil and electricity consumption) that their citizens' indoor environment should be 'ideal'. What he would have made of the indoor environment of citizens' homes in the industrialised nations in the twenty-first century we will never know, as the norms three-quarters of a century on have changed considerably. However, it is likely that he would consider today's greenhouse and health concerns as somewhat ironic.

Nearly a quarter of a century on from Markham's book, in the 1960s, all but the poorest of people in Britain had warm homes in winter. So much so that British biologists' concerns arising from climate were not so much focused on human health due to temperature but on climate and Britain's ability to feed its population; remember that rationing from World War II only completely ended in 1950. In the 1960s scientists were aware that the climate had changed in the past, so when in 1963 the Institute of Biology (the professional body for UK biologists) in its annual major symposium, looked at the biological significance of climatic changes in Britain, the human health dimension was not considered. Instead the Institute's focus was on the impact of climate change on natural and agricultural systems (Johnson and Smith, 1964). Apart from a few specialists, only since the late 1980s have health concerns from climate change (as opposed to weather) had serious attention.

Contemporary concerns as to climate change and health impacts have recently been summarised by the IPCC (2001a) and the World Health Organization (WHO, 1999, 2001a). The picture emerging is that some populations will experience health impacts from climate change due to:

- permanent or intermittent changes in disease range due to shifting climate zones and/or extreme weather events (such as the El Niño);
- flooding, be it due to sea-level rise or extreme weather events and
- changes in food security.

Finally, both the IPCC and WHO emphasise that the populations, and those within populations, most vulnerable to health impacts arising from climate change will be those that are poor. The question of how the developed and less-developed nations may respond to climate change will be explored in the next chapter, which will review climate change and sustainability policies.

7.3.1 Health and weather extremes

Weather extremes affect health in a number of ways. Besides affecting crops and facilitating outbreaks of disease, even the annual seasonal cycle of weather change outside of the tropics affects health. Human physiology can handle most variation in weather but within certain limits. Marked short-term fluctuations in weather that take the human body beyond these limits can cause adverse health effects that lead to a greater number of hospital admissions (so straining health services) and even to increased death rates. Depending on location, hence climate and the nature of typical weather extremes, regional mortality can deviate from the annual average particularly in winter and summer (the ends of the annual temperature cycle outside of the tropics), and in particular during extreme weather events.

An analysis of the Central England Temperature record (see Chapter 5) between 1665 and 1834 (when the British population was far less independent from the climate) shows that mortality increased in warm summers and also in cold winters above the annual average. The main mortality effect of extreme winter cold is immediate but for the summer the effect of excessive heat is delayed by a month or two. This is because in the winter the elderly died of influenza and bronchitis but in the summer a broader section of the population died from food poisoning, which took longer to have an effect. Broadly, in England in the seventeenth to nineteenth centuries, a 1 °C warming in winter reduced annual mortality by about 2%. Conversely, a 1 °C cooling in the summer reduced mortality by around 4%. There was also a synergistic correlation with the economy and prices, which suggests that poverty was a supplementary factor (Burroughs, 1997). However, care needs to be taken when correlating extreme weather events and mortality. An extreme event occurring once a decade among normal years will show a clear correlation, but a run of say three severe winters shows the greatest excess in mortality in the first year but there may even be no excess in the third consecutively bad winter (as happened in Britain in 1940–2). The correlation therefore is not so much between an extreme event and a population's mortality rate but the extreme event and the absolute number of susceptible individuals within that population. If that number is high then the mortality will be high but if that number is low then it is impossible for the mortality to be high, even if the extreme event is a record-breaking one.

The most susceptible people within a population are likely to be

- the elderly,
- the infirm (which will not only include the elderly),
- the poor (biologically due to poor diet, health care) and
- the poor (due to lack of insulation from environmental vagaries).

Because the most susceptible within a population will suffer, although the figures in terms of number of deaths may be high, the number of person-days of quality life lost may actually be low. This is another manifestation of the declining mortality from a run of extreme events described above that shows that only the susceptible tend to be vulnerable to extreme winters. If someone is susceptible then they could already be close to death, so the extreme event is bringing their death forward by weeks or months or a year or two. Those that would in any case die within a few days will die first, then later those that would otherwise have lived for weeks will die; and so forth.

Although it may seem to be humanly cold and calculating, this is how rigorous quantitative analysis needs to be conducted when attempting to calculate the costs and benefits of climate change. It is not mortality per se that should be counted but the number of person-days of quality life lost. For example, consider a population whose society can either afford air condition-ing for all its elderly citizens or bush fire-fighting equipment (or flood protec-tion or another such opportunity cost). Which opportunity should they choose assuming the numbers of anticipated deaths from heat stroke or fires (or whatever) were equal? The number of anticipated deaths does not help provide an answer but the number of person-days of quality life lost often provides a clear indication. Of course, in developed nations this dilemma need not arise (although sometimes it does with tight civil budgets). Conversely, such quand-aries are more common in less-developed nations.

Surprising as it may seem, climate change may cause health effects simply because the new climate is not within a population's perceptual history, and so the risks and avoidance measures may not be recognised. If they were, adverse health effects may be avoidable. This was brought home with the publication of research in 1998 by a team co-operating between London and Moscow. On balance, conventional wisdom had it that the more affluent Western European society would have a lower mortality than their Russian counterparts. However, when winter temperatures reached 0 °C excess deaths were not seen in the elderly cohorts of the Russian population, although they are known to exist in Western European populations. The reason given is that the Russian population was used to such winters and were prepared. Their houses were warmer and they wore more clothes. However, for many Western European population winters of 0 to −6°C were extreme events. The research suggests that *some* deaths from extreme events in affluent societies are avoid-able (Donaldson *et al.*, 1998).

Of the possibly avoidable deaths from extreme heat or cold events, it is the excess mortality above the norm in winter that can more easily be negated. Heating to combat cold need not rely on advanced technology and basic heating technology is as old as the camp fire. Conversely the least avoidable deaths will be those from excess summer heat, because to negate this usually does involve the higher-level technology of air conditioning. Before the use of air conditioning in the developed industrialised countries, major heatwaves would typically increase mortality in cohorts over 50 by several-fold. In 1936, one of the hottest summers on record in the USA, some 4700 people were estimated to have been killed directly by heat stroke. This represented an increase of several per cent above the numbers of deaths expected over the same period. Again, in Melbourne, Australia, in 1959, there was a four-fold increase in mortality (145 excess deaths) that was directly attributable to a prolonged heatwave, although most of these occurred in the first 4 days of the event, for reasons presumably akin to those where mortality from a successive run of bad winters occurs early in the run (see above). A similar pattern of early-run mortality occurred during the run of summer heatwaves in Los Angeles in 1939, 1955 and 1963 (McMichael, 1993). At the end of the twentieth century in the USA some 1000 died each year from extreme cold events, but twice as many died from extreme warm events. This last might in all probability be an under-estimate, as a proportion of death certificates made out during heatwaves cite respiratory disease or some other cause rather than heat stroke.

The UK provides an example of a policy response to the likelihood of increasing heatwaves. The August 2003 heatwave in north-west Europe is estimated to have caused some 15 000 extra deaths in France, 2000 in Spain, 1300 in Portugal and around 2000 in England and Wales. In 2004, from concerns that climate change will increase the frequency of heatwaves, the UK Government's Department of Health and the National Health Service published a *Heatwave Plan for England*. As outlined in Box 7.1, the plan has four levels of response for when threshold temperatures are forecast or exceeded. This plan is to be further developed over the years.

Climate change due to greenhouse warming for most (but not all) places on the planet will mean an increase in extremely warm summer events (above the mid twentieth-century average). It will also for most (but again, not all) places outside of the tropics mean a decrease in extremely cold winter events. The overall mortality of a population throughout the year may (or may not) increase, but what almost certainly will change will be the pattern of mortality within the year. There will in all likelihood be an above-average level of mortality in the summer and a lower level in the winter. Consequently hot countries are likely to see a net increase in annual mortality but countries with a mild climate (such as the UK) may see hardly any change in overall annual mortality and instead a decline in winter mortality and an increase in summer mortality. The winter exception to the above is that in a warmer world it is expected that the strength of severe storms would increase. Injury from debris and destruction could potentially increase public aware-ness, and enable local resource provision, of what to do in the event of severe storms.

Box 7.1   The Heatwave Plan for England

The plan details a heat-health watch from the beginning of June to mid-September each year. It consists of four levels of response based on threshold temperatures. These vary with region but are approximately 30 °C in the daytime and approximately 15 °C at night.

Level 1 (awareness)	The minimum constant level of heat-health awareness among heat-health stakeholders and the general public.
Level 2 (alert)	To be triggered as soon as the UK Meteorological Office forecasts that threshold temperatures are to be exceeded for 3 days or that there will be two consecutive days of extremely high temperatures near the threshold.
Level 3 (heatwave)	Triggered as soon as threshold temperatures are reached.
Level 4 (emergency)	When a heatwave is so severe that it threatens the integrity of health and social care systems.

It is the responsibility of the Government-funded health and social services to identify groups and individuals likely to be adversely affected by heatwaves, provide advice and implementing appropriate measures. At-risk groups include the elderly, babies and the very young, those with mental health problems, those taking certain medication, those with certain chronic conditions such as breathing problems, those with a disease/infection induced by high temperature and the physically active, such as manual workers and athletes. It is the responsibility of the UK Meteorological Office to provide the health and social services with 3-day forecast warnings.

The WHO recognises that global warming comes with health impacts and it has been working with the WMO and the UNEP (both of whom in turn formed the IPCC; see Chapter 5) since 1999. Its work in this area has three main strands: capacity building (informing to help nations develop the necessary services to address climate-change-related health impacts), information exchange and research promotion (WHO, 2001a).

Extreme events are also associated with climatic cycles. There are three or four principal climatic cycles: two encircling the poles (one each), a cycle related to northern mid latitudes and the El Niño Southern Oscillation (ENSO). The North-Pole cycle is the Arctic Oscillation or Northern Annular Mode, while the northern mid latitude cycle is known as the Pacific–North American pattern (Kerr, 2004). Of these cycles, the best studied is the ENSO. The ENSO has a period of 2–7 years and within these years warm water periodically forms in the eastern Pacific to create an El Niño event. The El Niño begins with a weakening of the prevailing winds in the Pacific and a change in rainfall patterns that create extreme-flood and drought events in countries surrounding the eastern Pacific and climate-related events further afield. Prolonged dry periods may occur in south-east Asia, southern Africa and northern Australia, as took place in 1991–2, together with heavy rainfall, and sometimes flooding in Peru and Ecuador. During a typical El Niño event the Asian monsoon usually weakens and is pushed towards the equator, often bringing summer drought to north-west and central India and heavy rainfall in the north east. The regions where the El Niño has a strong effect on climate are those with the least resources: southern Africa and parts of South America and south-east Asia.

The WHO has noted that the risks of natural disasters are often highest in the years during and immediately after the appearance of an El Niño, and the lowest in the years before. In the 1997 El Niño Central Ecuador experienced rainfall 10 times higher than normal, which resulted in flooding, extensive erosion and mudslides with the loss of lives, homes and food supplies. During the same El Niño, there were droughts in Malaysia, Indonesia and Brazil, exacerbating the extensive forest fires taking place at the time. The El Niño cycle is also associated with increased risks of disease range (see section 7.3.2).

The El Niño not only serves to illustrate the type of health impact an extreme weather event may impart, the impacts themselves may change with global warming. It has been suggested that the El Niño may become more intense and/or more frequent with warming and so will the potential health impacts (WHO, 1999, 2001b).

In addition to the direct physiological effects of heat, extreme climatic events also have indirect effects that are no less serious. A drought increases the likelihood of bush and forest fires which are not only dangerous in themselves but destroy homes and property, which can also have health consequences, including loss of life. For example, Portugal's heatwave of August 2003 resulted in that nation's worst forest fires for 20 years, killing six people. Meanwhile, at the same time Canada was experiencing the worst fires for 50 years, which necessitated the evacuation of some 8500 people. Consequently those concerned with the biological management of woodland are having to come to grips with climate change and fire hazards. For example, since 1990 the western USA has seen some extreme fires that have been notable for their size and severity. The annual US cost of fire suppression now exceeds S$1.6 billion (2004) and the figure is rising. In the absence of large fires, much of the western USA during the twentieth century has seen many forests fill with a dense understorey of shrubs and small trees that provide so-called ladder fuels that set crowns of trees alight: crown flares are among the most destructive of forest conflagrations. To address this the US Healthy Forest Restoration Act (2003) determined programmes of aggressive thinning, burning and replanting to create fire breaks and more open conditions by removing ladder fuels (Whitlock, 2004). However, there is evidence to suggest that this may not be appropriate as a generic strategy.

In 2004 Jennifer Pierce and colleagues from the universities of New Mexico and Arizona published results indicating that a one-size-fits-all woodland strategy may not be pragmatically effective. They came up with a way of looking at the long-term fire record in low-altitude pine forests of the western USA. These forests are prone to drying. There are also open forests that lie above the steppe and grasslands of the valleys and basins. At higher altitudes there is more moisture and the forest becomes enclosed and subalpine. The low-altitude forests have been affected by human land use (hence fires there have a potential health impact) and there is some controversy as to how best to return some areas to their 'natural' state. One question is whether the recent severe fires are atypical. Another concerns the role that climate change may (or may not) be playing. Conventional wisdom has also been challenged, in that fires have affected both open and closed forests: open forests were thought to be more fire-resistant.

Our current understanding of past low-altitude forest fires is based largely on two records. The first record is the fire scars of living and dead trees. As such, this only provides a record within about 500 years, and a record that is increasingly biased with age towards fires that scar but do not kill. Second is charcoal records from lake sediments. These can often provide a record going back a few thousand years but only in the main relating to montane and subalpine forests: these have a sufficient altitude difference from the lakes for watercourses for charcoal to be conveyed.

Conversely, Pierce and colleagues looked at fire-related sediments at low altitudes. Severe fires can remove surface litter and small plants, so enabling soil movement (especially following storms or snow melt) that result in debris-flow deposits. These accumulate in alluvial fans, and other places in valley floors, and can be carbon-dated. They found that over the past 8000 years in xeric (that is, characterised by dry conditions) ponderosa pine (*Pinus ponderosa*) forests, fire-related debris flows were associated with warm, dry periods such as the MCA around AD 1100–1300 (see Chapter 4). This contrasts with cooler, wetter periods such as the Little Ice Age (AD 1550–1850) that saw fires from the dense grass cover, but only minor episodes of sedimentation. Adding this knowledge to the previous picture it becomes clear that strategies of forest management need to recognise that the frequency and nature of fires varies with both location (hydrology and geology) and climate regimen. So, one-size-fits-all forest management may not be appropriate in that we may not be currently minimising risks to human settlements that are mainly associated with lower-, rather than higher-, altitude forests.

Forest fires also affect whole ecosystems and the humans associated with them. Conversely, the physiological effects of extreme heat (and cold) are a manifestation of effects on the whole organism, be it plant or animal (including humans), as discussed above. However, there are also effects of extreme hot weather at the cellular level.

Hot weather events tend to be associated with sunnier days, which combined with the human behavioural response of spending more time outdoors during such events can result in an increased exposure to ultraviolet (UV) radiation. Sunburn (erythema) and skin cancer are largely due to UV-B radiation, which has a spectral region centring around 300 nm. But there are other cellular degenerative effects more related to UV-A radiation, which has a spectral region centring around 340 nm.

The risk of skin cancer is greatest among fair-skinned populations that evolved in northern latitudes that are exposed to lower levels of intense sunshine. It is thought that the evolutionary driver for pale skin was to enable the skin to continue to produce sunlight-induced vitamin D. Further, sunlight-induced skin cancers are the commonest form of cancer among light-skinned populations: for example, they account for an estimated 70% of all skin cancers in the USA. Exposure to sunlight is the major environmental cause of these cancers. The least frequent, but most serious, of the three skin cancer types is malignant melanoma. If not detected and treated early, this cancer will spread via the bloodstream to internal organs and this is often fatal, with mortality of around 25% (McMichael, 1993). In the UK the Department of Health assessment is that by 2050 there will be a possible 5000 extra cases of skin cancer a year due to climate change (Department of Health, 2002).

UV light also damages eyes in a variety of ways, including encouraging cataract formation. The Department of Health estimate an increase of some 2000 cataract cases a year by 2050 (Department of Health, 2002).

As with extreme cold events discussed above, the health effect of increased surface UV on a population is strongly affected by behaviour. Avoiding sunbathing (and especially by avoiding erythema) greatly reduces risks, although even not sunbathing does not eliminate risk as skin exposure to UV takes place outdoors on nearly all uncovered parts of the body. Wearing glasses with glass lenses (as opposed to plastic-lensed glasses transparent to UV radiation) helps reduce the risk of cataract formation. However, small-lensed, dark sunglasses allow sunlight in around the rim while the dark lens allows the iris to dilate, so allowing in extra UV, even if it is not focused on the fovea of the retina, which is responsible for the detailed part of the human visual field. Large-lensed and only slightly tinted sunglasses give the best protection. (Wearers of glasses take note when next visiting your optician.)

Extreme hot-weather events can also affect microbial populations and this is discussed in section 7.3.2. Finally, extreme weather events can also result in flooding due to rain bursts and the synergies between sea-level rise and storm surges that are associated with global warming. These too have health impacts and these are covered in section 7.3.3.

7.3.2 Climate change and disease

Pathogens are disease-causing living organisms that are predominantly pro-toctistans (single-celled, eukaryotic, meaning with internal membranes), prokaryotes (such as bacteria, which have no internal membranes) and viruses (genetic material (DNA or RNA) with a protein and/or glycoprotein coat), although it has to be said that there is debate as to whether viruses should be considered as living. There are also metazoan (multi-celled animal) parasites. The ability of all these to flourish is affected by their environmental conditions, and hence climate change. Even those pathogens that rely on a vector (an organism capable of transmitting a disease from host to host) are affected by climate change, in that the vector itself will flourish best under certain environmental conditions. In short, just as a species' range and/or population will vary with a combination of climate change and other environmental factors (Chapter 5 and 6), so a disease's range will also change, as will the disease organism's population if not controlled.

As with humans and the annual temperature cycle outside of the tropics, so microbial populations vary. The incidence of reported cases of food poisoning in developed nations is illustrative of such annual cycles, with cases peaking in the summer months. In the UK the number of officially notified food-poisoning cases was rising in the late twentieth century, so that by 1998 it had reached about 100 000 per year. However, notifications do not always strongly correlate with monthly temperature and sometimes more strongly with the previous month's temperature. This is not due to incubation period as these are commonly short (typically less than 2 days), or due to delays in notification as the UK system is designed for prompt reporting. It is possibly to do with the level of infections in animals prior to slaughter, as meat is an important source of food poisoning. Consequently global warming does not just present a challenge to consumers but also to those responsible for the precursor food production, processing and transportation stages (Bentham, 1999).

The annual pattern of reported cases of *Salmonella enteritidis* by month in England and Wales is typical (see Figure 7.12). *Campylobacter* species show a

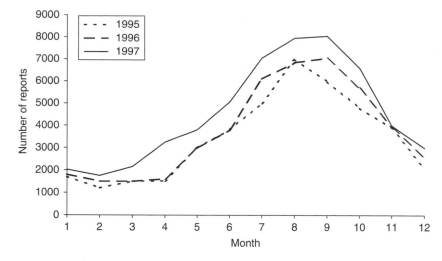

Figure 7.12 Reported cases of *Salmonella enteritidis* by month in England and Wales in 1995–7. After Stanwell-Smith (1999).

similar pattern with a summer maximum. *Campylobacter enteritis* has been linked to barbecues, particularly those involving chicken. Nonetheless this only explains a small proportion of UK outbreaks and a very few of the sporadic cases. However, that cases of *Campylobacter*-related illness are rising has made this species of interest to those engaged in climate-change and microbial-disease research in the UK.

This pattern of summer increase in temperate-zone incidence of food poisoning suggests that in a warmer world there will be more such cases. Worldwide, WHO and the UN Food and Agriculture Organization (FAO) estimate that there are at least 2 billion people affected by food poisoning each year and many die. For example, in Asia alone some 700 000 die due to food- and water-borne (contaminating food) disease (WHO/FAO, 2004).

Lyme disease is another disease whose seasonality is likely to be affected by climate change. It was first identified in the USA in 1975 and then noticed in much of Europe. Its name comes from a district in Connecticut, which is one of the many areas in which it is endemic. The disease is caused by the spirochaete bacterium *Borrelia burgdorferi*, with mammals acting as reservoirs, and is transmitted by ixodid ticks such as *Ixodes ricinus*. Humans are not the preferred host, but nonetheless Lyme disease is a public-health concern in a number of rural areas in Europe and especially the USA, where there are many thousands of cases a year. In the UK and Ireland the incidence of the disease has been sporadic with typically just a few scores of cases a year. It results in arthritis of large joints and may also inflame the brain and nervous system. Transmission occurs during the spring, summer and early autumn months, which is also when humans are likely to be outdoors in rural areas and exposed to the vectors. Because of climate change both the seasonality of the disease and also its geographic distribution may be extended.

Climate change not only affects the propensity of a disease to affect a human population but it can also change the species range of disease vectors. Of the major vector-borne diseases, malaria is the most significant with (in the 1990s) the potential to impact on over 2100 million people and successfully infecting some 300–500 million, of whom some 90% are in tropical Africa. Malaria currently kills more people than any other infection except tuberculosis, and has an annual death toll of 1–2 million people, mainly young African children. It is also one of the diseases whose range has changed, and will undoubtedly continue to change, with climate. The disease itself is caused by protoctistan parasite species of *Plasmodium*, namely *Plasmodium ovale, vivax, malariae* and *falciparum*, which are transmitted to (and between) humans by mosquitoes belonging to the genus *Anopheles*. *P. ovale* results in a less-severe form of malaria compared with the other three species, and Hypocrites (*c.*2500 years ago) divided malaria into three different types that, with the benefit of subsequent knowledge, largely appear to relate to three of the different *Plasmodium* species. As for the vector, there are some 380 species of *Anopheles* of which about 60 transmit malaria.

Genetic studies suggest that the virulent form of malaria evolved some 10 000 years ago just after the last glacial. This broadly coincides with the development of agriculture and the supposition is that water held for irrigation provided an ideal habitat for the *Anopheles* mosquito to evolve into its current form, but this has not been proven.

Malaria evolved in the topics and sub-tropics of the Old World (Africa) and it did not arrive in the New World (the Americas) until around 1500 with European settlers and African slaves. Malaria was not controlled in the USA until the 1940s with pesticides, but eradication has proven difficult and there are still sporadic outbreaks. Eradication programmes elsewhere have been more successful, such as in Brazil and Egypt.

The disease used to be found in Europe, including in southern England. Historically, in England malaria was known as ague and was frequently found around coastal salt marshes. Indeed, 200 years ago it was thought to be a leading cause of death in many marshland communities including the Fens (in Norfolk), the Thames estuary and further along the Kent and Essex coasts, as well as the Somerset Levels. It declined in the 1840s but there were still pockets of the disease in Kent and in other low-lying areas through to the early 1900s. The last major UK outbreak took place in Queenborough on the Isles of

Sheppey (in Kent) when soldiers carrying the malaria parasite who were returning from World War I were bitten by local mosquitoes that went on to transmit the disease to 32 people over several years. However, indigenous malaria remained and from 1917 to 1952 some 566 people contracted the disease in Britain. The indigenous disease died out in 1953 with two final cases in London: mosquitoes (probably *Anopheles plumbeus*) were found breeding in a tree hole near one of the victims' houses.

Twentieth-century England typifies the problem of malaria in many temperate zones in the twenty-first century. First, this century sees a far more internationally mobile population than any other. This increases the chance of re-introducing malaria. Indeed in 1971 Europe as a whole saw around 1400 cases of imported malaria. The subsequent years saw Europeans travel further outside the continent so that by 1986 cases of imported malaria had increased to around 6800 (Martens, 1999). By the turn of the millennium the number importing malaria into the UK alone approximated 2000. Fortunately, only 25% of these had *P. vivax* that used to be common.

Second, the disease's range is complex and the areas in which it may exist are determined by many factors. Just as climate change threatens many wildlife species (and note that the malaria parasites and vectors are wildlife species) so climate change may not necessarily serve to help promulgate the disease as might be thought. The actual theoretical areas where malaria may exist are determined by a number of factors of which climate is one. Just as human land use restricts wildlife species and impedes their migration/expansion (see Chapter 6), so it does with malaria. In the UK the extent of many wetlands near to large centres of population have declined. Climate itself is multifactorial and malaria, or rather the parasite in the *Anopheles* vector, not only requires a specific temperature range but also humidity and duration of these conditions, as well as the aforementioned land-use criteria. In 2000 David Rogers and Sarah Randolph showed, using climate-change models, that during the rest of this century, even under the more extreme climate-change scenarios *P. falciparum* malaria might move into new areas, but that more or less equally it would move out of old ones. The overall potential health effects of climate change appear to be marginal; it is just the distribution that will change. This is further complicated by whether the new areas contain more humans to infect compared to the areas vacated. The picture is complex and again is not untypical of the way ecosystems might respond to climate change: species assemblages may not always shift uniformly with climate change (see Figure 6.1b). The aforementioned forecast is also supported by limited observations in four high-altitude sites in East Africa where increases in malaria have been reported and then present and past conditions compared using

95-year climate records. Looking at the combination of required temperature, rainfall, humidity and number of months suitable for *P. falciparum*, conditions have not changed significantly either over the 95-year period or in the two decades of malaria increase (Hay *et al.*, 2002). In short, there is a natural variation in the disease in addition to anything that might be to do with climate change.

Does this mean that there are no future health impacts from malaria due to climate change? Certainly not. Notwithstanding surprises, the distribution of malaria will change but it may not result in huge increases in numbers of cases due to climate change alone, although the places where infections take place are likely to alter. However, climate change in combination with other factors, such as continued increasing international mobility, might result in sporadic localised outbreaks in parts of Western Europe and North America. By the end of the twenty-first century malaria might be found in some parts of mid South America, parts of southern Africa and further east of its existing range in Asia. Equally, parts of mid South America, eastern Africa, eastern India and parts of east Asia that are presently suitable for malaria may become unsuitable. Consequently, the deployment of anti-malaria measures will need modifying.

Here climate-change science is set to help. The marked increase in sophistication of computer models of the global climate, especially their regional-level spin-offs, are beginning to be used to develop forecast systems of malaria outbreaks. (Here, as elsewhere in this book, the term model means a mathematical simulation as opposed to how epidemiologists sometimes use the word as a simple correlation or statistical relationship between two things (Palmer, 2006).) One such system has already had practical success in semi-arid Botswana, adding up to 4 months' extra lead time in advance of existing warnings of outbreaks. It predicts regional climate patterns about 6 months in advance. In Botswana the rainy season, hence the malaria season, begins in December but health officials have to wait until February or March before they can make an accurate assessment. The new model enables them to make the prediction months before the rainy season starts (Thomson *et al.*, 2006).

Then again, some non-climatic factors can aggravate matters. For example, *Anopheles* pesticide resistance was first noticed in the 1950s and resistance to chloroquine (an anti-malaria pharmaceutical) was first noticed in the 1960s and has become an increasing problem since. Evidence of resistance to the modern anti-infective drugs atovaquone and proguanil was noticed in 2003. In China artemisinin-based combination drugs derived from the sweet wormwood plant (*Artemisa annua*) have yet to encounter resistance and they are increasingly being used in Africa. Consequently, demand for these drugs is rising but even with them the evolution of resistance will continue to be of concern, so increasing dependence on artemisinin-based pharmaceuticals. Indeed, in 2004 the WHO announced a shortage of artemisinin-based combination drugs and forecast a demand of 10 million treatment courses, predicting that this would rise to 60 million in 2005 (WHO, 2004).

Finally there is the question of how long artemisinin-based drugs will continue to avoid resistance by malaria parasites. One option to delay the onset of resistance would be to avoid the use of single-drug treatment (monotherapy) and instead use cocktails of pharmaceuticals or artemisinin combination therapies (sometimes known as ACTs). So, in January 2006 the WHO appealed to all pharmaceutical companies to stop marketing oral artemisinin monotherapies and 13 countries over the subsequent 3 months agreed to withdraw marketing authorisation for such monotherapies (WHO, 2006).

There are many other diseases whose ranges are likely to change with climate. Among those of major concern is Chagas' disease or American trypanosomiasis, a South American variation of the sleeping sickness (trypanosomiasis) that affects central Africa. Chagas' disease is named after the doctor who researched the affliction in northern Brazil during the building of a railway in 1909. About 18–20 million people in Central and South America and the USA are at risk of contracting Chagas' disease. Its geographical distribution ranges from the southern USA to southern Argentina and Chile. It is transmitted by the blood-sucking insect *Triatoma infestans*, and other species, which carry the protoctistan *Trypanosoma cruzi*. It originally affected rodents and marsupials but has adapted to human encroachment. Unlike African sleeping sickness, the trypanosomes do not multiply in the blood but become localised in various organs. The disease causes long-term, essentially untreatable debilitation and, by affecting heart muscle, it is a major cause of chronic heart disease and sudden death in apparently healthy young people.

Conversely, with the related African sleeping sickness the brain and the meninges (the covering of the brain and spinal cord) become infected with *Trypanosoma brucei* and, if treatment is delayed, it usually proves fatal. The WHO estimates that between 300 000 and 500 000 people are affected worldwide, and that the disease threatens 60 million people in the poorest underdeveloped countries. As for treatment, relatively drastic measures are required. In the acute stage of the disease pentamidine is given. If the parasite is also in the central nervous system, the drug melarsoprol is used, so exposing patients to serious risk, especially of arsenic encephalopathy, of which almost 1000 people die each year. Both substances are administered directly into the bloodstream. There is no vaccine or chemical prevention due to the relative toxicity of the drugs, so it is necessary to prevent bites from the tsetse fly (of the genus *Glossina*) and control the places where the fly lives.

Dengue, which covers both the milder 'breakbone fever' and the more serious but rarer haemorrhagic fever, is caused by arboviruses that are transmitted by *Aedes* mosquitoes. Dengue fever can be caused by any one of four types of dengue virus: DEN-1, DEN-2, DEN-3 and DEN-4. A person can be infected by at least two if not all four types at different times during their life time, but only once by the same type. Symptoms of typical uncomplicated dengue usually start with fever of up to 40.5 °C within 5–6 days of a person being bitten by an infected mosquito. Other symptoms include severe headache, retro-orbital (behind the eye) pain, severe joint and muscle pain, rash, nausea and vomiting. Of interest to clinicians is that most infected children never develop typical symptoms. There is no specific treatment for dengue fever, and most people recover completely within 2 weeks. Dengue is of particular concern because the vectors are associated with human environments and, unlike many anopheline mosquitoes, they prefer to feed on humans. Consequently, when epidemics take place they do so rapidly. Currently milder dengue has all-year transmission, but mostly during and shortly after the rainy season around the tropics, between about 25°N and 25°S, where the principal vector is *Aedes aegypti* (which also is the vector for yellow fever arbovirus, now largely eliminated from the USA, Central America and most of South America and still present in much of tropical Africa). The WHO estimates that there are 50 million cases of dengue infection each year. This includes 100–200 cases reported annually to the US Centers for Disease Control and Prevention (CDC), mostly in people who have recently travelled abroad. From 1977 to 1994, US health-care workers reported to the CDC 2248 cases of dengue that had been imported into the USA. Many more cases probably go unreported because some doctors do not recognise the disease. During the last part of the twentieth century, cases of dengue began to increase in many tropical regions. Epidemics also began to occur more frequently, and to be more severe. The concern for Europe is that with climate change another dengue vector, *Aedes albopictus*, is currently increasing its range and can already be found in Italy and parts of the Balkans.

In North America West Nile fever has come to be of concern in recent years. It is caused by a member of a group of viruses called flaviviruses, which are transmitted mainly by birds to humans via mosquitoes. It results in a mild, influenza-like illness that is often called West Nile fever. More severe forms of the disease, which can be life-threatening, may be called West Nile encephalitis, West Nile meningitis or neuroinvasive disease. Although many people are bitten by mosquitos that carry West Nile virus, most do not know that they have been exposed. Few people develop severe disease or even notice any symptoms at all. It is thought that only one in five infected people develop mild illness, and that only one in 100 or fever-infected people go on to develop brain inflammation (meningitis or encephalitis). By 2000 the disease was considered firmly established in the USA and it is of concern in Canada. In 2002 and 2003 in the USA it caused the death of 284 and 264 people respectively. By 2002 in Canada several populations of birds were found to have the disease. It was originally found throughout much of Africa, from the Cape of Good Hope to Cairo, and is also widespread in the Middle East and India. The disease occasionally appears in parts of southern Europe, such as Italy, and in Belarus, Ukraine and southern Russia. In 1997 some 500 in the Bucharest region of Romania caught the disease, with 50 people dying.

In 2004 there was sufficient concern over West Nile fever that Britain's Department of Health issued a short bulletin to alert clinicians to be vigilant and describing a plan to control key mosquito populations and a contingency plan for when cases were diagnosed (Department of Health, 2004).

That West Nile virus has recently hit the headlines in North America and Europe suggests that flavivirus zoonoses (virus diseases of animals that affect humans) have only recently presented a serious health impact on humans. However, Japanese encephalitis (found in Asia) is a virtually identical bird-borne disease caused by a sibling virus. It affects more than 50 000 people a year in Asia, and kills 15 000 of them. It currently has a far greater health impact, but is not of such concern in North America or Europe because it has not touched the western hemisphere (yet). But there are occasional incidents. For example, one such illness has already appeared in Britain, louping-ill fever. It affects a variety of animals and in sheep is known as ovine encephalomyelitis or infectious encephalomyelitis of sheep, for which the main vector is the sheep tick *Ixodes ricinus*. Other tick vectors include *Rhipicephalus appendiculatus*, *Ixodes persulcatus* and *Haemaphysalis anatolicum*. Infection has been demonstrated in other domestic species and wildlife; namely, cattle, horses, pigs, dogs and deer, as well as in a range of species of small mammals such as shrews, wood mice, voles and hares, the latter being considered an important reservoir species for the virus. It can decimate grouse populations but humans are only a tangential host and there have only been occasional isolated outbreaks of non-fatal meningo-encephalitis. Louping-ill is endemic in rough upland areas in Scotland, northern England, Wales and Ireland. A disease of sheep very closely related to louping-ill has been reported in Bulgaria, Turkey, the Basque region in Spain and Norway.

The aforementioned are just some of the diseases whose incidence is in part related to climate change. Of course, these diseases would still be with us without climate change and in one sense virtually all diseases are in some way or other influenced by climate and/or weather. Even common influenza viruses tend to have an annual cycle. As a very broad generality within the range of environments commonly colonised by humans, microbial organisms prefer warm, moist conditions as opposed to dry, cold conditions. Consequently, as global warming will result in an increase in precipitation globally, and as (again a generality) temperate-zone winters will become less cold, the balance is such that a warmer world is likely to see more, rather than less, microbial-caused disease. This increased likelihood is synergistically increased by non-climate factors such as the increasing global population and international mobility. Indeed, this last is causing a noticeable increase in tropical infections among travellers from temperate, developed nations returning home.

In December 1997, the first workshop for international agencies was held on climate change and human health monitoring when the UNEP, the WHO and Britain's Medical Research Council joined forces to assess the problem. This paralleled a WHO, UNEP and WMO initiative to establish an interagency network on climate-induced health impacts. The workshop helped set the inter-national research agenda and allowed agencies to pool resources to begin to tackle this issue. Worldwide health efforts to address these diseases have increased in recent years, although a lack of resources in many less-developed nations has hindered (and in some instances reversed) progress. Consequently, where climate change is a significant epidemiological factor, once again it is the wealthy countries generating the most greenhouse gases who can afford to take remedial action, while poorer nations tend to remain exposed.

7.3.3 Flooding and health

Sea-level rise (see Chapter 5) and increased precipitation on a warmer planet are likely to enhance the prospect of flooding. Importantly, the volume of water that can be held in the air increases with temperature. As this goes up so the possible maximum precipitation (PMP) increases. With higher PMPs in a warmer world, storm bursts can potentially become more extreme. In short, a warmer world can increase hydrological cycling.

The question to be asked is whether this in itself will seriously impact on people's lives, their well being and health? However, a brief consideration as to how floods have affected people worldwide in the last century does give us an indication as to their human impact in the more crowded and flood-prone twenty-first century. The WHO stated that 'almost 2 billion people – one-third of humanity – were affected by natural disasters in the last decade of the 20th century. Floods and droughts accounted for 86% of them.' Floods are the second most frequent cause of natural disaster after windstorms but affect more regions and more people than any other phenomenon (WHO, 2001b). Against this background any factor, such as climate change, likely to increase the frequency of floods (not to mention severe windstorms) needs to be examined so that the appropriate measures required can be anticipated.

WHO consider the most vulnerable victims of flooding globally to be the poor and the marginalised, most of whom live in low-quality housing, in flood- or drought-prone regions. Those fleeing floods often drink unclean water. If the drinking-water supply and sanitation systems are already inadequate, flooding poses a major health risk. Flooded areas that have industrial waste, such as used engine oil, and refuse dumps add to health risks. The WHO says that, 'people who have lost everything in the flood – their homes, their food, their livelihood – are all the more prone to disease'. It also noted that floods are becoming more frequent: from 66 major floods in 1990, the number rose to 110 in 1999. Further, the number of people who died in floods in 1999 was more than double that in any other year of the 1990s (WHO, 2001b).

In tropical countries receding waters provide an ideal breeding ground for mosquitoes and other insects, so creating an increased risk of diseases such as malaria and dengue (see the previous subsection) as well as Rift Valley fever viral infections that affect both humans and animals. Flood water combined with the effects of open sewage and reduced opportunities for good personal hygiene lead to cholera, diarrhoea and gastrointestinal viruses, and displaced rodent populations can themselves cause outbreaks of leptospirosis and hanta virus infections. Flooding may also encourage *Cryptosporidium* infections. *Cryptosporidium* is a widespread protoctistan found in both tropical and temperate climates, the infection of which causes cryptosporidiosis, a diarrhoeal disease. *Cryptosporidium* oocysts (the cyst formed around two conjugating gametes) can survive for long periods and are capable of penetrating groundwater sources. In addition *Cryptosporidium* spp. are readily spread from person to person. Infections have also been linked to swimming pools. *Campylobacter* is similarly associated with water.

As with disease impact and the resistance to anti-infectives (antibiotics and antivirals), the health impact from floods can be exacerbated by such, and other, non-climate-related factors. For example, regional development increased the impact in 2004 of heavy storms that hit the Philippines early in December, which resulted in exceptional floods along rivers. By 3 December over 650 were dead and some 400 were missing. Thousands more were made homeless. Extensive logging was blamed as a compounding factor. The next day the government announced that all logging was to be banned subject to a review.

Another example of a major flood imparting the aforementioned health impacts is that of the Bangladesh flood in July and August 2004. Aside from this low-lying area being one of a high population, it has also been identified by the IPCC (1990) as likely to have a higher summer rainfall in a warmer world. In 2004 the region experienced heavy monsoon rains, the worst on record. It was an extreme weather event, again of the kind that the IPCC anticipate. In this instance 1 million people were displaced from their homes, out of the 10 million in the nation's capital, Dhaka. Mosquitoes, escaped sewage, disrupted food supplies and disease (for example more than 270 000 cases of diarrhoea) compounded the problem. 2004 was a particularly bad year but floods in that region are common and in 1998 about 70% of the country was under water for nearly 3 months. With around 65% of the population deemed by the Red Cross to live below the poverty line, the people affected were typical of those the IPCC consider to be most vulnerable to extreme weather events and other climate-change impacts. The reasons why the IPCC cites Bangladesh as an example of a risk area are both because it is low-lying, hence exposed to sea-level rise, and because its climate naturally features weather extremes (often monsoon-related). Factors conspired synergistically to make the floods in the north-eastern region of Bangladesh in 2004 impact heavily on people. First, four flash floods came in quick succession and, second, unusually high tides slowed the water's discharge to the sea. But, in addition, population pressures, to farm and other land development, had constrained the waterways, so increasing their chance of failing during extreme events (*Nature* news, 2004). But are rain-fed floods actually getting worse?

The IPCC (2001) stated that there were no observable changes in tropical cyclones, and peak precipitation intensities were not found in the few analyses available to it. However, in the twenty-first century it forecast that increases would in the future be likely in some areas. Indeed, research is beginning to reveal some increases in rain-fed floods, although it is still unclear whether these will persist: a longer time frame is required as twentieth-century analyses reveal flooding periods of different frequencies (all be they at a lower level) that last for decades. For example, in 2001 a US team led by Stanley Goldenberg showed that the years 1995–2000 experienced the highest level of North Atlantic hurricane activity in the reliable record. Compared with the generally low activity of the previous 24 years (1971–94), the subsequent 6 years saw a doubling of overall North Atlantic heavy-storm activity, a 2.5-fold increase in major hurricanes (with wind speeds of over 50 m/s) and a five-fold increase in hurricanes affecting the Caribbean. The greater activity largely results from increases in sea-surface temperature that, the researchers note, has associated human well-being and property-damage costs.

The annual cost (1999) for the USA from cyclone damage was of the order of US$5 billion, but this is expected to rise in future years with increased population and wealth. It has been estimated that if the hurricanes of 1925 had happened in the wealthier and more populous 1990s, instead of a few billion, the cost would have been some US$75 billion after adjustment for inflation. In 1998, Hurricane Mitch killed at least 10 000 people in Central America, while the average economic loss in the Philippines was estimated at some 5% of gross national product (Bengtsson, 2001).

Floods have always taken place throughout human history. Noah's flood and similar stories that might date from pre-history appear to relate to the flooding of the Black Sea, which itself arose due to sea-level rise following the last glacial maximum and the onset of the current Holocene interglacial. Furthermore, there has always been a tension between the human desire to exploit the fertility of the floodplains of the world's greatest rivers and the fact that floodplains (by definition) flood.

In the late twentieth century the floods of the upper Mississippi in the summer of 1993, the winter floods in December 1993 and January 1995 of the Rhine and Bangladesh's chronic flood problem all served to illustrate the material costs and costs to human well-being. The 1993 Mississippi flood had an estimated cost of US$15 billion but only a small death toll, of 48. Following that flood there was concern expressed over the way the river watershed was managed and the flood-control measures that were in place, but irrespectively that flood was due to an extreme weather event following a season of exceptional snow in Iowa and the 1993 precipitation record was the greatest since records began in 1895. Meanwhile, both the 1993 and 1995 floods of the Rhine were also due to extreme weather events and parts of the region in 1993 had three times the average rainfall in December. The cost of the flood in Germany alone was around US$850 million in the money of the day. However, in 1993 lessons were learned so when in 1995 there were floods again, the damage costs were halved. Even so, there was disruption and in the Netherlands some 200 000 people were evacuated from their homes for safety. What is also clear is that in north-west Europe, as illustrated by England and Wales, instrumental measurements have shown that between 1765 and 1995 there has been a slight decrease in the intensity of winter drought years accompanied by a similar increase in the intensity of years with high winter rainfall, with average rainfall also increasing by about 17% (Osborn *et al.*, 2000). This trend is what is expected in that region with climatic warming. However, in the case of Bangladesh there is no historic meteorological evidence of increased rainfall (although future projections tell a different story) but deforestation in the Himalayas and isolated tropical storms appear to have caused much of the problem. Bring climate-change factors into the picture and in the future flood problems will undeniably be aggravated. If the recent climate events in Bangladesh are part of a longer-term future trend, then arguably this future may have already begun. Indeed, factor in the dimension of sea-level rise and matters will get worse still. As stated, the region is low-lying and has previously seen damage from sea flooding. In the case of the 1970 tropical storm some 300 000 perished when a tidal wave swept across the Ganges and Meghina deltas (Burroughs, 1997). The implications are clear.

With regards to UK climate trends, the autumn 2000 was then the wettest since records began in 1766. The heaviest rainfall that season was across England and Wales, with a total of 489 mm falling between September and November; the most extreme rainfall was in October, which resulted in extensive flooding. UK insurance claims arising from that season's floods came to around £1 billion. Many of the same areas of southern UK flooded again in early 2003. In the summer of 2004 a flash flood in Boscastle, Cornwall, was caused by around 20 cm of rain falling in just 4 hours. Two rivers burst their banks and a 3-m wall of water passed through the town's high street. More than 150 people had to be airlifted to safety and despite about 50–60 cars being washed away nobody died, but around £50 million of damage was caused. Another extreme event took place in January 2005, when 100 mm of rain fell around Carlisle in Cumbria. So intense was the rain that surface-water drainage could not cope and flooding began, quickly before river monitoring (which the UK Environmental Agency then relied on) generated a flood alert. The flood caused local power cuts and interrupted both the terrestrial and mobile phone systems; consequently, some people became trapped in their homes.

The same year, 2005, saw a similar event on the North Yorkshire moors. After a weekend heatwave in which temperatures reached 33.1 °C, the night of Sunday 19 June saw rainbursts over the north east of England that triggered regional flooding. North Yorkshire as a whole had the best part of a month's worth of rain in 3 hours. One fast-flowing flood that went through the villages of Thirlby, Helmsley and Hawnby was caused by 2.7 cm of rain that fell in just 15 min. The flood was in places 2 m deep and it carried cars along with it. Meanwhile, in river courses torrents washed bridges away. The highest rainfall in the UK on that day took place in Hawarden, Flintshire, in Wales, with 4.3 cm of rain. Two days after the event some 2500 homes were still without electricity. Indicative of the 'clumping' of rainfall patterns Friday 24 June 2005 saw some 2.8 cm fall on RAF Lyneham in Wiltshire. The usual rainfall for June in that area is 3–6 cm.

On continental Europe the most disastrous floods in recent times took place in August 2002 along the Elbe and Danube rivers. In the Czech

Republic some 200 000 had to leave their homes whereas in Germany 3600 patients had to be moved from hospitals threatened by waters. Rough cost estimates for the Elbe flood are US$3 billion in the Czech Republic and over $9 billion in Germany. Flood damage (which is not the same as flooding) of this magnitude had never happened in Europe before. There have been 10 extreme floods in Dresden, Germany, since the thirteenth century, with water levels peaking between 8.2 and 8.8 m, but the 2002 flood peaked at 9.4 m. However, the previous record-holder in 1845 had a higher flow rate than the flood in 2002. The reason why the 2002 flood was higher despite a lower flow rate is due to greater river containment as a result of floodplain development. This again exemplifies how non-climate factors can compound extreme weather events. Having said this, the rain that caused the 2002 flood exceeded most previously measured rainfall amounts and intensities, and in the core area of precipitation rainfall levels exceeded the 24-h record observed in Germany since records began and were close to the PMP (Becker and Grünewald, 2003).

The future Western European picture that is just beginning to emerge is therefore a complex mix of changes in seasonal precipitation (down in summer and up in winter), extreme events (similar frequency but slightly greater intensity) and watershed development (increased floodplain development). This is exemplified by historical analysis of rivers about which there is sub-stantial historical documentation. Among these are some European rivers. In 2003 German researchers led by Manfred Mudelsee looked at records for the middle Elbe and middle Olde rivers, primarily concentrating on events in the past 500 years, for which the records are more reliable. Discerning long-term trends in monthly run off was difficult. They found that there was no sub-stantial trend emerging above the natural variability in events. So, the histor-ical analysis did not provide evidence for a climate-change-related cause for increased European flooding.

How can this conclusion sit alongside model predictions? Leaving aside model reliability (which is continually improving), historical records look back in time while models look forward. The past has seen just a degree's worth of global warming since the Industrial Revolution, whereas the coming decades are expected to see a Business-as-Usual rise above 1990 temperatures of 2.0–4.5 °C (best-estimate all-model scenarios) with overall low and high estimates of 1.4 and 5.8 °C respectively. In short, retrospective reviews and prospective forecasts are not the same. So, the former may show hardly any change while the latter might suggest a definite future change. What it means is that the early twenty-first century sees us on the cusp. Even so, the actual changes in the future could be quite small, with just marginal excesses over previous instances. Here the problem is that such floods are catastrophic events. Small changes can have big consequences. Matters are fine until river banks are breached (such as with a peak rise of just a few millimetres), then there is a quick transition from no damage being incurred to considerable damage.

Finally, increased major floods could well happen in the summer despite European summers becoming drier. Seemingly paradoxically, computer models predict an increase in *intense* summer rainfall with global warming. Instead of the less rain being spread across summer months, there will a tendency for this precipitation to clump into extreme weather events (Christensen and Christensen, 2002). This is because in a warmer world there is more water vapour in the air. Of course, some air is warmer than other air. So when water-laden air (more than historically anticipated at a given latitude) meets cool air the resulting rainfall will be greater. Additionally, matters can even be worse. Consider two bodies of water-laden, warm air colliding. There is nowhere for the air to go but up. At a higher altitude it cools, again releasing water but even more so than in the previous scenario. The consequence of this for flooding is significant. If rainfall is spaced out over weeks then the water has the opportunity to sink into the ground and then into the subsurface geology. If a few weeks' or even a month's worth of rain falls at once, due to an atmospheric collision, then there is no time for the resulting precipitation to drain away and so it remains on the surface. Problems are compounded if the surface is corrugated with hills and valleys, as this volume of rain will run off the hills into the valleys as fast-flowing floods. Of course, there are parts of the Earth's surface that historically are warmer than temperate latitudes and which also have high rainfalls. However, these regions have seen waterways being gouged out and if they are inhabited they have the appropriate drainage systems in place. With climate change, a place that previously only had moderate, well-spaced rain with time becomes exposed to more extreme rainfall regimens; there will not be appropriate drainage systems in place and so flooding results, with property damage and possible loss of life.

Flooding from both sea-level rise and increased precipitation from greater marine evaporation in a warmer planet are expected symptoms of global warming and this is stated by the IPCC. However, the opposite of an excess of water, drought, is paradoxically also a problem and is also likely in other areas of a warmer world that have greater evaporation from the land. Further, the human impact of droughts in a more crowded future, and hence one that consumes more water, will be exacerbated. This is yet another example of a non-climate factor exacerbating impacts.

7.3.4 Droughts

As with floods, droughts are not new but have been increasing in their severity. Droughts not only affect natural and human-managed ecosystems (such as agricultural lands) but also damage property. The drying out of soil can damage housing, as exemplified by the 1976 drought in England with the cost of repairing domestic properties exceeding £100 million (US$160 million) in the money of the day (Burroughs, 1997).

In 2004 the south of China experienced the worse drought for half a century and low levels of rain continued in 2005 and 2006. Difficult choices then had to be made. In the autumn of 2004 100 hydroelectric plants were shut down so that water could be diverted to rice fields and farmland, some 730 000 ha of which were affected by the drought, including 36 000 ha abandoned due to lack of water. Widespread power cuts resulted. The level of one river, the Dongijang, fell by 80%, which dramatically reduced supplies to some 36 million people. Whereas one incident cannot be specifically blamed on climate change, as noted before, it is the pattern of events that is revealing and this instance is the sort of event that one might expect with global warming. The greatest health impacts arising from climate change are those that adversely impact on the increasing global population's ability to feed itself. Water availability, be it a deficit or a surplus, is one such factor that links climate change to food security. This is the subject of the next section, in which we will discuss drought in a little more detail.

7.4 Climate change and food security

7.4.1 Past and present food security

Humanity's food security improved greatly in the latter quarter of the twentieth century. This has been due to the mechanisation of agriculture and the use of chemical fertilisers and pesticides, which together have formed the Green Revolution that has dramatically increased agricultural productivity (which is defined as output per unit area). The success of the Green Revolution is evident by the late twentieth-century growth of the global population (see the beginning of this chapter), which in turn was built on previous population growth. As previously noted, the global population (Figure 7.4b) began to increase following the Renaissance and continued to rise with the Industrial Revolution. The rate of growth (as opposed to population itself) peaked in the twentieth century. This determined the size of the population in the following decades (Figure 7.4a), which itself is expected to peak during this century. All these extra mouths required, and will require, feeding.

Yet, there has been a price for this success. First, the success did not come without its failures. New agricultural practices were not always sustainable. Concerns such as soil erosion, ecotoxicological impacts from chemical inputs and resistance to pesticides, have meant that techniques have had to be refined. Furthermore, the post-agricultural component to food supply has come to dominate the economics of food supply so that, for example, in the 1990s in the UK while agriculture contributed to a little over 1% of its GDP, post-agricultural processing and commerce added more than another 2% (see section 7.1.4 on food). This post-agricultural dominance can be seen in virtually all countries. Consequently, post-agricultural commercial concerns have shaped much of the way the non-subsistence proportion of the global population feeds itself. Today those in the developed nations are used to eating out-of-season produce and products that have a high energy intensity (again see section 7.1.4). Irrespective of climate change there is a very important question (worthy of a textbook in its own right) as to how the global population of the twenty-first century will feed itself.

In summary, and remembering the backdrop of growing population, the position as revealed by FAO data to date since the middle of the last century is as follows.

- Annual grain production (which includes wheat, maize (corn) and rice) has increased from around 500 million t to around 1800–1900 million t in the 1990s, when it peaked and levelled off. Conversely per-capita grain production has stabilised at around 290–330 kg/year since the 1970s and indeed the trend has been downwards since a mid 1980s peak.
- Annual meat production (over 90% of which is pork, poultry and beef) has increased from 44 million t in 1950 to 237 million t in 2001, whereas average individual consumption has risen from 17.2 kg in 1950 to 39.0 kg in 2002.
- The above has been fuelled by fertilisers, global production of which has increased from 14 million t in 1950 to 120–46 million t in the 1990s.
- The global annual fish catch (as opposed to aquaculture; fish farming) has increased from 19 million t in 1950 to around 92–6 million t in the latter half of the 1990s. However, against a backdrop of growing global population, the annual catch per person despite growth since 1950 has levelled off since the mid 1970s to around 15–17.5 kg. Fortunately the global fish supply has been augmented since the 1980s with aquaculture, whose production has increased from around 7 million t in the mid 1980s to some 36 million t in 2000.

Those in industrialised countries consume more grain than non-industrialised ones via animal feed for the increasing meat in their diet. For example, a North American consumes nearly 10 times as much grain as someone from a sub-Saharan nation. Middle-income countries have also seen a twentieth-century increase in per-capita meat consumption. Not irrelevant is the increasing incidence of obesity and diabetes in these countries. Despite this inefficiency in healthy nutrition, globally those facing chronic hunger have decreased from 956 million (approximately 26% of the global population) in 1970 to 815 million (approximately 13% of the global population) in 2002.

There are two key dimensions undermining the future sustainability of food supply. First, on land much of the most productive ecosystems are already heavily exploited. Much of the most suitable land has already been turned to agriculture while other land has gone to urban sprawl and related infrastructure. Worldwide from 1700 to 1980 crop lands increased by 466.4% from 265 million ha to 1501 million ha. Most of this has been at the expense of forest and woodland that has declined by 18.7% from 6215 million to 5053 million ha (Turner *et al.*, 1990). At sea the price of human success has been the collapse of fisheries, such as in the North Sea. There have also been effects elsewhere in the marine food web. For example, around Alaska the tripling of the pollack (*Pollachius* spp.) catch between the mid 1980s and the end of the century saw the stellar sea lion (*Eumetopias jubatus*) population that feed off of them collapse by around 90% (Starke, 1999).

The second factor is that the growth in human population, and its modern fossil-energy-intensive culture, appears not have been sustainable, not just in food-security terms, but also in terms of the sustainable harvesting of ecosystems. For example, non-sustainable fossil fuels are the main source of urea that is used for agricultural fertiliser. Again, regarding fresh water (another essential agricultural input), the growth in demand for water in many parts of the planet is not sustainable. In other words, not only are we quantitatively over-exploiting ecosystems so that they collapse, we are also doing this qualitatively, in a non-sustainable way.

However, while both these dimensions are real, combating hunger in modern times has been arguably hindered by an enabling factor of economic development. True, while much of the most productive land around centres of population has been exploited, it is often possible to irrigate marginal lands, develop sustainable agricultural systems or buy in food from elsewhere, etc. However, this takes both finance and a sound commercial framework within which to operate. This development dimension is cited by the FAO as being one of the current key factors determining food security. Unfortunately international trade barriers and development agreements have in some instances undermined sustainable development. Whether twentieth-century-style economic development will remain in a more crowded future with climate change remains to be seen.

7.4.2 Future food security and climate change

As discussed, without future climate change we know (from current human longevity records and birth and death rates) that the global population will continue to rise through to the middle of the twenty-first century. Given that we are continuing to push for increased food production, what does the future hold? To begin to answer this we need to know where we are now.

The 1996 World Food Summit set the goal to reduce the number of people undernourished globally by half by 2015. Since then the UN Food and Agricultural Organization (FAO) has produced annual reports on *The State of Food Insecurity in the World*. In summary, its early reports point to a declining number of people who are undernourished. In 1970 there were approximately 960 million undernourished. The FAO estimate that in the period 1997–9 this figure had declined to 815 million. Of these, over 95% were in developing nations, around 3% in transitional nations and 1.4% in developed industrial nations. The FAO's 2001 report noted that the (then) current decline in undernourished was 6 million a year, which in the 1990s represented a reduction in the rate of decline in undernourishment compared to the 1980s. In part this might be attributable to the increase in global population (see earlier in this chapter) but nonetheless by 2001 if this reduced trend continued then it would take 60 years to reach the 1996 Summit's target number, which was supposed to be achievable 14 years hence, by 2015. In short, to achieve the original target the annual reduction in undernourished people would have to increase from 6 million to 21 million a year. This target, the 2001 report concluded, was unlikely to be reached without 'rallying political will and resources'. This was prophetic.

The publication of the FAO's 2004 report announced a turning point. The numbers of undernourished people had stopped declining and in fact had increased, while the world population also continued to grow. The 815 million undernourished for the period 1997–9 increased to 842 million for 1999–2001. Consequently the FAO reported that the world was further away from meeting the 1996 World Food Summit target. This could now be reached only if annual reductions were to be accelerated to 26 million per year, more than 12 times the pace of 2.1 million per year achieved on average from the 1996 summit to 2004. The 2004 report also pointed to Africa as the continent with the greatest nourishment problems.

With regards to climate change as a factor affecting food security, the FAO food security reports do not specifically raise this as a concern but do touch upon it tangentially. The closest the 2001 report came to mentioning climate change was when it briefly cited weather extremes – specifically droughts, floods, cyclones and extreme temperatures – as threatening progress towards food security. By the time of the 2004 report the FAO gave weather extremes a higher profile, although developmental concerns (including human conflict) were (rightly) of greater current concern. With reference to both these issues the reports said: 'Many countries that are plagued by unfavourable weather but enjoy relatively stable economies and governments have implemented crisis prevention and mitigation programmes and established effective channels for relief and rehabilitation efforts. But when a country has also been battered by conflict or economic collapse, programmes and infrastructure for prevention, relief and rehabilitation are usually disrupted or destroyed.' The 2004 FAO data demonstrated that the trends in either increasing or decreasing hunger were both related to a nation's growth in GDP, with those countries with the lowest economic growth faring the worst.

The FAO's 2004 food-security report did acknowledge the importance of weather-related events in causing food crises, but it noted that the number of crises caused by weather, as well as jointly weather and human conflict, had decreased from 86% of crises in the period 1986–91 to 63% for 1992–2004. Conversely problems due to human activities (mainly conflict) had increased. This explains the FAO's current focus away from climate change.

Despite the aforementioned immediate-crisis caveat, from a climate-change perspective the 2004 report did highlight the importance of water to long-term food security. It observed that agriculture is by far the biggest user of water, accounting for about 69% of all withdrawals worldwide and over 80% in developing countries. Reliable access to adequate water increases agricultural yields, providing more food and higher incomes in the rural areas that are home to three-quarters of the globe's hungry people. So, not surprisingly, countries with better access to water also tend to have lower levels of hunger. Drought ranks as the single most common cause of severe food shortage in developing countries. For the three most recent years for which data were available to the 2004 FAO report's authors, weather-related causes of hunger increased year on year, with drought listed as a cause in 60% of food emergencies. Even where overall water availability is adequate, erratic rainfall and access to water can cause both short-term food shortages and long-term food insecurity. Floods are another major cause of food emergencies. Sharp seasonal differences in water availability can also increase food insecurity. In India, for example, more than 70% of annual rainfall occurs during the 3 months of the monsoon, when most of it floods out to sea.

Conversely, farmers who lack irrigation facilities must contend with water scarcity through much of the year and the threat of crop failures when the monsoons fail. Where water is scarce and the environment fragile, achieving food security may depend on what has been called virtual water – foods

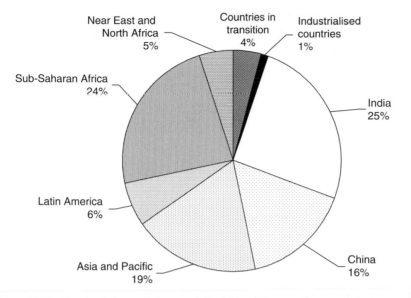

Figure 7.13 The location of the world's 842 million undernourished people (1999–2001). Based on data from FAO (2004).

imported from countries with an abundance of water. It takes 1 m³ of water to produce 1 kg of wheat. Extrapolating from those numbers, the FAO calculated that to grow the amount of food imported by Near-Eastern countries in 1994 would have required as much water as the total annual flow of the Nile at Aswan. In such conditions, it may make sense to import food and use limited water resources for other purposes, including growing high-value crops for export. As we have seen (Chapter 6), in a globally warmed world not only can we expect extremes in precipitation (more evaporation of seas, more rain) but also greater evaporation on land and through plant photorespiration. With global warming water availability will become a key issue and, against a backdrop of an increasingly populous planet, a driver of food insecurity.

With regards to where food security is currently the greatest problem, as can be seen from Figure 7.13, Africa is the continent bearing the brunt. In 2005 it was the UK's turn to lead the G8 group of most economically active nations and for its part it made improving conditions in Africa a priority theme. With regards to the contribution of science to this priority, the UK's theme was climate-change science and Africa. As for the impact that this initiative will have, time will tell.

Perspectives on the prospects for improved food security extend beyond the FAO. An analysis by Lester Brown of the Worldwatch Institute is not encouraging (Brown, 1995; incidentally, the Worldwatch Institute contributed to the aforementioned 2001 FAO food-security report). Lester Brown noted that two of the four largest grain producers (the USA, China, the former Soviet block and India) that historically in 1950 either were net exporters or were virtually self-sufficient, had by 1990 became net importers, with only the USA increasing its production to a level that more than offset the decline in the others. However, over the same period this US surplus was also offset by declines in production by the nine other most-populous countries. Factor in the anticipated future growth in population and it becomes clear that we cannot continue to feed the world as we now do.

Contrary to initial impressions, this is not a doom-and-gloom conclusion. In 1972 the report for the Club of Rome, *The Limits to Growth*, was prophetic (Meadows *et al.*). Back then it predicted that resources would run out *unless* we changed the way they were exploited and used. Fortunately, this happened. For example, since its publication the efficiency of extraction of copper has improved, so remarkably the spent tailing mounds of previous extractions have been able to be mined, while at the other end of the production cycle, material recycling has boomed (see Chapter 8). Such is the wastage, similar improvements in efficiency could happen with global food. However, if this does happen then western consumers would have to accept changes. To take just three examples, food miles would need to be reduced dramatically so as to free agri-energy to improve productivity elsewhere as well as to reduce transition times during which spoiling can take place. This last would mean a return to the seasonal availability of many foods and a reduction in year-round food choice (although not choice across the year). A second possibility might be a change in animal protein and fat consumption, which through much of the twentieth century has grown ahead of population and which arguably is one of the factors behind obesity-related health impacts that have begun to dominate the health agendas of a number of developed nations. Finally, there would need to be an increase in (sustainable) agricultural intensity. This rise in production per unit area (productivity) in turn means a decrease in biological resources available to wildlife within that agricultural area. There would be biodiversity implications both within and outside of these agricultural areas (see the discussion on agriculture, biodiversity and landscape earlier).

Whereas improvements in food-supply efficiency may be a note of hope for our species, what it does mean if we are to avoid increasing starvation is that humanity will feed itself differently 30 years from now compared to today, and probably as differently as it was fed 50 years ago. However, even with this it is hard to see pressures on natural systems decline; indeed, it is almost inevitable that pressures on them will increase. Either way, greater resource management worldwide will be required.

The above summary does not emphasise the climate-change factors. So, what are the food-security implications of global warming? We have already seen that past climate change has had a detrimental impact on civilisations such as with the Incas in the New World and the people affected by the Little Ice Age in the Old World. It is therefore difficult to see why future climate change should not have impacts on future food supply. With past change agricultural belts moved and the problems arose because the civilisations did not move with them and exploit new potentials. The same problem is likely to affect future climate-related changes in agricultural production. The big difficulty is that the present and future are far more crowded than the past. There are increasingly limited options for opening up land that does not already have a claim on it. A compounding matter is that a significant proportion of the human population is concentrated along land-mass coastlines, some of which will be under threat from climate-induced sea-level rise compounded by occasional intense run off from the land.

Climate–food models of future global agricultural production vary. Much depends on both which model is used and the agricultural products that will be grown. This is in part because, as with climate models, their resolution is too coarse. Also, because climate and agriculture are so closely related, while climate models are not (yet) sufficiently sophisticated for precision use in regional and local planning, so climate–food models cannot be used for regional and local agricultural planning.

Regarding food security, the IPCC (2001a) concludes that while plant growth will benefit from the increased atmospheric carbon dioxide, agricultural belts will shift with climate. This means that farmers in existing agricultural areas will need to change their crops or, in come cases, cease farming or move. The IPCC also notes that in areas that become favourable for agriculture there will be new opportunities.

The IPCC raises the concern that extreme weather events will become increasingly problematic. Indeed, examples from food production subsequent to their latest report (2001) already abound. For instance, the Western European heatwave of 2003 scorched enough of the potential wheat harvest to make a noticeable impact on bread prices. Most of the harvest was lost in southern Europe compared to the north. France lost about a fifth of its anticipated harvest while Italy and Britain lost over 10%. Moldova was one of the worst hit countries, losing about three-quarters of its anticipated harvest. In August 2003, Rank Hovis, the UK's largest flour miller, announced that the price of its milling flour had increased from £35 a tonne to around £245. This in turn increased bread prices by around 10–15%. In part this modest rise was due to flour costs being a minor component of bread's retail price. The modest increase was also due to being able to import wheat from elsewhere, as these losses were not similarly reflected worldwide, with the year's global grain (including rice and maize (corn)) harvest being down on expectations by less than 2%. But other crops were affected in Europe; for example, potatoes also saw a price increase.

A major future challenge for global agriculture, the IPCC predict (2001a), will be the degradation of soil and also water resources. Drier summers in many temperate areas but with bursts of more intense rainfall will necessitate greater water management and the ability to capture downpours for storage. Heavier winter rainfall again will necessitate greater watershed management to minimise soil erosion and flooding.

The IPCC's overall conclusion is that modest climate change that might be expected in the first quarter of the twenty-first century will result in agricultural winners and losers. However, more significant annual temperature increases of 2.5 °C or more could prompt global food prices to rise. Again, the IPCC note that the greatest impacts will be on the poorer farmers and consumers.

So why would increases of annual temperatures upwards of 2.5 °C have a net detrimental effect? After all, a warmer, carbon dioxide-rich planet should mean that there are some (new) places that will favour agriculture even if some other places become unsuitable. Here the concern is that higher temperatures shorten the life cycle of many cereals, hastening senescence and so reducing the length of the growing season. The world's staple cereal crops tend to tolerate narrow temperature ranges, which, if exceeded during the flowering stages, can damage production of fertile seeds and reduce yields. Given that global warming is to increase the number of extreme weather events, crop yield is likely to be impaired. This would also be compounded by increases in ozone in the troposphere (which is low down in the atmosphere; Porter, 2005). This is likely to be more of a factor in parts of the globe where there are fewer effective controls on air pollution, such as Asia.

Given the almost certain demographic pressures anticipated for the rest of this century, agricultural research needs to become a priority and this needs to be related to high-resolution climate models that include the biological dimensions (which they are now beginning to do). This is worth stressing because many developed nations have been markedly reducing their investment in whole-organism agricultural research due to past successes – reflected in the agricultural productivity (production per unit area) rises – and past European and American food surpluses. Consequently a number of governments have felt that with past success achieved, there is little new challenge on the horizon. In the UK governmental investment in policy-driven agricultural research from the department responsible for farming and fisheries has declined in real terms year on year since the mid 1980s to the present (2006/7).

Internationally, one of the recent initiatives to help co-ordinate research into food security and environmental change is Global Environmental Change and Food Systems (or GECAFS). GECAFS was established in 2001 with formal partnerships with the UN FAO and WMO. It is due to run for a decade.

Then there are broader policy concerns beyond those of science research: policies that enable us to both reduce the speed of climate change and its extent, as well as to adapt to likely impacts will help enable our global society to cope with the warming world of the twenty-first century.

7.5 The biology of reducing anthropogenic climate change

Just as it is a combination of human short-circuiting of the deep carbon cycle (through fossil-fuel burning) as well as altering the fast carbon cycle (through land-use change such as deforestation) that is increasing atmospheric carbon dioxide, so it is possible for us to take carbon-cycle actions that reduce climate change or, if not that as such, at least the rate of change. Improving the efficiency with which fossil fuels are consumed and switching reliance towards non-fossil energy resources will be briefly discussed in the next chapter. However, in terms of human ecology (our own species' demographics and relationship with other species) we might manipulate the fast carbon cycle through land-use, and perhaps marine, management. The aim would be to use the fast carbon cycle to offset carbon dioxide from fossil carbon (part of the deep carbon cycle) by increasing fast-cycle carbon sinks as well as partly replacing dependence on fossil carbon with fast-cycle carbon (or biofuels). However, care would be needed to ensure that we did not build up ecological reservoirs of carbon that would subsequently be released through climate change.

In terms of tapping into the deep carbon cycle, the IPCC (2001a) broadly estimates that the total fossil carbon released between the onset of the Industrial Revolution (1880) and almost 1998 was around 300 GtC. This compares with a broad upper estimate of conventional fossil carbon resources (past and the next two centuries) of ∼1700 GtC. Alternatively, if one includes unconventional fossil reserves (such as tar sands and shale oils, etc.), then this increases to a total fossil carbon inventory that might be exploited of almost 5000 GtC. (And this last excludes carbon fuels that we have as yet not even begun to consider as unconventional resources, such as marine hydrates (clathrates), which the IPCC estimate at 12 000 GtC.) This estimate of approximately 300 GtC for historic fossil releases compares to the forecast total historic *and* future releases (1880–2100), which the IPCC (2001a) use based on its Special Report on Emission Scenarios (SRES), of between 1000 and

2150 GtC for the twenty-first century. In other words, the IPCC thinks it likely – depending on various economic futures and compared to the 12 decades up to the end of the twentieth century – that twenty-first-century carbon releases may be double to over six times that released historically.

The aforementioned data give us a rough upper estimate for *average* annual fossil carbon emissions for the twenty-first century of around 18 GtC/year for a high Business-as-Usual (BaU) scenario (though more is most likely to be released annually at the end of the century than the beginning). For biological sequestration to have a significant impact it is going to have to be extremely effective. So what is its likely potential?

7.5.1 *Terrestrial photosynthesis and soil carbon*

The above twenty-first-century 18-GtC BaU annual fossil release is small compared to the broadly 50 GtC photosynthetically captured by the Earth's terrestrial ecosystems (i.e. excluding ocean productivity) each year *after* allowing for respiration. In theory therefore it should be possible to harvest 18 GtC of terrestrial vegetation (say as wood) each year and bury it (effectively removing carbon from the atmosphere) or burn it as fuel (effectively part-offsetting fossil carbon emissions). This would negate human average annual twenty-first-century carbon emissions. However, to do this we would need to harvest the equivalent of all new tree growth globally each year and this is clearly impractical. (Note: notwithstanding in addition to the approximately 50 GtC photosynthetically captured there is also a carbon stock from previous years' growth and this last forms the bulk of the carbon stocks depicted in Figure 7.14.)

However, the theoretical possibility of photosynthetically drawing down such a large amount of carbon can perhaps be better understood by looking at terrestrial–atmospheric carbon exchange. Such is the present rate of carbon removal from the atmosphere that at current rates the entire atmospheric volume of carbon in theory might be exchanged within about a decade. Of course, there are mixing problems and just as carbon is drawn down in the form of carbon dioxide it is also respired back as well as returned in other ways and places (such as the oceans), not to mention turned into other forms of carbon, such as methane. Another theoretical problem of this hypothetical situation would be that actually drawing down that much carbon would serve to draw carbon out of the oceans. This is one reason why the atmospheric residence time of adding a single molecule of carbon dioxide to the atmosphere is effectively between 50 and 200 years (see Table 1.1). Nonetheless, the idea of trapping (sequestering) carbon in biomass and soils is intriguing. Indeed, we

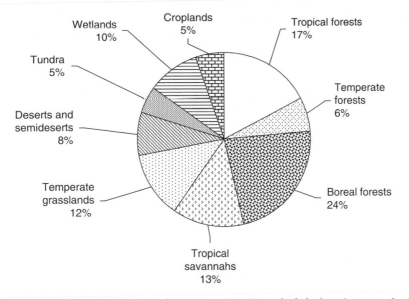

Figure 7.14 The proportions of some 2.3 Tt of total global carbon stocks in soil and vegetation in different terrestrial biomes. After Royal Society (2001).

know that roughly half the carbon released into the atmosphere through human action each year remains in the atmosphere. The other half is absorbed by vegetation, soils and the oceans, so sequestration already takes place somewhere in the biosphere.

The possibility that forests and soils might be deliberately manipulated to mitigate carbon dioxide emissions was recognised by the UN's Framework Convention on Climate Change (FCCC) in 1992 and the Kyoto Protocol (1997). (The convention and protocol will be discussed further in the next chapter.) Both these emphasise natural terrestrial sinks rather than marine ones, largely because land carbon is easier to manipulate as well as due to issues of ownership.

There is very roughly 2.3 Tt of carbon in vegetation and soils (and a further 1 Tt in the oceans' surface layers, which we will return to later): this is very much a ball-park figure for discussion purposes (see Figure 1.3). However, carbon is not distributed equally across terrestrial biomes (categories of regional communities of species). This is not only because some biomes contain more carbon per unit area than others but because some biomes are more extensive globally than others. Figure 7.14 provides a proportional breakdown showing in which terrestrial biomes the carbon can be found.

Forests and forest soils contain about 47% of terrestrial-biome carbon and so this form of carbon storage is attracting considerable attention. However, broadly speaking forests that have been in existence for thousands of years (climax forest communities) tend to be greenhouse-neutral. ('Broadly speaking' because a forest's carbon balance is climate-dependent and will change with temperature and water availability, see Chapters 1, 5 and 6.) It is the establishment of new forests that has a net, short-term effect sequestering carbon from the atmosphere. If you like, this is the opposite of deforestation, which is the second anthropogenic factor contributing to the build-up of atmospheric carbon dioxide (see Table 1.3). This new forest necessity places a severe constraint on forestation as a form of carbon sequestration. Not only is land itself finite, it is also required for other uses. Furthermore, not all the Earth's land is suitable for forestation. The IPCC's second assessment report (1995) estimated that by 2050 some 60–87 GtC could be conserved or sequestered in forests. Compared to the above-estimated mean annual twenty-first-century BaU fossil-fuel emissions of 18 GtC, this represents an annual saving of around 1.2–1.7 GtC, or around 6–9% of annual emissions.

Soils can also be managed to increase their carbon content (although it has to be said that we have a lot to learn about soil-carbon mobility and fixation). The potential here might at first appear quite considerable, especially as roughly three times the amount of carbon is stored in soils than as vegetation above ground. Indeed, previously noted is that sub-arctic soils already hold a considerable volume of carbon (albeit a small proportion of terrestrial carbon in both vegetation and soils). Of the various soil types, peatlands represent a huge store of carbon (see Chapter 4). As previously noted, high-latitude peatlands (which include parts of tundra, boreal and semi-desert soils and wetlands), with between 180 and 455 GtC, represent up to about a third of the global soil carbon pool. Some 70 GtC has been sequestered since the last glacial maximum (LGM) but, since the LGM was over 15 000 years ago and with the current interglacial Holocene only lasting 11 500 years so far, this represents a small annual sequestration rate which would be difficult to enhance to meaningful levels. So, despite being a large carbon pool, high-latitude soils do not have a commensurate potential for further sequestration. If anything, there is concern for the opposite. As with forests these high-latitude peatlands and permafrost soils are climate-sensitive. It is quite likely that they might release their store of carbon if warmed. In short, with global warming they could become a carbon source.

Conversely, agricultural soils do have more potential. Soils that are currently used for agriculture are regularly ploughed, so bringing organic carbon to the surface and destroying plant root networks that physically trap carbon compounds. Ploughing also greatly aerates soils, so facilitating oxidation of carbon compounds. On the other hand natural and semi-natural grassland and forest systems are not subject to regular ploughing but additionally provide the soil with carbon compound inputs. The IPCC (2001a) notes that some agricultural land, such as set-aside (which represents some 10% of agricultural land in the EU), could be managed to enhance soil carbon. In total the IPCC's second assessment report (1995) estimated that up to 2050 some 23–44 GtC could be sequestered by agricultural soils, or around 2.6–5% of estimated mean annual twenty-first-century BaU emissions. This excludes the potential saving of fossil-fuel emissions (mitigation) from biofuels.

The IPCC's 2001 estimate for carbon mitigation from *both* vegetation and soil management up to the year 2050 is 100 GtC. This is equivalent to 10–20% of fossil-fuel emissions estimated for that period. The IPCC (2001a) also notes that 'hypothetically' if all the carbon released due to historic land-use change (forest clearance and such) could be reversed then projected 2100 atmospheric concentrations (approximately 800 ppm under BaU scenarios) could be reduced by 40–70 ppm. This last would be a one-off gain from the finite land area historically affected by land use to release carbon dioxide (see Chapter 5), for once a new forest is fully established it ceases to be a net absorber of carbon.

Managing the terrestrial short-term carbon-cycle carbon can offset a small, albeit significant, proportion of carbon likely to be emitted from twenty-first-century fossil-fuel burning. However, it is not risk-free. The more carbon built up in an ecosystem the more the likelihood of possible carbon leakage. Clearly, if an ecosystem is devoid of carbon then there is none to leak. On the other hand, a system rich in carbon, be it through peaty soils or woodland, has plenty of carbon that can escape. For instance, a forest fire can release nearly all the above-ground carbon extremely quickly, whereas ploughing grassland managed to maximise soil carbon can undo much of the benefit in a single season. Leakage can also take place in soils whose carbon pools have adjusted to the comparatively stable Holocene climate of the past 10 000 years or so, but which could well release their carbon with warming anticipated for the twenty-first century that takes them beyond temperatures seen in this past time (see Chapter 1).

The potential for carbon leakage should be of concern to policy-makers concerned with climate change. As we shall see in the next chapter, some countries (typically nations that use a lot of fossil fuels) have insisted that it be possible to trade, using permits, the right to emit carbon dioxide for carbon-sequestering ecosystem-management schemes. Trading in such greenhouse permits at best can have a short-term mitigating effect. However, it comes with a risk, and there is the question as to what would happen in this permit game if after a number of decades a forest – created, say, by funding from a fossil-fuel station – burned down? Or if its soil carbon was released with global warming? How would the trading permit be paid back and the permit's environmental impact nullified? These questions have not been resolved. Indeed the IPCC (2001a) warn that 'larger carbon stocks [in ecosystems] may pose a risk for higher CO_2 emissions in the future'. It says that 'if biological mitigation activities are modest, leakage is likely to be small. However, the amount of leakage could rise if biological activities became large and widespread'.

Boreal and tundra soils hold considerable carbon stocks, and of these, as noted in Chapters 1 and 4 high-latitude soils currently have up to a third of the global soil carbon pool. These, as per the IPCC warning, pose a risk for higher carbon dioxide emissions with warming. Already there is some experimental evidence suggesting that warming will result in carbon loss from these high-latitude soils and indeed carbon loss from temperate and tropical soils (see Chapter 1). Yet soil carbon loss may also increase another way as the climate warms.

The amount of carbon held in a soil should really be viewed as a balance between the rate of carbon entering the soil and the rate at which it leaves. Such a consideration should include the living plant dimension, as it is the photo-synthetically driven primary productivity (plant growth) that largely draws down carbon, with plants creating roots within the soil and plant remains lying on the soil surface, from which carbon compounds can leach into the soil beneath. Therefore, if something were to hinder primary productivity then the balance of carbon flows would shift. In 2005 it was reported that this indeed happened in Europe during the 2003 heatwave. Then the July temperature in places was up to 6 °C above long-term means and rainfall showed deficits of up to 30 cm/year. This extreme weather event impeded primary productivity and reversed these ecosystems' former net carbon sequestration.

The work (Ciais *et al.*, 2005) was conducted as part of the EU CarboEurope research programme and, though not as detailed an assessment as many ecologists would like, as a partial snapshot it gives cause for thought. The researchers looked at the programme's monitoring of carbon dioxide, water and energy from one grassland and 14 forest sites for 2003 (the heatwave year) and 2002 (the control year). They also analysed crop-harvest data at country level and compared 2003's figures with the 1998–2002 annual averages. Finally, they linked these field data to a sophisticated ecosystem computer model that necessitated supercomputing facilities (which were provided by the French Commissariat à l'Energie Atomique). The results suggested that the 2003 heatwave months resulted in a 30% reduction in gross primary produc-tivity for the year and that carbon release across Western Europe was some 500 GtC. Up to then it had (tentatively) been thought that Europe's soils had slowly been accumulating carbon. This carbon loss roughly equalled some 4 years of such accumulation. What was not known is the knock-on effect of this on the following year's carbon balance. However, what is known is that the 2003 European heatwave temperatures, while those of an extreme event early in the twenty-first century, are destined to be average summer temperatures later in the century (and that heatwave summers then will be correspondingly warmer; see Chapter 6). In short, Western Europe's 2003 carbon loss was not a one-off event.

All of this does not bode well for a system of trading of greenhouse permits that uses carbon sequestration in natural systems, because it assumes that these systems are known and can be taken for granted given a certain management regime in the future. As a way of mitigating fossil emissions, it can at the very best only be as good as however long is the commitment to ecosystem management. ('At the very best' because, as noted, a warmer world facilitates ecosystem carbon release regardless of commitment and present carbon savings could easily turn into future carbon losses). The success of such trading schemes also depends on our knowledge (currently not sufficiently complete) as to agricultural sinks and especially soils, and our ability to monitor them, as well as the schemes' robustness lest they be open to gamesmanship by fossil-energy producers and consumers. Nonetheless, in the 1990s and early 2000s many policy-makers welcomed such permit-trading schemes.

However, whereas permit-trading schemes that rely on soils as carbon sinks may not function properly per se, carbon sequestration can still play a role in helping (albeit temporarily) slow the build-up of atmospheric carbon dioxide. This last could well help slow the rate of climate change and may (marginally) help both ecosystems and human systems adapt even if a hectare of ecosystem cannot be equated with the consumption of a specific amount of fossil fuel.

7.5.2 Manipulating marine photosynthesis

Compared to a very broad figure of around 50 GtC for annual net terrestrial biological productivity for a biotic carbon stock of 550 GtC, the oceans have a small biotic carbon stock of only around 3 GtC. Much of the 1 Tt of carbon in the oceans, that was noted in the previous subsection, is in inorganic (non-biotic) forms. The other big carbon difference between the terrestrial and marine biotic reservoirs is that, whereas terrestrial ecosystems have an annual carbon turnover of roughly 20% of the terrestrial biotic stock, the carbon turnover of the marine carbon stock is in excess of 1000%.

In other words, the terrestrial biota represent a large mass of biotic carbon with more modest carbon flows to and from this volume. Conversely, marine biota represent a small mass of carbon with a comparatively very high carbon turnover whereby the carbon ends up in the large inorganic carbon ocean reservoir. Managing the marine carbon cycle to offset atmospheric increases in carbon dioxide from fossil-fuel burning depends on this naturally high turnover and indeed this is the attraction of such management.

As noted in Chapter 1, there have been attempts to encourage algal blooms with the IronEx experiments. However, while successful in forming blooms, the energy required to make, transport and distribute the iron used to fertilise the ocean more or less balanced the fossil energy that was saved in carbon sequestration. Furthermore, the ecological implications should an IronEx type scheme be implemented on a large scale are unknown. With regard to marine ecosystem manipulation, the IPCC (2001a) conclude that much remains 'unresolved' and that such schemes 'are not, therefore, ready for near-term application'.

7.5.3 *Biofuels*

Biofuels arise from living species and are greenhouse-neutral in that they sequester atmospheric carbon dioxide when grown and return it when burnt. However, if the products of biofuel consumption are captured (for example, from the flue of a biofuel power station) and stored, then there is a net transference of carbon from the fast to the deep carbon cycle (see section 8.2.4). This is one way in which carbon dioxide can be usefully sequestered from the atmosphere. Biofuels can also be considered to have a positive effect offsetting fossil carbon emissions when biofuels displace oil, gas and coal consumption. Here the transference of carbon from the deep carbon cycle to the atmosphere is reduced.

There is a continuum of agricultural biological resources from which biofuels may be generated. This continuum is essentially founded within the agriculture sector but also includes forestry. At one end of this spectrum the biofuel resources are derived almost by happenstance from agricultural wastes such as animal slurry (from piggeries, poultry wastes, etc.) and vegetative wastes (such as straw from cereal production). At the continuum's other end, biofuels can be derived from existing agricultural products such as sugars and oils. In between these there are both dedicated and dependent biofuel sources. Dedicated biofuel sources are those whose planting solely provides biofuels and nothing else of economic value. The tropical grass *Miscanthus sinensis*, for example, has no other real value as a crop in the USA or Europe (although it is used as an ornamental plant in the European domestic sector). Conversely, dependent biofuel sources are those that are dependent on another economic gain (commonly food-product-related) from growing the crop. However, the reality is that few biofuel sources fall strictly into either category, even though some may be predominantly one or the other. *Miscanthus*, for instance, is also being investigated as a possible UK crop for fibre production. Birch (*Betula* spp.), willow (*Salix* spp.) and alder (*Alnus* spp.) can contribute to biofuel production while continuing to provide stock for paper and fibre-board production. Since these are broad-leaved, woody species, their appropriate cultivation adds substantially to the biodiversity of associated flora and fauna and offers visually attractive amenities during production.

As to the potential contribution of biofuels, taking the UK as an example, if all 500 000 ha of UK set-aside land (this is equivalent to 3% of the land *within* agricultural holdings) was planted with, for instance, *Miscanthus*, and assuming a median dry matter production of 15 t/ha per year and a broadly typical 40% conversion to oil in energy-content terms, then some 3 mtoe (million t of oil equivalent) of biofuel would result. This would be equivalent to nearly 4% of the UK annual oil demand. If *Miscanthus* oil production could be increased to an optimistic 25 t of dry matter/ha, then over 6% of UK oil demand would be met. However, *Miscanthus* does not reach its peak yields until the third or fourth season, even though absolute peak yields for *Miscanthus sacchariflorus* of 44 t of dry matter/ha per year have been reported from Denmark. This peak, if possible across all such potential UK *Miscanthus* production, would result in 14 or 15 t of oil/ha per year. However, such high yields are highly theoretical and cannot be considered a realistically obtainable national average at this time (Cowie, 2003), although they might with appropriate breeding or genetic modification if environmental concerns could be addressed.

A similar calculation for oilseed rape would take into account the approximately 3.2 t harvested per hectare (in the UK), the 37% of recoverable oil from the rapeseed and the 95% conversion to diesel. This gives us about 0.4 mtoe/year (0.5% of UK oil demand). The advantages of rape are that the plant's entire life cycle takes place in the UK, but has the disadvantages that methanol (equivalent to about 12% of fuel yield) is required during the diesel-making process: however, this methanol could come from biological sources too. It should be noted that there are rotational constraints on rapeseed production: rapeseed is usually grown one year in five in the UK.

Miscanthus and rapeseed oil are taken as just two examples of how currently unused set-aside land could contribute to make the UK more energy-secure through increasing energy self-sufficiency. However, it may be that some other crops, or crop mix, would be preferable. Climate conditions do not favour *Miscanthus* rhizome production in Europe as they do in the tropics.

Nonetheless, rhizome production is possible in the UK (which is warmed by the Gulf Stream) and in some other parts of Europe. Consequently, the examples are illustrative and the production levels of *Miscanthus* and rapeseed oil are not atypical in energy terms of other biofuels.

It should be noted that in addition to the 3% of land within agricultural holdings being set aside, 97% is not. (Of this 24% is rough grazing and 11% unspecified. Indeed, of the 97% of agricultural land that is not set aside, only 25% is used for crops.) In short, there is room for some considerable increases in the land that might be available for biofuels. Assuming, solely for purposes of example, that 6% of current UK oil consumption was met through the use of specific biofuel crops, then there is the additional possibility of a biofuel contribution from dependent sources: that is, crops that are grown for other purposes; for example, straw (as mentioned above) or crop trimmings, such as from field vegetables. This last is currently on land covering over eight times the size of set-aside land. Indeed, much of the above-ground biomass that is not harvested for food or export (the waste biomass) from these species could be used for biofuel production. Finally, there are the biofuels available from forestry and woodland, such as from short-rotation coppicing.

In 2002 the Institute of Biology ran a biofuels workshop in London as part of an energy series run also by the Institute of Physics and Royal Society of Chemistry, which was sponsored by the Department of Trade and Industry. Workshop participants, all associated with either the biofuel industry or biofuel research, wcre asked in advance as to the potential for biofuels to offset UK oil demand. The modal answer was around $20 \pm 5\%$ of UK oil demand or about 7% of total UK primary energy consumption. If this estimate is realistic then it represents some 16 million t of UK domestic annual oil consumption and a roughly a £1.3 billion gain to the UK's domestic economic output at the point of refinement/production (not point of sale) in terms of crude-oil savings. This 7% figure of UK primary energy goes a long way towards meeting the Government's medium-term renewable targets, although the UK will require greater displacement of fossil fuel than that if it is to meet its Kyoto targets. Even so, it is difficult to see any fossil-fuel-reducing strategy in the medium-to-long term in the UK that does not include biofuels (Cowie, 2003).

The estimate for a practical biofuel contribution to the UK of some 15–25% of current oil consumption, or 7% of UK primary energy, is not at all unrealistic. True, while a greater proportion may well be feasible, equally it might perhaps incur significant opportunity costs (probably either to strategic food production and/or the environment). The importance of biofuels to the UK would be best appreciated if undertaken as part of a coherent energy strategy, along with improved energy efficiency as well as changes of behaviour favouring lower energy consumption. The question is whether the UK Government would consider a firm drive for biofuels to be a worthy goal in economic, environmental and strategic terms. It is also a question other governments will need to face. The concerns raised by the workshop that the UK was not coherently exploring biofuel options were subsequently echoed by a House of Commons Select Committee report (EFRA Select Committee, 2003; See also the UK policy case history in Chapter 8).

Finally, the 2002 Institute of Biology biofuels workshop looked at costs. The range of figures presented did result in an end biofuel cost per barrel that was of the same order of magnitude as that for crude oil (which still has refining costs to bear). Unfortunately the current vagaries of small-scale production, lack of distribution and other factors made a firm costing difficult. Nonetheless, biofuels remain a useful option as part of a suite of measures to counter fossil-fuel emissions of carbon dioxide.

Globally, aside from non-commercial local use of biofuels (see section 8.2.3) a little over 1% of commercially traded energy currently comes from biomass. Of all countries, Brazil arguably leads in biofuel production and consumption with biofuels contributing some 20% of its primary energy supply. (However, questions need to be asked as to how much of this comes from biologically sustainable sources that do not contribute net greenhouse emissions from climate change). While biomass could theoretically make a major contribution to offsetting fossil-fuel consumption it is more likely to be used to meet chemical feedstock demand as oil becomes more expensive, especially as there are competing land pressures for food production. It has been calculated that if, for example, the USA were to displace 10% of its car petrol consumption with biofuel then 12% of US cropland would be required, but a further 38% would be needed (bringing the total to nearly 50% of currently available cropland) once the energy requirements for processing, harvesting and transportation are taken into account (Kheshgi *et al.*, 2000). Consequently it is only likely that commercial biofuels could possibly make a small but significant contribution to meeting future energy needs within an early to mid twenty-first-century window of opportunity (see Table 8.2) before the preference tends to food production.

7.6 Summary and conclusions

Human ecology, as opposed to the ecology of any other species, is different in that sentience has enabled a shaping of *H. sapiens'* relationships with other species. This shaping has often been with disregard to possible ecological and environmental side effects (anthropogenic global warming itself being just one instance). This sentient dimension to the shaping of human ecology has been enhanced since the Renaissance and the subsequent Industrial Revolution and this in turn is reflected in the growth in human numbers and the quantity of energy harnessed by our species, both globally and on a per capita basis. Yet the environmental resources commensurately consumed in the process have impacted, and will continue to impact, markedly on natural systems on which our species continues to depend. Human numbers and projected population growth for the twenty-first century are such that, *irrespective* of climate change, a more rational and efficient use of resources is required. This can in part be achieved with an understanding of the carbon cycle and related ecology. Climate-change policies coincide with more rational and efficient environmental resource use so that irrespective of global warming it would be prudent to pursue such policies (Cowie, 1998). The history of international environmental policies, the possible measures that might be taken to address climate change concerns, as well as international climate-change policies, will be summarised in the next chapter.

7.7 References

Balmford, A., Bruner, A., Cooper, P. *et al.* (2002) Economic reasons for conserving wildlife. *Science*, **297**, 950–3.

Becker, A. and Grünewald, U. (2003) Flood risk in Central Europe. *Science*, **300**, 1099.

Bengtsson, L. (2001) Hurricane threats. *Science*, **293**, 440–1.

Bentham, G. (1999) Direct effects of climate change on health. In Haines, A. and McMichael, A. J., eds. *Climate Change and Human Health*, pp. 35–44. London: Royal Society.

Bentley, R. W. (2002) Global oil and gas depletion: an overview. *Energy Policy*, **30**, 189–205.

British Petroleum (2002) *BP Statistical Review of World Energy*. London: British Petroleum.

British Petroleum (2005) *BP Statistical Review of World Energy*. London: British Petroleum.

Brown, L. (1995) *Full House: Reassessing the Earth's Population Carrying Capacity*. London: Earthscan.

Burroughs, W. J. (1997) *Does The Weather Really Matter? The Social Implications of Climate Change*. Cambridge: Cambridge University Press.

Cabinet Office (1994) *Sustainable Development: The UK Strategy*. Cm 2426. London: HMSO.

Cardillo, M., Mace, G. M., Jones, K. E. *et al.* (2005) Multiple causes of high extinction risk in large mammal species. *Science*, **309**, 1239–41.

Ceballos, G., Ehrlich, P. R., Soberón, J., Salazar, I. and Fay, J. P. (2005) Global mammal conservation: what must we manage? *Science*, **309**, 603–7.

Chichilnisky, G. and Heal, G. (1998) Economic returns from the biosphere. *Nature*, **391**, 629–30.

Christensen, J. H. and Christensen, O. B. (2002) Severe summertime flooding in Europe. *Nature*, **421**, 805.

Ciais, P., Reichstein, M., Viovy, N. *et al.* (2005) Europe-wide reduction in primary productivity caused by the heat and drought in 2003. *Nature*, **437**, 529–33.

Coates, S. and Cohen, T. (2003) Weird and dangerous: the side-effects of a heatwave. *The Times*, 8 August.

Cohen, J. E. (1995) *How Many People Can the Earth Support?* New York: Norton.

Costanza, R., d'Arge, R., de Groot, R. *et al.* (1997) The value of the World's ecosystem services and natural capital. *Nature*, **387**, 253–60.

Cottrell, F. (1955) *Energy and Society*. Westport, CT: Greenwood Press.

Cowie, J. (1998) *Climate and Human Change: Disaster or Opportunity?* London: Parthenon Publishing.

Cowie, J. (ed.) (2003) *Fuelling the Future 3: Biofuels*, 2nd edn. London: Institute of Biology and the British Council for Crop Protection.

Department of Health (2002) *Health Effects in the UK*. London: Department of Health.

Department of Heath (2004) *West Nile Virus: a Contingency Plan to Protect the Public's Health*. London: Department of Health.

Donaldson, C. G., Tchernjavskii, V. E., Ermakov, S. P., Bucher, K. and Keatinge, W. R. (1998) Winter mortality and cold stress in Yekaterinburg, Russia interview survey. *British Medical Journal*, **316**, 514–18.

Ehrlich, P. R. and Holdren, J. P. (1971) Impact of population growth. *Science*, **171**, 1212–17.

Environment, Food and Rural Affairs Select Committee (2003) *Biofuels*. Norwich: House of Commons EFRA Select Committee, Stationery Office.

Food and Agriculture Organization (2001, 2003, 2004) *The State of Food Insecurity in the World* [various annual reports]. Rome: UN Food and Agriculture Organization.

Food and Agriculture Organization (2002) *FAOSTAT Statistical Database*. Rome: UN Food and Agriculture Organization.

Goldenberg, S. B., Landsea, C. W., Mestas-Nuñez, A. M. and Gray, W. M. (2001) The recent increase in Atlantic hurricane activity: causes and implications. *Science*, **293**, 474–9.

Hallam, T. (2004) *Catastrophes and Lesser Calamities*. Oxford: Oxford University Press.

Hay S. J., Cox, J., Rogers, D. J. *et al.* (2002) Climate change and resurgence of malaria in the East African highlands. *Nature*, **415**, 905–9.

Imhoff, M. L., Bounaoua, L., Ricketts, T. *et al.* (2004) Global patterns in human consumption of net primary production. *Nature*, **429**, 870–3.

Intergovernmental Panel on Climate Change (1990) *Climate Change: the IPCC Scientific Assessment*. Cambridge: Cambridge University Press.

Intergovernmental Panel on Climate Change (1995) *Climate Change 1995: the Science of Climate Change*. Cambridge: Cambridge University Press.

Intergovernmental Panel on Climate Change (2000) *Emission Scenarios – IPCC Special Report*. Geneva: WMO.

Intergovernmental Panel on Climate Change (2001a) *Climate Change 2001: Impacts Adaptation and Vulnerability – a Report of Working Group II*. Cambridge: Cambridge University Press.

Intergovernmental Panel on Climate Change (2001b) *Climate Change 2001: the Scientific Basis – Summary for Policymakers and Technical Summary of the Working Group I Report*. Cambridge: Cambridge University Press.

Johnson, C. G. and Smith, L. P. (1964) *The Biological Significance of Climatic Changes in Britain*. London: Institute of Biology and Academic Press.

Keilman, N. (2003) The threat of small households. *Nature*, **421**, 489–533.

Kerr, R. A. (2004) A few good climate shifters. *Science*, **306**, 599–600.

Kheshgi, H. S., Prince, R. C. and Marland, G., (2000) The potential of biomass fuels in the context of global climate change. *Annual Review of Energy and the Environment*, **25**, 199–244.

Markham, S. F. (1944) *Climate and the Energy of Nations*, revised edition. London: Oxford University Press.

Martens, P. (1999) Climate change impacts on vector-borne disease transmission in Europe. In Haines, A. and McMichael, A. J., eds. *Climate Change and Human Health*, pp. 45–53. London: Royal Society.

Maugeri, L. (2004) Oil: never cry wolf – why the petroleum age is far from over. *Science*, **304**, 1114–16.

McMichael, A. J. (1993) *Planetary Overload: Global Environmental Change and the Health of Human Species*. Cambridge: Cambridge University Press.

McMichael, A. J. and Haines, A. (1997) Global climate change: the potential effects on health. *British Medical Journal*, **315**, 805–9.

McMullan J. T., Morgan, R. and Murray, R. B. (1977) *Energy Resources*. London: Edward Arnold.

Meadows, D. H., Meadows, D. L., Randers, J. and Behrens, III, W. W. (1972) *The Limits to Growth*. London: Pan.

Meehl, G. A. and Tebaldi, C. (2004) More intense, more frequent, and longer lasting heat waves in the 21st century. *Science*, **305**, 994–7.

Mooney, H., Cropper, A. and Reid, W. (2005) Confronting the human dilemma. *Nature*, **434**, 561–2.

Mudelsee, M., Börngen, T. and Grünewald, U. (2003) No upward trends in the occurrence of extreme floods in central Europe. *Nature*, **425**, 166–9.

National Health Service (2004) *Heatwave Plan for England: Protecting Health and Reducing Harm from Extreme Heat and Heatwaves*. London: Department of Health.

Nature News (2004) Reform of land use urged as floodwaters rise across Asia. *Nature*, **430**, 596.

Osborn, T. J., Hulme, M., Jones, P. D. and Basnett, T. A. (2000) Observed trends in the daily intensity of United Kingdom precipitation. *International Journal of Climatology*, **20**, 347–64.

Palmer, T. (2006) Making the paper: how climate data are helping to predict malaria outbreaks in Africa. *Nature*, **439**, xi.

Pierce, J. L., Meyer, G. A. and Jull, A. J. T. (2004) Fire-induced erosion and millennial-scale climate change in northern ponderosa pine forests. *Nature*, **432**, 87–90.

Porter J. R. (2005) Rising temperatures are likely to reduce crop yields. *Nature*, **436**, 174.

Raleigh, V. S. (1999) World population and health in transition. *British Medical Journal*, **319**, 981–3.

Rogers, D. J. and Randolph, S. E. (2000) The global spread of malaria in a future, warmer world. *Science*, **289**, 1763–5.

Rojstaczer, S., Sterling, S. M. and Moore, N. J. (2001) Human appropriation of photosynthesis products. *Science*, **294**, 2549–51.

Royal Society (2001) *The Role of Land Carbon Sinks in Mitigating Global Climate Change*. Policy document 10/01. London: Royal Society.

Ruddiman, W. F. (2003) The anthropogenic greenhouse era began thousands of years ago. *Climate Change*, **61(3)**, 261–93.

Schroter, D., Cramer, W., Leemans, R. *et al.* (2005) Ecosystem service supply and vulnerability to global change in Europe. *Science*, **310**, 1333–7.

Simon, J. (1996) *The Ultimate Resource 2*. Princeton, NJ: Princeton University Press.

Stanwell-Smith, R. (1999) Effects of climate change on other communicable diseases. In Haines, A. and McMichael, A. J., eds. *Climate Change and Human Health*, pp. 55–69. London: Royal Society.

Starke, L. (ed.) (1999) *Vital Signs: 1999–2000*. London: Earthscan.

Sustainable Development Commission (2006) *The Role of Nuclear Power in a Low Carbon Economy*. London: Sustainable Development Commission.

Thomson, M. C., Doblas-Reyes, F. J., Mason, S. J. *et al.* (2006) Malaria early warnings based on seasonal climate forecasts from multi-model ensembles. *Nature*, **439**, 576–9.

Turner, II, B. L., Clark, W. C., Kates, R. W. *et al.* (1990) *The Earth as Transformed by Human Action: Global and Regional Changes in the Biosphere over the Past 300 Years*. Cambridge: Cambridge University Press.

UN Population Division of the Department of Economic and Social Affairs (2002) *World Population Prospects: The 2002 Revision*. New York: United Nations Secretariat.

Vitousek, P. M., Mooncy, H. A., Lubchenco, J. and Melillo, J. M. (1997) Human domination of Earth's ecosystems. *Science*, **277**, 494–9.

Whitlock, C. (2004) Forest, fires and climate. *Nature*, **432**, 28–9.

World Health Organization (1999) *El Niño and Health: Report of the Task Force on Climate and Health*. Geneva: World Health Organization. [Summarised in WHO Fact Sheet 192 (2000)].

World Health Organization (2001a) *Climate and Health*. Fact Sheet 266. Geneva: World Health Organization.

World Health Organization (2001b) *Water – Too Much or Too Little – The Foremost Cause of Natural Disasters*. WHO Feature no. 203. Geneva: World Health Organization.

World Health Organization (2004) *Surge in Demand Leads to Shortage of Artemisinin-based Combination Therapy for Malaria*. Media release WHO/77. Geneva: World Health Organization.

World Health Organization (2006) *WHO Announces Pharmaceutical Companies Agree To Stop Marketing Single-Drug Artemisinin Malaria Pills*. News release WHO/23. Geneva: World Health Organization.

World Health Organization/Food and Agriculture Organization (2004) *Minimizing Food-borne Illness Requires Co-ordination Right up the Food Chain*. Joint news release WHO/FAO/71. Geneva/Rome: World Health Organization/Food and Agriculture Organization.

Worldwatch Institute (2003) *Vital Signs 2003–2004*. London: Earthscan.

8

Sustainability and policy

Having an understanding of climate change is one thing, relating this to real-world development is another. In the latter half of the twentieth century it became apparent to politicians that human impacts on the environment were sufficiently detrimental that they undermined the sustainability of human well-being and relatedly environmental quality. Human well-being is a catch-all term relating to material and cultural standards as well as quality of life.

Many of these terms, while having a clear meaning to western politicians and policy-makers, have no strict definition or individual basis of quantitative indexing in the strict scientific sense, although in some instances attempts have been made. Sustainability itself, as we shall see, does have a specific definition that is enshrined in international agreements and sustainability is affected by climate change. Consequently the history of what politicians mean by sustainability, and how current climate-change issues affect it, are central to developing climate change and human ecology from topics of academic interest to those of application.

8.1 Key developments of sustainability policy

8.1.1 UN Conference on the Human Environment (1972)

The 1972 Stockholm Conference on the Human Environment was the first of a series of UN conferences that arguably developed international environmental policy significantly. It was attended by 113 nations, but due to a dispute on the status of East Germany the USSR and all the Eastern European nations (many of which now are considered Central European), bar Romania, refused to participate.

The idea for the conference had been around since the end of the 1960s and Sweden played a leading role in much of the preparatory work. Sweden was at the time suffering the effects of acid rain and so realised that many

Table 8.1. *Summary of Stockholm's Declaration of Principles.*

1. Human rights should be asserted, and apartheid, colonialism, etc., should be condemned.
2. Natural resources must be safeguarded.
3. The Earth's capacity to produce renewable resources must be maintained.
4. Wildlife must be safeguarded.
5. Non-renewable resources must be shared and not exhausted.
6. Pollution must not exceed the environment's capacity to clean itself.
7. Damaging oceanic pollution must be prevented.
8. Development is needed to improve the environment.
9. Developing countries therefore need assistance.
10. Developing countries need reasonable prices for exports to carry out environmental management.
11. Environment policy must not hamper development.
12. Developing countries need money to develop environmental safeguards.
13. Integrated development planning is needed.
14. Rational planning should resolve conflicts between environment and development.
15. Human settlements must be planned to eliminate environmental problems.
16. Governments should plan their own appropriate population policies.
17. National institutions must plan development of states' natural resources.
18. Science and technology must be used to improve the environment.
19. Environmental education is essential.
20. Environmental research must be promoted, especially in developing countries.
21. States may exploit their natural resources as they wish but must not endanger others.
22. Compensation is due to states thus endangered.
23. Each nation must establish its own standards.
24. There must be cooperation on international issues.
25. International organizations should help to improve the environment.
26. Weapons of mass destruction must be eliminated.

environmental problems could only be tackled by international action. Not surprisingly the theme of 'Only One Earth' arose again and again throughout the conference.

Three things emerged from the conference: a proclamation, a list of principles (Table 8.1) and an action plan.

The proclamation is memorable for one sentence included by the head of the Chinese delegation, Tang Ke: 'We hold that of all things in the World, people are the most precious'.

The declaration of principles consisted of 26 in all. The UN Environment Programme (UNEP) that arose as a consequence of the conference later adopted some of these. UNEP, you will recall (Chapter 5), was to be the UN agency under whose auspices (along with the UN World Meteorological Organization; WMO) the Intergovernmental Panel on Climate Change (IPCC) was established 16 years later. Of relevance to anthropogenic climate change the declarations included: nations must be free to exploit their own resources; no nation should endanger others; compensation is due if they do; each nation should plan its own population policy; non-renewable resources must be shared by all and environmental concerns must not infringe human rights. Today anthropogenic climate change is being caused mostly by the developed nations and this is endangering others, through both climate change and sea-level rise. Fortunately, for a number of the participating states, the UN conference's principles are not enshrined in international law, for if they were there would undoubtedly be severe economic repercussions. Conversely for others, at the sharp end of anthropogenic climate change, the principles provide no protection.

The conference's action plan consisted of 109 recommendations covering, albeit it broad and loosely, all areas of environmental concern. Eighteen recommendations (1–18) related to human settlements, 51 recommendations (19–69) to natural resources, 16 recommendations (70–85) to pollution generally, nine recommendations (86–94) to marine pollution, seven recommendations to education and culture (95–101) and eight recommendations (102–109) to environment and development. Climate change was covered mainly by the general pollution recommendations and much of the science to address this was either reviewed or conducted by the UN WMO following the conference. However, recommendation 70, which called for countries to consult with other nations before doing anything that might affect the climate, has been an abject failure. This was recognised by many (well before the 1990 IPCC report), including the International Institute for Environment and Development, whose briefing a decade on from Stockholm said that 'provisions for this have become *seriously unstuck*' (the italics represent the source's own emphasis; Clarke and Timberlake, 1982). In July 1980 UNEP wrote to all nations submitting the provisions for co-operation in weather modification. A year later only nine replies had been received, of which six were simply acknowledgements of receipt of correspondence; just three accepted the proposals.

The decade following the conference saw the development of some climate models that were basic by today's standards. It has to be remembered that at that time many large computers used by industry and academia had *less* computer power than today's home PCs. Indeed, up to 1982 PCs were extremely rare in homes and indeed prior to 1980 virtually unheard of domestically. Those that became available early in the 1980s had memory capacities that were measured in kilobytes, not the gigabytes of today. Similarly, the models that academics used were very elementary. Nonetheless, back then most models predicted a rise of between 1.5 and 3 °C for a doubling of carbon dioxide (Clarke and Timberlake, 1982), which compares quite favourably with the IPCC's first assessment in 1990. All of this demonstrates that it takes a leap of orders of magnitude in climate-model complexity and data processing to realise relatively modest improvements in detailed output and reductions in uncertainty.

8.1.2 The Club of Rome's Limits to Growth (1972)

In 1968 30 individuals – scientists, educators, economists, humanists, industrialists and civil servants – gathered in the Accademian dei Lincei in Rome. It was the first meeting of the Club of Rome and they were there to discuss the present and future predicament of humans.

Jay Forrester of the Massachusetts Institute of Technology (MIT) had devised a computer model of the global human population and its resource consumption, which incorporated not just population levels, but industrial output, food availability, pollution and finite resource levels. Again by today's standards it was a very simple model. Then in 1972, the same year as the Stockholm Conference, an MIT team led by Dennis Meadows produced a report based on runs of the Forrester World Model for the Club of Rome. This was *The Limits to Growth* report (Meadows *et al.*, 1972).

The Limits to Growth report made headline news in many countries and in turn this spawned numerous documentaries and books. Yet it also came under severe criticism from a number of academics and was by many misunderstood. In essence it reflected the idea of the Reverend Thomas Robert Malthus (1798) that arithmetic (exponential) growth in population and resource consumption tends to outstrip additive (linear) growth in the ability to harvest resources, so that eventually checks come into play that reduce population. These checks include famine, pestilence, disease and war, but also include 'moral restraint'. The *Limits to Growth* report argued that human population growth and per-capita resource consumption cannot continue to increase indefinitely, so that if we were to avoid some sort of collapse then 'entirely new approaches are required to redirect society towards goals of equilibrium rather than growth'.

The Limits to Growth's many critics point to the fact that the decades since the report have not shown the collapse in resource use and society that were predicted. This is certainly true. To take just one example from the report, copper was predicted to run out in 36–48 years' time and gold in just 11–29 years from when the report was written, in 1972. This meant that by today gold should have run out while copper would be on the point of being exhausted. This has not happened, so clearly, the critics say, *The Limits to Growth* analysis must be wrong.

However, such criticisms completely miss the point. *The Limits to Growth* report pointed to a mismatch between resource consumption and supply that would come about unless circumstances changed, hence the need for new approaches. Indeed, in part what we got were new approaches. To continue with the examples of copper and gold, the situation regarding the development and use of these resources has improved. Change has taken place. Within a decade after the publication of *The Limits to Growth*, copper mining and ore-refining techniques had improved so much that the spent tailing mounds from past copper mining were themselves being mined routinely for copper. Similarly, technological improvements had been made that enhanced the extraction of gold and also the uses to which gold had been put changed; indeed, its importance as an international basis for monetary exchange diminished, so freeing the metal for alternative use.

Twenty years on from the original report its authors published a follow-up (Meadows *et al.*, 1992) that included a rebuttal of many of the criticisms levelled at them. This included pointing out that they did not seek to provide 'a warning about the future [that] is a prediction of doom' but 'a warning . . . to follow a different path'. They also emphasised that they were not making a judgement as to whether 'growth' per se was all good or all bad but that what was required was 'sustainable development'. The 20-year follow-up pointed out that the future held a range of options, but that considerable change was still required in the management of a number of key resources (including fossil fuels, water and food supply). They also pointed out that at the time of the UN's Stockholm conference, the number of nations with environment ministries numbered no more than 10, whereas by 1992 it was around 100. In short, the follow-up report urged for continued efforts to develop sustainably. To be fair to the critics of *The Limits to Growth*, the first report did have (rightly or wrongly) an alarmist tone, whereas the 20-year follow-up was more measured. However, in equal fairness to the authors, the first report did attract considerable media attention, something at which the 20-year follow-up was not nearly so successful.

8.1.3 *World Climate Conference (1979)*

Several international conferences on climate and global warming were held following Stockholm in the 1970s. Arguably one of the most significant was the 1979 World Climate Conference organised by the UN WMO. It concluded, echoing Stockholm, that the carbon dioxide problem deserved the most urgent of attention by the international community of nations. Nonetheless, following the conference, little discernable action was taken by politicians.

8.1.4 The World Conservation Strategy (1980)

In 1980 the International Union for the Conservation of Nature (IUCN), the World Wildlife Fund (WWF) and UNEP came together to produce the *World Conservation Strategy*. It was a usefully slim document, hence to the point, that outlined the central tenets of sustainable development from an ecological perspective. It took the Stockholm Conference's recognition of economic endeavours and environmental conservation as being two sides of the sustainable development coin, but went on to cite key principles and current priorities to realise sustainable development. Its three core principles were:

- to maintain ecological processes and life-support systems,
- to preserve genetic diversity and
- to ensure the sustainable utilisation of species and ecosystems.

The strategy warned, similar to *The Limits to Growth*, that the planet's ability to support its growing human population was being undermined through destruction of (biotic) resources. Importantly, the strategy stressed the fundamental need for 'genetic diversity' and this was a precursor to the popular concept of 'biodiversity' that was not formaliscd (strictly in policy terms) until the 1987 Brundtland Report (see section 8.1.6). However, it was the marrying of biological conservation ideas (and not just broad environmental concerns, as with Stockholm) with development issues that made the report stand out. This fact, and that it sprang from three major international bodies, meant that it attracted significant press coverage worldwide. For example, in the UK *The Times* ran a major feature on aspects of the strategy each day of the week of its launch, while the final programme of David Attenborough's *Life on Earth* TV series, devoted to the future of wildlife and humans, specifically cited the *Strategy* and its conclusions.

There was in 1991 a follow-up to the *World Conservation Strategy* called *Caring for the Earth: a Strategy for Sustainable Living* (IUCN, WWF and UNEP, 1991). Yet despite a simultaneous launch in several capital cities (the one in the UK had the Duke of Edinburgh as the lead speaker) it did not capture the attention of the media as did the original *Strategy*. The reasons for this are not entirely clear. However, *Caring for the Earth* was longer than – and hence lacked the punch of – the *World Conservation Strategy*. It also lacked a proper summary and key points; these had focused the reader's attention in the original. This may have been because *Caring for the Earth* was subject to extensive consultation and revision but either way it failed to make anywhere near as big an impact as the *World Conservation Strategy*.

Yet, while the *World Conservation Strategy* made the headlines, and even caught some political attention, it had little immediate effect on the ground, although it did pave the way for future policy documents. Had the *Strategy* been properly implemented then each nation would produce its own national strategy and implement it. A few countries did develop their own national response on paper but fewer still went on to implement these. The UK was one such. The Prince of Wales, three governmental agencies and two non-governmental bodies as well as the WWF launched its national plan, The Conservation and Development Programme for the UK, in the summer of 1983. Its 10-point action plan was never implemented. Yet the idea that an international policy document might spawn national strategies was one that caught on. As we shall see, more formal international climate and biodiversity policies of the 1990s (with political leadership from the UN as opposed to being led by non-governmental organisations), among others, also called for individual national strategies. Here such strategies, or national action plans were developed.

8.1.5 The Brandt Report – Common Crisis North-South (1980)

In the same year as the *World Conservation Strategy* a report came from a commission of 18 politicians of international repute (some of them former premiers) led by West Germany's Willy Brandt. The report (Brandt Commission, 1980) was primarily concerned with the growing 'Third World' problem (or 'developing nations' as they are more commonly referred to today). The North and South in the title referred to the fact that most developed nations were in the north while those developing or undeveloped were in the south. The report was concerned that the divergence between the north and south was increasing.

The second report of the Brandt Commission, *Common Crisis* (1983), pointed out that the problem of wealth divergence affected both the north and the south and that both depended on each other. The south depended on the north for financial institutions and technology, while the north depended on the south for raw materials and tropical and semi-tropical cash crops.

Whereas the *World Conservation Strategy* looked at global problems from a biological-conservation and human-ecology perspective, the 1980 Brandt report looked at developmental problems from a political perspective. As such it focused on international trade and monetary agreements. However, pointedly it did mention the need 'to integrate conservation with development'. Also, like the *World Conservation Strategy*, the report attracted considerable media coverage. Again like the *Strategy*, because politicians actively in power at the time did not produce the report, Brandt's impact on the ground was muted although it did help establish a tone of political thought. Three years later (1983) the follow-up report was very much along the same lines.

8.1.6 The Brundtland, World Commission on Environment and Development Report (1987)

The World Commission on Environment and Development was the first body to produce an international environment and development report that had real political credibility. Not only was it created as a consequence of a UN General Assembly resolution (38/161), but it was chaired by the premiere of Norway, Gro Harlem Brundtland, and had on it those who were either currently active senior politicians or who worked for major organisations such as the Organization for Economic Cooperation and Development (OECD), as well as leading scientific organisations and environmental agencies.

In effect the Brundtland report, called *Our Common Future* (World Commission on Environment and Development, 1987), was a synthesis of both the *World Conservation Strategy* and the report of the Brandt Commission. On the one hand it recognised the fundamental need to conserve biological resources and have sustainable development, but on the other it was aware that institutions, financial systems and international economic relations needed a radical overhaul.

With regards to biology and (separately) climate change it made some key (for international policy-makers at the time) observations and recommendations. On the biological front it called for a 'Species Convention' and the preservation of genetic diversity. It recognised the need to preserve ecosystems and have sustainable development. As such it re-iterated the arguments of the *World Conservation Strategy*. Here the Brundtland report has been cited as one of the first popularisers of the term biodiversity, although the term actually used in the chapter on species and ecosystems and in the report's summary is biological diversity. (Walter G. Rozen, of the US Academy of Science, used the term biodiversity in 1985 in the run up to the National Forum on Biodiversity held in Washington DC the following year; later the ecologist Edward O. Wilson used it as the title of his 1988 book.)

As for energy, the Brundtland report recognised the need for energy for economic growth and human well-being but also that much was being used inefficiently and that global warming and acid rain were major problems to be avoided. It called for improved energy efficiencies and alternative energy sources, and a 'safe, environmentally sound, and economically viable energy pathway that will sustain human progress into the distant future'. It said that this was 'clearly an imperative'. However, the report was not clear as to how this might be achieved. This was largely because no country had a significant long-term energy strategy (although a few – such as France and Norway – had short- and medium-term plans that were being implemented).

8.1.7 United Nations' Conference on the Environment and Development – Rio de Janeiro (1992)

In 1992 in Rio de Janeiro the UN held a conference on the Environment and Development (known as UNCED for short). It was convened on much the same basis as the Stockholm conference but unlike Stockholm had two distinct foci: biodiversity and climate change. Indeed, at the conference two international conventions were signed relating to each of these subjects. The biodiversity convention was in essence the realisation of recognition in international policy terms of the *World Conservation Strategy*'s and the Brundtland report's aspiration of the importance of biodiversity. The climate-related outcome was the UN Framework Convention on Climate Change (FCCC).

In the run up to UNCED over a score or so of nations had declared targets for the reduction of carbon emissions. The EU had a target of carbon emissions to be no higher than they were in 1990 by the year 2000, whereas the UK's target was emissions no higher than 1990 by 2005 and a reduction in all greenhouse gases by 20%. Not all these targets, and those of nearly all other nations, were met. The UK did manage to reduce its carbon emissions to below 1990 by 2004 and had even greater success with its total greenhouse emissions; however, as we shall see below, these savings owed much to a switch to natural gas (especially for electricity generation), with its higher energy content per carbon atom than coal or oil. Meanwhile, the US pre-UNCED had no official targets, although it did sign up to the FCCC.

The convention lacked any clearly binding, short-term commitments to reducing greenhouse emissions. What it did was to establish a process for negotiating further measures and introduced mechanisms for transferring finance and technology to the less-developed nations. One of the UNCED's successes was that in addition to the nations that already had adopted some form of greenhouse policy, the USA signed the convention.

The UNCED had its failures too. These included the removal of much of the potential from the FCCC's original draft but also proposals with climate implications outside of the FCCC. For instance, the UNCED Convention on Biodiversity was not signed by the USA. This convention would have increased the financial value of standing tropical forests, in that there would have been equitable sharing (between developed and less-developed nations) of the benefits that flow from using genetic resources (of which tropical forests abound). President George Bush Senior, perceiving that signing would have a cost for the USA in employment terms, and with an eye on the forthcoming presidential election, declined to sign. Another failure with greenhouse implications was a third convention on Sustainable Forestry. Here the international gulf between the developed and less-developed nations was so great that before the UNCED began the convention had been watered down to a set of so-called Forest Principles.

The UNCED was a summit at which nations took two steps forward and one step back: progress was made only in that nations agreed that progress needed to be made. Nonetheless, there was a framework for development. The next major step down this road from a climate perspective was taken at Kyoto in 1997.

8.1.8 The Kyoto Protocol (1997)

The World Climate Conference was held in Kyoto, Japan, in December 1997 and was a direct result of the UN's FCCC. The problem during Kyoto was that Russia and, arguably more importantly, the USA, dragged their feet. Russia did so because its economy was behind that of Europe and the USA, and although it sought to develop its industrial base had gone into recession after the 1988–90 devolution of many of the former Soviet Bloc nations. Russia did eventually sign up in 2004. The USA was noticeably unenthusiastic and had much at stake, as North America has less than 5% of global population yet around 24% of global emissions. Here the problem was not so much that President Clinton's administration did not accept the urgency of climate-change issues, but that the administration was gridlocked between the Senate and Congress. There was little point in the USA agreeing policies that would not be ratified back home. However, the USA did call for carbon trading between nations, whereby efficient users of carbon could trade permits with inefficient users. It also called for trading to take into account a nation's planting of forests. The problem with this last, as noted in Chapter 7, is that whereas new forest growth does sequester carbon, as do some soils, quantifying this is difficult. There are also problems in ensuring that carbon so sequestered remains sequestered.

The Kyoto Protocol set an international target for there to be a reduction of emissions (of all anthropogenic greenhouse gases) by 5% of the 1990 level (although a 1995 base year was allowable for some nations' accounting) by the Protocol's first implementation period of 2008–12. Nationally, though, individual nation's commitments to savings varied. The OECD nations generally had to make the most savings (an exception being Australia and some of the smaller nations) and Britain and Germany agreed to make greater savings so as to lessen the burden on some other EU states. Britain's target is to reduce its greenhouse emissions by 12.5% below the 1990 base year by 2008–12. This is greater than the collective EU target (by the then 1997 EU states) of 8%: individual EU nations contribute to this collective goal by agreement subject to their economic circumstances. Importantly, the Protocol established three 'flexible mechanisms' to achieving mechanism reductions: Joint Implementation (JI), the Clean Development Mechanism (CDM) and international emissions trading. Together these allow countries to achieve part of their legally binding targets by actions in other countries that are party to the FCCC. Emissions trading enables countries which achieve greater reductions than needed to meet their own target to sell the surplus to other countries. JI and CDM allow countries with targets to receive a credit for project-based activities that reduce emissions in other nations. Even so, under the Protocol, these three mechanisms must be in addition to action to curb emissions taken by a nation itself with such domestic action contributing a significant amount to the country's total effort to meeting its target.

The problem with the Protocol is that the enthusiasm of individual nations for it differs. Although the Kyoto conference took place at the end of 1997, Russia did not sign up to it until 2004. Worse, the USA, which is responsible for about a quarter of global emissions, withdrew from the Protocol in 2001 on the grounds that it would harm its economy even though it negotiated hard for carbon trading and for carbon sinks to be taken into account. Further, even many of the nations that did sign up to Kyoto had failed to make any savings by 2004. Not surprisingly then, globally carbon emissions have risen and this is reflected in increased atmospheric accumulation of carbon dioxide. Nonetheless, the Kyoto Protocol to date represents the most serious international policy attempt to curb global warming. Its first commitment period ends in 2012 and the UN has been holding a number of preliminary meetings to develop the Protocol beyond that date and also to encourage nations that are not on board to join. The first of these was held in Montreal in November 2005 (of which more shortly, in section 8.1.10).

So what effect would the Kyoto Protocol have on the typical range of IPCC BaU scenarios for the twenty-first century? If it were implemented by itself with no further carbon emission cuts then the twenty-first century would see global emissions reduced to 5% less than those of 1990 by 2012 and then remain steady. Under such a scenario one might presume that the developed nations would continue with further cuts (beyond 5%) to offset the increase in emissions from developing countries. In short, the remainder of the twenty-first century would see a static level of emissions at 5% less than those in 1990. At the end of the century global warming would be delayed by just a few years (optimistically, 2–5 years) compared to most of the IPCC's 2000 BaU-type scenarios. To actually stabilise atmospheric concentrations at 420 ppm CO_2 (50% above pre-industrial levels of 270–80 ppm) and to avoid even further warming beyond what this will incur, the IPCC say we need to cut emissions to 50% of 1990 levels by 2050 (the 2005 atmospheric concentration was 379 ppm). Conversely, more drastically, if we were to have stabilised the atmospheric concentration at its 1990 level of around 353 ppm, then the IPCC (1990) state that we would have needed to have made an immediate cut of carbon emissions of 60–80%. In short, the Kyoto Protocol by itself (assuming its goals are fulfilled) is very much just a beginning in an attempt to manage fossil carbon emissions. Much depends on the continuing Protocol negotiations.

Having said all of this, it is important to note the assumption in the above conclusion, namely that the Protocol, within its own terms, succeeded: many nations agreed to sign up to climate goals. As such it would be a beginning but would necessitate further development so that emissions could be meaningfully curbed to reduce (if not eliminate) anthropogenic climate forcing. So, is the Protocol really succeeding in terms of this broader goal? Two-thirds of a decade on from Kyoto, and half way to 2010 (the mid year between 2008–12), the 2004 global emissions figures were 26% higher than in 1990. Consequently, only an undaunted optimist could consider Kyoto to be working. Yet it should not be dismissed. Kyoto was undoubtedly a spur to the limited switching of energy away from fossil carbon that is occuring. Had this not taken place then emissions would be higher still and so make subsequent reductions that much harder. Furthermore, those Kyoto countries that have developed more non-fossil energy sources may well find future economic benefits in being more independent from fossil fuels and this could act as a spur to further, more ecologically meaningful, cuts in the longer term.

8.1.9 Johannesburg Summit – UNCED+10 (2002)

The UN World Summit on Sustainable Development in Johannesburg was also known as UNCED+10 and this last more accurately reflected its purpose. This was to chart the progress made since the UNCED in Rio de Janeiro in 1992 and to take things forward. Alas, despite the presence of many premier politicians and G8 leaders, UNCED+10 was hardly considered a success. It transpired that on the ground little progress had been made since 1992. Further, the future commitments were not binding and the wording was loose. This was also reflected in those issues that have a major focus on biology, human ecology and climate change.

Food security had deteriorated so it was agreed at Johannesburg that this must be improved and that the numbers suffering from hunger in Africa should be halved by 2015. It also had a goal that 'on an urgent basis and where possible by 2015, [to] maintain or restore depleted fish stocks to levels that can produce maximum sustainable yield'. With regards to the UNCED Biodiversity Convention, it was recognised that little progress, if any, had been made. Those at Johannesburg resolved to 'achieve by 2010 a significant reduction in the current rate of loss of biological diversity'. This signalled that they accepted that biodiversity was going to continue to deteriorate and so only planned to slow the rate at which this happened. By 2005, with just half a decade to the deadline, progress on even this modest goal had yet to show signs of success. With regards to human mortality it reaffirmed a goal the UN made at the turn of the millennium (the Millennium Development Goal) to 'reduce by 2015 mortality rates for infants and children under 5 by two-thirds, and maternal mortality rates by three quarters, of the prevailing rate in 2000'. It also affirmed to reduce HIV prevalence in the most affected countries in people of 15–24 years of age by 25% and globally by 2010. This too, as of 2006, appears to be difficult to realise as deaths from AIDS (all ages) had increased from under a third of a million a year in 1990 to nearly 3 million in 2000. HIV/AIDS in many nations (especially some of those with weak economies) continues to be a growing problem.

The commitments of UNCED+10 to climate change were limited, primarily because politicians were leaving this to the development of the Kyoto Protocol taking place at the same time (and which itself was having its own difficulties, especially with the recent loss of the USA from negotiations). However, energy did feature, and here the energy goals, though lacking in specificity, were in line with Kyoto goals. These included diversification of global supply and increasing the contribution from renewable energy, tackling energy poverty, freeing markets and removing harmful subsidies that might exacerbate the greenhouse effect and improving energy efficiency. The UNEP, though, did launch a new initiative, called the Global Network on Energy for Sustainable Development, to promote research and energy-technology transfer to developing nations. Meanwhile, the EU announced a $700 million partnership initiative and the USA said that it would invest up to $43 million in 2003. However, a fair proportion of these monetary pledges were already earmarked for international aid, and another proportion have yet (2005) to be provided. UNCED+10 never lived up to the promise of the UNCED itself.

8.1.10 Post 2002

Subsequent to UNCED+10 among the most significant international climate-change policy developments was the 2005 G8 Summit in Gleneagles, Scotland. This had two principal items on its agenda: Africa and climate change. However, the communiqué that was released was a watered-down version of the earlier drafts, saying that fossil fuels 'contribute a large part to increases in greenhouse gases associated with the warming of our Earth's surface'. This carefully crafted sentence (and rest of the communiqué) left considerable room for interpretation and political manoeuvrability but was viewed by some as signalling that US President George W. Bush was beginning to recognise the science behind the issue. The communiqué was accompanied by a list of 38 action points including the promotion of low-energy vehicles and systems. In essence, whereas Kyoto imposed carbon limits, the US view (and that of the G8 communiqué) was that although climate change was a problem, carbon limits were not essential, rather the ability to switch to a low-fossil economy. The USA sought to increase the economic activity produced per unit of carbon dioxide released into the atmosphere. However, Kyoto did score in one area. The G8 meeting also committed to the rapid adequate funding of the Kyoto CDM, whereby rich countries or industries can buy rights to emit carbon dioxide provided that they supply the technology and expertise to reduce emissions in developing countries. The success or not of national and international climate policies over subsequent years will play their part in the next UN environment and development conference set for around 2012.

At their Gleneagles summit the G8 countries also agreed to a new Dialogue with the leaders of China, India, Brazil and other emerging economies. Together they are due to hold dialogue climate meetings annually up to 2008, when they will report to the Japanese G8 Presidency. Even though some of the countries involved are not part of the Kyoto agreement, the goal of these meetings is to try to achieve real progress in stabilising greenhouse-gas emissions before the Kyoto 2012 deadline.

The 2005 G8 summit also saw international discussion on the further development of the Kyoto Protocol in Montreal: this might be considered the opening discussions in the aforementioned dialogue. The meeting consisted of two sets of parallel talks. One track of discussion consisted of nations who had signed up to the Kyoto Protocol. They agreed to further future discussions for greater carbon emission cuts as well as to find ways to help developing countries beyond 2012, up to when the original Kyoto Protocol relates. The other track consisted of parties who had signed the 1992 UN FCCC in Rio de Janeiro. Importantly these included nations like the USA, who had signed the FCCC but not the subsequent Kyoto Protocol (and who had made it clear at the above mentioned G8 summit earlier that year that they could not at that time join the Protocol). These non-Kyoto-nation talks enabled non-Kyoto countries to participate in climate-related policy discussion and keeps the door open for them to potentially join the Kyoto nations later.

A pessimistic perspective might be that the 2005 Montreal meeting achieved little, with no clear decision for further cuts in greenhouse-gas emissions beyond 2012. Optimistically it did commit to future discussions that could potentially lead to ecologically meaningful cuts in greenhouse gases.

8.2 Energy sustainability and carbon (global)

An overview of the history of human energy-resource production and consumption is given in section 7.2. Figure 7.8 shows how the growth in global energy consumption roughly quintupled in the latter half of the twentieth century and that this energy provision was by far dominated by fossil-fuel consumption. For this reason alone, it is easy to see how any greenhouse impact from the addition of carbon dioxide became more marked at the century's end. It also suggests that should such trends continue into the future that climate-change impacts will continue to increase. However, in terms of possible resource consumption and availability (as opposed to biological impact), how likely is it that we could continue trends in a BaU way?

The IPCC has in part answered this with their BaU scenarios, but these only cover the twenty-first century. As we have seen, in geological and biological terms a century is nothing. So, what are the prospects for continued profligate fossil carbon consumption given the likely exploitable geological base?

The Kyoto Protocol calls for a reduced overall level of carbon emissions but, to make the maths easier, and allowing for just some increase in consumption by developing nations, let us assume that carbon emissions stabilise in the near future. How long can we continue to consume fossil carbon at near-current rates? From Chapter 7 we can see that we have over two centuries of coal, and while oil and gas will run out long before then (at present rates of consumption), future economically recoverable fossil reserves (i.e. reserves not *currently* economically recoverable) could extend the potential fossil fuel era at current levels by another century or so, to over 300 years. Indeed, there are a number of resources, such as low-grade coal and orimulsion (a high-sulphur water/oil mix), that could be harnessed if the price were high enough to cover the extra processing costs.

Currently (using data from 2001) we are releasing some 8 Gt/year of fossil carbon into the atmosphere, of which some 3.5 GtC accumulates. Of course there is also the deforestation factor to take into account. Deforestation, though, will not last at the current rate for 200 years as the forests would have gone by then, so this dimension of land use may not be a carbon source in the long term. Then again, there are other carbon sources and the possible atmospheric accumulation from soils as the boreal climate warms (and that of other areas with high soil carbon). There is also the confounding factor of reduced ocean absorption once the oceans have absorbed a critical amount of carbon dioxide. So future guessing for the long term is a complex business. Nonetheless, a-back-of-an-envelope calculation (which if good enough for Arrhenius is good enough for us) suggests that if total fossil-fuel consumption continues as it is now (1995–2005) for up to two (or more) centuries, to use up our currently commercially available reserves (and without using some form of carbon-capture technology), then we will easily increase the atmospheric concentration of carbon dioxide to over three or four times pre-industrial levels, as opposed to around 33% above pre-industrial levels, which is where we are today, early in the twenty-first century. This carbon emission for two or more centuries at current levels is likely to raise global temperatures by easily over 6 °C and so have a palaeoanalogue earlier in the Cenozoic than a Pliocene palaeoanalogue, which itself might be expected somewhere towards the end of the twenty-first century with just the IPCC's anticipated 2.5–4.5 °C rise (Chapter 6). Whether enough of this heat of 6 °C or more will sufficiently penetrate the oceans to disrupt methane hydrates is not known with certainty (yet), but it is almost certain to release some of the soil's carbon, so compounding warming. For the palaeobiologist the nightmare scenario is that we might release enough carbon directly (through fossil-fuel and land use) and indirectly (due to soil carbon and possibly methane hydrates) to incur an extinction event. This may sound dramatic, and it is, but remember equally that we are already in the midst of a human-induced (anthropogenic) extinction event through biodiversity loss that is distinct from extinction due to anthropogenic climate change. The majority of the population do not perceive this as serious as the human species continues to flourish, as do the major commercial species on which we depend.

It is important to remember that the above discussion is based on rates of emission in 2001. As was noted in the previous chapters, the twin drivers of increasing per-capita energy demand as well as increasing global population synergistically conspire to drive up emissions beyond the 2001 level. (Indeed, in

2007, it looks like this is happening and will continue to do so unless we switch our energy dependence away from carbon; clean-carbon technologies notwithstanding.) In short, should emissions continue to rise this increase in mean global temperature of 6 °C or more will happen long before two centuries are up.

However, if a major (6 °C+) climate-driven event were to occur, what would be the consequence? Well, we can with a fair degree of confidence rule out turning our Earth into a Venus-like planet. Yet this does not mean that we can be excused from striving for greater energy sustainability that relies less on fossil carbon, or from adopting carbon-mitigation strategies. The palaeo-analogue of continued substantial atmospheric fossil carbon accumulation is likely be the carbon isotope excursions associated with the initial Eocene thermal maximum (IETM) and, worse, the Toarcian event (see Chapters 3 and 6).

Should we see a future event analogous to the IETM (let alone the Toarcian event) then this would have a profound effect on most of the planet's ecosystems. As it stands the IETM did have a considerable ecological effect some 55 mya, even though globally much biodiversity was maintained (the marine extinction notwithstanding). Yet this maintenance of global biodiversity only took place because of biome (hence ecosystem) shift. However, such an ecological response would not be analogous to the situation today. To begin with, present-day ecosystems have (up to the so-called Anthropocene era) arisen in a cooler Pliocene and ice-age Pleistocene of alternating glacials and interglacials. Further, as mentioned in previous chapters, current human ecosystem management to our species' own ends means that options for natural ecosystems to shift and evolve with climate are restricted due to human fragmentation of the landscape, which was not the case back in the Eocene. Furthermore, the ecological disruption of the IETM resulted in subsequent opportunities for new evolutionary paths, this again would be limited in the human-dominated global environment. Certainly the diversification and expansion of primates began around the time of, or shortly after, the IETM so in theory at least there could be evolutionary opportunities. It would also be a little ironic if our species – one current end-point of this primate diversification – triggered a climatic event analogous to the IETM that arguably marked the beginning of a new diversification for some other species.

The ecological disruption of an IETM-type event today would in all probability be more severe (because of our current ecosystem management, which itself is resulting in biodiversity loss). Equally, as with the IETM itself, there would be winners and losers. One of the losers would be humanity – more specifically our global culture – for although we as a species would probably not be completely wiped out, many of our 6 billion-plus population (and growing) would face dislocation, as parts of the planet become too hot, and famine, unless we radically adapted our grain crops (including rice) that currently dominate human calorie provision.

So, how realistic is this nightmare scenario? This scenario may well be a longer-term one than the IPCC's 100-year scenarios. Equally, as the IPCC warn, being wary of surprises, depending on natural systems' critical thresholds the scenario might even be manifest sooner. There has so far only been informal discussion about this among researchers but, to date, no analysis published in a high-impact, peer-reviewed journal. Nonetheless, given that globally, the IPCC (1990) say, we need to reduce carbon emissions by some 50% of their 1990 value by 2050 to stabilise CO_2 at 50% above pre-industrial levels (resulting in a low probability of an eventual IETM-type event) – which in turn necessitates both cutting fossil-carbon emissions per capita today and planning for even greater per-capita reductions for tomorrow's larger population – and that after 15 years since the first IPCC report we have singularly failed to halt growth in emissions, it would be foolish to dismiss this IETM-analogue nightmare scenario out of hand.

As things stand, irrespective of whether or not this nightmare scenario comes about, we are facing the very real prospect of there being climate change to a regime not seen at least since the Pliocene 3 mya or more. As such we are likely to lose much of the fauna and flora that have evolved over this time and see some major changes in the location and type of many of today's ecosystems.

So, what then are the theoretical prospects of reducing fossil emissions? This question relates more to the abiotic aspects of human ecology rather than biology per se and has been covered elsewhere (such as in Cowie (1998a) and by the IPCC SRES scenarios; IPCC, 2000) and so is only very briefly summarised in the following subsections and only for the short term to 2025.

8.2.1 Prospects for savings from changes in land use

As noted towards the end of the last chapter, the IPCC's second assessment report (1995) estimated that by 2050 some 60–87 GtC could be conserved or sequestered in forests. Compared to the above-estimated mean twenty-first-century BaU fossil-fuel emissions of 18 GtC, this represents an annual saving of around 1.2–1.7 GtC, or around 6–9% of annual emissions. More optimistic estimates (a number are cited in Cowie, 1998a) increase this saving to over 2 GtC per annum. However, even this degree of saving would not last indefinitely, for once forests have become firmly established (reached their climax community) they become (climate change itself notwithstanding) carbon-neutral. Nonetheless, as a fossil-carbon-mitigation strategy it has the potential to make a significant contribution for at least the first half of this century. A round-figure equivalent to an average figure of 2 GtC/year for a major global forestation strategy is not unreasonable but, because of the aforementioned reasons, unlikely to be of significant long-term benefit.

8.2.2 Prospects for savings from improvements in energy efficiency

Since the early 1970s there has been a considerable body of work looking at the potential for improvements in energy efficiency that, of course, have direct implications for reducing energy use and hence possible fossil-carbon demand and emissions. Much of this energy-efficiency work was originally prompted by the 1973 oil crisis and some of this research is referred to in the US and UK case history sections that follow (section 8.3; and again many are cited in Cowie, 1998a). Prior to 1973 the predominant policy view on both sides of the Atlantic was that a nation's energy consumption was proportional to its GDP. After the oil crisis this changed as it was recognised that different types of economic activity consume different amounts of energy per unit of economic turnover. In other words, people did not want to consume energy per se but to receive the goods and services that can be realised by harnessing energy resources.

It is never possible to achieve a theoretical 100% of energy transference from fuel to user application due to the laws of thermodynamics and physics, so it is always theoretically possible to improve energy efficiencies of generation, transfer and use. Indeed, because historically efficiencies have been so low, there has been an overall historical trend of improved efficiencies. The situation is further complicated because over time consumers have found new ways to use energy. For example, two decades ago (early 1980s) few western homes had home computers, no homes had DVD players and only a few had video recorders (VHS had only just become established as the surviving format out of four vying for the commercial privilege). Then again, household sizes change (Chapter 7) and that also affects energy efficiencies or energy consumed per capita.

Thermodynamically, energy efficiency from a heat engine or generator can be mathematically expressed in the following equation:

$$\text{Carnot efficiency} = \frac{\text{The heat converted into work}}{\text{Initial heat intake}} = \frac{T2 - T1}{T2}.$$

$T1$ is heat discarded in degrees Kelvin and $T2$ is heat input in degrees Kelvin. Efficiency trends for generation have improved. Energy generation that relies on a boiler or the thermal expansion of gases is restricted by the Carnot cycle that relates the temperature (in Kelvin) of the burning fuel to the temperature of the outgoing coolant (see above). Because of engineering constraints it is difficult to have commercial boilers operating much above 550 °C (823 K), whereas coolant temperatures (in temperate countries) tend to be a little above the average environmental temperature and so are around 10–20 °C (283–293 K). This gives a theoretical maximum practical efficiency of around 65%. This applies to both fossil- and nuclear-power generation. Such early twentieth-century power stations had an efficiency of around 25% while at the century's end, in the more developed nations, efficiencies of new generating plant had risen to 30–7%. One trick around this limitation is not to solely transform the fuel energy into another form (electricity) but to give the station's heat output direct to the consumer. So, instead of the station inefficiently generating electricity for consumers' electric heaters, heat from the plant goes straight to their houses. This is the theory behind combined heat and power stations. Another way around this limitation is not to use a boiler at all but some alternative way of transforming energy. Hydroelectric plants capture the energy of falling water while solar photovoltaic cells capture that of the Sun and both do so with far higher efficiencies than power stations. However, there are other limitations, here respectively the availability of water and sunshine, as well as limitations of cost. These costs can change with time, as the downward trend in the cost of photovoltaic cells demonstrates.

Aside from the cost of new, more-efficient generating plants, there is the cost (and indeed the carbon-energy cost) of discarding older, less-efficient plants. To take an extreme illustration, consider what might happen if a new technology became available tomorrow that doubled efficiency in construction and running in terms of both financial costs and energy costs (in terms of oil and cement). It would actually probably be less efficient in both these terms to discard recently built plants that used the less-efficient, old technology if the old plants had not been running long enough to pay off their construction costs (in terms of energy and money).

There is therefore a very real time lag involved in introducing new energy-efficient technology. Regardless of development time, new efficient technologies, even if theoretically practical, cannot be introduced instantaneously.

This same time-delay factor also applies to introducing energy-consuming technology with improved efficiency. A motor car needs to have paid off its construction energy and finance costs before it is replaced by a more efficient car. True, we only tend to think of such factors in financial terms, but the principle applies equally to the energy costs of technology, be it the energy used to refine (or recycle) the metal or the actual carbon in the said technology (plastics and so forth).

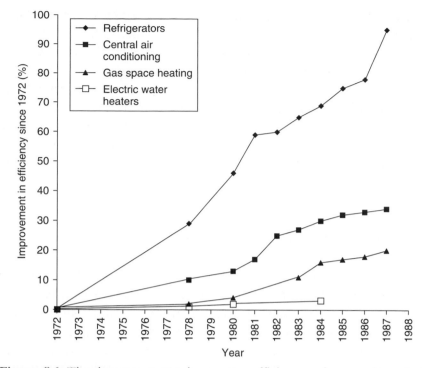

Figure 8.1 The improvements in energy efficiency of some household electrical equipment between 1972 and 1987. Note that this does not mean, for instance, that by 1987 refrigerators were 90% efficient, rather that their typical efficiency had improved by 90% over their efficiency in 1972. Data based on Schipper and Hawk, (1991).

Again, as with energy generation, efficiency trends in consumption have also tended to improve, especially since 1950, and progress was marked in the years following the 1973 and 1980 oil crises (see Figure 8.1). However, these efficiency gains in consumption have been confounded in a number of ways.

First, we actually might use the improved efficiency savings to maintain consumption comparable to old inefficient levels rather than lower consumption itself. The classic example is space heating. Over the years improved household insulation has reduced the loss of warmth from homes. However, the trend has not been one of reducing the energy that nations use for domestic space heating. Leaving aside the question of reduced household occupancy (Figure 7.3), home owners have tended to recoup the energy savings by being able to have warmer homes. This maintaining of dwellings at higher temperatures offsets potential savings. Similarly, whereas there has been an overall trend in improved fuel efficiency of cars (in terms of fuel consumption/unit of car mass per kilometre travelled) so this has been more than offset by the distance travelled per year by car owners. (This is one reason why the green lobby is against new motorway construction as it leads to increased car use, hence the need to build even more motorways; Tyme, 1978).

Second, we might devote our increased affluence (see the previous chapter) to using energy in less efficient ways. An example might be the use of less-efficient (in terms of common occupancy energy per passenger kilometre), larger sports utility vehicles (or SUVs) than traditional smaller and lighter family cars. This trend was manifest in developed nations early in the twenty-first century where some SUVs began to be used solely in urban settings and not off-road, for which they were notionally designed. The motivation here is largely psychological in that SUVs are sometimes perceived as conferring status for the driver/owner.

These two trends of improved efficiency and increased consumption serve to offset each other. Indeed, in some cases they can cancel each other out. Here an example would be the energy used in watching television in the UK. In the 1950s televisions were still a luxury and very rare but the number households in developed or industrialised nations with television sets was increasing. For example, in the UK in 1955 television sets consumed about 830 GWh. By 1970 more households had television sets and the energy consumed that year was around 5000 GWh. However during that decade semiconductor technology was being introduced to replace the old thermo-electronic valves and so the electricity consumption per set began to lower. In the late 1970s colour tele-visions began to replace the former black-and-white models and energy con-sumption per set rose once more. However, by 1984, despite almost universal use in the UK, these sets too had become more efficient and so annual consumption decreased almost to the 1970 level (Evans and Herring, 1989). The 1990s saw an increase in the number of television sets in the UK and with the advent of video (and later still DVD) increased total UK energy consump-tion by the sets themselves being left on standby rather than being turned off (as well as of course the ancillary video and DVD equipment). Domestic equipment left on standby accounts for approximately 10% of total domestic electricity consumption in the UK today. Finally, the new generation of high-definition, wide-screen plasma sets consume almost five times as much elec-tricity as the sets they are replacing. This pattern of counteracting trends is common to virtually all areas of energy use, although on balance one tends to dominate over the other so that they rarely negate each other exactly. For these reasons, since the 1970s to the early 2000s electricity consumption by UK domestic appliances and lighting increased at a rate of around 2% a year (Parliamentary Office of Science and Technology, 2005).

Finally, there has been one misunderstanding about energy efficiency that has re-surfaced in the literature over the decades. This is that energy efficiency per se is a substitute for energy generation. This is not strictly true. While it is eminently important to improve energy consumption and production efficiencies, improved efficiency by itself is meaningless: *something* needs its efficiency improving. Arguments that *the bulk of* energy investment should be directed towards improving energy efficiency are not practical. Investment will *always* be required *first* for energy production (although hopefully this will be efficient production). Calling for investment to be almost exclusively directed to energy conservation will fail because a nation needs power production as well as consumption (Greenhalgh, 1990). Given this, and the offsetting twin trends in improved energy efficiency and increased consumption, it is unwise to consider improved energy efficiency by itself as a way of countering increased greenhouse-gas emissions. Instead, a package of measures is required of which increased efficiency of energy production and consumption is but one. The others include low (or zero) fossil options such as nuclear and renewable energy, and carbon-capture technologies as well as pricing. These are summarised briefly in the following subsections.

The bottom line with regard to the savings that might be realised through energy efficiency – should a broad quantified estimate be required – might amount to around 20% of consumption over 1990 efficiencies by 2025. (For a more detailed overview see Cowie (1998a), which is in line with the IPCC (1990, 2001a) and the IPCC emission scenarios (2000).) We will return to this broad estimate later in the context of other potential savings that might accrue from possible energy strategies, and, with regards to efficiency, low-energy strategies.

8.2.3 *Prospects for fossil-carbon savings from renewable energy*

Renewable (flow) energy resources are currently (using 2000–3 data) contributing to 6–7% of global commercially traded energy (BP Economics Unit, 2005). This excludes wood and other biofuels gathered locally and consumed. (Estimates for global non-commercial local-biomass energy production are around 12 600 TWh/year (Royal Society of Chemistry, 2005), or about 1000 mtoe (which itself is close to about 10% of commercially traded energy).) Among renewable commercially traded fuels, hydroelectric power (HEP) dominates (at about 6–8% of global commercial energy traded in terms of coal-plant equivalents). However, there are also wind, wave, geothermal and solar (thermal and photoelectric) forms of renewable energy. These are not biological in nature and will not be described in this book save to say that of these wind is the renewable energy currently being significantly developed but which currently (2006) only contributes around 0.5% of global commercial energy. Biofuels are biological, though, and a summary account appeared in the previous chapter: they make a small but significant global contribution. Biofuels not only have the potential to offset additions of fossil carbon to the atmosphere but, when combined with carbon-capture technology described below, can take carbon out of the atmosphere.

As with fossil-fuel consumption and nuclear power, renewable energy production does have environmental impacts, but these are largely unrelated to climate – hence are not covered in detail in this text – and the exceptions to this only have a local (not global) impact. An example might be those hydroelectric schemes that create a substantially increased surface area of water over that of the previous river course, such as with the Iron Gates scheme on the River Danube between Romania and Serbia. The valley in which the scheme is located has acquired a more Mediterranean (as opposed to its formerly continental) climate.

There is a substantial body of literature as to the potential of various renewable energy resources to be found. My previous review (Cowie, 1998a) and potential for the year 2025 is in line with the proportions of total non-fossil energy contributions from a number of the IPCC (2000) scenarios including the principal scenarios within its A1, A2, B1 and B2 groups. One reason for this is that the various IPCC scenarios see their greatest divergence in emissions after 2025 through to 2100. Another reason is that the most economically attractive sites that lend themselves to renewable energy will be exploited first, which means that there is greater uncertainty associated with estimates further into the future. Finally, because bulk exploitation of renewable energy other than HEP is still in its infancy there is uncertainty associated with these technologies. Broad working upper and lower estimates for potential non-fossil renewables for the year 2025 should there be a drive to increase renewables energies are depicted in Table 8.2. These straddle the International Energy Authority's renewable potential forecast for 2030. The low estimate of 1296 mtoe roughly equates with a little over 1 GtC/year whereas the high estimate of 1960 mtoe equates with around 1.7 GtC/year. (For comparison, in 2000 some 8 GtC was released globally from fossil fuel.)

8.2.4 Prospects for carbon-capture technology

Improving energy efficiency is one way to increase the economic effectiveness of fossil-energy resources without increasing fossil consumption and developing renewable energy is another. However, as indicated in the previous sections, both these ways have their limitations in contributing to a low-fossil energy strategy. With enough coal for over two centuries of global consumption the question arises as to whether there is a way of continuing to use

Table 8.2. *Possible potential contribution of renewable energy for the year 2025.*
(Source: Cowie, 1998a, and checked to see if currently compatible against IPCC SRESs.
Further, the table's total low/high estimates of 1296–1960 mtoe for 2025 fit in with the
International Energy Authority's (2004) estimate of some 2000 mtoe possible from
renewable energy for the year 2030.) The mtoe data assume that the fossil fuel it displaces
would have been burned in a power station with an efficiency of around 35%.

	Potential contribution (mtoe)	
2025 resource	Low estimate	High estimate
Existing (2000) HEP	600	600
New HEP	300	500
Photovoltaic cells	5	45
Wave	20	220
Tidal	50	125
Wind	50	90
Geothermal	20	40
Biomass	250	340
Total	**1295**	**1960**
Of which is new	695	1360

fossil-carbon energy sources yet reduce carbon emissions to the atmosphere. The comparatively new idea of carbon capture and storage (CCS) has recently come to the fore.

The idea behind CCS is that instead of fossil-fuel consumption short-circuiting the deep carbon cycle (Chapter 1), through the burning of geological deposits of carbon-releasing carbon dioxide to the atmosphere, the carbon is captured and stored geologically. It could also apply to the burning of biofuels, in which case carbon from the fast carbon cycle is transferred to the deep carbon cycle as opposed to the other way around. This means that there would be a net reduction in atmospheric carbon dioxide.

There are other storage options but geological storage is both the most viable and has the least environmental risk (although the technology is currently far from proven). Other options include precipitating the carbon out as carbonate, so in effect creating limestone, or pumping it into the deep ocean. The chemical commercial viability of the former is still a little way off. The possible environmental impact of the latter on abyssal ecosystems is unclear, as is its stability (especially should the oceans warm up).

The capture of carbon dioxide takes place directly after the fossil fuel, or even biofuel, is burnt, by taking it out of the exhaust flue of a power station and pumping it into geological formations. The use of depleted oil and gas fields, or alternatively deep saline aquifers (water-laden porous rock formations), are the principal options. There are advantages with the former. Primarily, oil and gas fields have a proven stability and storage ability or tightness, whereas saline aquifers do not. Tightness is an important factor as the carbon must remain trapped for longer than the time in which fossil fuel is used at biosphere-affecting levels. Given that this last is likely to be between four and eight centuries, and given that the half life of carbon dioxide in the atmosphere is of the order of a couple of centuries, ideally the geological structure's tightness needs to be such that the carbon dioxide is secure for over a thousand years. One problem is that exploratory drills and the oil-/gas-extraction processes serve to undermine tightness. Well-capping technology is reasonably developed but may need optimising. Estimates vary, and further work needs to be done, but optimistically it is thought that leakage could be 0.004–2.4% on a 1000-year horizon. Further, some 0.03% leakage might take place during transportation and injection (DTI, 2003). Less optimistically, the *IPCC Special Report on Carbon Dioxide Capture and Storage* (2005) estimates that the technology could be reasonably assumed to be capable of storing 80–99% for 100 years and some 60–95% for 500 years.

There is another key advantage to using oil and gas fields and that is the actual process of storing carbon dioxide can be used to help extract remaining oil and gas. Pumping in carbon dioxide at the base of one end of the field drives remaining oil and gas further away and upwards. This is known as enhanced oil recovery (EOR). It is particularly attractive because it helps extend an oil or gas field's economic life and offset the cost of the carbon-capture exercise. It also means that the field provides more energy, which is itself important as the carbon dioxide from power-station flue gases needs to be compressed, and then pumped from the power station before being injected into the geological structure. This takes energy and so can offset (in energy terms) the utility of carbon capture.

Estimate costs per tonne of carbon dioxide emissions reduced by carbon capture vary both due to the uncertainties associated with the new technology and because much depends on the locations of the geological fields and the power stations. Storage costs without EOR could be about £30–£90/t (2003 prices) and with EOR significantly less. (For comparison it is interesting to note that the price of carbon on the EU trading scheme in the latter half of 2005 was around £22/t.) It should be noted that if EOR was always economically viable then the oil companies would always be using it. That they do not means that governments are going to have to provide a long-term economic and regulatory framework to ensure the process's economic viability. Although it is early days, the suggestion is that fossil-fuel power stations with such carbon-capture methods have costs comparable with nuclear power and off-shore wind energy. However, because carbon capture is not immediately economic the prospects for wide-scale adoption of the technology in the short term are unrealistic. What is more easily possible is to ensure that any new fossil-fuel power stations built are built as 'capture-ready' so that CCS can be added later. This is particularly important for less-developed nations, whose future fossil-fuel consumption is likely to rise markedly. A number of nations, such as India and China, are already thinking about CCS and are participating in the Carbon Sequestration Leadership Forum. Developed western nations have yet to fund CCS in developing nations but this is a likely option in the future.

It is important to note that there are limitations to CCS technology above those already cited. Chief of these is that, as currently envisioned, CCS systems only apply to power stations where the amount of carbon dioxide being captured is of sufficient quantity to be practically feasible (even if it is still not generally economic without some support). This limits the ability to reduce carbon emissions. This will be looked at more closely in section 8.3.2.

The argument against CCS is that it facilitates continued reliance on fossil fuels. There is also a sustainability argument in that in the long term (many decades and centuries), as fossil fuels dwindle, other longer-lived energy resources will need to be exploited. The nuclear option is currently one of the few technically proven greenhouse-friendly options available, although it has its own associated environmental concerns. Because of these some green campaigners view CCS as the preferable investment option.

The counter argument to this is that because carbon capture is predominantly related to fossil-fuel consumption, and because finite fossil-fuel consumption is not sustainable in the long term (beyond a couple of centuries as humanity's main energy resource), its climatic impact, although very significant, will be eventually curtailed. CCS could therefore have a limited role in providing time while we transfer to a sustainable, long-term energy strategy. Nonetheless, if considerable quantities are to be stored then the tightness of the geological formation needs to be greater than 60–95% for 500 years, as the IPCC's 2005 special report estimates, if we are to completely prevent a long-term climatic impact.

The IPCC (2001a) SRES scenarios for the twenty-first century assume 1000–2200 GtC consumption and one might presume approximately double this if one included the twenty-second century. This last greatly exceeds the estimated carbon release that triggered the IETM. This thermal maximum (and indeed the earlier Toarcian event) lasted well over 100 000 years.

Consequently, from a palaeo-environmental perspective if CCS is used to store hundreds or thousands of gigatonnes of carbon then significant leakage needs to be minimal over this kind of 100 000-year time frame if a similar thermal event is to be avoided. (It is interesting to note how this perspective compares with that for storage of nuclear waste – see below.) Upon further technology appraisal it may be that such storage tightness is not achievable. However, this does not mean that the technology is without value. It would appear from both the Toarcian and Eocene carbon isotope excursion (CIE) events that the build up of atmospheric carbon took tens of thousands of years. We know that the biosphere coped with both these events (ecosystem disruption, biome latitude migration, greatly increased global precipitation and erosion notwithstanding). What we do not know, even loosely, is how the biosphere will react to at least 1000 GtC injected into the atmosphere in a century. Even if carbon-capture technology could help reduce our twenty-first- (and twenty-second-) century anthropogenic impact to 'only' that of an Eocene event, that would confer major benefits. It would give more time for the biosphere to adjust and contribute to minimising human impacts. The IPCC's 2005 report on carbon capture cites the security (longevity) of storage sites as one of the unknowns requiring further investigation.

8.2.5 Prospects for nuclear options

Nuclear fission is a non-biological energy resource and so will not be described in this book other than to note its possible contribution to a sustainable energy strategy. Nor does nuclear power pose a serious threat to the global commons in the way that fossil-fuel carbon release does. As with renewable energy there are potential local impacts and among these the potential for radiological ecotoxicology is in the fore, especially (for politicians) if humans are affected. The biggest radiological burden on natural systems does not come from the power stations themselves (other than accidents primarily arising out of poor design, such as the Chernobyl nuclear-power-plant disaster in 1986) but the creation of radioactive isotope-enriched fuels, fuel reprocessing and waste disposal. For example, in the UK the greatest radiological burden imposed on natural systems comes from the Sellafield processing and reprocessing plant (formerly Windscale), a burden which greatly exceeds the sum total of radiological burden from all British nuclear power stations. Even so, this burden is not so great as to be considered a public health risk. What is of concern is the question of the long-term disposal of radioactive waste. In both the UK and the USA (and virtually all other nuclear countries) this question has not been resolved and both nations have been engaged in lengthy investigations and consultations as to viable options. No country has yet fully resolved the question of high-level radioactive waste but some countries, such as Finland, have excavated repositories for low- and intermediate-level waste. The major concern is that high-level waste depositories need to be kept isolated from the biosphere for many thousands of years. This, by definition, will involve an unproven method, and hence incur a risk that needs to be anticipated if it is to be minimised. There is also a cost involved. Not surprisingly, this dimension to the nuclear option makes the question of nuclear fission power in a number of countries politically sensitive.

Although the waste problem has yet to be solved by most nuclear energy countries there is a far greater understanding at the moment of the concerns and related science than for carbon capture. Whereas there is not space in this text to go into the issues in any depth, it is interesting to note the timescales involved and compare these with carbon burial (carbon capture; see earlier in this chapter). One of the most considered schemes is that proposed for the Yucca Mountain in the USA. In 2001 the Environmental Protection Agency (EPA) issued standards for the radiological dose one would receive near the proposed repository of 15 milli-rems/year for 10 000 years. However, a court ruling in July 2004 found that the EPA, by limiting its compliance assessment to 10 000 years, had not heeded the recommendations (as it is statutorily obliged to do) of the National Academy of Sciences. They had earlier concluded that 'peak risks might occur tens to hundreds of thousands of years or even further into the future'. In response to this court ruling the EPA issued a consultative statement for public comment in August 2005 that there be a two-tiered protection regime: a maximum dose of 15 milli-rems/year for the first 10 000 years and 350 milli-rems/year for up to 1 million years thereafter (Carter and Pigford, 2005). This means that the civil-engineering considerations for leakage for nuclear power and carbon capture (to avoid leakage contributing to an Eocene-like warming event) are in timescale terms broadly similar.

In terms of rate of growth nuclear power proved to be the most rapidly developed of the major energy resources in the twentieth century. That it does not involve fossil carbon (the moderate carbon use of gas-cooled reactors' excepted) makes it from both climate-impact and energy-sustainability perspectives very attractive. In 1975 the global production of nuclear power was just under 100 mtoe (assuming the oil was burnt in a modern 38%-efficient station) and this rose to over 600 mtoe by the early twenty-first century, so contributing around 6.5% of global commercially traded energy (BP Economics Unit, 2005). However, because of the aforementioned environmental concerns, the growth of nuclear power capacity (as opposed to supply) has slowed since the Chernobyl disaster in 1986. Nonetheless, of the non-fossil carbon resources, it is the only one that has seen significant growth and contribution to the global energy budget in the latter half of the last century. It is therefore the only non-fossil energy resource that can demonstrably contribute to meaningfully cutting future carbon emissions without a decline in energy supply this century.

However, from an environmental-science and human-ecology perspective the question of whether we should adopt nuclear fission as part of a move towards a sustainable-energy strategy depends on the balance or biological risk and hazard of nuclear as opposed to fossil fuels. What should not be an issue in climate terms, although early in the twenty-first century it has featured in much public discussion, is whether the choice of energy investment is one of investing in either nuclear or renewable energy. Such a question simply displaces investment between developing non-fossil-carbon energy resources. The key question is whether nuclear investment should be included in investment displacing fossil-carbon consumption. Given the time it is taking to develop significant amounts of renewable capacity, it is very difficult (though not entirely impossible) to dismiss the nuclear option. Indeed, some of the public may be surprised at this human-ecology and environmental-science perspective, which differs markedly from that of the green movement.

From a political and sociological perspective, there are other (non-scientific) problems of nuclear power to do with proliferation: that of nations acquiring nuclear military capability through civil nuclear programmes. However, these might be surmounted if it is possible to keep nuclear-power generation quite separate from the other parts of the nuclear epicycle (of fuel enrichment, processing and disposal). This is especially important if the nuclear generation takes place in less stable countries.

It is not often noted, but as with finite fossil fuels there are limits to nuclear fission that depend on the finite supply of fissile uranium. As such, some fission can be considered a finite or fund resource. The size of this fund depends on the rate of use, which has seen periods of rapid growth up to the 1970s and which slowed markedly after 1986 (and Chernobyl). Consequently the estimates for the global supply of uranium vary from over three centuries at broadly current use, to less than a few decades assuming that uranium consumption grows to meet some 40% of anticipated global electricity demand by around 2030. There is, though, a property of nuclear physics which can extend this resource's lifetime, by which a more energetic reactor can be used to produce more nuclear fuel from certain isotopes. These reactors are called fast breeders; fast because they actively depend on an energetic type of neutron (a fast neutron) and breeder because they breed fuel through fast-neutron interaction with appropriate isotopes in the reactor's core. However, because the process relies on high-energy fast neutrons, the resulting high thermal energy produced needs to be quickly carried away from the core and so typically such reactors use liquid metal (sodium) coolant to conduct the heat away. This presents significant engineering challenges that make such reactors inherently more expensive than non-breeder reactors. Nonetheless, such breeder reactors would extend the global uranium resource by around some 4–6-fold.

The other problem with any major global use of nuclear power to offset greenhouse emissions is that of nuclear proliferation. Even without breeder reactors it is possible to manipulate conditions in reactors and to increase reactor core dwell times to produce transuranic elements such as plutonium that can then be used in warheads. Further, those countries with nuclear-epicycle capability (primarily isotope enrichment) for enriched nuclear-fuel production can produce materials for nuclear warfare. Such considerations, though beyond the scope of this climate-biology and human-ecology text, are of global socio-political consequence. Indeed nuclear war, no matter how limited, is not without its own biological impact.

Nuclear fusion is the process by which light elements are built up to more nuclear-stable heavier elements (the most stable being iron) with the release of energy. As such it is a counterpart to fission whereby heavy elements are split to more nuclear-stable lighter elements (again the most stable being iron). Fusion is the process that takes place in stars (including the Sun). The attractions of fusion include that far more energy per light atom (such as hydrogen) is released than with an atom of fissile uranium or plutonium. Also, that while fission converts the fuel to radioactive isotopes, fusion can convert hydrogen to helium without the production of long-lived radioactive isotopes. However, both fusion and fission do lead to the transmutation of their reactors' non-fuel structural components to radioactive isotopes. (In fusion, it is the neutral fast neutron which is not contained by the magnetic 'bottle' that does the transmuting.) So, both fusion and fission produce waste, albeit of markedly different natures, and it is considered that a fusion reactor would result in far less radio-active waste per unit of energy produced than fission.

The other problem with fusion is that it is technically difficult in engineering terms to contain a plasma of hydrogen in a magnetic bottle at a sufficient temperature and density for fusion to take place, let alone to extract the heat on an on-going basis. The first serious proposal for harnessing fusion came in 1946 with the Thomson and Blackman toroidal design: a toroidal has no ends from which a plasma may leak. Famously the 1957 ZETA (Zero Energy Thermonuclear Assembly) was predicted to be the forerunner of electricity that would be too cheap to meter. It failed. Over subsequent decades various designs increased plasma density, temperature and duration, including the Russian Tokamak and the Joint European Torus (based in Culham in England). The next major experimental fusion reactor that it is hoped will demonstrate sustained net energy output will be expensive, at over $10 billion, so necessitating international co-operation. It will be the International Experimental Thermonuclear Reactor (ITER) that France is hosting. The final decision on constructing this was made in 2005 and it is thought that the first ITER plasma should be possible by 2015.

In terms of energy sustainability and carbon the big attraction of fusion is two-fold. First, it imposes a low greenhouse burden. (There will always be some greenhouse burden if only because of the carbon dioxide generated in the cement-making process, even if reduced through carbon capture.) Second, compared to human energy needs there is an abundance of hydrogen and even its more rare deuterium isotope. If all humanity's commercial energy production came from deuterium fusion (at current levels) then theoretically there would be enough fuel for well over 1 billion years. Indeed, should fusion become sufficiently economical (i.e. only marginally cheaper than oil's real-term price for much of the twentieth century) then, aside from the radioactive waste issue, a real environmental concern might become the thermal waste from profligate energy production and consumption. (For example, consider far more pronounced current urban thermal islands.) However, we still have at least decades to go before the first commercial fusion reactor. ITER has to prove itself theoretically and then a fully commercial reactor needs to be designed and produced along with a fusion epicycle plant to produce fuel and process waste. Fusion is therefore unlikely to make a significant contribution to meeting global energy before the mid twenty-first century. It is, though, a worthy long-term goal.

8.2.6 Overall prospects for fossil-carbon savings to 2025

Such are the errors of estimation it is not worthwhile presenting an overall picture for potential carbon dioxide emissions for the present (taken here as 2000) and those possible for the near future (taken as 2025) to greater than one or two significant figures. Table 8.3 below depicts the overall current and near-future carbon scenarios with the latter considered both in a BaU way (i.e. continuing as we did in the 1990s without adopting any greenhouse strategies) and what might realistically be done with greenhouse measures. Note the total carbon emissions include all those associated with human activity and so include fossil fuels, cement manufacture and land-use change.

The second and third columns of Table 8.3 tie in with many of the IPCC SRESs. In part this is a cheat on my part because most of the IPCC's scenarios do not diversify until the second quarter of the twenty-first century

Table 8.3. *Potential B-a-U carbon emissions against maximum likely possible savings for the year 2025. Non-fossil energy includes renewable energies and fission (hence is higher than the estimates for solely non-nuclear renewables given in Table 8.2).*

	Human-related CO_2 emissions/savings in GtC per year		
Carbon source or sink	2000	2025 B-a-U	2025 with savings
Total human activity	8	14	(savings not realised) 14
Decline in deforestation and some reforestation	–	0	Save ~2
Improvement in energy efficiency	–	0	Save ~20%
Offset with non-fossil energy	–	0	Save ~2.2%
Total	**8**	**14**	**7 (low) – 8.5 (high)**

onward: after then, things not only get difficult to summarise but the outcomes and scenarios are so varied that any attempt to include them would have limited meaning. However, although calculated independently, the final column of savings happens to approximate broadly the IPCCs B1 SRES optimistic scenario, especially the high with-savings total. The low with-savings total also ties in with the extreme low-fossil SRESs. Although this scenario could in theory be realised, it would be foolish to assume that it would, for the global political and economic will for this would need to be considerable.

It should be noted that the non-fossil contribution includes elements of both renewable energies and nuclear fission. It may be possible to invest solely in non-nuclear renewable energy but as some of this would be new unproven technology (or established technology operating in untried circumstances), individual nation's policy-makers have to decide whether to go down the nuclear route. Globally speaking, from the 1970s to the Chernobyl incident in 1986 nuclear-generating capacity was growing at between 25 and 30 GW/year. After Chernobyl capacity typically grew between 1 and 5 GW/year. However, this belies a more complex picture revealed by the number of nuclear reactors starting to be constructed. These peaked *before* the Three Mile Island nuclear-plant incident in 1979 and declined markedly after it (Worldwatch Institute, 2003). There is little doubt that these two incidents undermined global confidence in nuclear power. Consequently, the contribution nuclear is likely to make in the future is uncertain, although the potential is there.

Nonetheless, even the broad-brush picture that Table 8.3 portrays provides a couple of illuminating points as to our species' near-future relationship with the carbon cycle and climate. First, without taking any action to combat carbon emissions and to continue as we are then anthropogenic carbon emissions are set to approximately double between 2000 and 2025. Second, even with the adoption of major greenhouse measures (developing non-fossil fuels, enhancing energy efficiency, reducing deforestation and engaging in some reforestation) we will barely stabilise emissions.

The IPCC 2001 report, based on the then-peer-reviewed science, is quite clear as to what a BaU future without greenhouse-combating measures will look like. By 2025 global temperatures will have increased by 0.75–0.9 °C above 1990 temperatures and by the century's end temperatures will have risen by 1.4–5.8 °C. At the high end this would take us well beyond Quaternary climates and the likelihood of the IPCC's warning of climate surprises becomes very pertinent (see Chapter 6). With greenhouse measures by the end of the century the temperature is still likely to exceed anything seen during our own time (that of *H. sapiens*) and especially the past 11 500 years (the Holocene). Our world will change.

8.3 Energy policy and carbon

It is the human influence on the carbon cycle that is changing our world above and beyond our species' physical presence and its direct impacts. Of course these last, through land-use change, are affecting the global carbon cycle, but beyond this it is fossil-fuel emissions that are by far the dominant factor (see Table 1.3). Consequently no exploration of the human ecology (the way our species relates to others) of climate change can be complete without looking at the energy dimension and energy policy in carbon (and non-carbon) terms. This itself is a huge subject worthy of a number of books in its own right. However, it is perhaps worth looking at three case histories. The first relates to a country whose inhabitants emit the most carbon when taken as a combination of a per-capita basis *and* as a nation – the USA. (Although of course if looking purely at a per-capita basis there are smaller nations who consume more.) The second is the UK, whose citizens along with a number of their Western European peers (and a number of other developed nations such as Japan), emit a little over half that of their US counterparts. Finally, we briefly examine India and China. Their citizens each release less carbon than North Americans and Europeans but because they are developing and because India and China's populations are 1.1 and 1.3 billion respectively (CIA, 2005) together they represent around 38% of the Earth's total population and consume (in 2004) 18.7% of fossil carbon produced globally. It is these countries that will generate much of the new climate forcing expected this century.

All told, the USA, UK, India and China account for nearly half (just over 44% in 2004) the planetary fossil carbon consumption (BP Economic Unit, 2005) and coincidentally together have nearly the same proportion of the global population. However, the average US and UK citizen individually consumes more energy than their Chinese and Indian counterparts.

Turning away from purely carbon concerns, energy policy may for some seem to have little to do with human ecology. Indeed, in the past many countries had little regard for the environmental impact of energy-resource utilisation. This has slowly changed, at least among the most developed (OECD) nations. Nonetheless non-fossil energy policy decisions do affect fossil energy use (consider France's nuclear energy policy in the latter half of the twentieth century). And even fossil energy policy decisions can affect the degree of a nation's anthropogenic climate impact (consider the late twentieth century's so-called dash for gas). Both non-fossil and fossil energy policy decisions have implications for environmental sustainability. These then are the reasons why anybody trying to understand the biology of climate change needs to have an appreciation of the energy policy implications for human ecology.

Each of the four nations in the three case histories below has a fossil carbon and energy graph (Figures 8.2–8.5) that conveys its country's carbon energy relationship for at least the past quarter century. Each depicts the nation's fossil-fuel energy consumed (the total of gas, oil and coal) in million tonnes of oil equivalent (mtoe) and its fossil-fuel energy produced. If the production line (dashed or dotted lines) is below the consumption line for fossil fuels (light solid lines) then the nation is not managing its needs self-sufficiently (this is just one, limited, perspective of sustainability). The separation between these two lines indicates the amount of imports required. Conversely, if the production line is above the consumption line then the nation is a net exporter of fossil fuel. The bold solid lines denote the nation's total energy requirements. Therefore the gap between that and the line for fossil fuel consumed denotes the amount of non-fossil energy that the country produces (mainly nuclear power and HEP but also including wind and other renewable energy). This gap can be viewed (the finite nature of global uranium notwithstanding) as indicative of the proportion of sustainable energy that a nation has in its energy mix.

8.3.1 Case history: USA

As noted in the last chapter, the USA, with less than 5% of the planet's population, generates nearly a quarter of the Earth's anthropogenic greenhouse gases. In part this is due to the individual high-energy consumption of US citizens. Such per-capita consumption been growing steadily since the Industrial Revolution and in no small part is due to the nation's indigenous resources: initially of wood and then coal and oil. However, the finite nature of US oil soon became apparent, as long ago as the late 1940s.

If any single thing in the latter half of the twentieth century could have shaken US citizens from profligate energy consumption and demonstrated the need for a successful energy policy then it should have been the 1973 oil crisis. The Organization of the Petroleum Exporting Countries' (OPEC) oil then increased from US$1.77 to $7.00 per barrel at a time when nearly half of the USA's annual oil demand (818 mtoe) was met by imports. However, the 1973 crisis did focus some political attention on to energy concerns. Then the western hemisphere's energy gap (between production and consumption) was some 9 million barrels per day, or around 13% of global refinery throughput.

In 1973 US President Ford proposed a series of plans to ensure that by the end of the decade the USA would not have to rely on any source of energy beyond its own. These proposals were packaged in what the President called Project Independence 1990. However, by 1990 40% of US primary energy needs were still being met by imported oil. With the benefit of hindsight few today might be surprised at this failure. Yet back in the 1970s some energy-policy experts had their doubts. David Rose was one such, and in January 1974 he condemned US energy policy as unrealistic and unworkable without the 'application of Draconian measures'. Indeed, the question energy-policy experts were asking at the time was whether the President's policy ideals would be formally adopted and implemented by Congress. So how did matters turn out?

In 1990 the USA produced some 417 million t of oil and 561 mtoe of coal. Yet because it (only) consumed 481 mtoe of coal it had a coal surplus of some 80 mtoe (17%). US oil consumption was 782 million t, so resulting in an oil deficit of 365 million t. Natural gas consumption, 486 mtoe, just exceeded production, by 23 mtoe (or 5%). In total in 1990 the USA saw fossil-carbon consumption exceed production by 21% or 308 mtoe. In terms of overall fossil sustainability (combining oil, gas and coal production and consumption figures) the USA has been importing increasingly more fossil carbon to meet its needs. Notwithstanding the climate dimension, given the finite nature of fossil reserves this fossil-carbon trade deficit is clearly not sustainable (see Figure 8.2.).

Today the US fossil self-sufficiency situation has deteriorated further. In 2004 the USA consumed the following primary fossil resources: oil 937.6 mtoe, coal 564.3 mtoe and gas 582.0 mtoe, a total of 2083.9 mtoe, whereas it produced

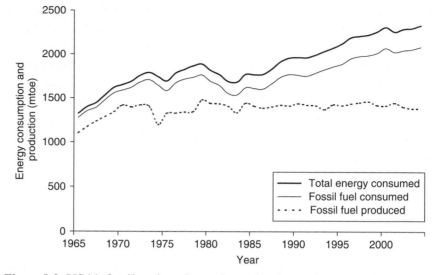

Figure 8.2 USA's fossil-carbon domestic production and consumption trends. US domestic fossil-carbon production remained broadly constant since 1975 but overall fossil consumption has increased. Further, non-fossil energy has been increasingly relied upon to meet the nation's total energy needs.

329.8 mtoe of oil, 567.2 mtoe of coal and 488.6 mtoe of gas, a total of 1385.6 mtoe. In short, the fossil deficit between 1990 and 2004 of US fossil-carbon consumption over production more than doubled, increasing to just over 50% of its fossil-fuel production level. This growing fossil deficit has been a hallmark of the USA's energy-resource profiles throughout the post-World War II years.

The finite nature of US fossil-energy resources had been predicted by energy experts as long ago as the late 1940s. As mentioned, the 1973 oil crisis served to highlight the problem. Today, with the benefit of hindsight, we can see that President Ford's Project Independence 1990 failed. Indeed, just one and a half decades after Rose's 1974 condemnation of US energy policy, Raymond Seiver, in an introduction to the 1980 reprint of Rose's paper, noted that 'when Rose's article was written, in late 1973, there was no USA energy policy. In mid-1979, there was still none. How long will we wait?'

Regrettably, a decade later still in the early 1990s, the target decade of President Ford's initiative, the answer remained effectively unchanged. This despite the Energy Policy Act in 1992 that encouraged renewable energies and natural gas (with its higher energy content per carbon atom than coal or oil), and the provision of more research and development as well as the extension of tax credits. Even so, the US Department of Energy estimated that by 2000 annual fossil savings would be of the order of just 2.5% of the USA's 1993 primary energy consumption. Such a saving, though significant, is small. As it was, the Act ignored car efficiency and lacked major economic incentives proposed by energy-efficiency groups (Eikeland, 1993).

As can be seen from Figure 8.2, since then trends have still continued more or less the same: US fossil consumption has grown, although its domestic production has remained static so that the US fossil deficit has increased commensurably. More recently, in 2001 the National Energy Policy Development Group under the chairmanship of Vice-President Dick Cheney reported back to President Bush, who had instructed it to assess America's energy position and policy at the outset of his administration's first term. Then the President had said, 'America must have an energy policy that plans for the future, but meets the needs of today. I believe we can develop our natural resources and protect our environment.' The report, entitled *National Energy Policy*, highlighted the need for some 38 000 miles of new gas pipelines along with 255 000 miles of distribution lines as well as increasing oil-refinery capacity, and a significant upgrading of the nation's electrical transmission grid. It estimated that over the next 20 years the nation's oil consumption would increase by 33%, natural gas by 50% and electricity by 45%. Yet it noted that both domestic oil and gas production was failing to grow to meet increases in demand. Conversely it observed that in terms of its economy's energy intensity – energy consumed per US$ of GDP – it was becoming more effective, in that since 1980 it was getting more dollars per unit of energy consumed. Above all it called for increased energy production by, not just renewable energy and nuclear power, but all fossil fuels including natural gas. This last was particularly interesting in terms of energy sustainability because estimates for the ratio of US reserves to (economic) production for gas were then of the order of just a decade. However, the Policy Development Group anticipated this criticism by calling for a vigorous domestic fossil-energy exploration programme, part of which would be in an opened part of Alaska's Arctic National Wildlife Refuge. It also called for the USA to build strong relationships with its overseas energy providers, namely the Middle East. In short, it assumed that the nation's fossil-carbon demand needed to increase along with increases in alternative energy so as to meet continued growth in energy consumption. In 2005 the essence of the report was enshrined in an Energy Bill, for a brief summary of which see Kintisch (2005).

In light of the above, it was not surprising that George Bush's administration also refused to sign the Kyoto Protocol. Which meant that as a nation it continued not to have any targets for cutting fossil-carbon use. 'As a nation' because some US city mayors and State Governors have introduced local targets. In 2005 nine north-eastern US states – Connecticut, Delaware,

Maine, Massachusetts, New Hampshire, New Jersey, New York, Rhode Island and Vermont – planned to cap some 600 of their power stations' carbon dioxide emissions at roughly 150 million t (about that year's level) starting in 2009. Further, between 2015 and 2020 those states plan to reduce power station emissions by 10%. The grouping represents about 3% of global fossil-energy emissions. The proposal came after 2 years from the Regional Greenhouse-gas Initiative. The same year California (which as a state represents 12.19% of the US population and 13.26% of its GDP) pledged to reach 1990 levels of emissions by 2020. Although these US regional policy goals were welcome, and contrast sharply with US national policy, they are only policy goals. As we shall see, and has been noted in the earlier subsection on the Kyoto Protocol, policy goals are one thing, realising them quite another. Notwithstanding this, even if these states, or any country for that matter, did manage to lower their carbon emissions, care is needed to ensure that there has been no carbon leakage. Carbon leakage takes place when an area supposedly has reduced its emissions but in fact has ceased high-carbon-emitting activities and imported goods and/or energy whose production elsewhere involved high-carbon emissions. (This is not to be confused with carbon leakage from soils, to which some IPCC reports refer.) In short, the savings the US regions are laudably proposing need to be genuine savings without carbon leakage.

Regarding the science underpinning the policy the 2006 report *Temperature Trends in the Lower Atmosphere* from the US Climate Change Science Program also signaled a change in which the USA perceived climate change science. Up to 2006, the Bush administration appeared reluctant to accept the IPCC's conclusions. This report, the first of a number from the Climate Change Science Program, recognised some of the apparent discrepancies of past climate data (such as from satellites). It concluded that 'the most recent versions of all available data sets show that both the surface and troposphere have warmed, while the stratosphere has cooled'. Which is what you would expect with a world warming due to the effects of increased greenhouse gases. While the report fell short of explicitly spelling this last out, it did call for more research and assessment of climate-related data. The importance of this report is not that it said anything new as far as many US climate-related scientists were concerned, but that it carried weight with US policy-makers and so represented the beginnings of a shift in the way President Bush's administration was likely to view climate science.

As a nation, the first years of the twenty-first century very much saw the USA on the same energy track as it had been for more than the last 50 years. It continues to increase its fossil consumption and enables this through fossil-energy imports and so, on a per-capita basis, continues to make a disproportionately large contribution to global carbon emissions. In terms of Kyoto, by 2004 the USA had increased its fossil-energy consumption by over 18% above 1990 (the Kyoto base year). The 2005 US administration had policies in place that would allow for emissions to rise by 30% over the nation's 1990 emissions by the target window of 2008–12.

8.3.2 Case history: UK

The average UK citizen produces just over half as much carbon dioxide compared to his or her US counterpart (see Figure 7.1), but far less as a nation, as its population is just less than a quarter of the USA's (Figure 8.3). In fossil-fuel terms the UK consumes (in 2004) around 200 mtoe compared to the US consumption of 2000 mtoe. As such, in carbon terms it is (loosely speaking) broadly representative of Western European nations, though using slightly more fossil fuel per capita than most others. By 2000 in Europe only Germany as a nation consumed more fossil carbon than the UK. The UK is also similar in fossil-carbon terms to Japan and so can be considered to be part of a group of nations that represent a halfway house between our US case history and the large developing nations.

Perhaps the single reason why the UK is less fossil-carbon-intensive per capita than the US is due to oil tax. Notwithstanding this, the USA is far larger than the UK and so expends more energy on transport, and historically it has a higher proportion of energy devoted to air conditioning. Factors such as historically poor car fuel efficiency have also played their parts in determining the intensity of fossil-fuel use. Automobile petrol has throughout the

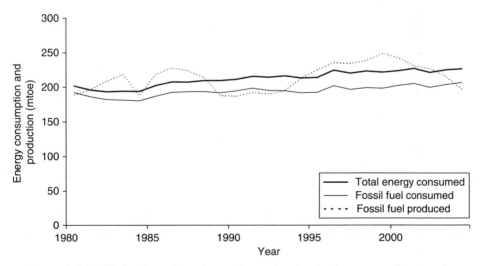

Figure 8.3 UK's fossil-carbon domestic production and consumption trends.

Table 8.4. *Prices for car petrol in the money of the day in mid 1992 and mid 2005 (prior to the 2005 sharp rise in crude oil). Note: fluctuations in the exchange rate mean that caution is advised when making strict comparisons and hence an allowance of around ±5% is advisable. Nonetheless, this price comparison between nations is fairly representative.*

Year	Automobile fuel	Price in UK (London)	Price in USA (east)
1992	1 US gallon	£1.75 ($3.24)	$1.30 (£0.74)
1992	1 UK gallon	£2.10 ($3.88)	$1.56 (£0.84)
2005	1 US gallon	£3.50 ($5.95)	$2.04 (£1.20)
2005	1 UK gallon	£4.20 ($7.14)	$2.45 (£1.44)

post-World War II period been cheaper in the USA than the UK by, very broadly (allowing for exchange-rate fluctuations), some 38% (see Table 8.4).

The UK saw considerable fuel switching throughout the twentieth century and towards the century's end it strove for energy sustainability (meaning in this case self-sufficiency), albeit with only limited success as far as the very early twenty-first century is concerned.

Following World War II UK energy policy (such as it was) focused on recovery from the war's rationing years and then in the 1960s in reducing acid emissions from coal burning (mainly in domestic homes and then in electricity power stations). However, it was the 1973 oil crisis that prompted much debate over energy sustainability, possibly more so than in the USA, which had not seen its imports threatened as had Britain during the war. In the years that followed 1973 the UK Department of Energy produced a number of Energy Papers, number 22 of which specifically dealt with national energy policy (Department of Energy, 1977). It defined the energy-policy objective as to secure the nation's energy needs at the lowest cost in real resources, but consistent with national security, environmental, social and other objectives. It highlighted needs to develop guidelines for the future expansion of coal, to ensure that reactor technology was sound for post-1990 expansion if required, to have regard for the long-term role of North Sea oil and gas, to secure all cost-effective savings of energy and to ascertain the viability and potential of renewable energy. It also noted that 'There [was] an argument for devoting some of the benefits to come from North Sea oil to investment in energy production and conservation'. This energy paper was particularly important as it preceded a governmental energy-policy Green Paper. (UK Green Papers are discussion documents and usually accompany a public consultation as a precursor to a White Paper that formally outlines forthcoming national policy and legislative intent.) The subsequent 1978 Green Paper (Department of Energy, 1978) was a landmark in UK energy policy for it represented the next most significant UK energy-policy official statement since the previous White Paper on fuel policy in 1967.

The 1978 Green Paper outlined the criteria for the formulation of a possible long-term energy strategy and covered much of the ground presented in the aforementioned 1977 Energy Paper. It addressed each of the primary fuels (gas, oil, coal, HEP and nuclear). It concluded that there was an urgent need to allow for the dwindling potential of global oil reserves to sustain economic growth and hence the need to switch to other fuels. It said that:

This change was in no way comparable with previous changes in the predominant source of World energy – from wood to coal, from coal to oil – since these changes were not in general dictated by an insufficiency of supply of the old fuel but took place as and when the new fuel appeared more advantageous. By contrast the change away from oil will be an enforced change and subject to the constraints in timing. We may have to accept that the available alternatives will be less convenient, more expensive or less environmentally attractive than the use of oil. Careful planning will be required if we are to effect the transition successfully and without damage to the global economy.

In terms of manifesting the 1978 Green Paper's policy aspirations, UK politicians have both succeeded and failed. They succeeded in that UK oil consumption largely stabilised between 1979 and 2004 to 74–84 mtoe (1984 and its coal-industry problems excepted). It failed because there was no long-term decline in oil consumption. Indeed, spending on cutting fuel use through energy-efficiency measures, for example, was cut: the Department of Energy energy-efficiency budget fell from £24.5 million in 1985 to £15 million in 1990.

In 1982 the then UK Secretary of State, Nigel Lawson, referred to what was to be governmental policy through to the 1990s, in that the energy market was to be liberalised and made competitive. Since then UK Conservative governments have privatised gas, electricity and coal. Electricity privatisation was a success in that it met the policy goals of its time and reduced prices. However, the cost was in cutting the strategic development of new supply, closing the former Central Electricity Generating Board's research laboratories and shortening the industry's longer-term perspectives. This was to lead to supply concerns in the early years of the twenty-first century.

Yet UK politicians were concerned over global warming issues from fossil-carbon consumption, and in the run up to the 1992 UNCED (see section 8.1.7) the UK was for a while leading Europe in calling for a carbon tax. (The problem being that unilateral action by a single country would disadvantage that country economically as it would be burdened with higher energy costs while its competitors would not.)

Following the 1992 UNCED, in 1994 the UK published *Climate Change: The UK Programme* (HMSO, 1994), which set out its proposals to combat anthropogenic climate change. The UK was the first country to fulfil this obligation made under the UNCED FCCC. It featured several measures that were considered by some as weak, and implementing it was not helped by the closure of the government's Department of Energy (which was then subsumed within the Department of Trade and Industry). However, *The UK Programme* (1994) did explore the aforementioned possibility of a carbon tax (as did some other Western European countries at the time). To date this has not come to pass. However, a tax (value added tax) was added to energy sold to consumers. Originally it was proposed to be at a level of 17.5% but this proved so unpopular that it was only implemented at a reduced rate. Indeed, it continued to be lowered so that by 1997 it was 8% and by 2005 just 5%. Another concern with this energy tax was that it did not discriminate between fossil-carbon and non-fossil-carbon energy sources. The UK strategy strongly relied on voluntary measures and what was known as the dash for gas: natural gas (methane) having a higher energy content per carbon atom than either coal or oil and so being more greenhouse friendly.

During the mid 1990s it became clear that the original *Climate Change: The UK Programme* was not enabling the UK to meet its climate policy goals and so other measures were introduced. First there was a levy on electricity from fossil-fuel power stations. However, because of EU Charter restrictions this could not continue beyond 1998. Second, there was a series of Non-Fossil Fuel Obligations (NFFOs) that guaranteed the sales of electricity from a variety of non-fossil-fuel sources. These too were criticised as weak and the third NFFO in 1995 related to just 1% of UK electricity production.

In 1997 the Labour Party took office. It published a number of policy documents, including *Climate Change: The UK Programme* (Department of Environment, Transport and the Regions, 2000), which is not to be confused with the 1994 document of the same title. It affirmed the UK's obligation under the Kyoto Protocol (unlike the USA), stating the UK's commitment to reducing greenhouse-gas emissions to 12.5% below 1990 levels by 2008–12, and stating its 'domestic goal' of a 20% reduction in carbon dioxide by 2010. The means by which this was to be achieved included a continued move to gas-fired power stations ('subject to cost'), increased contribution from renewable energy and a target to double combined-heat-and-power stations. (Such stations have a lower efficiency of electricity production but as waste heat is used to warm buildings, the overall energy efficiency from fuel consumption to total energy (electricity and heat) delivered is greater.) It also formalised the UK commitment to the Kyoto measures of Joint Implementation (JI) and the

Clean Development Mechanism (CDM). Climate-related criticisms of the new *UK Programme* included the cost restrictions on renewable energy and that it side-stepped the question of replacing existing UK nuclear plants that were anticipated to come to the end of their generating life. Nonetheless, its aspirations to address greenhouse-gas emissions were commendable.

In 2003 the same government published the UK's first energy White Paper for over two decades. The *Energy White Paper: Our Energy Future – Creating a Low Carbon Economy* was published under the auspices primarily of the Department of Trade and Industry, as well as the Department for Transport and DEFRA. It was one of a number of Government and Parliamentary documents that used the phrase 'low-carbon economy'. Technically this was misleading as there was still an emphasis on switching from coal to gas (which contains carbon) and encouraging biofuels (which also contain carbon).

The 2003 *Energy White Paper* was the first energy White Paper to have as one of its core themes climate-change concerns. The other core themes were energy diversity (for security purposes), maintaining a competitive energy market and tackling energy poverty (to prevent the poor from being unable to afford energy). Laudably – from a climate-change policy perspective – it recommended, above and beyond its existing Kyoto and domestic goals, that the UK should reduce carbon dioxide emissions from the current (2003) levels by 60% by 2050. It also affirmed commitment to the policy goals of *Climate Change: The UK Programme* (2000) regarding renewable energy and combined-heat-and-power stations. However, on cost grounds it did not commit to any nuclear build, although it did 'keep the option open'. Another problem area identified by the White Paper was that of transport, especially road transport, and the dominance of the use of cars for personal transport. It said that the solutions lay with hydrogen and biofuels.

The White Paper's greenhouse-emission aspirations were lauded, although a number of commentators and policy analysts were not so confident that the paper had identified the means to deliver these. Climate-policy-related criticisms included that it side-stepped the question of nuclear build (given the forecast decline in UK nuclear power which meant that a modest increase in renewable energy offset nuclear as a non-fossil-carbon fuel rather than fossil fuel), and that the road-transport goals could not be fulfilled by domestic biofuel production (where even optimistic forecasts were a power of 10 too low for required levels) and hydrogen (which is an energy vector and not a fuel, as it takes energy to make the hydrogen in the first place). Such criticisms reflect those by others that the UK was not striving as it might to develop alternatives (for instance, see section 7.5.3). The other criticism, less vocally made at the time, was that the UK was leaking carbon (in the economic as opposed to the ecological sense). Since the 1980s it had been shutting down some of its energy-intensive industrial processes such as steel manufacture and so was importing such energy-intensive goods instead. This meant that the carbon emissions were being generated overseas.

Over the 25 years up to 2005 the UK has struggled to reduce its reliance on fossil imports and for much of this time, from the mid 1980s onwards, also to address greenhouse-gas emissions. Unlike the USA it has had some success in that total energy consumption has risen less than the USA's. Further, in a broad sense, it managed to roughly equate fossil-fuel consumption with domestic fossil production. Again this was unlike the USA, but it did rely heavily on its North Sea oil and gas reserves, whose production peaked around the turn of the millennium. However, longer-term energy-policy concerns focus on the slow development of non-fossil-carbon renewables as well as the lack of a nuclear strategy. Given the absence of the former, addressing the latter in some way was becoming an issue. Here, in Parliamentarian minds, concern over nuclear power related to public perceptions, as well as the uncertain long-term environmental impact of nuclear waste, counted against the technology. (Here it is perhaps worth being reminded of the comparability of past natural carbon (CIE) releases into the biosphere of the Eocene and Toarcian with that of the timescale needed to secure nuclear power's radioactive waste, as noted earlier on this chapter. Up to this time neither policy-makers nor the public were making such comparisons between carbon emissions' and radioactive waste's environmental impact.)

By 2005 energy-policy concerns over UK security of supply, sustainability (mainly meaning medium-to-long-term national self-sufficiency) and climate concerns came together. This was perhaps most strongly articulated by the UK learned community in a 2-day discussion symposium, 'Challenges & Solutions: UK Energy to 2050', followed by a half-day conclusion launch by a clutch of learned and professional societies led by the Geological Society that included the Royal Society of Chemistry, Institute of Physics, Institution of Civil Engineers and the Energy Institute and with the support of UK research councils. Only the absence of biological learned societies prevented nearly all of the major disciplines being represented. One of the main conclusions was the need for Government to seriously consider a new generation of nuclear power stations (the current stations then coming to the end of their generating life-times). A month later, on 29 November 2005, Prime Minister Tony Blair announced an energy review based on widespread consultation that included the question of whether the UK should embark on a new nuclear programme. This review was due to be completed in 2006. Before then, though, there were two other documents published that throw an intriguing light on the use of science to underpin UK climate and energy policy and to meet UK climate-policy goals.

The first of these was a consultation launched in December 2005 on *Proposals for Introducing a Code for Sustainable Homes* (produced by the Office of the Deputy Prime Minister). The idea was that the Government was to introduce a code of practice that would apply to all government-funded home construction and be voluntary for private builds. This code would help ensure that new homes were more environmentally sustainable and, importantly, impose a smaller carbon burden. But how committed was the Government in involving the UK science community in achieving this policy goal? The document's introductory section (page 7) stated that 'sustainability is not an exact science and is still developing', which is a fair comment. What perplexed some involved in science policy was that the list (appended) of those to whom the consultation was sent did not include the chartered bodies for biology, chemistry, geology or physics, or the Royal Society (Britain's 'academy' of science), even though the document appeared to recognise the need for science to underpin sustainability. The consultation document itself said that the proposed code was to include the need for homes to have a lifetime in excess of 60 years. Yet the proposed code was that the homes considered to be most sustainable would be heated by natural gas. Another concern from the greenhouse perspective was that the lowest level of the proposed code was the same as the then existing legal energy-conserving requirements. In short builders, should they wish, need not do anything more than they had been already in energy-efficiency terms.

The reason for citing this policy instance is that it is illustrative of one of the main reasons why successive UK governments since 1990 had difficulty in meeting climate-policy goals. The second document actually recognised this difficulty and it was the Governments' *Climate Change: the UK Programme* (DEFRA, 2006) (this was the successor to the *Climate Change: the UK Programme* 2000 and 1994 policy documents). Although some of the UK Government's 2006 policy documents had yet to be published *The UK Programme 2006* did predict some of their content. For example, it stated that the then forthcoming code for sustainable homes would raise carbon standards above the mandatory level. Its forecast for UK electricity generation depicted a constant contribution from nuclear power between 2000 and 2020 that clearly suggests that policy-makers were actively considering new nuclear build to replace old reactors scheduled to be decommissioned. A number of policy analysts consider that the reason for *The UK Programme 2006* document anticipating forthcoming carbon-saving measures was because *The UK Programme 2006* itself predicted that the Government's previous goal to reduce by 2010 carbon dioxide emissions by

20% (over the 1990 base level) would not be achieved: instead, a new forecast of a 15–18% reduction was given. Nonetheless, the Government remained guardedly optimistic that its original long-term aspirations were still in theory achievable.

From a purely policy-analysis perspective it is also worth noting that, reminiscent of its earlier contribution to the discussion of a possible European carbon tax, the UK has taken an international lead to encourage all nations to invest together in addressing greenhouse-gas concerns. This was not a purely altruistic action. British politicians did not want to make their own nation's markets uneconomic through unduly high (that is, significantly higher than Western European nations) energy prices so as to pay for renewable development. In short, the UK is in favour of multilateral action to address greenhouse emissions and so it only undertakes limited unilateral action. The two leading UK Parliamentary parties have maintained this approach for over two decades.

Economics lies at the heart of politics, and also at the heart of climate-change policy issues. That successive UK governments have recognised climate change as a problem since the 1980s, yet have failed since 1990 to set in train strategic policies to meaningfully reduce UK emissions of greenhouse gases, exemplifies political weakness as far as the environment is concerned. To help address this, in July 2005 the Government (through the Treasury's Chancellor) asked the former World Bank Chief Economist Sir Nicholas Stern to conduct an independent review of the economics of climate change. The review was taken forward jointly by the Cabinet Office and Her Majesty's Treasury. It reported both to the Prime Minister and the Chancellor on 30 October 2006 with an electronic version placed on the Internet, and a mass market paper version published in 2007 (Stern, 2007).

Its conclusions were that climate change will incur real costs, both in terms of reducing the causes (reducing greenhouse emissions) and in terms of dealing with its impacts. Stern warned that costs could be as high as 20% of the world economy. However, by taking 'strong action' now we could 'avoid the worst impacts'. Action could cost as little as 1% of the world economy if emissions were reduced. However, delay would exacerbate matters. Importantly Stern's report said:

Action on climate change will also create significant business opportunities, as new markets are created in low-carbon energy technologies and other low-carbon goods and services. These markets could grow to be worth hundreds of billions of dollars each year, and employment in these sectors will expand accordingly. The World does not need to choose between averting climate change and promoting growth and development. Changes in energy technologies and in the structure of economies have created opportunities to decouple growth from greenhouse gas emissions. Indeed, ignoring climate change will eventually damage economic growth.

This argument was not new. Some scientists and the green lobby had noted previously that climate change would have severe costs. Further, some had said that adopting low-fossil carbon measures would be advantageous through freeing an economy from its dependence on dwindling finite fossil resources. Indeed, both these arguments were detailed together nearly a decade earlier in *Climate and Human Change: Disaster or Opportunity?* (Cowie, 1998). What was new from Stern was that this message had come from an independent, yet Government-commissioned, review led by a senior economist. What remains to be seen is whether such measures will be adopted, especially as at the time of publication the UK was still struggling to keep on track to meet its targets to reduce emissions.

With regards to any climate-stabilisation target, the UK view has been influenced by EU policy. Since 1996 EU policy documents have frequently referred to a carbon dioxide stabilisation level of around 550 ppm, which (depending on the greenhouse-gas mix) has been associated with a 2 °C rise above pre-industrial temperatures, although the EU itself has no official target. One reason for this figure is that European crop yields are expected to begin to fall when warming reaches 2 °C. More recently evidence suggests that to ensure that the temperature does not rise above this 2 °C limit that levels will need to be kept below 400 ppm. As carbon dioxide levels are currently (2006) nearing 380 ppm and rising at around 1.5 ppm per year it is virtually certain that the 400 ppm limit will be reached well before 2010 and, given that carbon dioxide levels are expected to rise faster as the century progresses, the 550 ppm limit around, or shortly after, the middle of the twenty-first century. The UK's current aspirational target (a 60% cut in carbon dioxide emissions below 1990 levels by 2050) is currently broadly consistent with the UK's overall contribution to global emission reductions required for stabilisation at 550 ppm. Because curbing UK and global emissions is proving so difficult it therefore looks as if the 2 °C limit will be reached.

8.3.3 Case history: China and India

China and India have many similarities in terms of energy and greenhouse emissions. Further, on the spectrum of energy used by citizens, they are at the opposite end (currently low energy per capita) to North America (high energy per capita) in the amount of energy that their respective citizens use. However, representing some 38% of the global population China and India both make up for their low per-capita consumption in high capita numbers (CIA, 2005). Further, as both are developing rapidly and so increasing their per-capita energy consumption, these nations are likely to dominate emissions from before the middle of the twenty-first century. This is especially inevitable because, in addition to their per-capita energy consumption increasing, their respective populations are also growing. Of the two countries, India is likely to see proportionally more population growth, for unlike China (whose population has a comparatively low fecundity of 1.78) India has an average of 2.78 children born per woman (CIA, 2005). This is despite China having the larger population, with some 1.3 billion (2005 estimate), or just over 20% of the global population, and is due to that nation having a policy that actively discourages parents from having a second child.

Of the two, China has the higher per-capita energy consumption: over 1 t of oil equivalent a year in 2004 (inclusive of minority non-carbon fuels) compared to an Indian citizen, who consumes just over a third of a tonne of oil equivalent. Of China's domestic energy consumption only around 6% comes from non-fossil carbon sources. Consequently China makes a considerable contribution to global anthropogenic carbon emissions and as a nation is currently (2005) the second largest carbon dioxide emitter after the US. As it is India and China (some 38%) of the global population together account for nearly 19% (2004 estimate) of global human carbon emissions from fossil fuels (BP Economics Unit, 2005).

China's fossil-fuel consumption, its fossil-fuel deficit (domestic production less consumption) and total energy consumption (fossil and non-fossil consumption) have steadily increased over the past quarter of a century, with an exception around the turn of the millennium when its coal production and domestic consumption declined. This dip was in part caused by a decrease in coal supply and also a move towards increased electrification combined with the transference of electricity regulation from the former State Power Corporation. This rise in energy consumption has fuelled its economy, and its GDP quadrupled between 1978 and 2004. Its major energy-intensive industries include iron and steel making, aluminium refining, cement manufacture and coal mining.

As can be seen from Figure 8.4, China has gone from being a marginal exporter of fossil fuel prior to the early 1990s to a marginal (currently slowly increasing) fossil-fuel importer. It does produce (hence consume) a small amount of non-fossil energy. This carbon and energy picture over the past quarter of a century is shown in Figure 8.4. The small amount of non-fossil energy can be seen in the difference between the total energy consumed and the total fossil-fuel consumed, and this non-fossil contribution has been up to now clearly a minority energy resource. Both China's growing energy demand, and its small but increasing fossil deficit, has caused increasing governmental unease that has been detected internationally. This last is not a trivial point

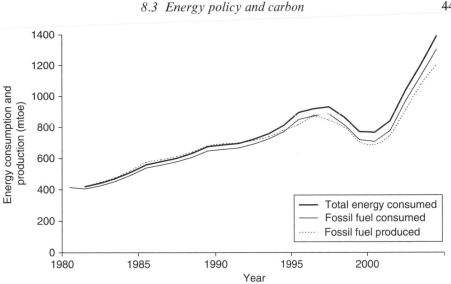

Figure 8.4 China's fossil-carbon domestic production and consumption trends.

as China has seen much governmental secrecy in the latter half of the twentieth century. However, China's recent opening up of its economy internationally has revealed particular governmental concern over the nation's growing dependence on imported fossil carbon (largely in the form of oil) and it is known that China is seeking to improve its energy security.

In November 2002 the Chinese government announced an oil strategy. It included new fuel-efficiency standards for cars and lorries sold there. The first phase of the standards came into effect in 2005 and the new standards range from 38 miles/US gallon for the lightest cars to 19 miles/US gallon for heavier trucks. In 2008, the standards will increase to 43 and 21 miles/US gallon, respectively.

On a broader economic basis, China aims to quadruple its GDP from 2000 to 2020 while only doubling energy use. This is a challenge to address its aforementioned energy security, as well as social welfare and environmental, concerns. It markedly contrasts with the years up to 2000 when energy use had been rising faster than GDP. As for the future, China's energy demand has been greatly increasing and is expected to grow at about 5.5% per year to 2020. By 2030, the International Energy Agency predicts the country will account for one-fifth of the total global annual energy demand. At the 2005 Kyoto next-phase meeting in Montreal, China declared an intent to double its use of renewable energy to meet 15% of its electrical demand by 2020.

There is no one body responsible for the energy in China; the government's energy responsibilities are shared mainly between the State Development Planning Commission, the State Economic and Trade Commission and the Ministry of Land and Natural Resources. They respond to a set of objectives provided by China's Parliament, the National People's Congress. These objectives are set down in national Five-Year Plans. In China's tenth Five-Year Plan (2001–5) it aims to increase the share of HEP and natural gas in its energy budget. Its commitment to nuclear power is considerable and it hopes to more than treble this. So currently there is much emphasis on the construction of gas pipelines and terminals, as well as many more hydroelectric schemes and nuclear plants. However, the impression given overseas is that China's priorities lie with economic development first and environmental and climate concerns second.

In 2006 the International Energy Agency (IEA) published a report, *China's Power Sector Reforms: Where to Next?* It noted that every 2 years, China adds as much power-generation capacity as the total in France or Canada. Also, that the country is now the biggest electricity consumer in the world after the USA and its needs are still growing. It concluded that China needs to strengthen its institutional and governance framework so as to use energy far more efficiently. Importantly, in greenhouse-gas and climate-change terms, this is needed so as to tackle the environmental consequences of coal that fuels 70% of China's electricity. To date China has adjusted some of its internal energy prices but much still needs to be done if it is to minimise its greenhouse impact. Development (as opposed to sustainable development) appears to be the overriding priority.

India consumes roughly a third of China's national energy consumption. Yet, like China it has seen considerable growth in both consumption and production, albeit at a far slower rate and over a longer time period, since the late 1970s (see Figure 8.5). However, like China it has been almost exactly self-sufficient in gas, which it uses as a minority primary fuel, and of which it has (according to 2004 r/p ratios) over three decades worth of reserves. Further, again as with China, India mainly relies on coal but it is not quite self sufficient and so is a net coal importer. This combined with its oil deficit means that India has had a significant, and growing, fossil deficit for more than the past quarter century. It has just under the estimated size of economical coal reserves of China but, because it produces (mines) so much less, it has a higher estimated r/p ratio of over two centuries. This means that India, should it wish, could sustain its increasing coal production for much of the twenty-first century.

With regards to India's oil deficit, it produces just a third of the oil it consumes. Its oil consumption (in energy terms) is equivalent to just over half

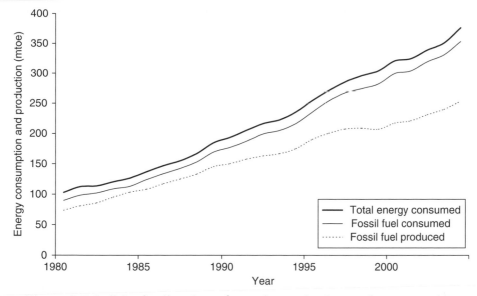

Figure 8.5 India's fossil-carbon domestic production and consumption trends.

of its coal consumption and around four times its gas consumption. As with China, India has some non-fossil energy (represented in Figure 8.5 by the difference between total fossil-fuel consumed and the total energy consumed). This non-fossil contribution is increasing but not nearly as fast as its fossil-energy demand and in turn its domestic production of fossil fuel is increasing at a slower rate than consumption. Therefore, proportionally speaking, India has a greater fossil sustainability problem than China in that the gap between what it can produce domestically and its consumption is already steadily increasing.

With regards to energy policy, India's energy-policy formulation and implementation is split between several ministries: one for each fuel, including electricity, and also one for energy planning. Since the 1980s India's policy has barely changed and this is reflected in the way it produces and consumes energy and fossil carbon (Figure 8.5). Foreign investment, especially from Japan in the mid 1990s, has enabled the nation to develop its coal resources. However, around the turn of the millennium it was considered that the lack of a comprehensive energy policy had been a barrier to significant further investment and so the government published its *Hydrocarbon Vision 2025* report. *Hydrocarbon Vision 2025* states that India should revise its foreign ownership regulations for refinery operations to allow 100% foreign ownership but be more restrictive of foreign control of petroleum trading. It also established a goal of India supplying 90% of its own petroleum. However, attempts to increase domestic oil production have since the mid 1980s had minimal success while demand has increased. So, like China, India has a net oil deficit but unlike China, India's domestic production has effectively remained static since the mid 1980s.

In common with many other developing countries whose economies in the 1990s and 2000s are seeing considerable growth, part (though not all) of their increase in carbon emissions is due to carbon gain from developed nations. This is the opposite of carbon leakage. Such rapidly developing countries are tempted to take on energy-intensive industrial processes to make goods or raw materials for export to developed nations.

With regard to twenty-first-century carbon emissions, neither China nor India are capable of sustainably accessing fossil carbon within their own borders to fuel their ongoing economic growth for many decades. This means that they will become increasingly subject to external fossil limitations, other than (for much of this century) for coal. In theory, should oil petroleum imports become sufficiently expensive, both India and China could synthesise liquid fuel from coal. Both China and India have a policy of increased fossil-energy consumption to drive growth. However, should the developed (OECD) nations provide low-fossil emitting technology at no extra cost to what they would otherwise have had to invest in conventional energy technology (what economists call at no extra opportunity cost), then they would explore this option. What they are not prepared to do is to halt their economic development.

This technology transfer from the developed nations is enshrined in the Kyoto Protocol. Participation of developing countries in carbon-emission mitigation is not specifically quantified by the Protocol, although it is a goal: India and China are not therefore required to limit their carbon emissions by 2012. However, the commitment to mitigate per se is conditional on the fulfilment of developing countries' own commitments to transfer enabling finance and technology. Furthermore China and India, as well as other Kyoto nations whose economies are rapidly developing, agreed, at the next-phase Kyoto meeting in Montreal in 2005, to discussing how to further develop the Protocol beyond 2012.

So far in this chapter we have looked at carbon drivers and carbon trends that have affected the anthropogenic manipulation of the carbon cycle. This manipulation has in one sense been inadvertent since manipulation of the carbon cycle was not intended, only the harnessing of an energy resource. These drivers have to date shaped our species' manipulation of the biosphere's carbon cycle, and will in all likelihood continue to do so in the future. So, given current carbon trends in developed (currently high-carbon-emitting) and developing (potential future high-carbon-emitting) nations, can we now manage our effect on the carbon cycle with a view (for example as the Kyoto Protocol aims) to reduce our forcing of climate? Also, can we also do this to meet sustainability concerns, in the economic sense, given the finite nature of comparatively easy-to-access economic fossil carbon? (Once again, note that resource-sustainability and climate-change concerns coincide.) Indeed, what are our prospects for successful carbon management?

8.4 Possible future energy options

8.4.1 Managing fossil-carbon emissions – the scale of the problem

As noted, to stabilise the global climate at present levels an immediate reduction in human emissions of greenhouse gases of 60–80% would be needed (Chapter 5). Indeed, in order not to exceed an atmospheric carbon dioxide concentration of 50% above pre-industrial levels – 420 ppm – then we would have needed to start reducing emissions in the 1990s, or make greater cuts later, so that they would be reduced by 50% of their 1990 value by 2050 (IPCC, 1990). Such is their magnitude, both these reduction options seem exceedingly unlikely, especially given the current drivers and trends outlined in this chapter. But exactly how great is the problem of managing carbon emissions? To get an idea of the scale of the problem we can look at the energy demands of the past 40 years, the present and the likely demands of the future. Energy demand is the main contributor to atmospheric carbon dioxide, the principal anthropogenic greenhouse gas. To date, how we have met this energy demand?

Forecasting the future is not possible with any certainty but it is possible to make some reasonable assumptions to underpin a scenario. In this case the scenario is depicted in Figure 8.6. Here three fundamental assumptions are made. One is that future global population growth follows the median UN forecast (UN Population Division of the Department of Economic and Social Affairs, 2002). Second, that future energy demand is a middle IPCC SRES (IPCC, 2000) and so close to many of the IPCC scenarios. Finally, at least part of the energy contribution in 2044 will be similar to that in 2004. In other words, that we will be consuming at least a similar amount of coal, generating at least as much from HEP, etc. and then additional energy to allow for the economic, and also population, growth due to take place between now and then.

In short, the assumption is that both global per-capita energy consumption and population will grow over the next 40 years up to 2044 but that this growth will be *less* than that which actually took place since 1964. Under this

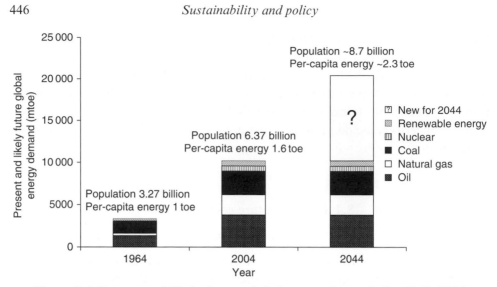

Figure 8.6 Present and likely future global energy demand, for 1964, 2004 and 2044.

scenario global energy consumption approximately doubles to 2044. Indeed, the International Energy Agency in 2005 forecast that global energy consumption would be even higher: it forecast an energy increase between 2005 and 2030 of over 52%. So, the assumption of 50% growth by 2044 can be considered a conservative one.

Given the aforementioned facts, then if fossil emissions in 2044 are to be held at 2004 levels then a similar additional contribution of non-greenhouse-impacting energy to the total energy used in 2004 needs to be found. The options, of course, are renewable energy, nuclear power and fossil fuels with carbon capture. Figure 8.6 visually encapsulates the scale of the problem by suggesting the likely amount of new fuel capacity required. This is not a trivial quantity.

The climate-related question here is whether fossil emissions can be kept at present levels or whether new fossil consumption (without carbon capture) will take place? When addressing such questions it is important to note that the limited ability of trees and soils to absorb carbon (assuming that a successful global sequestration programme is possible) can only at best (given likely boreal soil carbon release) offset a comparatively small proportion of any extra greenhouse emissions (see Chapter 7). Consequently, the biological implication is that we do run a considerable risk of increasing atmospheric carbon dioxide above 420 ppm (50% above pre-industrial levels). Indeed, unless radical action is taken we will exceed this 420-ppm limit, so incurring consequential biological and human ecological impacts.

8.4.2 Fossil futures

Virtually all the multi-national oil companies and energy consultants forecast (as much as anyone can forecast) that oil production outside of the 11 nations of OPEC will peak before 2015: in 2004 they met 60% of international demand. After 2015 OPEC nations, especially its Middle Eastern members, will have to meet the vast majority of international oil demand. OPEC production itself might peak as early as 2018 or as late as 2025 (Kerr, 2005). So by 2044 the Middle East will in effect have strategic control of the global oil supply. This brings us on to considering fossil-energy considerations beyond our 2044 scenario. Here the one bit of good news is that both the UN low and medium global population forecasts see the population begin to stabilise; indeed, the low scenario sees the beginning of a decline. Only the UN high-population scenario sees continued growth. Conversely, other than coal, *currently economically recoverable* reserves of oil and gas are likely to be largely depleted. This means either that the price of these fuels will markedly rise in real terms due to scarcity (as per the law of supply and demand) or the price of these fuels will rise due to the exploitation of *currently* uneconomic reserves. In short, by the middle of the twenty-first century it is almost certain that we will be paying significantly more in real terms for these fuels. This will make currently uneconomic technologies such as many types of renewable energy and nuclear power more economic, though far more expensive in real terms than those commonly utilised today. These price rises will herald the end of the post-Industrial Revolution era of cheap fossil energy with implications for the utilisation of non-fossil energy resources. Importantly, though, it will not necessarily significantly impact on the further perturbation of the carbon cycle or climate due to our past, existing and near-future fossil-carbon-related activities.

In 2005 an independently researched report was published that had been undertaken for the US Department of Energy. The study for the report was led by Robert L. Hirsch and so it is commonly known as the Hirsch Report. It noted that many estimates for the date at which peak conventional oil production will take place are before 2025. It noted that to most easily survive the economic disruption that will follow the peak, policies to both facilitate production and curb oil consumption (through improved efficiency and the use of other sources of energy) would need to be instigated two decades in advance, i.e. now.

On a purely economic basis real-term price increases in natural gas and oil will again make liquid- and gas-fuel conversion from coal more attractive. However, there will be economic repercussions in the twenty-second century as economic reserves of coal, if heavily exploited, will themselves become depleted. Consequently, in the very long term (from a human perspective), all fossil- (including coal-derived) fuel costs will be subject to a real-term marked rise. On a climate basis, without some form of carbon capture (which at this time is not easy to envisage), extensive use of coal as predicted over such a scale of the coming century will mean that climate change is likely to follow the high IPCC forecasts but into the twenty-second century (which has not to date been the subject of principal IPCC focus). What it also means is that by then the human perturbation of the carbon cycle will be of a comparable order to that of the early Eocene CIE, and hence increase the likelihood of setting in train some similar climatic event.

8.4.3 *Nuclear futures*

In part because of the above economics, and in part because it is a low- (though not entirely zero-) fossil-carbon resource, nuclear fission is likely to become increasingly more attractive in future. Although there are, and will likely continue to be, concerns over radioactive waste and potential nuclear proliferation, it is difficult not to see many countries adopting nuclear power, or increasing their existing nuclear-generating capability by the middle of the century. The scale of fission development (and its nature, such as developing or not fast-breeder reactors) will influence the rate at which high-grade uranium ore reserves are depleted. At some stage (somewhere from the middle of the century, but it is unclear when), as high-grade uranium ore deposits are depleted, the cost of fission will rise. Increased fission costs would be bearable if cheaper fossil fuel ran out and/or if fossil fuel combined with carbon capture was at least as expensive. This would enable lower-grade uranium ore be extracted and processed.

Ore extraction and processing currently has a carbon cost in that oil is used to transport the ore and energy is used to process it as well as to isotopically enrich the fuel. This processing and enrichment energy in our high-fossil-using present day currently necessitates some fossil-carbon consumption (although nuclear power still offsets more carbon than it consumes; see Chapter 7). However, in a potentially low-fossil (with high use of nuclear or renewable energy) future this nuclear-fuel manufacturing energy could conceivably come from renewable energy and nuclear power itself: such are the complexities that make energy futures difficult to predict. Currently part of this carbon cost is offset by nuclear electricity generation displacing electricity generation that would otherwise have been undertaken by fossil fuels. However, if lower-grade uranium ore is mined, then more oil is used in its transportation and more energy for processing and enrichment. Of course, if we were by then in a largely renewable- and alternative-energy future then conceivably fossil carbon would not be required to contribute to a nuclear infrastructure.

Increased future (beyond the mid twenty-first century) fission and fossil costs are likely to make fusion economically attractive: assuming, of course, that the current International Experimental Thermonuclear Reactor (ITER) project demonstrates fusion viability. It is unlikely to be cheap, hence there are fossil-carbon-release implications, but *if* fusion does prove technically viable, and economic in a more energy-expensive end-of-twenty-first century, then fusion combined with a hydrogen economy could be a long-term, climate-friendly hope. There would, though, still be a cost in the form of radioactive waste and we would still have to live with the climate impacts already in train and also likely due from whatever carbon emissions are to come. Finally, like fission, fusion, with its fast-neutron flow, also has nuclear-proliferation consequences for any nation with sufficiently advanced technology.

8.4.4 Renewable futures

The aforementioned higher price of future energy will make the majority of currently expensive renewable energies increasingly attractive and so they are an option for nations that can afford to develop these resources. Large-scale hydroelectric schemes are the cheapest of large commercial energy sources. However, there are a limited number of sites available for developing these without incurring (currently?) unacceptable environmental impact, not to mention socio-economic consequences. Currently (using 2004 data), HEP accounts for some 595 mtoe which is about 6% of global fossil- and non-fossil-energy consumption. It may well be possible to double this contribution by our 2044 snapshot date in terms of tonnes of oil equivalent but not in terms of its percentage contribution to the global energy market. This means that new HEP development will not be a factor that keeps energy prices down. Furthermore, HEP in itself does not offset fossil consumption for personal transport. Were there enough of it then it could theoretically be possible via a hydrogen economy but other forms of energy generation will be required to take that option forward.

That currently expensive renewable energies will become economically attractive in the future, not to mention climate-change concerns, will mean that these technologies will become increasingly developed. Yet because most renewable energies, such as solar, wind and tidal power, exploit a low-density energy resource they take up considerable land or sea-surface area. This in turn confers a significant local environmental impact. If renewable energies were extensively developed worldwide then this impact would be extended and begin to have a global effect. In addition, given the existing human domination of the terrestrial landscape and the current impacts that this has (most notably the post-Industrial Revolution increase in Holocene biodiversity loss), renewable energies are likely to compound current impacts and so not be the totally environmentally benign energy resources that they are currently popularly believed to be.

Having said the above, there is scope for a significant contribution from a mix of renewable energies. Further, there is scope for considerable micro-generation, by solar and wind power, or on individual dwellings in the urban setting. Here cost and individual motivation are implementation factors.

8.4.5 *Low-energy futures*

Low-energy strategies are strategies that maximise energy efficiency and so, in theory, have lower energy demand. As noted earlier this chapter, there have been continual improvements in the efficiency of energy generation by power stations that use boilers. Mileage (distance travelled per unit of fuel) of the most efficient family cars has also improved. However, variations in family preference regarding type of car and in use mean that there is still much wastage, so improvements are still possible.

As also previously noted, there are other behavioural difficulties with low-energy strategies. Improvement in home insulation resulted in some home-owners preferring warmer houses to lower energy bills. Another example from the UK was the 1990s fad for adding glass conservatories to the sides of houses. These were thought to help reduce energy bills by providing a green-housed thermal buffer. However, many home owners decided to treat con-servatories as extensions to the all-year living space and so heated these in winter, increasing energy loss.

Yet, low-energy strategies are, in terms of reducing energy-driven climate impact, a necessary part of the energy equation. Nobody in normal circum-stances would consider carrying water in a leaky bucket and even if leaky buckets were all that there were to be had, one would be likely to choose the ones with the fewest leaks. However, people do not tend to think like that when it comes to energy. Nonetheless, using energy efficiently is an important part of reducing energy consumption, and hence lowering greenhouse-gas emissions.

Low-energy strategies are most effective when included in settlement plan-ning and building construction rather than when introduced later. In the developed world ensuring that planning and construction standards are such that energy efficiency is greatly enhanced is undermined by a lack of what the industry calls a level playing field. If firms and commercial interests are not obliged for their products to meet the same specific minimum energy-efficiency standards then a competitor is able to save on efficiency measures and so undercut the prices of suppliers who have paid to meet standards. Consequently, legally determining efficiency standards and monitoring these standards in the market are both essential if maximum energy efficiencies are to be realised. Again, human psychology is a factor. Many consumers prefer to pay less for a product today even if over several years they may pay more through higher energy bills. Surveys have also shown that managers in the workplace often see less value in addressing energy-efficiency concerns as they save little money (and often cost more) in the short term even if long-term savings are realisable and ongoing. I have addressed this in more detail elsewhere (Cowie, 1998a).

Looking at energy efficiency and the prospect of low-energy strategies on an international basis, historically the developed nations used to export their old and inefficient power-generating plant, and technology which used that power, to less-developed and developing nations. This is a cousin to carbon leakage (whereby developed nations do not produce, but import, products whose manufacture is likely to result in high carbon emissions). For this reason the Kyoto technology-transfer mechanisms are important to help developing nations develop more cleanly (efficiently). Without such technology transfer developing nations, whose overriding priority is growth, will develop in a more energy-wasteful way, so raising emissions more than they otherwise would have from such economic growth.

It is therefore generally agreed that what is required, be it domestically or internationally, is highly energy-efficient and low-energy-consuming market controls and standards that are monitored and enforced. Within this regulatory regime the free market would be allowed to operate. However, as mentioned earlier in this chapter, such measures have to go hand in hand with greenhouse-friendly energy-production strategies.

8.4.6 Possible future energy options and greenhouse gases

Given the above, and the scale of the likely future energy demand to meet an evermore energy-intensive and growing human population, it is a virtual certainty (barring some cataclysm) that fossil-fuel consumption will continue to increase beyond present values as we approach the middle of the twenty-first century (and beyond) – see Figure 8.6. Carbon capture, even if the technology works on the necessary timescale, is only applicable (as far as can be currently foreseen) to large, point-source carbon dioxide emitters, namely carbon-fuel power stations (be they fossil or biofuel) and carbon dioxide-emitting industries such as cement manufacturers, and here only really effectively if they are located near, or within pipeline reach of, the appropriate geological strata to act as a carbon store. Consequently, for the medium-term future continuing to use fossil fuels for the most part means continuing to emit carbon dioxide and at levels higher than at present.

Nobody can predict the future with certainty, which is why the IPCC have their various futures from the Special Report Emissions Scenarios (SRESs). Not surprisingly, given the aforementioned range of pressures for increased carbon dioxide emissions, all the SRESs forecast increased emissions until 2050 and only a minority predicted reductions after that. However, such is the scale of reduction needed to stabilise climate change, let alone return the global temperature to its 1990 level, that all the SRES forecasts result in global warming and none see carbon dioxide at less than 500 ppm through to the end of the twenty-first century. See how this compares with the discussion of climate and carbon dioxide levels in the earlier section on the Kyoto Protocol (section 8.1.8).

The alternatives to fossil fuel, (renewable energy and nuclear) power, are currently more expensive than fossil fuel. Market regulation (different to that currently in place) will therefore be required. Having said that, some countries may adopt civil nuclear power for other (military-related) reasons. The future therefore seems to be one of increased energy prices. Increased prices impact inherently most on the poor and poorer nations. This in turn will have biology-related impacts on poor-nations' population health and longevity as well as reducing their ability to pay for climate-change impacts. Even so, such is the magnitude of likely energy demand, even before the middle of this century (see Figure 8.6), that the required growth in non-fossil-energy generation will be unprecedented. This is not to say that it is not possible but, as with climate change itself, it is new territory.

With regards to the developing and developed nations, carbon emissions and economic growth have been the subject of some discussion about compromise, and indeed it can be found within the Kyoto Protocol. This is the idea of contraction and convergence. The developed, wealthy nations would switch to a low-fossil, high-energy-efficiency economy and so contract or reduce their carbon emissions. Meanwhile, the developing nations would be allowed some leeway and so increase emissions. In this way the developing and developed nations, on a per-capita basis, would see their respective levels of emissions converge. The problem at the moment (since Kyoto) is that the wealthy developed nations en masse are not contracting emissions but increasing them, and the developing nations are similarly raising emissions. At the moment (at least) our species is firmly on a high-emission IPCC scenario and, despite international rhetoric by the Kyoto nations, there seems little prospect of this changing. Climate change and climate-change impacts are therefore not only inevitable but could be more towards the extreme ends of the range discussed in this book.

The consequence of this is that not only will we have to adapt to our currently slightly warmer world but, we will need to make greater adaptations to an even warmer world in the mid twenty-first century and beyond. Of importance to policy-makers, the question is almost certainly no longer one of expenditure on either adaptation *or* greenhouse-emission mitigation, but expenditure on both, or paying for the consequences (namely the range of climate impacts and fossil-fuel scarcity).

8.5 Future human and biological change

Anthropogenic (human-induced) climate change is undoubtedly taking place. We can measure greenhouse-gas concentrations in the atmosphere and their respective heat-absorbing properties in the laboratory. We have detected how their atmospheric concentrations have fluctuated in the past: for example, between significant climate (glacial and interglacial) swings. We know that other factors also affect the global climate, within certain margins of error, depending on the factor being considered. So the case for human activity causing climate change is virtually conclusive, even if the exact detail is unclear (Chapter 1).

What is a little less certain is whether we are in the process of triggering a biosphere event like the Initial Eocene Thermal Maximum (IETM). Here there is less evidence and considerable uncertainty as to the environmental thresholds that would have to be crossed to initiate such an event. Only in recent years has there been concern that the amount of carbon being released into the atmosphere is of a similar magnitude to that involved in the Eocene event. Yet this similarity exists. Second, computer models as yet do not include the necessary biospheric components needed to simulate such events with sufficient detail/confidence, but they probably will in time. However, the evidence that does exist seems to point to an Eocene-maximum-type possibility. This suggests that future global-warming impacts are likely to be very significant, especially in the long term from the point of view of humans.

Here the language and perceptions of scientists and policy-makers varies. Geologists commonly view matters in thousands to tens of millions of years and longer. Such timescales do not figure in policy-makers' operational perceptions. For them 2 or 3 years is short term and a decade or more is long term.

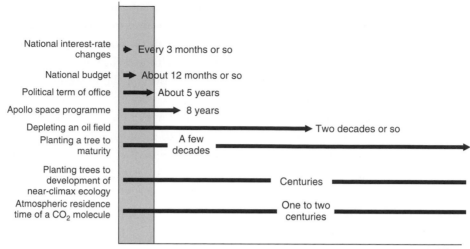

Figure 8.7 Time horizons of various human activities, natural processes and atmospheric carbon dioxide. The political time horizon is primarily dictated by the length of term of office which is a few orders of magnitude less than many energy-related natural processes; this time horizon (the shaded area in the figure) is also the typical period of *direct* policy involvement (responsibility). From Cowie (1998).

Even the timescale of short-term climate-science processes tends to be greater than that which policy-makers usually consider (see Figure 8.7).

Reconciling these two sets of perceptions is not easy. Consequently biosphere scientists need to recognise that their long-term concerns will not be readily (if at all) appreciated by policy-makers and so the scientists will need to adjust their presentations of science to politicians. Yet climate change and its associated impacts will be very significant on *both* these levels.

On the decade level we will see significant change and impacts. For example, over the coming years parts of the Earth (be it parts of the USA or the UK or South Africa) will become markedly drier, with profound economic and ecological effects. Over the coming couple of decades past weather extremes, such as the heatwaves experienced in 2003 by London and Paris, will be normal for the summer and what will become heatwaves at that time will be that much warmer and their associated biological impacts greater (Chapters 6 and 7).

On the 1000-year level we will see several metres of sea-level rise. Large parts of Greenland will become ice-free. Year-round sea ice in the North Pole will have long gone and we will see carbon cascade between various biospheric pools via the atmosphere, which will maintain global temperatures even if the global economy has by then switched away from fossil carbon. This will persist for many thousands of more years unless some means of sequestering atmospheric carbon is found. Over this timescale the oceans will rise scores of metres. All of Greenland and parts of Antarctica will become ice-free.

As said, many thousands of years, let alone 1000 years, hardly figures in politicians' or civil servants' thinking. Yet 1000 years, very nearly the time from the Domesday Book to the present, is in human terms just 40 generations!

As I have previously argued, if environmental scientists, or green lobbyists (two quite distinct groupings), wish to focus people's minds then they should learn from biology (Cowie, 1998a). Here the biggest bio-psychological imperative is associated with reproduction. Our bioclade (the sum total of life from a common primordial nucleic acid ancestor to date) has had over 3.5 billion years (Chapter 2) of Darwinian evolution dictating successful reproduction, enabling the selective passing on and evolution of genes. We are therefore programmed to care for our offspring. The intergenerational arguments of a climatically changed Earth, as well as the exhaustion of cheap fossil-carbon fuel over the coming decades, need to be brought to the fore. Anyone born today will witness many of the climate-change impacts discussed in this book and will almost certainly see the peak in consumption of global oil and gas as well as witnessing their decline to scarcity. This is not a shock statement but one of simple virtual certainty and something very pertinent to all parents.

Yet if to help get the message across we recognise the biological imperative to nurture offspring then equally we should recognise other biological imperatives that serve to counter moves to curb greenhouse-gas emissions. While it is difficult on moral grounds to expect fast-developing nations (such as China and India) with low standards of living to forego economic growth (and hence the energy needed to drive such growth), recognition of biological imperatives that encourage profligate resource consumption needs to be accepted. Affluent North Americans and Western and Central Europeans consume energy (and fossil carbon) not for essentials (basic warmth, clean water and so forth) but for luxuries (the SUV in an urban setting, overly warm homes and such). From a biologist's perspective this too can be viewed as part of the reproductive imperative and can be seen in many species. It is the driver behind the spectacular tail-feather displays of male peafowl such as the blue peacock (*Pavo cristatus*) or the large antlers of male fallow deer (*Dama dama*): the examples are legion. These signal to females that the males have the necessary genes to secure resources for such displays and so are worthy mates. The problem is that it can seem to be a never-ending race. As soon as one individual raises the stakes to a level above those of his or her peers, so the peers are forced to raise their own stakes if they are to successfully compete.

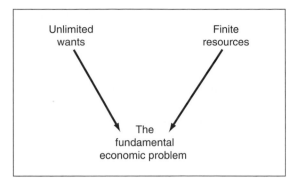

Figure 8.8 The fundamental economic problem.

In humans such resource-control signals include being able to command resource consumption so as to secure the big car or being able to afford the latest fashions. It leads to what economists call the fundamental economic problem (see Figure 8.8). Countering it is possible through changing perceptions of what is desirable. Humans can do this being both a sentient and a social species but just because it is possible does not necessarily make it easy.

So how is it possible to get around the fundamental economic problem? In biology and in the wild such displays have their limitations. Too large a set of display feathers, or antlers, can impede flight from predators. There are balances. With humans such checks and balances include limitations of access to resources (availability), which is where sustainability comes in. Failure to curb resource consumption leads to over-consumption and this applies to resources common to life on the planet. These common resources are principally the main biogeochemical sources and sinks in the biosphere and over-consumption leads to what is known by environmental scientists as the tragedy of the commons (Hardin, 1968). Global commons include the oceans and atmosphere. The over-consumption comes about through competition up to and beyond the sustainability limit. A simple model of grazing sheep on common land illustrates how this comes about. Each sheep herder is motivated to get the maximum grass for the flock. That the common can become over-grazed is a secondary consideration, for if a herder takes some sheep away from the common then sheep from other herders can have that grass. The result is that herders keep their sheep on the common, which then becomes overgrazed. The parallels between this and perturbing the fast carbon cycle, hence atmospheric carbon concentrations, should be clear. What the tragedy of the commons illustrates is the difficulty in modifying motivations so as to preserve commons. This brings us back to employing other powerful biological drivers such as the desire to nurture offspring (the next generation).

8.5.1 *The ease and difficulty of adapting to future impacts*

It is now clear that at best we might marginally slow global warming in the twenty-first century; it is most unlikely that it will be stopped, and virtually unrealistic to consider the Earth's system returning to the climatic regimen of the early twentieth century. Consequently twenty-first-century policies addressing climate change, in addition to any mitigation, are going to have to either include adaptation or pay for the consequential costs of climate impacts.

Leaving aside temperature, in terms of human global distribution, sea-level rise is arguably one of the factors that will incur both impacts and adaptation costs. Worldwide some 70% of the global population live in coastal areas or near rivers. A significant proportion of these could potentially be affected by sea-level rise and/or possible flooding due to increased (seasonal?) precipitation. The IPCC (2001b) estimate that in Western Europe alone in 1990 some 25.5 million were exposed to coastal flood risk and that of these roughly one in 1000 actually experienced flooding. Assuming no adaptation the IPCC consider that between half as much again to 90 times that number might be affected by the 2080s. The degree of uncertainty reflects the combination of uncertainty in the proportion of the population with a degree of flood-proofing and the uncertainty in the IPCC's forecasts. However, either way this represents a significant cost.

Adapting to climate-change-related impacts, such as flooding, does not mean reducing risks or flooding to zero. This is for two reasons. First, to eliminate all risks and flooding would be the most expensive option and is likely to be too expensive for society/communities to condone. Second, not all flooding does significant damage or incurs a permanent cost. For example, you may be aware of a street near you that becomes difficult to traverse after a heavy sustained downpour but then drains comparatively quickly. There are also land areas that are naturally prone to, and might even benefit from, occasional flooding. Marshes and other wetlands are obvious examples. Finally, extreme flooding for some will be a very rare event and proofing everyone against these may be too expensive. In these cases 'soft' measures (such as allowing fields and parks to flood) as opposed to physical infrastructure (flood barriers) may be more economical. It is a question of assessing costs and risk.

There are two key problems of assessing costs and risks with climate change. First there is the fundamental assumption that the risk and cost assessments are accurate. The 2005 example of Hurricane Katrina and New Orleans is illustrative. The 1998 proposed protection plan *Coast 2050* had a potential cost of some US$14 billion. It did not get implemented as nobody was prepared to pay; the risk or hazard was not considered to be worth the cost. It was a gamble that such an event would not happen. Yet following Katrina the economic losses were in excess of US$125 billion. (We will return to this shortly.) The second problem is that climate change is by definition dynamic and not static: past experience cannot be used as a guide to the future. For example, most of the New Orleans levees breached by Katrina were designed to deal with floods that occur once every 30 years or so. Now, notwithstanding whether this itself was an appropriate judgement of the degree of protection required, given the change in strength and frequency of extreme weather events due to global warming (Chapter 6), basing safety decisions on past criteria will be unreliable whereas including forecasts of future change would be prudent. That there is and will be global warming is not a gamble, even if the exact rate and extent of this warming are not exactly certain.

One of the countries that has historically spent the most on sea protection per length of coastline is The Netherlands. Quite simply this is because it is a low-lying country, with some 60% below sea level, and it is also a wealthy Western European nation. In addition, some 70% of its gross national product is earned in its low-lying areas and this provides a certain financial motivation. Indeed, the country is due to spend nearly £600 million (US$1 billion) on a programme called ARK (which translated from the Dutch stands for Adaptation Programme for Spatial Planning and Climate) to generate a comprehensive agenda to deal with climate change across several sectors of society and economy. As with other low-lying areas, The Netherlands faces a number of climate adaptation questions. First, how to identify and cope with increased risk. Second, on what timescale will the adaptations (identified from answering the first question) need to be implemented: when does acting later become too late? Third, what are the comparative costs and benefits of alternative strategies? Such adaptations are not just restricted to enhancing conventional sea defences (dykes etc.) For example, the 2003 summer was the hottest in Europe for some 500 years but it is thought that it will represent a normal summer in the middle of the century. During the 2003 summer the low discharge of the Rhine resulted in sea-water seepage into the ground water that in turn affected agricultural and horticultural activities. One option therefore will be new canals to bring in extra fresh water. However, with low river flow securing an extra volume of fresh water is likely to prove difficult. Another option therefore would be to increase the area of land devoted to fresh-water reservoirs. However, this will take land out of alternative use and income generation. A compromise therefore may be to build the reservoirs but to house large floating rafts, or hydrometropoles, on them to provide an area for, say, greenhouses (Kabat *et al.*, 2005).

Of course, just as overall in a warmer world there will be more ocean evaporation and hence precipitation, so on land there will be more evaporation and some areas will see significant water shortage. So, global warming will incur costs to address water shortages as well as costs of surplus (floods). Even some wealthy countries, whose politicians recognise climate change, ignore the full implications of the costs associated with such change, especially if such costs fall outside of government departments whose remit relates to climate and energy matters. The UK provides a typical example. Although the UK Government has plans to build homes for some 500 000 people, mainly in the area of Milton Keynes and north Kent, it ignored many of the major implications of climate change of which its own agencies warned (see Hulme *et al.*, 2002). In 2005 it launched a public consultation into *Proposals for Introducing a Code for Sustainable Homes* that would be used in the construction of these new (and any other Government-funded) homes (Office of the Deputy Prime Minister, 2005). However, the proposals did not take into account the broader landscape and regional issues for the areas in which the homes were to be built. (Nor were the independent UK learned scientific (biology, chemistry, geology and physics) societies on the initial consultation distribution list.) Yet the areas in which nearly all the new build was to take place are already experiencing water shortages in the form of lowering water tables, and stream and river flow, and are forecast to see less rainfall with twenty-first-century climate change. Further, there were (at the time) no plans to enable water to be routinely transferred from parts of England and Wales predicted to see an increase in rainfall. Even the Kyoto-supporting government of one of the planet's leading economies was ignoring costs.

As indicated, climate-adapting measures have a substantial price tag. Even if the cost of adaptation is far smaller than the economic activity protected, such costs can only be paid if there is the finance both available and in place. In 2003 the international news media expressed surprise that the wealthiest nation had not protected itself from, and then was slow to address, Hurricane Katrina. As mentioned above this was because the estimated US$14 billion was not available to implement a protection scheme even though the resultant cost from failing to make this investment was some nine times greater. Clearly the USA had the economic resources to make that investment but, for whatever reason (let's say policy-implementation inertia) it did not. However, the proportion of the global population living in wealthy countries is a distinct minority (indeed, even as is the proportion of wealthy within these).

An example of a far less wealthy nation to be significantly affected by sea-level rise is Bangladesh. Sea-level rise of just 100 cm (which is predicted by virtually all the IPCC SRESs for the twenty-second century) could potentially flood nearly 30 000 km^2 of Bangladesh, representing over 20% of its land area. This will expose some 14.8 million people, or some 13.8% of its current population, to dislocation. The problem is compounded because the nation is not only one of the most densely populated but most of its population relies either directly or indirectly on site-dependent agriculture: over a third of Bangladesh's GDP comes from agriculture (IPCC, 2001b). Yet despite the threat of floods, its lack of wealth is impeding it from developing strategies to address future sea-level rise. This is evident if only because it currently suffers from regular monsoon flooding which it is equally unable to address.

This problem of the poor adapting to climate change is not restricted to sea-level rise but includes, as has been mentioned earlier (Chapter 7), the human-ecological aspects of climate change, especially with regards to health and food security. Yet, as has been emphasised in this chapter, it is both the wealthy and those nations striving to develop that are dominating greenhouse-gas emissions. So, while adapting to climate change will be easier (more affordable) for the wealthy, it is not an option available to (or within control of) the poor. Further, it is an unpaid environmental cost (or environmental externality) imposed by the wealthy and those accruing wealth.

If the poor will have a difficult time adapting, then what of natural systems? Semi-natural systems of woodland and agriculture will at least see management, as they have a recognised economic value. Of course, the most vulnerable components of such semi-natural systems will be those species and ecosystems within the semi-natural biome that have no immediate economic value attached to them, i.e. wildlife species (see section 8.5.3).

As noted in Chapter 7, even natural ecosystems with no apparent immediate economic worth have a value, in that their ecosystem function enables biosphere processes on which we rely to take place. In theory it may be possible to attach a value to these ecosystems for performing these functions. However, acting practically on such environmental economic theory is not an affordable option for poor nations and, without assistance, their options for adaptation remain limited.

Finally, climate change often results in a significant step change. This is because many of the factors relating to climate forcing (see Chapter 1) and impacts (Chapter 6) interact with positive feedback until a new equilibrium is reached. Again, wealthy nations can afford to adapt or relocate climate-sensitive economic activity. And again this is not an option for poorer countries. For example, near-subsistence communities in the sub-Sahara can do little in the face of dune-stabilising vegetation dying due to climate change, hence consequential dune movement and the covering by sand of other vegetation.

In short, the difference in the ability of societies to adapt to climate change is both wealth-related and likely to exacerbate wealth differences. The question of difficulty in adapting to climate change then becomes one of managing international development. This is an entire topic in itself, but the struggle the global community has had in meeting the UN's millennium goals (primarily to reduce poverty and increase human well-being and environmental sustainability) is a testimony to how successfully it will manage international development so as to reduce climate-change impacts on human well-being.

8.5.2 Future climate change and human health

As future climate change takes the planet beyond the climate regime experienced in the Holocene (Chapter 5), there will be climate-related health impacts new to a number of regions that have been inhabited over past centuries. Indeed, later in the twenty-first century as the global climate regime becomes one not experienced since our species evolved, these will become more pronounced. That this will occur when the planet's population is more than four times greater than it has been at any time since our species evolved up to 1900 will serve to compound impacts on human well-being, and hence health. Global warming and this human demographic maximisation are related because it is human activity that is driving atmospheric concentrations of greenhouse gases beyond previous Quaternary levels (Chapter 7).

As with other aspects of climate change and human interactions, there will be winners and losers. Again, the wealthy are likely to fare far better than the poor, and again the disparity between the two will increase. An ironic illustration is that of atmospheric particulates (a cause of respiratory impairment) that are commonly associated with industrial and urban areas. In the past this association was universal but since the late twentieth century urban atmospheric particulates have declined markedly in OECD countries. Yet atmospheric particulates have served to provide a countering force through global dimming (Chapter 5). If they decline elsewhere as developing countries progress with the cleaning of their industries and economies, then this brake will be lessened and global warming enhanced. In short, one set of human health problems will be exchanged for another.

Finally, generally speaking (and this is very generally speaking), a warmer and wetter world is a world that will tend to favour terrestrial pathogenic microbes over a cooler and drier world. However, there are still many unknowns. For example, how will domestic and urban water use change? Will climate change and associated ecosystem changes affect microbial mutation rates and successful speciation? How would an increase in tempo of extreme weather events and natural disasters affect infectious disease occurrence? (McMichael, 2005). Combining such climate developments with the forthcoming human demographic maximisation, as well as the continued post-Industrial Revolution trend in increased international transportation, the likelihood is that health risks will increase.

8.5.3 *Future climate and human-ecology implications for wildlife*

As we have seen, climate and climate change have discernable, indeed marked, impacts on species and biological systems. Consequently, through a variety of ways (a number of which were covered in Chapter 2) we have managed to discern many of the principal global climatic regimes and changes of the past couple of hundred million years, although, of course, we know far more about relatively recent climate change than that in distant deep time. The picture that each tool provides is necessarily incomplete. However, taking evidence together enables the bigger picture to emerge, and it is still emerging with ongoing research. This picture is coherent as both the palaeo and current evidence is very much corroborative (Chapters 3 and 4), which in turn lends confidence to the conclusions drawn.

It is therefore certain that current climate-change influences (Chapter 5) and likely future change (Chapter 6) would without the presence of humans cause species to migrate and ecosystems to transform. Ecotones (the boundary zones between habitats) will be a focus of profound biological change: one that in no small part reflects the changes to take place within bordering ecosystems. Yet, in all likelihood, without humans terrestrial species would largely survive if (as we expect will happen) a similar amount of carbon were added over the next century to the atmosphere to that added since the Industrial Revolution (of course, there would be a genetic fingerprint; Chapter 4). Calcareous marine species, though, would be susceptible to extinction through increased ocean acidity, as happened with the Eocene thermal maximum (Chapter 3). Other than this, most groups and families of species would survive and the biosphere would recover. However, this is theoretical. The reality is that we do currently have a climbing population of over 6 billion humans that already is exerting considerable pressure on natural systems (Chapter 7), even without additional impacts due to anthropogenic climate change. So, whereas wildlife has adapted to the series of considerable glacial–interglacial climatic fluctuations over the past 2 million years (Chapter 4), and even major carbon-cycle perturbations earlier still, the past is not the present. First, the current anthropogenic perturbation of the carbon cycle is far more rapid than previous carbon excursions (consider the Eocene thermal maximum), as is the commensurate rapidity of global climate change (even if in the past some regional climate change has been as fast). Second, competition with humans has seen the landscape fragment: a landscape through which species adapting to change would have to migrate.

That we do have anthropogenic climate change in addition to an extinction already taking place will only serve to enhance threats to species. That current human pressures include the fragmenting of many natural terrestrial systems, and management and conversion of others to semi-natural systems, will impede species' ability to adapt to change. Such negative factors tend to interact synergistically. For example, a forest subject to climate change is one thing: species may migrate. Alternatively, a forest that becomes a large island surrounded by highly managed land becomes a nature reserve. However, impose significant climate change on isolated wildlife reserves, from which species cannot migrate, and extinctions virtually become inevitable.

That this anthropogenic climate change will, beyond the twenty-first century, take us to a global climate regime not seen since before the Quaternary means that species that have evolved since then, such as in the Pleistocene (or even the Pliocene), will in all likelihood be most adversely affected. This will be especially so for those species accustomed to the cooler end of the global climatic spectrum. Indeed, this can already be seen with some boreal and sub-polar species.

We are therefore going through an extinction event of a magnitude only last exceeded by the Cretaceous/Tertiary extinction 65.5 mya (Chapter 3). Yet, unlike that event, much of which took place rapidly, the current extinction arguably began when (some consider) the Anthropocene began with the extinction of Eurasian megafauna: that is to say, an extinction over thousands of years. This extinction is slowly but steadily continuing today and the size of the species affected is getting smaller. Future climate change will exacerbate matters.

Can this extinction be halted? In all probability, no. Can it be slowed and can some species be conserved? In all probability, yes, although it is impossible to estimate how many species that will be.

If we are to minimise extinctions then, aside from investment in biological conservation and management, investment will be needed to continue improving our understanding of biology and climate interactions. Computer models of global climate are now increasingly including biological dimensions and they are also becoming increasingly detailed. However, much still needs to be done. To date much of our present-day climate-change understanding (hence computer modelling) comes from our knowledge of Holocene and glacial climates discerned from actual biological and geological evidence.

Notwithstanding this, we have still much to learn as to how carbon moves between various biosphere sources and sinks. Filling this knowledge gap is fundamentally important, especially as the twenty-first century will see the Earth move into a climatic regime not seen for a few million years. Nonetheless, it is possible to foresee, before the end of this century's first quarter, that in some areas ecological biogeographic mapping will be sufficiently detailed, and the climate models sufficiently accurate, to foresee where and what species are best likely to flourish. This will greatly facilitate not just wildlife management but management of natural systems, especially agricultural and forestry systems, and hence generate a financial return on such investment in research, especially as we have a burgeoning human population to feed. In short, there is a very real role for whole-organism biological, ecological and biosphere expertise to address such issues for the foreseeable long-term future.

8.5.4 *Reducing future anthropogenic greenhouse-gas emissions*

Curbing anthropogenic carbon dioxide emissions, as the principal greenhouse gas, is the main, though not the sole, priority (Chapter 1). Yet such is human ecology (Chapter 7), and especially human energy requirements, that reductions are proving difficult. Even if the Kyoto Protocol were implemented globally in full then the curbing of our current growth in emissions will only amount to saving a few years' worth of warming and climate forcing by the end of the twenty-first century (Chapter 8). It is therefore difficult to envisage, at least from our early twenty-first-century perspective, that international political measures will have any significant effect in halting global warming. The best that may perhaps be hoped for is that measures above and beyond the Kyoto Protocol might begin to meaningfully slow matters down.

   Given this, is it worth investing in the curbing of carbon emissions? The future may be one of continued global warming. Reducing its rate would certainly help but there is (from a purely human materialistic and well-being perspective) another motivation for cutting fossil-carbon emissions. This stems from two facts. First, we know that fossil-carbon reserves are finite. Second, we know that the human demand for energy to power our technological global society is also certainly set to increase (barring some global human catastrophe). Consequently the need to curb emissions should not be perceived as an economic threat but an opportunity to develop alternative ways of generating and using (more efficiently) energy. Those nations that succeed most in doing this will be more secure from fossil-fuel shocks as well as see more economic benefit per unit of energy consumed and so be more competitive globally (Cowie, 1998a). This economic benefit will also help protect us from some of the adverse effects of climate change and help conserve wildlife. It is up to us, and nobody else, to seize this opportunity. In short, distinct from purely climate-change arguments, there are a raft of (largely sustainability-based) reasons for reducing our consumption of fossil carbon. That these chime perfectly with climate-change arguments for reducing greenhouse-gas emissions should be positively welcome (Cowie, 1998b). Yet as global greenhouse-gas emissions deviate from the Kyoto path such a welcome is far from being practically realised.

The null-hypothesis approach is equally compelling. Suppose some nations do not adapt to a low-fossil-carbon economy? Well, before the end of the twenty-first century these nations will be competing (in a world with a far larger population) for dwindling reserves of oil and gas. One of these factors alone (be it the larger population or the dwindling supply) would drive up the price of fossil fuel; both factors together will have a synergistic effect. This is due to what economists call the law of supply and demand. Economists have long known that if supply decreases but demand stays the same then price rises. They also know that if demand increases and supply stays the same the price also rises. Economists equally recognise a worse situation, that of supply decreasing while demand simultaneously increases. This prospect is the one which prudent national leaders need to take into account when framing policies to develop their respective nations' economies. The conclusion on purely economic grounds, irrespective of climate changes' biological and environmental impacts, is that renewable and alternative energy (arguably including nuclear power) needs to provide the vast bulk of society's energy needs.

From a human ecological perspective, if human societies are to be environmentally sustainable then one of the fundamental factors is the need to rely on sustainable energy and not the planet's finite fund of fossil photosynthetic energy.

8.5.5 A final conclusion

One final conclusion. It is actually more an observation, and it is a personal one at that. Yet as the author of this broad review and appraisal, it is something I feel entitled to make after some 200 000 words. If we are to comprehend and address climate-change concerns and the implications for species, including our own, then this understanding needs to come from science. It will not come from anywhere else. Science in many nations, including a number of highly developed ones, is not as valued as it might be and in some instances is positively under attack. Indeed, even where it is supported it is increasingly being constrained (no western university today would allow a scientist a few years out to write a book like Newton's *Principia*, and work to develop bold new ideas outside of science institutions (such as a monk growing peas) are an anathema (so much for any modern Mendel influencing genetics)). Yet if we are to see our global society survive the coming century (albeit transformed) without considerable suffering, let alone denuded of many biological and ecological assets, then we are going to have to change the way we relate to both resource use and (scientific and technological) knowledge. These are equally formidable, but not impossible, challenges. If we fail...well, someone else can narrate doom and disaster, at least as far as humans and many wildlife species are concerned. As for the rest of the biosphere, it will go on. Of that I am very confident.

8.6 References

BP Economics Unit (1992) *BP Statistical Review of World Energy*. London: British Petroleum Corporate Communications Services.

BP Economics Unit (2005) *BP Statistical Review of World Energy*. London: British Petroleum Corporate Communications Services.

Brandt Commission (1980) *North-South: a Programme for Survival*. London: Pan Books.

Brandt Commission (1983) *Common Crisis – North-South: Co-operation for World Recovery*. London: Pan Books.

Carter, L. J. and Pigford, T. H. (2005) Proof of safety at Yucca Mountain. *Science*, **310**, 447–8.

CIA (2005) *CIA: The World Factbook – 2005*. Washington, DC: Central Intelligence Agency.

Clarke, R. and Timberlake, L. (1982) *Stockholm Plus Ten: Promises, Promises? The Decade since the 1972 UN Environment Conference*. London: Earthscan – International Institute for Environment and Development.

Climate Change Science Program (US) (2006) *Temperature Trends in the Lower Atmosphere*. Washington, DC: U.S. Climate Change Science Program.

Cowie, J. (1998a) *Climate and Human Change: Disaster or Opportunity?* London: Parthenon Publishing.

Cowie, J. (1998b) The other reasons for cutting carbon dioxide emissions. *Science in Parliament*, **55(4)**, 21.

Department of Energy (1977) *Energy Policy Review – Energy Paper 22*. London: Department of Energy and HMSO.

Department of Energy (1978) *Energy Policy: a Consultative Document*, Cmnd 7107. London: HMSO.

Department of Environment, Food and Rural Affairs (2006) *Climate Change: the UK Programme*. London and Norwich: DEFRA and Stationery Office.

Department of Environment, Transport and the Regions (2000) *Climate Change: the UK Programme*. London and Norwich: DETR and Stationery Office.

Department of Trade and Industry (2003) *Review of the Feasibility of Carbon Dioxide Capture and Storage in the UK*. London: DTI.

Department of Trade and Industry, Department for Transport and DEFRA (2003) *Energy White Paper: Our Energy Future – Creating a Low Carbon Economy*, Cmnd 5761. Norwich: Stationery Office.

Eikeland, P. O. (1993) US energy policy at a crossroads? *Energy Policy*, **21**, 987–98.

Evans, R. D. and Herring, H. P. J. (1989) *Energy Use and Energy Efficiency in the UK Domestic Sector up to the Year 2010*. London: Department of Energy and HMSO.

Greenhalgh, G. (1990) The comforting illusion of energy conservation. *Atom*, **393**, 6–7.

Hardin, G. (1968) The tragedy of the commons. *Science*, **162**, 1243–8.

Hirsch, R. L. (2005) *Peaking of World Oil Production: Impacts, Mitigation, & Risk Management*. Washington, DC: Department of Energy.

HMSO (1994) *Climate Change: The UK Programme*, Cmnd 2427. London: HMSO.

Hulme, M., Turnpenny, J. and Jenkins, G. (2002) *Climate Change Scenarios for the United Kingdom: The UKCIP02 Briefing Report*. Norwich: Tyndall Centre.

Intergovernmental Panel on Climate Change (1990) *Climate Change: the IPCC Scientific Assessment*. Cambridge: Cambridge University Press.

Intergovernmental Panel on Climate Change (2000) *Emission Scenarios – IPCC Special Report*. Geneva: WMO.

Intergovernmental Panel on Climate Change (2001a) *Climate Change 2001: the Scientific Basis – Summary for Policymakers and Technical Summary of the Working Group I Report*. Cambridge: Cambridge University Press.

Intergovernmental Panel on Climate Change (2001b) *Climate Change 2001: Impacts Adaptation and Vulnerability – a Report of Working Group II*. Cambridge: Cambridge University Press.

Intergovernmental Panel on Climate Change (2005) *IPCC Special Report on Carbon Dioxide Capture and Storage*. IPCC, Geneva.

International Energy Agency (2005) *World Energy Outlook 2005*. Paris: IEA.

International Energy Agency (2006) *China's Power Sector Reforms: Where Next?* Paris: IEA.

International Energy Authority (2004) *Key World Energy Statistics*. Paris: IEA.

International Union for the Conservation of Nature, World Wildlife Fund and United Nations Environment Programme (1980) *The World Conservation Strategy*. Gland, Switzerland: IUCN.

International Union for the Conservation of Nature, World Wildlife Fund and United Nations Environment Programme (1991) *Caring for the Earth: a Strategy for Sustainable Living*. London: Earthscan.

Kabat, P., Vellinga, P., van Vierssen, W., Veraat, J. and Aerts, J. (2005) Climate proofing the Netherlands, *Nature*, **438**, 283–4.

Kerr, R. A. (2005) Bumpy road ahead for World's oil. *Science*, **310**, 1106–8.

Kintisch, E. (2005) US Energy Bill promises some boosts for research. *Science*, **309**, 863.

Malthus, T. R. (1798) *Essay on the Principle of Population as it Affects the Future Improvement of Society With Remarks on the Speculation of Mr Godwin, Mr Condorcet and Other Writers*. Reprinted in *On Population* (1960) (Himmelfarb, G., ed.). New York: Modern Library.

McMichael, A. J. (2005) Environmental and social influences on emerging infectious diseases: past, present and future. In McLean, A. R., May, R. M., Pattison, J. and Weiss, R. A., eds, *SARS: a Case Study in Emerging Infections*. Oxford: Oxford University Press.

Meadows, D. H., Meadows, D. L., Randers, J. and Behrens, III, W. W. (1972) *The Limits to Growth*. London: Pan Books.

Meadows, D. H., Meadows, D. L. and Randers, J. (1992) *Beyond the Limits: Global Collapse or a Sustainable Future*. Godalming and London: World Wide Fund For Nature and Earthscan.

National Energy Policy Development Group (2001) *National Energy Policy*. Washington, DC: National Energy Policy Development Group.

Office of the Deputy Prime Minister (2005) *Proposals for Introducing a Code for Sustainable Homes*. London: Office of the Deputy Prime Minister.

Parliamentary Office of Science and Technology (2005) *Household Energy Efficiency*, POST Note 249. www.parliament.uk/parliamentary_offices/post/pubs2005.cfm.

Rose, D. J. (1974) Energy policy in the US. *Scientific American*, **230**, 20–9. Reprinted in *Energy and Environment: Readings from Scientific America* (1980) (Siever, R., ed.). San Francisco: Freeman & Co.

Royal Society of Chemistry (2005) *Chemical Science Priorities for Sustainable Energy Solutions*. London: Royal Society of Chemistry.

Schipper, L. and Hawk, D. V. (1991) More efficient household electricity use. *Energy Policy*, **19(3)**, 244–62.

Seiver, R. (1980) *Energy and Environment: Readings from the Scientific American – Energy Policy in the US*. San Francisco: Freeman & Co.

Stern, N. (2007) *Economics of Climate Change: the Stern Review*. Cambridge: Cambridge University Press.

Tyme, J. (1978) *Motorways Versus Democracy*. London: Macmillan Press.

UN Population Division of the Department of Economic and Social Affairs (2002) *World Population Prospects: The 2002 Revision*. New York: United Nations Secretariat.

Wilson, E. O. (1988) *Biodiversity*. Harvard, CT: Harvard University Press.

World Commission on Environment and Development (1987) *Our Common Future*. Oxford: Oxford University Press.

Worldwatch Institute (2003) *Vital Signs: 2003–2004*, London: Earthscan.

Appendix 1
Glossary and abbreviations

Glossary

The following are just some of the main biological, geological and climate terms used. It is not a complete list but focuses on terms used in more than one chapter. Although the terms are common to biology, geology or climatology, they are provided to help readers from one discipline understand passages of this book that relate to another discipline.

abyssal (zone)	The zone of greatest ocean depth, frequently defined as deeper than 1000 m.
aerosol	A colloidal system such as a mist or fog in which the dispersion medium is a gas. In climatology the gas is the atmosphere.
albedo	The fraction of light reflected from a surface. It includes the proportion of solar energy reflected from a surface on the Earth or from the planet in its entirety.
Allerød and Bølling	A warm phase lasting about 1000 years towards the end (deglaciation period) of the last glacial and prior to the Holocene.
allopatric	Species having completely separate geographical distributions (*see also* sympatric).
anthropogenic	Human-generated.
biome	The largest ecological category of a bio-geographical region (such as tundra or tropical rainforest).
Bond cycle	A cycle of waxing and waning of the Arctic glacial ice sheet.
boreal	The northern continental biogeographic zone of short summers and long winters.
Calvin cycle	The cycle of reactions in which chloroplasts make two key biochemicals (ATP and NADPH) through the light-driven process of photosynthesis.
Carnivora	Flesh-eating mammals.
dendrochronology	Past climate-change reconstruction from tree rings.
ecotone	A transitional zone between two habitats (e.g. forest bordering grassland).
Eemian	The previous interglacial (in some texts referred to as the Ipswichian, Sangamon or Eem-Sangamon interglacial) some 125 000 years ago.
endosymbiont	Organism(s) living inside the cell of another, in a symbiotic (close and mutually beneficial) relationship.

Eocene	A 22-million-year epoch beginning 55.8 mya (*see* Appendix 2).
eustatic	Changes in global sea level due to changes in the size of the Earth's ocean basins, as well as in local sea level due to tectonic movement.
evapotranspiration	The total loss of water from the soil by evaporation and by the transpiration of plants growing in the said soil.
fecundity	The number of offspring produced by an individual or breeding pair.
Fennoscandia	The north-western part of the Eurasian continent and commonly used in association with the glacial ice sheets covering the north of that continent.
fertility	The ability to reproduce (irrespective of realisation). (Note: some human demography texts do not distinguish between fertility and fecundity. Ecologists do, and this text uses the ecological definition to avoid confusion.)
folivorous	Leaf-eating.
forcing	A factor that affects the climate, warming or cooling it. Climate-forcing factors include aerosols and greenhouse gases.
glacial	A cool period within an ice age with expanded ice caps and lower sea levels. However, often in common parlance glacials are known as ice ages and this can lead to some confusion.
global warming potential	The global warming over a *specific* time period resulting from the addition of a greenhouse gas compared to the same mass of carbon dioxide.
Gondwana (Gondwanaland)	The southern-hemisphere supercontinent that existed 200 mya but which has since fragmented.
Heinrich events	Short periods of glacial Arctic ice rafting.
Holocene	The current interglacial, which began 11 500 years ago (*see* Appendix 2).
Holocene climatic optimum	A warm period some 8000–5000 years ago.
ice age	A period of time with glacials.
interglacial	A warm period in an ice age between glacials.
interstadial	A cool period within a (cool) glacial.
Ipswichian	*See* Eemian.
Iron Age neoglaciation	A minor glaciation that occurred 4500–2500 years ago.
Laurentide	Commonly used to describe the glacial ice sheet over North America.
Little Ice Age	A short cooling period particularly associated with, but not restricted to, the northern hemisphere in AD 1550–1850.
medieval climatic anomaly	A short warming period in approximately AD 1100–1300, also known as the medieval warm period or the medieval climatic optimum.
medieval warm period	*See* medieval climatic anomaly.
Metazoan	A multi-celled animal.

mitochondrion	An organelle within a cell that consumes carbohydrate and releases energy. Contains DNA that is inherited via the female line through generations.
nunatak	A mountain peak poking through an ice sheet.
Oligocene	The geological epoch lasting between 34 and 23 mya (*see* Appendix 2).
opportunity cost	An economist's term for the lost opportunity of employing one investment (be it cash or a resource) option over another (this book considers resource options). For example, some land used to grow biofuel would have an opportunity cost of not being available to grow food. Both options can also be quantified in product and cash terms.
Pangaea	A supercontinent existing in the deep geological past that subsequently fragmented into Gondwanaland to the south and Laurasia to the north.
Phanerozoic	The eon since the beginning of the Cambrian (542 mya), with multi-celled species (*see* Appendix 2).
phenology	The study of plant and animal development in response to the seasons.
photoperiodism	The response of an organism to the day/night cycle as well as its changes in the course of the year.
photosynthesis	The (sun)light-powered process by which plants (including algae) synthesise carbohydrates from carbon dioxide and water.
Pliocene	The epoch 5.3–1.8 mya (*see* Appendix 2).
pluvial	A geologically short time of high rainfall that is usually associated with Milankovitch or other climatic cycles such as those related to Heinrich events.
production	The biomass fixed over a period of time (usually a year), expressed in terms of either mass or energy. Primary production is the production due to photosynthesis (also called gross primary production). Net primary production is the production due to photosynthesis after respiration losses are taken into account. Secondary production relates to the production of species living off of photosynthetic species. Productivity relates to production in a unit area over a period of time (again, usually a year).
Quaternary	The time from the present to nearly 2 million years ago that saw a number of deep glacials and interglacials (*see* Appendix 2).
refugium	An ecological area where species can survive surrounding or nearby environmental change (plural, refugia).
Sangamon	*See* Eemian.
Snowball Earth	The Earth, either nearly or completely, covered with snow and ice.
speciation	The processes by which new species evolve.
stoma	A minute opening on the leaf of a plant (usually on the underside) though which gas and water exchange takes place (plural, stomata).

sympatric	Species occupying the same, or overlapping, geographical area (*see also* allopatric).
tundra	The treeless polar regions with permanently frozen subsoil (although surface soil may be thawed for some parts of the year).
Younger Dryas	A short period of cooling at the end of the last glacial, 12 900–11 600 years ago.
zonation	The occurrence of distinct bands of species across an environmental gradient such as a seashore or mountainside.

Abbreviations

ACC	Antarctic Circumpolar Current
ACR	Antarctic cold reversal
BAP	Biodiversity Action Plan
B-a-U scenario	Business-as-Usual scenario
Be	Bering, a unit of water flow (equivalent to 10^3 km^3)
bya	billion years ago
CBD	UN Convention on Biological Diversity (1992)
CCD	calcite compensation depth
CCM	carbon-concentrating mechanism
CCS	carbon capture and storage
CDC	US Centers for Disease Control and Prevention
CDM	Clean Development Mechanism
CEH	UK Centre for Ecology and Hydrology (formerly ITE and the NERC River Laboratory)
CET record	Central England Temperature record
CFB event	continental flood basalt (event)
CIE	carbon isotope excursion
DEFRA	Department of the Environment, Food and Rural Affairs (England and Wales; formerly DETR and MAFF)
DETR	Department of the Environment, Transport and the Regions (England and Wales; formerly the Department of the Environment)
DIC	dissolved inorganic carbon
DoE	Department of Environment (England and Wales)
ECM	electrical conductivity measurement (here applied to ice cores)
ENSO	El Niño Southern Oscillation
EOR	enhanced oil recovery
EPICA	European Project for Ice Coring in Antarctica
EU	European Union
FAO	UN Food and Agriculture Organization
FCCC	UN Framework Convention on Climate Change (1992)
G5	The emerging economic powers: China, India, South Africa, Mexico and Brazil
G8	Group of Eight, consisting of Canada, France, Germany, Italy, Japan, the UK, the USA and Russia.
GDP	gross domestic product
GECAFS	Global Environmental Change and Food Systems
GSA	Great Salinity Anomaly

GtC	giga-tonnes of carbon
GWh	giga-Watt hour
GWP	global warming potential
Ha	hectare ($100\,m \times 100\,m = 10^4\,m^2$ or approximately 2.47 acres)
HEP	hydroelectric power
HMSO	Her Majesty's Stationery Office (now just the Stationery Office)
IETM	initial Eocene thermal maximum (also known as the PETM)
IETR	International Experimental Thermonuclear Reactor
IPCC	Intergovernmental Panel on Climate Change
ITCZ	Intertropical Convergence Zone
ITE	UK Institute for Terrestrial Ecology (now CEH)
IUCN	International Union for the Conservation of Nature (and Natural Resources; now also known as the World Conservation Union)
JI	Joint Implementation
LAA	leaf area analysis (used to approximate rainfall)
LGM	last glacial maximum (approximately 20 000 years ago)
LMA	leaf margin analysis (used to approximate temperature)
MAFF	UK Ministry of Agriculture, Fisheries and Food (now part of DEFRA)
MAR	mass accumulation rate
MCA	medieval climatic anomaly (also known as the medieval warm period)
MIT	Massachusetts Institute of Technology
MONARCH	Modelling Natural Resource Responses to Climate Change
mtDNA	mitochondrial DNA
mtoe	million tonnes of oil equivalent ($= 42\,PJ = 42 \times 10^{15}\,J = 12\,TWh = 12 \times 10^{12}\,Wh$; excluding station efficiency factors)
MWP	medieval warm period (also known as the medieval climatic anomaly)
mya	millions of years ago
NAO	North Atlantic Oscillation
NERC	UK Natural Environment Research Council
NFFO	Non-Fossil Fuel Obligation
NHG	northern-hemisphere glaciation
NOAA	US National Oceanic and Atmospheric Administration
OECD	Organization for Economic Cooperation and Development (Austria, Australia, Belgium, Canada, Czech Republic, Denmark, Finland, France, Germany, Greece, Hungary, Iceland, Republic of Ireland, Italy, Japan, Luxembourg, Mexico, New Zealand, Netherlands, Norway, Poland, Portugal, Slovakia, South Korea, Spain, Sweden, Switzerland, Turkey, UK and USA)
OPEC	Organization of the Petroleum Exporting Countries (Iran, Iraq, Kuwait, Qatar, Saudi Arabia, United Arab Emirates, Algeria, Libya, Nigeria, Indonesia and Venezuela)
PETM	Palaeocene–Eocene thermal maximum (also known as the IETM)
PMP	possible maximum precipitation
ppb	parts per billion
ppm	parts per million
ppbv	parts per billion by volume
ppmv	parts per million by volume

ppt	parts per trillion
r/p ratio	(fossil-fuel) reserve/production ratio
SRES	IPCC special report on emission scenario
tC GWh^{-1}	tonnes of carbon per giga-Watt hour
TGS	thermal growing season
toe	tonnes of oil equivalent
UKCIP	UK Climate Impacts Programme
UN	United Nations
UNCED	UN Conference on Environment and Development (1992)
UNEP	United Nations Environment Programme
USGCRP	US Global Change Research Program
UV	ultraviolet (light)
W	Watt, the SI unit of power ($= 1\,\mathrm{Js}^{-1}$)
WAIS	West Antarctic Ice Sheet
WHO	World Health Organization
WMO	World Meteorological Organization
WWF	World Wildlife Fund (now World Wide Fund for Nature)

Appendix 2
Bio-geological chronology

Age (mya)	Eon	Era		Period	Epoch	Major events	Time beginning (mya)
0.0115	Phanerozoic	Cenozoic	Neogene	Quaternary	Holocene	Current interglacial (much of which is sometimes called the Anthropocene)	0.0115 (11 500 years ago)
1.806					Pleistocene	Series of interglacials and glacials	1.806
5.332				Tertiary	Pliocene		5.332
23.03					Miocene	35–15 mya temperatures 3–4 °C warmer than present but cooling	23.03
33.9			Paleogene		Oligocene	Gets cooler from Eocene; Antarctic ice from 34 mya	33.9
55.8					Eocene	Early Eocene warm	55.8
65.5					Paleocene		65.5
99.6		Mesozoic		Cretaceous	Upper	End-Cretaceous extinction	99.6
145.5					Lower		145.5
199.6				Jurassic		CO_2 levels begin a slow, variable trend down to Quaternary levels	199.6
251				Triassic		CO_2 levels rise to ~5 times present level	251

Eon	Era	Sub-era	Period	Epoch/System	Notes	Date (Ma)
Precambrian	Palaeozoic		Permian		End-Permian extinction	299
			Carboniferous	Pennsylvanian	CO_2 levels low	320
				Mississippian		359.2
			Devonian		CO_2 levels peak at over 22 times present level	416
			Silurian			443.7
			Ordovician			488.3
			Cambrian		'Age of invertebrates'	542
	Proterozoic	Neoproterozoic		Ediacaran		630
				Cryogenian		850
				Tonian		1000
		Mesoproterozoic		Stenian		1200
				Ectasian		1400
				Calymmian		1600
		Palaeo-proterozoic				2500
	Archean				Fossil evidence of bacteria and algae.	>3600 (lower limit not defined)

The dates given are those designated by the International Commission on Stratigraphy (2004), or ICS. The breakdown of nomenclature includes commonly used terms such as Quaternary, which the ICS has made a move to discontinue, but debate goes on. Yet because of common use and for student ease (especially by non-geologists) such terms are retained in this chart.

Appendix 3
Calculations of energy demand/supply and orders of magnitude

Calculations of energy demand/supply

From Chapter 7 onwards there are a number of depictions of global energy supply and demand. The following definitions have been applied in constructing these graphs of energy statistics.

Where statistics are given in millions of tonnes of oil equivalent (mtoe) for quantities of nuclear and renewable energy (such as hydro electric power) they are expressed in terms of the oil equivalent as if the energy was generated by an oil-fired electricity-generating plant with an efficiency of around 33%. Consequently it is possible to make some comparison and to see how these non-fossil energy sources (could) displace oil sources.

Historic proportional energy-supply estimates prior to 1950 are taken from Morgan and Murray (1976). Subsequent estimates were taken from the US Government's Energy Information Administration www.eia.doe.gov in autumn 2004 and spring 2005 as well as the *BP Statistical Review of World Energy* (1990, 2000, 2005) and adapted where necessary, as per previous power-station efficiency assumptions. The Worldwatch Institute's *Vital Signs* series (2003), which in turn draws on additional sources, was used as a check on statistics post-1950. The UK Department of Trade and Industry's (www.dti.gov.uk) *Energy Trends* was similarly used as a check for UK data.

Data from 1950 onwards strictly relates to commercially traded fuels and does not include locally gathered or produced domestic fuels in less-developed nations, such as burning of rural wood or animal dung. The proportional estimates prior to 1900 for the UK and USA may include some from such categories although the elements of change in pattern of energy consumption depicted in this book prior to the twentieth century are thought to capture the essential features.

- 1 t (metric) of oil = 1.1023 short tons = 256 gallons (imperial) = 308 US gallons
- 1 mtoe (million tonnes of oil equivalent) produces around 4 TWh = 4×10^{12} Wh in a power station of about 33% efficiency.
- 1 mtoe, when burnt, releases approximately 4.25×10^{13} Btu (British thermal units) or 4.48×10^{16} J.
- A machine doing work consuming energy at the rate of $1 \, \mathrm{Js}^{-1}$ consumes 1 W each second. If this continues for an hour then (60 s × 60 min) 1 kWh of energy is consumed, which is equivalent to 3600 J.

Orders of magnitude

Other than fossil-fuel energy content (above), other non-biological units and nomenclature are based on Baron (1988).

Factor	Prefix	Symbol
10^1	deca	da
10^3	kilo	k
10^6	mega	M
10^9	giga	G
10^{12}	tera	T
10^{15}	peta	P
10^{18}	exa	E
10^{21}	zetta	Z

Sources

Baron, D. N. (1988) *Units, Symbols and Abbreviations: a Guide for Biological and Medical Editors and Authors*. London: Royal Society of Medicine Services.

BP Economics Unit (1990, 2000, 2005) *BP Statistical Review of World Energy*. London: British Petroleum Corporate Communications Services.

Morgan, R. and Murray, R. B. (1976) *Energy Resources and Supply*. London: Wiley Interscience.

Worldwatch Institute (2003) *Vital Signs 2003–2004: the Trends that are Shaping Our Future*. London: Earthscan.

Appendix 4
The IPCC 2007 report

In Paris on 2 February 2007, while this book was at its page-proof stage prior to press, the IPCC launched the first part of its Fourth Assessment Report. The announced summary report was itself yet to be checked editorially, but its core message was clear: it gave a more assured statement, compared to the previous assessments, that human activity was in all probability affecting the global climate. Its summary statements included the following.

The understanding of anthropogenic warming and cooling influences on climate has improved since the Third Assessment Report (TAR), leading to *very high confidence* that the globally averaged net effect of human activities since 1750 has been one of warming, with a radiative forcing of $+1.6 [+0.6$ to $+2.4]$ W m $^{-2}$.

Warming of the climate system is unequivocal, as is now evident from observations of increases in global average air and ocean temperatures, widespread melting of snow and ice, and rising global mean sea level.

At continental, regional, and ocean basin scales, numerous long-term changes in climate have been observed. These include changes in Arctic temperatures and ice, widespread changes in precipitation amounts, ocean salinity, wind patterns and aspects of extreme weather including droughts, heavy precipitation, heat waves and the intensity of tropical cyclones.

Anthropogenic warming and sea level rise would continue for centuries due to the timescales associated with climate processes and feedbacks, even if greenhouse gas concentrations were to be stabilized.

The 2007 report (from the February announcement) gives a range of 1.1–6.4°C, including the high–low envelopes for all the scenarios for warming from 1990 to the end of the twenty-first century. The 2001 IPCC assessment high and low estimates from the 2001 report's IPCC SRES future scenarios were just 1.4–5.8°C over the same 110-year period (see Table 5.1 and Figure 5.6). This was also reported in *Nature's* coverage of the report's release, which also said that where we end up on that scale will largely depend on how much fossil fuel the world burns (Giles, 2007). In short, the 2007 estimated warming range broadened. However, its upper limit went up (0.6°C) twice as much than the bottom of the range went down (0.3°C). In short, while the IPCC was less certain as to the warming we might expect in the twenty-first century, the summary report reflects the view that warming might possibly be greater than previously thought.

Contrary to what might be expected from the above more uncertain but possibly greater warming, was the 2007 assessment's estimate for sea-level rise from 1990 to

2100. In the 2001 assessment the full range of sea-level rise to 2100 was 9–88 cm (best estimates, 22–70 cm). The 2007 assessment's revised estimate was 18–59 cm. This narrower range implies that the IPCC is more certain of possible sea-level rise, yet the lowering of the upper limit does not seem to chime with the possibility of greater warming and greater uncertainty as to the range of warming.

There remains considerable uncertainty as to possible sea-level rise (see sections 5.3.2 and 6.5.3). Indeed, just prior to the IPCC's February 2007 announcement a paper was published in the renowned journal *Science* based on the proportionality of past warmth and sea-level rise suggesting that 1990–2100 sea-level rise might be between 50 cm and 1.4 m (Rahmstorf, 2007). For the reasons given in this book, sea-level rise is likely to remain a source of considerable controversy within the scientific community for at least the next few years.

As far as readers of this book are concerned, there is nothing in the IPCC's 2007 initial statement that alters the substance of discussion, or the broader conclusions reached, in any of this work's chapters.

References

Giles, J. (2007) From words to action. *Nature*, **445**, 578–9.
IPCC (2007) *Climate Change 2007: the Physical Science Basis – Summary for Policymakers*. Geneva: IPCC.
Rahmstorf, R. (2007) A semi-empirical approach to projecting future sea-level rise. *Science*, **315**, 368–70.

Index